Symplectic techniques in physics

Symplectic techniques in physics

VICTOR GUILLEMIN
Professor of Mathematics
Massachusetts Institute of Technology

SHLOMO STERNBERG
George Putnam Professor of Pure and Applied Mathematics
Harvard University
and Permanent Sackler Fellow
University of Tel Aviv

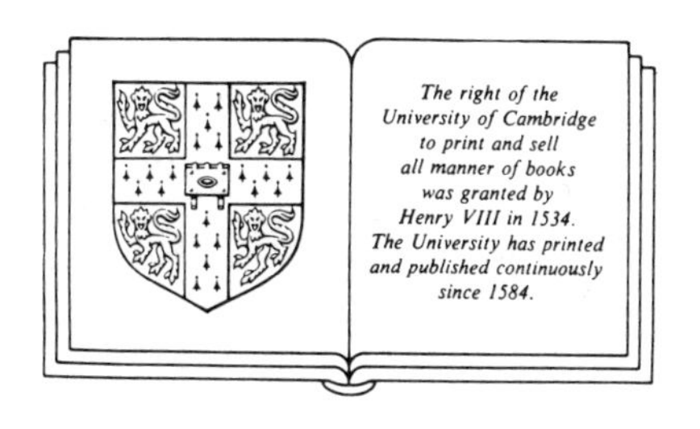

CAMBRIDGE UNIVERSITY PRESS

Cambridge
New York Port Chester
Melbourne Sydney

Published by the Press Syndicate of the University of Cambridge
The Pitt Building, Trumpington Street, Cambridge CB2 1RP
40 West 20th Street, New York, NY 10011, USA
10 Stamford Road, Oakleigh, Melbourne 3166, Australia

First published 1984
Reprinted 1986
First paperback edition 1990
Reprinted 1991

Printed in the United States of America

Library of Congress Cataloging in Publication Data

Guillemin, Victor, 1937–
Symplectic techniques in physics.
Bibliography: p.
Includes index.
1. Geometry, Differential. 2. Mathematical physics. 3. Transformations (Mathematics).
I. Sternberg, Shlomo. II. Title.
QC20.7.D52G84 1984 530.1′5636 83–7762

ISBN 0-521-24866-3 hardback
ISBN 0-521-38990-9 paperback

Contents

Preface

This book is based on lectures on symplectic geometry that we have given over the past few years at MIT, Harvard University, and the University of Tel Aviv. Symplectic geometry – especially under its old name, "the theory of canonical transformations" – is a venerable topic in mathematical physics. It has recently experienced a great rejuvenation and is currently an active area of research. Our purpose in this book is twofold: to provide an introduction to the subject and to present the central results of the subject from a modern point of view. There is, accordingly, a difference in style and tone between Chapter I and the rest of the book.

Chapter I is directed at the general reader interested in mathematics or physics. The mathematical prerequisites are quite modest. A knowledge of calculus and the rudiments of linear algebra suffice for most of the chapter. Although some mention of more sophisticated topics is made from time to time, the passages containing such material are inessential and can be glossed over without loss of continuity. The mix of mathematics and physics is homogeneous. We have tried the genetic approach. Using the various theories of light as our paradigm, we have attempted to explain the development of the mathematical and physical ideas involved in symplectic geometry. Despite its elementary character, most of the key ideas of the book are to be found, albeit in embryonic form, in Chapter I.

Chapter II presents the key mathematical results of the book and describes several of the important physical applications. The style is more formal and the mathematical demands are greater than those of Chapter I. The reader is expected to have some familiarity with the basics of differential geometry and a degree of mathematical sophistication. The mix between the mathematics and the physics is less homogeneous here. We have indicated those sections that can be skipped by a reader more interested in the physical applications than in the mathematical proofs.

Chapters III, IV, and V are practically independent of one another. Chapter III is mainly concerned with the use of symplectic geometry as a tool for formulating (possible) laws of physics. The main theme is the principle of general covariance due to Einstein, Infeld, and Hoffman in a form recently explained and recast by Souriau. Section 41 contains one of the main mathematical results of the book.

Chapter IV is devoted to the use of the interaction between group theory and symplectic geometry for the solution of various mechanical systems. We have tried to explain how a broad variety of integrable mechanical systems can be explained using overt or hidden symmetries. We have, in the main, restricted attention to finite-dimensional systems, although we do discuss some evolution equations of the Korteweg–de Vries type. We have not included the very interesting work involving the Kac–Moody algebras.

Chapter V serves two purposes. The first part presents some of the key results in the theory of Lie algebras that are needed and used throughout the book. Although these results, and our treatment of them, are completely standard, we provide them here for the convenience of the reader. We also illustrate these results in computations of physical interest. The second part is a discussion of the deformation theory of symplectic group actions. The results are all due to Don Coppersmith, and much of the last few sections is taken directly from his unpublished Harvard thesis. We thank him for his kind permission to reproduce these results here. Much work remains to be done on this highly interesting and important topic.

The diagram of logical dependence is thus roughly as follows:

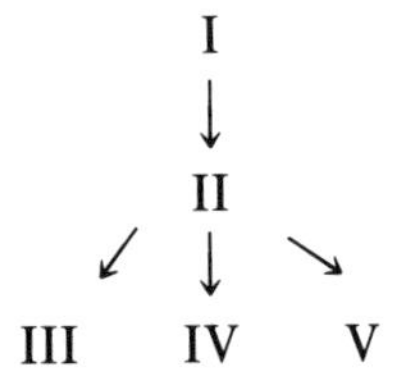

We should pay tribute to some of the people who originated the ideas and techniques that are the subject matter of a large part of the last three chapters of this book. The coadjoint representation and coadjoint orbits appear for the first time in Kirillov's work on classification of the irreducible representations of the nilpotent Lie groups. The "orbit method" in its symplectic setting was extensively developed by Kostant in the late sixties and early seventies. The moment mapping also plays an important role in Kostant's work at this time, and its physical implications were emphasized by Souriau in his beautiful monograph (1970). The idea of reduction occurs for the first time in the short paper of Marsden and Weinstein (1974) and was subsequently used by them and their students and collaborators for a wide range of applications in mathematics and physics. The recent work, described in Chapter IV, on complete integrability owes much to the work of Arnold and Moser in the late sixties and early seventies. In particular, it was from their work that the idea emerged that the complete integrability of most of the classical examples of completely integrable systems could be accounted for by "hidden symmetries." The material in the first part of Chapter IV, on Lagrangian fibrations and the geometric structure of completely integrable systems, is largely adapted from Duistermaat's beautiful paper (1980); and the material in Section 33, on stationary phase, is based partly on work of Duistermaat and Heckman and partly on work (some of it unpublished) of Bott and Berline and Vergne.

We wish to thank Professor G. Emch for a careful reading and many constructive comments on Chapter I. We also wish to thank Bert Kostant for long and fruitful collaboration on many of the topics covered in this book and for his permission to reproduce many joint results, and Jerry Marsden and Alan Weinstein for intense but friendly competition. We would also like to express our appreciation to many of our friends who provided illuminating conversations, unexpected insights, and moral support during the three or four years that this book was in gestation (as course notes for graduate courses at Harvard and MIT). Among them are Michael Atiyah, Bob Blattner, Raoul Bott, David Kahzdan, Richard Melrose, John Rawnsley, J. M. Souriau, Joe Wolf, and Alejandro Uribe.

V. G.
S. S.

December 1982

I

Introduction

Not enough has been written about the philosophical problems involved in the application of mathematics, and particularly of group theory, to physics. On the one hand, mathematics is created to solve specific problems arising in physics, and, on the other hand, it provides the very language in which the laws of physics are formulated. One need only think of calculus or of Fourier analysis as examples of this dual relationship.

We are all familiar with the exploitation of symmetry in the solution of a mathematical problem. On the other hand, the very assertion of symmetry is often the most profound formulation of a physical law or the key step in the development of a new theory.

Another example is provided by Hamiltonian mechanics. The mathematical theory underlying Hamiltonian mechanics is currently called symplectic geometry. We briefly recall the basic definitions and early history. A symplectic vector space is a real vector space equipped with an antisymmetric, nondegenerate bilinear form. For example, on $\mathbb{R}^2$ we can define the form (,) by $(u_1, u_2) = q_1 p_2 - q_2 p_1$, where $u_1 = \begin{pmatrix} q_1 \\ p_1 \end{pmatrix}$ and $u_2 = \begin{pmatrix} q_2 \\ p_2 \end{pmatrix}$. It is not hard to prove that such a vector space must be even-dimensional (if finite-dimensional) (see Section 21 below). A linear isomorphism of a symplectic vector space V (or more generally of V onto W) is called symplectic if it preserves the bilinear form. (In two dimensions, a linear transformation is symplectic if and only if its determinant is 1. In higher dimensions, the condition is more restrictive.) A differentiable map of an open subset of a symplectic vector space V into W is called symplectic if its (Jacobian) matrix of partial derivatives is symplectic at every point. A symplectic manifold M is an even-dimensional manifold that locally has the structure of a symplectic vector space. This means that one has local charts ψ_i mapping open sets U_i of M onto open subsets of some fixed symplectic vector space V and such that the change of coordinates maps $\psi_i \circ \psi_j^{-1}$ defined on $\psi_j(U_i \cap U_j)$ are symplectic. (Alternatively, thanks to a theorem of Darboux (see Chapter II, Section 22), a symplectic manifold is a manifold together with a closed 2-form of maximal rank.) One has the obvious definition of symplectic diffeomorphism (i.e., one-to-one smooth transformations with smooth inverses, which are locally symplectic in the above sense). In the older literature, symplectic diffeomorphisms were called canonical transformations. Symplectic geometry is the study of symplectic manifolds and diffeomorphisms. The relation with mechanics is usually expressed by saying that the *phase space* of a mechanical system is a symplectic manifold and that the time evolution of a (conservative) dynamical system is a one-parameter family of symplectic diffeomorphisms. The role of the symplectic structure had first appeared, at least implicitly, in Lagrange's work on the variation of the orbital parameters of the planets in celestial mechanics. Its central importance, emerged, however, from the work of Hamilton.

At the age of eighteen, Hamilton submitted a paper entitled "Caustics" to Dr. John Brinkley, then the first royal astronomer for Ireland, who, as a result, is said to have remarked "This young man, I do not say *will be* but *is* the first mathematician of his age." Brinkley presented the paper to the Royal Irish Academy. It was referred, as usual, to a committee, whose report, while acknowledging the novelty and value of its contents recommended that it should be further developed and simplified before publication. Five years later, in greatly expanded form, the paper finally appeared, entitled "Theory of systems of rays," published in 1828 in the Transactions of the Royal Irish Academy.

The gist of Hamiltonian optics, in modern language is as follows: One is interested in studying the geometry of rays of light as they pass through some optical system. Suppose our system is aligned along some axis, and we study rays that enter the system at the left and emerge from the right. The portion of the rays to the left of the system are straight line segments. One needs four variables, locally, to specify such a line – two variables to specify the point of intersection of the line with a plane perpendicular to the optical axis and two additional angular variables giving the inclination of the line to this plane. The problem is to relate the incoming line segments, to the left of the system, to the outgoing line segments to the right. The first basic assertion is that if we use the proper coordinates (which involve the index of refraction of the ambient space), the transformation from the incoming to the outgoing coordinates is a symplectic diffeomorphism. Thus geometrical optics is reduced to symplectic geometry. Hamilton showed that if the graph of a symplectic transformation satisfies an appropriate "transversality" condition, then the transformation determines and is determined by a function of one-half the number of incoming and outgoing variables – the so-called generating function of the symplectic transformation. As this function is determined solely by the physical properties of the optical system, Hamilton called it a characteristic function. Depending on the transversality assumptions made, it can be a function of the points of intersection of the incoming and outgoing rays with the transversal planes – the point characteristic, the incoming points of the intersection and the outgoing angles – the mixed characteristic, or the incoming and outgoing angles – the angle characteristic. These functions are of use in combining optical systems, that is, in composing the corresponding symplectic transformations. They are also extremely useful in describing the deviation of the symplectic transformations from linearity – the *geometric aberrations* of the optical system. Finally, they are closely related to the *optical length* of the light rays themselves, and these light rays can be characterized as being extremals for optical length (Fermat's principle).

Some years later, Hamilton realized that this same method applies unchanged to mechanics: Replace the optical axis by the time axis, the light rays by the trajectories of the system, and the four incoming and four outgoing variables by the $2n$ incoming and outgoing variables of the phase space of the mechanical system. Hamilton's methods, as developed by Jacobi and other great nineteenth-century mathematicians, became a powerful tool in the solution or analysis of mechanical problems. Hamilton's analogy between optics and mechanics served as a guiding beacon to the development of quantum mechanics some one hundred years later.

The main purpose of this chapter is to discuss the relation between linear optics, geometric optics, and wave optics, stressing Hamilton's point of view and the corresponding relations between classical and quantum mechanics.

We first make a few comments on the relation between various theories of optics. In the history of physics it is often the case that when an older theory is superseded by a newer one, the older theory still retains its validity – either as an approximation to the newer theory, an approximation that is valid for an interesting range of circumstances, or as a special case of the newer theory. Thus Newtonian mechanics can be regarded as approximation to relativistic mechanics, valid when the velocities that arise are very small in comparison to the velocity of light. Similarly, Newtonian mechanics can be regarded as an approximation to quantum mechanics, valid when the bodies in question are sufficiently large. Kepler's laws of planetary motion are a special case of Newton's laws, valid for the inverse-square law of force between two bodies. Kepler's laws can also be regarded as an approximation to the laws of motion derived from Newtonian mechanics when we ignore the effects of the planets on one another's motion.

The currently held theory of light is known as quantum electrodynamics. It describes very successfully and very accurately the interaction of light with charged particles, explaining both the discrete character of light, as evinced in the photoelectric effect, and the wavelike character of electromagnetic radiation. The triumph of nineteenth-century physics was Maxwell's electromagnetic theory, which was a self-contained theory explaining electricity, magnetism, and electromagnetic radiation. Maxwell's theory can be regarded as an approximation to quantum electrodynamics, valid in that range where it is safe to ignore quantum effects. Maxwell's theory fails to explain a whole range of phenomena that occur at the atomic or subatomic level.

One of Maxwell's remarkable discoveries was that visible light is a form of electromagnetic radiation, as is "radiant heat." In fact, since Maxwell, optics is a special chapter of the theory of electricity and magnetism that treats electromagnetic vibrations of all wavelengths: from the shortest γ rays of radioactive substances (having a wavelength of one hundred millionth of a millimeter) up through x rays, ultraviolet, visible light, and infrared to the longest radio waves (having a wavelength of many kilometers).

Maxwell's theory dealt with the source of electromagnetic radiation as well as its propagation. Before Maxwell, there was a fairly well developed wave theory of light, due mainly to Fresnel, which dealt rather successfully with the propagation of light in various media, but had nothing to say about the production of light. Fresnel's theory did account for three physical effects that could not be explained by earlier theories – diffraction, interference, and polarization. Diffraction has to do with the behavior of light in the immediate vicinity of surfaces through which it is transmitted or reflected. A typical diffraction effect is the fact that we cannot produce an absolutely straight, arbitrarily narrow beam of light. For example, we might try to produce such a beam by lining up two opaque screens with holes in them to collimate light arriving from the left, as shown in Figure 1.1. When the holes are made very small (of the order of the wavelength of the light) we find that the region to the right of the second screen is suffused with light, instead of there being a narrow beam. Interference refers to those phenomena where the wave character of light

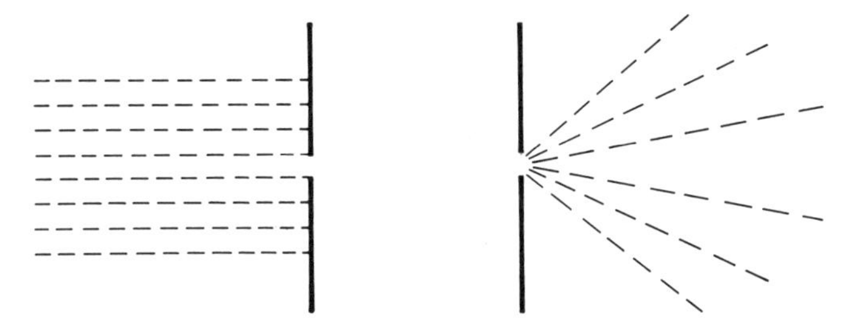

Figure 1.1.

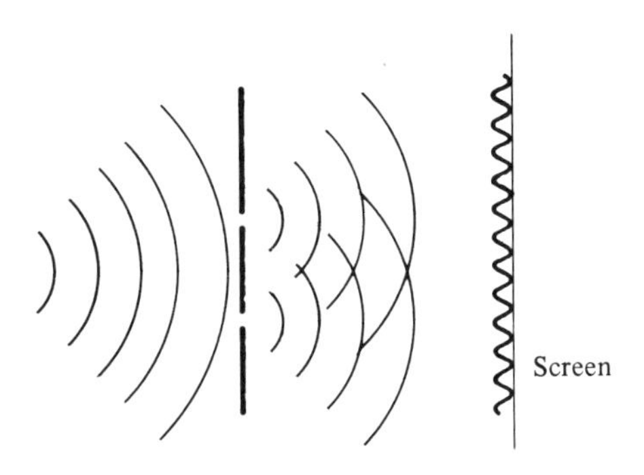

Figure 1.2.

manifests itself by the constructive or destructive superposition of light traveling different paths. Typical is the famous Young interference experiment illustrated in Figure 1.2. Polarization refers to the fact that when light passes through certain materials it appears to acquire a preferred direction in the plane perpendicular to the ray; such effects can be observed, for example, by using Polaroid filters.

Geometrical optics is an approximation to wave optics in which the wave character of light is ignored. It is valid whenever the dimensions of the various apertures are very large when compared to the wavelength of the light and when we do not examine too closely what is happening in the neighborhood of shadows or foci. It does not account for diffraction, interference, or polarization.

Linear optics is an approximation to geometrical optics which is valid when the various angles that enter into consideration are small. In linear optics one makes the approximation $\sin \theta \doteq \theta$, $\tan \theta \doteq \theta$, $\cos \theta \doteq 1$, etc.; that is, all quadratic (or higher-order) expressions in the angles are ignored. For example, in geometrical optics, *Snell's law* says that if light passes from a region whose index of refraction is n into a region whose index of refraction is n', then $n \sin i = n' \sin i'$, where i and i' are the angles that the

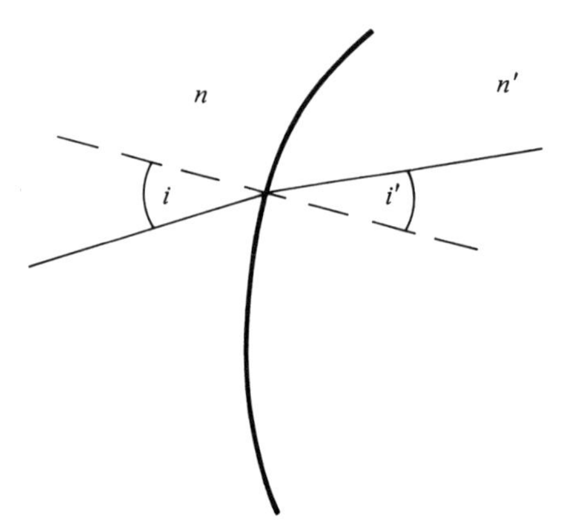

Figure 1.3.

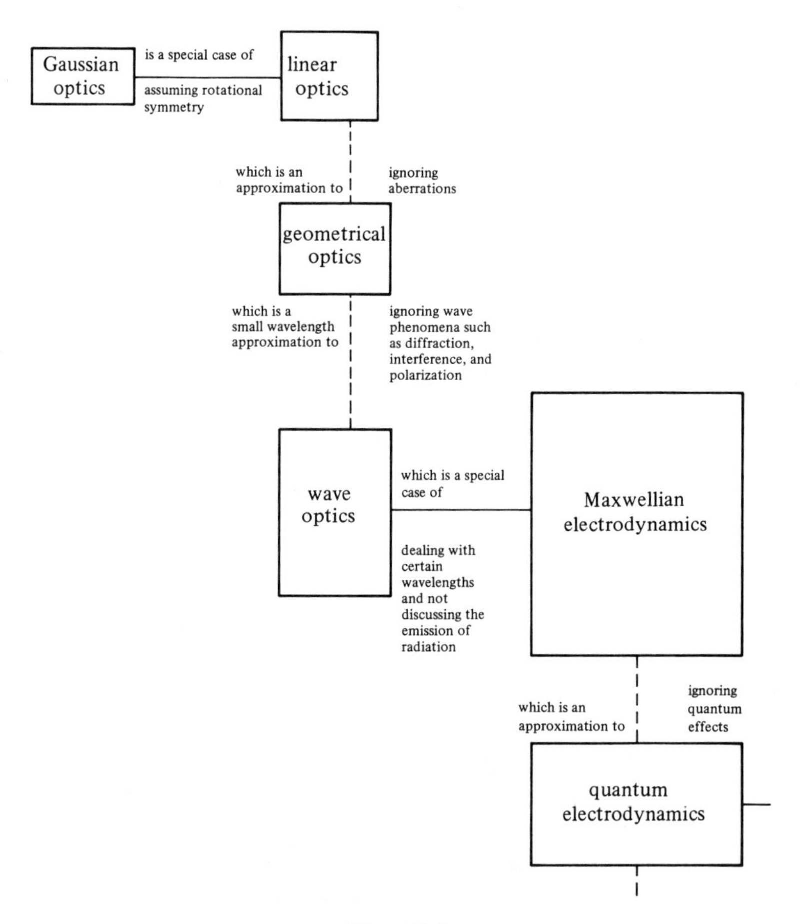

Figure 1.4.

light ray makes with the normal to the surface separating the regions (Figure 1.3). In linear optics, we replace this law by the simpler law $ni = n'i'$, which is a good approximation if i and i' are small. (This approximate law was known to Ptolemy.) The deviations of geometrical optics from the linear-optics approximation are known as (geometrical) aberrations.

Gaussian optics is a special case of linear optics in which it is assumed that all the surfaces that enter are rotationally symmetric about a central axis. This is a very important special case, since all ground lenses and most polished mirrors have this property. We summarize our discussion in Figure 1.4.

1. Gaussian optics

In linear optics we will make the approximation that the index of refraction is constant between refracting surfaces. (In our study of geometrical optics we will make the opposite approximation: that the index of refraction is a smoothly (possibly rapidly) varying function.) In Gaussian optics we are interested in tracing the trajectory of a light ray as it passes through the various refracting surfaces of the optical system (or is reflected by reflecting surfaces). We introduce a coordinate system so that the z axis (pointing from left to right in our diagram) coincides with the optical axis (i.e., the axis of symmetry of our system).

We shall study rays which are coplanar – ones that lie in one plane with the z axis; we shall prove later on that, in this approximation, the study of rays that do not lie in a single plane with the optical axis can be reduced to the study of coplanar rays.

By rotational symmetry, it is clearly sufficient to restrict attention to rays lying in one fixed plane. The trajectory of a ray as it passes through aperatures which are large compared to the wave length consists of pieces of straight lines. Our problem is to relate the straight line of the ray after it emerges from the system to the straight line that entered. For this we need to have a way of specifying straight lines. We do so as follows: We choose some fixed z value. (This amounts to choosing a plane perpendicular to the optical axis, called the *reference plane*.) Then a straight line is specified by two numbers, its height q above the axis at z, and the angle θ that the line makes with the optical axis. The angle will be measured in radians and will be considered positive if a counterclockwise rotation carries the positive z direction into the direction of the ray along the straight line (see Figure 1.5).

It is convenient to choose new reference planes, suitably adjusted to each stage in the calculation. Thus, for example, if light enters our optical system

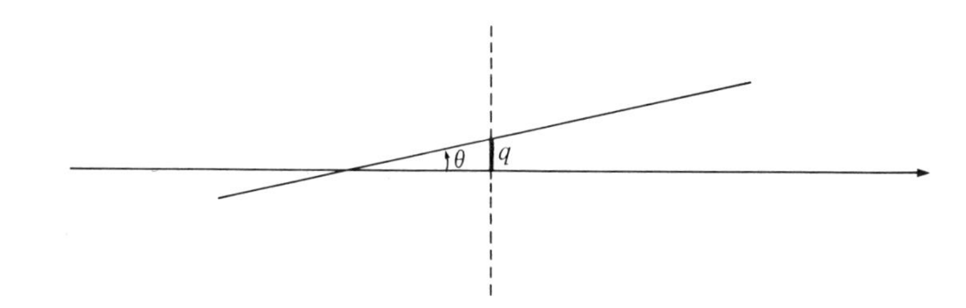

Figure 1.5.

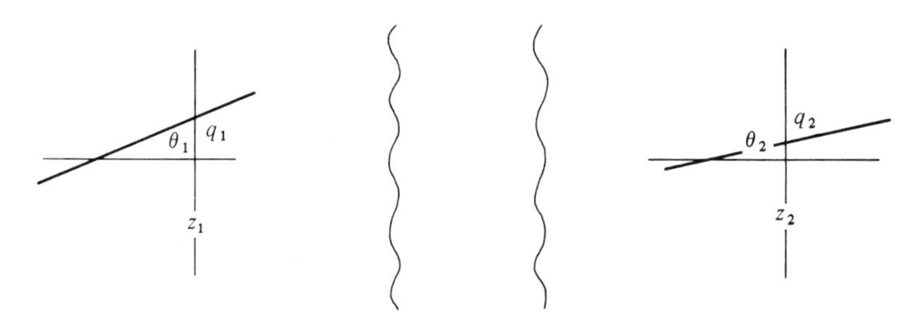

Figure 1.6.

from the left and emerges from the right, we would choose one reference plane z_1 to the left of the system of lenses and a second reference plane z_2 to the right. A ray enters the system as a straight line specified by q_1 and θ_1 at z_1 and emerges as a straight line specified by q_2 and θ_2 at z_2 (Figure 1.6). Our problem, for any system of lenses, is to find (q_2, θ_2) as a function of (q_1, θ_1).

Now comes a simple but crucial step of far-reaching significance that is basic to the geometry of optics and mechanics: Replace the variable θ by $p = n\theta$, where n is the index of refraction of the medium at the reference plane. (In mechanics, the corresponding step is to replace velocity by momentum.)

We thus describe a light ray by the vector $\begin{pmatrix} q \\ p \end{pmatrix}$, and our problem is to find $\begin{pmatrix} q_2 \\ p_2 \end{pmatrix}$ as a function of $\begin{pmatrix} q_1 \\ p_1 \end{pmatrix}$. Since we are ignoring all terms quadratic or higher, it follows from our approximation that $\begin{pmatrix} q_2 \\ p_2 \end{pmatrix}$ is a linear function of $\begin{pmatrix} q_1 \\ p_1 \end{pmatrix}$; that is, that

$$\begin{pmatrix} q_2 \\ p_2 \end{pmatrix} = M_{21} \begin{pmatrix} q_1 \\ p_1 \end{pmatrix}$$

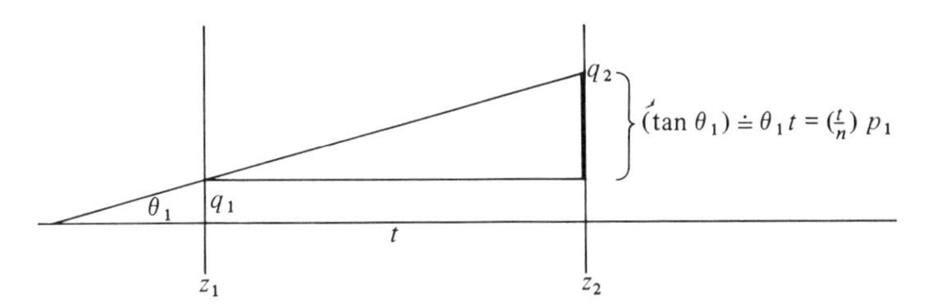

Figure 1.7.

for some matrix M_{21}. The key effect of our choice of p instead of θ as variable, is the assertion that

$$\det M_{21} = 1;$$

in other words, that the study of Gaussian optics is equivalent to the study of the group $Sl(2, \mathbb{R})$ of 2×2 real matrices of determinant 1. To prove this, observe that if we have three reference planes, z_1, z_2, and z_3, situated so that a light ray going from z_1 to z_3 passes through z_2, then by definition

$$M_{31} = M_{32}M_{21}.$$

Thus, if our optical system is built out of two components, we need only verify $\det M = 1$ for each component separately. To simplify the exposition we assume that our system does not contain mirrors. Then any refracting lens system can be considered as the composite of several systems of two basic types:

(a) A translation, in which the ray continues to travel in a straight line between two reference planes lying in the same medium. To describe such a system we must specify the gap t between the planes and the refractive index n of the medium (Figure 1.7). It is clear for such a system that θ and hence p does not change, and that $q_2 = q_1 + (t/n)p_1$. We write $T = t/n$ (called the *reduced distance*) and see that

$$\begin{pmatrix} q_2 \\ p_2 \end{pmatrix} = \begin{pmatrix} 1 & T \\ 0 & 1 \end{pmatrix} \begin{pmatrix} q_1 \\ p_1 \end{pmatrix},$$

$$\det \begin{pmatrix} 1 & T \\ 0 & 1 \end{pmatrix} = 1.$$

(b) Refraction at the boundary surface between two regions of differing refractive index. We must specify the curvature of the surface and the two indices of refraction, n_1 and n_2. The two reference planes will be taken immediately to the left and immediately to the right of the surface, respectively.

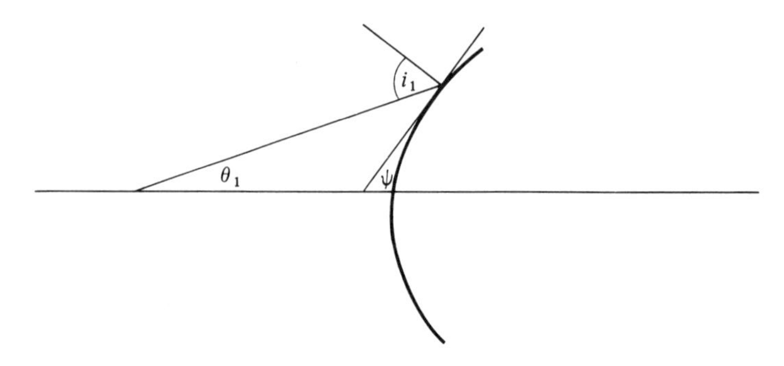

Figure 1.8.

At such a surface of refraction the q value does not change. The angle and hence the value of p change according to (the linearized version of) Snell's law. Now Snell's law involves the slope of the tangent to the surface at the point of refraction. In our approximation we are ignoring quadratic terms in this slope and hence terms of degree 3 or higher in the surface. We may thus assume that the curve giving the intersection of this surface with our plane is a parabola:

$$z - z_1 = \tfrac{1}{2}kq^2.$$

Then the derivative of z with respect to q is $z'(q) = kq$, which is $\tan(\pi/2 - \psi)$, where ψ is the angle in the figure. For small angles θ, that is, for small values of q, ψ will be close to $\pi/2$, and hence we may replace $\tan(\pi/2 - \psi)$ by $\pi/2 - \psi$, if we are willing to drop higher-order terms in q or p. Thus $\pi/2 - \psi = kq$ in our Gaussian approximation. On the other hand, if i_1 denotes the angle that the incident ray makes with this tangent line (Figure 1.8), then the fact that the sum of the interior angles of a triangle adds up to π shows that $(\pi - \psi) + \theta_1 + (\pi/2 - i_1) = \pi$ or

$$i_1 = \theta_1 + kq,$$

and similarly,

$$i_2 = \theta_2 + kq,$$

where $q = q_1 = q_2$ is the point where the rays hit the refracting surface. Multiplying the first equation by n_1 and the second equation by n_2 and using Snell's law in the approximate form $n_1 i_1 = n_2 i_2$ gives

$$\begin{pmatrix} q_2 \\ p_2 \end{pmatrix} = \begin{pmatrix} 1 & 0 \\ -P & 1 \end{pmatrix} \begin{pmatrix} q_1 \\ p_1 \end{pmatrix},$$

where $P = (n_2 - n_1)k$ is called the *power* of the refracting surface.

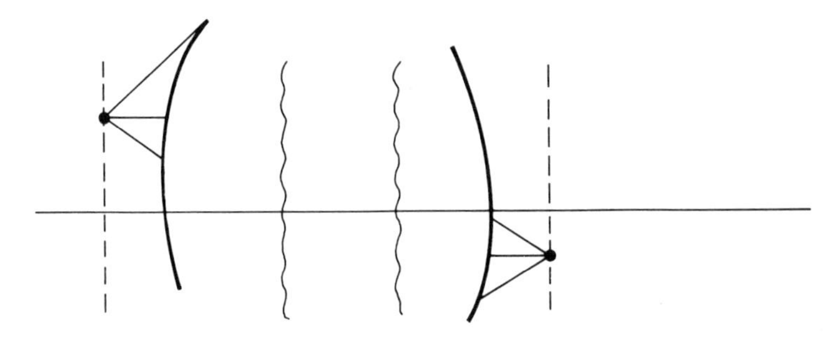

Figure 1.9.

Thus each Gaussian optical system is composed of two reference planes corresponding to a matrix

$$M = \begin{pmatrix} A & B \\ C & D \end{pmatrix} \qquad AD - BC = 1,$$

and one can set up a "dictionary" that translates properties of the matrix into optical properties.

For instance, the two planes are called *conjugate* (or in focus with one another) for any q_1 at z_1 if all the light rays leaving q_1 converge to the same point q_2 at z_2 (Figure 1.9). This of course means that q_2 should not depend on p_1; in other words, that $B = 0$.

Notice that the product of two matrices of the form $\begin{pmatrix} 1 & 0 \\ -P & 1 \end{pmatrix}$ again has this same form

$$\begin{pmatrix} 1 & 0 \\ -P_1 & 1 \end{pmatrix}\begin{pmatrix} 1 & 0 \\ -P_2 & 1 \end{pmatrix} = \begin{pmatrix} 1 & 0 \\ -(P_1 + P_2) & 1 \end{pmatrix}.$$

This gives the equation for the so-called thin lens, consisting of refracting surfaces with negligible separation between them (Figure 1.10). In this case, the reference planes z_1 and z_2 can be conveniently both be taken to coincide with the plane of the lens. The plane z_1, relates, of course, to rays incident from the left, while z_2 relates to rays that emerge from the lens and continue to the right.

The matrix for the left refracting surface is

$$\begin{pmatrix} 1 & 0 \\ -\dfrac{n_2 - n_1}{R_1} & 1 \end{pmatrix}.$$

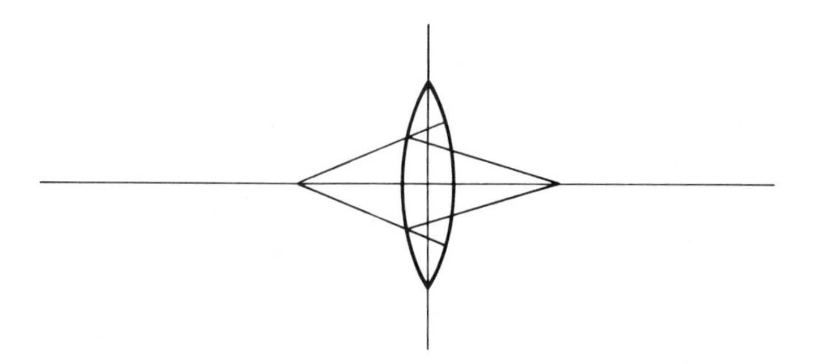

Figure 1.10.

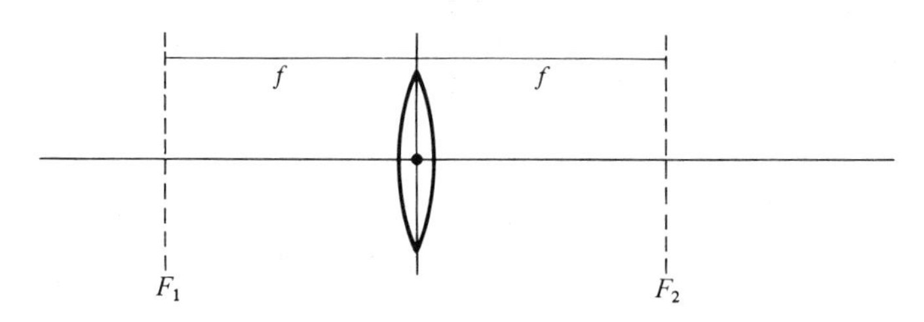

Figure 1.11.

The matrix for the right refracting surface is

$$\begin{pmatrix} 1 & 0 \\ -\dfrac{n_1 - n_2}{R_2} & 1 \end{pmatrix}.$$

(Note that R_2 is negative in the diagram.) Multiplying these matrices, we find that the matrix for the thin lens is

$$\begin{pmatrix} 1 & 0 \\ -1/f & 1 \end{pmatrix}, \qquad \text{where} \quad \frac{1}{f} = (n_2 - n_1)\left(\frac{1}{R_1} - \frac{1}{R_2}\right).$$

We shall assume that the lens is in air, so $n_1 = 1$ and $n_2 > 1$. In the case where R_1 is positive, R_2 is negative, and $n_2 - n_1 > 0$ (a double convex lens), the *focal length* f is positive. If we calculate the matrix of the thin lens between a reference plane E_1, located a distance f to the left of the lens and a reference plane F_2 located a distance f to the right, we find

$$\begin{pmatrix} 1 & f \\ 0 & 1 \end{pmatrix}\begin{pmatrix} 1 & 0 \\ -1/f & 1 \end{pmatrix}\begin{pmatrix} 1 & f \\ 0 & 1 \end{pmatrix} = \begin{pmatrix} 0 & f \\ -1/f & 0 \end{pmatrix}$$

(see Figure 1.11). The plane F_1 is called the *first focal plane*. If a ray, incident on the lens, passes through this plane at $q_1 = 0$ with slope p_1, then the outgoing ray has

$$\begin{pmatrix} q_2 \\ p_2 \end{pmatrix} = \begin{pmatrix} 0 & f \\ -1/f & 0 \end{pmatrix} \begin{pmatrix} 0 \\ p_1 \end{pmatrix} = \begin{pmatrix} fp_1 \\ 0 \end{pmatrix};$$

that is, it has zero slope and so is parallel to the axis. Conversely, if the incident ray has zero slope, the outgoing ray has

$$\begin{pmatrix} q_2 \\ p_2 \end{pmatrix} = \begin{pmatrix} 0 & f \\ -1/f & 0 \end{pmatrix} \begin{pmatrix} q_1 \\ 0 \end{pmatrix} = \begin{pmatrix} 0 \\ -q_1/f \end{pmatrix};$$

that is, it crosses the axis in the second focal plane. More generally, we can see that p_2 is independent of p_1, so that incident rays passing through a given point in the first focal plane emerge as parallel rays, all with the same slope. Furthermore, q_2 is independent of q_1, so that incident rays all emerge to pass through the same position in the second focal plane.

As a simple illustration of the use of matrix methods to locate an image, suppose that we take reference plane z_1 to lie a distance s_1 to the left of a thin lens, while z_2 lies a distance s_2 to the right of the lens. Between these planes, the matrix is

$$\begin{pmatrix} 1 & s_2 \\ 0 & 1 \end{pmatrix} \begin{pmatrix} 1 & 0 \\ -1/f & 1 \end{pmatrix} \begin{pmatrix} 1 & s_1 \\ 0 & 1 \end{pmatrix} = \begin{pmatrix} 1 - s_2/f & s_2 + s_1 - s_1 s_2/f \\ -1/f & 1 - s_1/f \end{pmatrix}.$$

The planes are conjugate if the upper-right entry of this matrix is 0 (i.e., if $s_2 + s_1 - s_1 s_2/f = 0$). Thus we obtain $1/s_1 + 1/s_2 = 1/f$, the well-known *thin-lens equation.* We shall write this as

$$s_1 + s_2 - p s_1 s_2 = 0,$$

where $p = 1/f$.

We can solve this equation for s_2 so long as $s_1 \neq f = 1/p$. Thus each plane other than the one corresponding to $s_1 = f$ has a unique conjugate plane. At the first focal plane, where $s_1 = f$, all light rays entering from a single point q emerge parallel, so the conjugate plane to the first focal plane is "at infinity." A similar discussion (with right and left interchanged) applies to the second focal plane.

For $s_1 \neq f$ and s_2 corresponding to the conjugate plane, the magnification is given by

$$\frac{q_2}{q_1} = 1 - \frac{s_2}{f} = 1 - s_2\left(\frac{1}{s_1} + \frac{1}{s_2}\right) = -\frac{s_2}{s_1}.$$

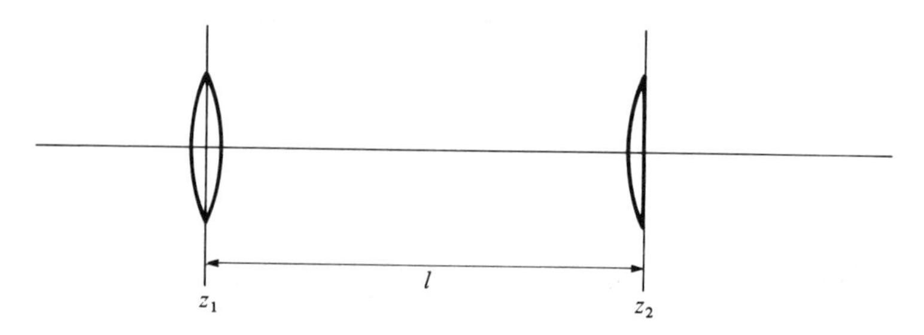

Figure 1.12.

If s_1 and s_2 are both positive (object to left of lens, image to right), then the magnification is negative, which means that the image is inverted.

By multiplying matrices, it is straightforward to construct the matrix for any combination of thin lenses. For example, in the case of thin lenses with focal lengths f_1 and f_2, separated by distance l in air, we find the matrix

$$\begin{pmatrix} 1 & 0 \\ -1/f_2 & 1 \end{pmatrix} \begin{pmatrix} 1 & l \\ 0 & 1 \end{pmatrix} \begin{pmatrix} 1 & 0 \\ -1/f_1 & 1 \end{pmatrix} = \begin{pmatrix} 1 - \dfrac{l}{f_1} & l \\ \dfrac{l}{f_1 f_2} - \dfrac{1}{f_2} - \dfrac{1}{f_1} & 1 - \dfrac{l}{f_2} \end{pmatrix}$$

between the reference plane z_1 (first lens) and z_2 (second lens), as is shown in Figure 1.12. A particularly interesting situation arises when $l = f_1 + f_2$, for then the matrix takes the form

$$\begin{pmatrix} A & B \\ 0 & D \end{pmatrix};$$

that is, $C = 0$. This means that $p_2 = Dp_1$; in other words, that the outgoing directions depend only on the incoming directions. The condition $l = f_1 + f_2$ is satisfied in the *astronomical telescope*, which consists of an objective lens of large positive focal length f_1 and an eyepiece of small positive focal length f_2 separated by a distance $f_1 + f_2$. Parallel rays from a distant star that enter this optical system are "converted" into other parallel rays that are presented to the eye.

The *angular magnification* of such a telescope is the ratio of the slope of the *outgoing* rays to the slope of the *incoming* rays, which equals

$$D = 1 - \frac{l}{f_2} = 1 - \frac{f_1 + f_2}{f_2} = -\frac{f_1}{f_2}.$$

This magnification is negative (the image is inverted) and its magnitude is the ratio of the focal length of the objective to that of the eyepiece

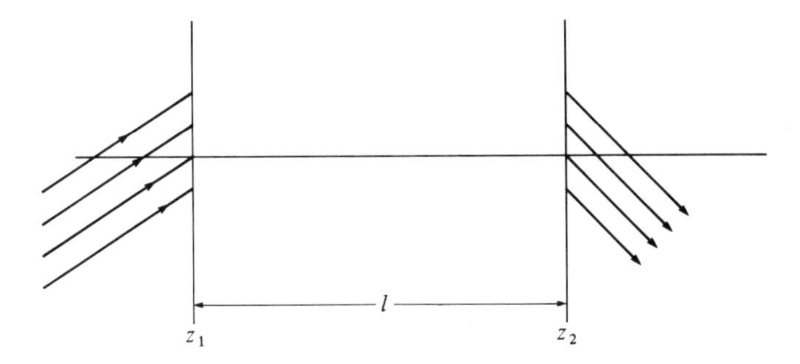

Figure 1.13.

(Figure 1.13). For further details of practical applications to optics, the reader should consult Gerrard & Birch (1975).

We now want to show that *any* 2×2 matrix with determinant 1 can arise as the matrix of some optical system. First of all, suppose that the matrix is telescopic ($C = 0$). Then $A \neq 0$, and, if $P \neq 0$, then

$$\begin{pmatrix} 1 & 0 \\ -P & 1 \end{pmatrix}\begin{pmatrix} A & B \\ 0 & D \end{pmatrix} = \begin{pmatrix} A & B \\ -PA & D - PB \end{pmatrix}$$

has $PA \neq 0$, so is not telescopic. We shall show that every matrix $\begin{pmatrix} A & B \\ C & D \end{pmatrix}$ with $C \neq 0$ can be written as

$$\begin{pmatrix} A & B \\ C & D \end{pmatrix} = \begin{pmatrix} 1 & -s \\ 0 & 1 \end{pmatrix}\begin{pmatrix} 1 & 0 \\ C & 1 \end{pmatrix}\begin{pmatrix} 1 & -t \\ 0 & 1 \end{pmatrix}, \qquad (*)$$

and thus arises as an optical matrix. If $C = 0$ then we need only multiply $\begin{pmatrix} A & B \\ -PA & D - PB \end{pmatrix}$ by $\begin{pmatrix} 1 & 0 \\ P & 1 \end{pmatrix}$ on the left to get $\begin{pmatrix} A & B \\ 0 & D \end{pmatrix}$, so it too is an optical matrix. To prove (*), consider

$$\begin{pmatrix} 1 & s \\ 0 & 1 \end{pmatrix}\begin{pmatrix} A & B \\ C & D \end{pmatrix}\begin{pmatrix} 1 & t \\ 0 & 1 \end{pmatrix} = \begin{pmatrix} A + sC & t(A + sC) + B + sD \\ C & Ct + D \end{pmatrix}.$$

Since $C \neq 0$ we can choose s so that $A + sC = 1$ and then choose $t = -(B + sD)$. The resulting matrix has 1 in the upper left-hand corner and 0 in the upper-right-hand corner. This implies that the lower-right-hand corner is also 1 so that the matrix on the right has the form $\begin{pmatrix} 1 & 0 \\ C & 1 \end{pmatrix}$ and this proves our assertion. Notice that s and t were uniquely determined. Thus, for any nontelescopic optical system, there are two unique planes

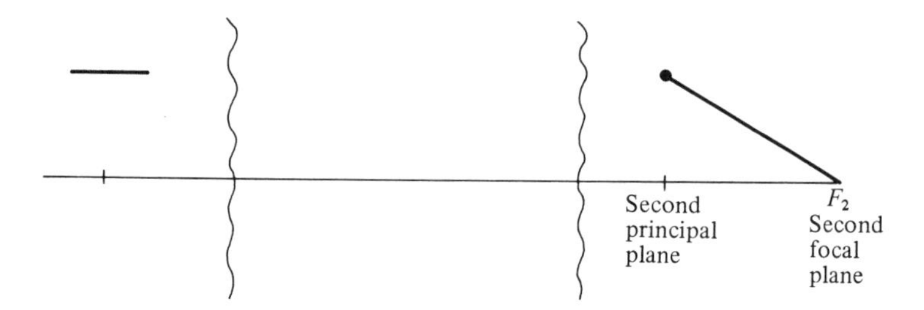

Figure 1.14.

such that the matrix between them has the form $\begin{pmatrix} 1 & 0 \\ C & 1 \end{pmatrix}$. These planes are conjugate to each other and have magnification 1. Gauss called them the principal planes. If we start with the optical matrix $\begin{pmatrix} 1 & 0 \\ C & 1 \end{pmatrix}$ between the two principal planes, we can proceed exactly as for the thin lens to find the plane conjugate to any other plane. All we have to do is write $C = -P = -1/f$. For instance, the two focal planes are located f units to the right and left of the principal planes:

$$\begin{pmatrix} 1 & f \\ 0 & 1 \end{pmatrix}\begin{pmatrix} 1 & 0 \\ -1/f & 1 \end{pmatrix}\begin{pmatrix} 1 & f \\ 0 & 1 \end{pmatrix} = \begin{pmatrix} 0 & f \\ -1/f & 0 \end{pmatrix}.$$

Gauss gave the following interpretation of the decomposition we derived above for the most general nontelescopic system in terms of ray tracing. Suppose a ray $\begin{pmatrix} q \\ 0 \end{pmatrix}$, parallel to the axis, enters the system at z_1 (Figure 1.14). When it reaches the second principal plane it is at the same height but is bent into $\begin{pmatrix} q \\ -q/f \end{pmatrix}$ and is focused on the axis at the second focal point, F_2. Similarly, a ray emerging from the first focal point is bent at the first principal plane into a ray parallel to the axis and arrives at z_2 still parallel to the axis and at the same height above the axis as it was at the first principal plane.

We see that the most general nontelescopic optical system can be expressed simply in terms of three parameters: the location of the two principal planes and the focal length. (We know that there should be three parameters, since there are only three free parameters in the matrix, the fourth matrix coefficient being determined by the fact that the determinant must equal 1.)

We can summarize the results of this section as follows: Let $Sl(2, \mathbb{R})$ denote the group of all 2×2 matrices of determinant 1. We have shown that there is an isomorphism between $Sl(2, \mathbb{R})$ and Gaussian optics. Each matrix corresponds to an optical system; multiplication of matrices corresponds to composition of the corresponding systems.

We now turn to Hamilton's ideas, in embryonic form.

2. Hamilton's method in Gaussian optics

Suppose that z_1 and z_2 are planes in an optical system but are *not* conjugate. This means that the B term in the optical matrix is not 0. Thus, from the equations

$$q_2 = Aq_1 + Bp_1$$

and

$$p_2 = Cq_1 + Dp_1,$$

we can solve for p_1 and p_2 in terms of q_1 and q_2 as

$$p_1 = (1/B)(q_2 - Aq_1)$$

and

$$p_2 = (1/B)(Dq_2 - q_1).$$

This has the following geometrical significance. Given a point q_1 on the z_1 plane and a point q_2 on the z_2 plane, there exists a unique light ray joining these two points. (This is exactly what fails to be the case if the planes *are* conjugate. For conjugate planes, if q_2 is the image of q_1, there will be an infinity of light rays joining q_1 and q_2; in fact all light rays leaving q_1 arrive at q_2. If q_2 is not the image of q_1, then there will be no light ray joining q_1 and q_2.) Let $W = W(q_1, q_2)$ be the function

$$W(q_1, q_2) = (1/2B)(Aq_1^2 + Dq_2^2 - 2q_1q_2) + K,$$

where K is a constant. Then we can write the equations for p_1 and p_2 as

$$p_1 = -(\partial W/\partial q_1) \qquad \text{and} \qquad p_2 = \partial W/\partial q_2.$$

Hamilton called the function W the *point characteristic* of the system. In the modern physics literature this function is called the *eikonal.* Suppose that z_1, z_2, and z_3 are planes such that no two of them are conjugate, with $z_1 < z_2 < z_3$ and such that z_2 does not coincide with a refracting surface. Let W_{21} be the point characteristic for the $z_1 \leftrightarrow z_2$ system and let W_{32} be the point characteristic for the $z_2 \leftrightarrow z_3$ system. We claim that (up to an additive constant) the point characteristic for the $z_1 \leftrightarrow z_3$ system is given by

$$W_{31}(q_1, q_3) = W_{21}(q_1, q_2) + W_{32}(q_2, q_3),$$

where, in this equation, $q_2 = q_2(q_1, q_3)$ is taken to be the point at which the ray from q_1 to q_3 hits the z_2 plane. To see why this is so, we first observe that since the z_2 plane does not coincide with a refracting surface the direction of the ray does not change at z_2. Thus

$$p_2 = (\partial W_{21}/\partial q_2)(q_1, q_2) = -(\partial W_{32}/\partial q_2)(q_2, q_3).$$

Now apply the chain rule to conclude that $\partial W_{31}/\partial q_1 = -p_1$ and, similarly, that $\partial W_{31}/\partial q_3 = p_3$ at (q_1, q_3).

The function W is determined by the above properties only up to an additive constant. Hamilton showed that, by an appropriate choice of the constant, we can arrange that $W(q_1, q_2)$ be the "*optical length*" of the light ray joining q_1 to q_2, where the optical length is defined as follows: For a line segment of length l in a medium of constant index of refraction n, the optical length is nl. A *path* γ is defined as a broken line segment where each component segment lies in a medium of constant index of refraction. If the component segments have length l_i and lie in media of refractive index n_i, then the optical length of γ is

$$L(\gamma) = \sum n_i l_i.$$

Let us prove Hamilton's result within the framework of our Gaussian optics approximation. Our approximation is such that terms in p and q of degree higher than 1 are dropped from the derivatives of W. Thus in computing the optical length and W we must retain terms up to degree 2, but may ignore higher-order terms. We will prove Hamilton's result by establishing the following general formula for the optical length (in the Gaussian approximation) of a light ray γ whose incoming parameters (at z_1) are $\begin{pmatrix} q_1 \\ p_1 \end{pmatrix}$ and whose outgoing parameters (at z_2) are $\begin{pmatrix} q_2 \\ p_2 \end{pmatrix}$:

$$L(\gamma) = L_{\text{axis}} + \tfrac{1}{2}(p_2 q_2 - p_1 q_1),$$

where L_{axis} denotes the optical length from z_1 to z_2 of the axis ($p_1 = q_1 = 0 = p_2 = q_2$) of the system. Notice that once this is proved, then, if we assume that z_1 and z_2 are nonconjugate, we can solve for p_2 and p_1 as functions of q_1 and q_2. That is, substituting $p_1 = (1/B)(q_2 - Aq_1)$ and $p_2 = (1/B)(Dq_2 - q_1)$ into the above formula gives our expression for W with $K = L_{\text{axis}}$.

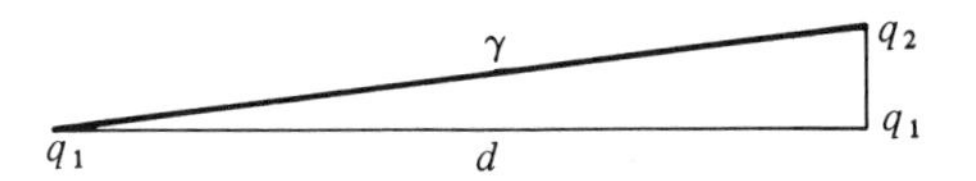

Figure 2.1.

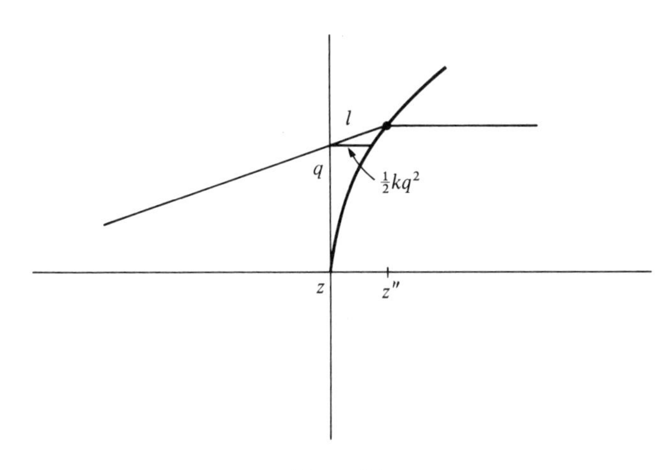

Figure 2.2.

To prove the above formula for $L(\gamma)$, we observe that it behaves correctly when we combine systems: If we have z_1, z_2, and z_3, then the length along the axis certainly adds, and $\frac{1}{2}(p_2q_2 - p_1q_1) + \frac{1}{2}(p_3q_3 - p_2q_2) = \frac{1}{2}(p_3q_3 - p_1q_1)$. So we need only prove the formula for our two fundamental cases:

(i) If n is constant,

$$\begin{aligned} L(\gamma) &= n\left(d^2 + (q_2 - q_1)^2\right)^{1/2} \\ &\doteq nd + \tfrac{1}{2}\frac{n}{d}(q_2 - q_1)^2 \\ &= nd + \tfrac{1}{2}\left[\frac{n}{d}(q_2 - q_1)\right](q_2 - q_1) \\ &= nd + \tfrac{1}{2}p(q_2 - q_1), \end{aligned}$$

where $p_2 = p_1 = p \doteq (n/d)(q_2 - q_1)$ is the formula that holds for this case (Figure 2.1).

(ii) At a refracting surface, $z' - z = \frac{1}{2}kq^2$ with index of refraction n_1 to the left and n_2 to the right (Figure 2.2). Here the computation must be understood in the following sense. Suppose we choose some point z_3 to the left and some point z_4 to the right of our refracting surface. If n_1 were equal to n_2, the optical length would be $n_1 l_3 + n_2(l + l_4)$, where l_3 is the portion of the ray to the left of our plane and $l + l_4$ is the portion to the right and where l_4 is the portion to the right of the surface. (We have drawn the figure with $k > 0$, but a similar argument works for $k < 0$.) If $n_2 \neq n_1$ then $n_2 l_4$ will be different but can be calculated by the methods of case (i) from z to z_4. In addition, an effect of the refracting surface is to replace $n_2 l$ by $n_1 l$ in the above expression, which serves to modify the optical length by $(n_1 - n_2)l$. This is the contribution at the refracting surface. Now

$$l = (z'' - z) \csc \theta_1,$$

where z'' is determined by the pair of equations

$$z'' - z = \tfrac{1}{2}kq''^2,$$
$$q'' = (\tan \theta_1)(z'' - z) + q.$$

It is clear that up to terms of higher order we may take $z'' = z' = \frac{1}{2}kq^2 + z$ and replace $\csc \theta_1$ by 1 so

$$\begin{aligned}(n_1 - n_2)l &= \tfrac{1}{2}k(n_1 - n_2)q^2 = -\tfrac{1}{2}Pq^2 \\ &= \tfrac{1}{2}[k(n_1 - n_2)q]q \\ &= \tfrac{1}{2}(p_2 - p_1)q,\end{aligned}$$

since $q_2 = q_1 = q$ and $p_2 = p_1 - Pq$ at a refracting surface, where $P = k(n_2 - n_1)$. This completes the proof of our formula.

3. Fermat's principle

Let us consider a refracting surface with power $P = (n_1 - n_2)k$ located at z. Here P might be 0. Consider planes z_1 to the left and z_2 to the right of z. We assume constant index of refraction between z_1 and z and between z and z_2. Let q_1 be a point on the z_1 plane, q_2 a point on the z_2 plane, and q a point on the z plane. Consider the path consisting of the light ray from q_1 to q and the light ray from q to q_2 (see Figure 3.1). This path will not, in general, be an optical path, since q can be arbitrary. However, its optical length

$$n_1 l_1 + n_1 l + n_2 l_2$$

is given, in the Gaussian approximation, by the sum of three terms, as we saw in the last section:

$$L(q_1, q, q_2) = L_{\text{axis}} + \tfrac{1}{2}\left(p_1(q - q_1) + p_2(q_2 - q) - Pq^2\right).$$

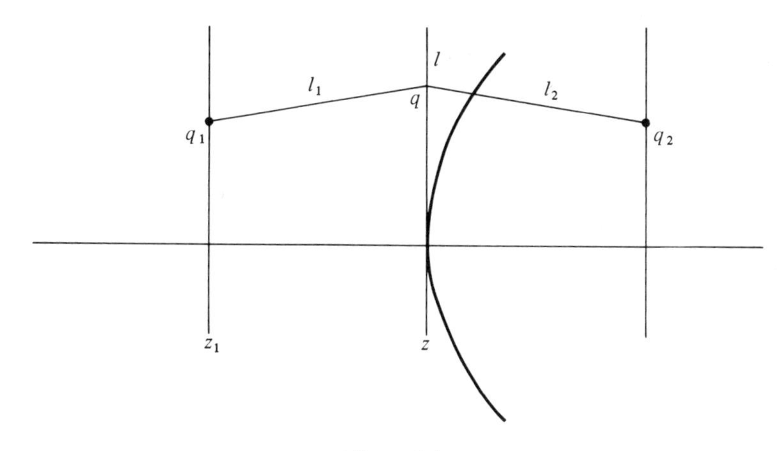

Figure 3.1.

In this expression

$$p_1 = \frac{n_1}{d_1}(q - q_1)$$

and

$$p_2 = \frac{n_2}{d_2}(q_2 - q)$$

so we can write

$$L = L_{\text{axis}} + \frac{1}{2}\left(\frac{d_1}{n_1}p_1^2 + \frac{d_2}{n_2}p_2^2 - Pq^2\right).$$

Suppose that we hold q_1 and q_2 fixed, and look for that value of q that extremizes L; in other words, we wish to solve the equation $\partial L/\partial q = 0$ for fixed values of q_1 and q_2. Substituting into the last expression for L, and knowing that $\partial p_1/\partial q = (n_1/d_1)$ and $(\partial p_2/\partial q) = -n_2/d_2$, we obtain the equation

$$p_1 - p_2 - Pq = 0.$$

In other words,

$$p_2 = Pq + p_1.$$

But this is precisely the relation between p_1 and p_2 given by the refraction matrix at z.

We have thus proved the following fact:

Let us fix q_1 and q_2 and consider the set of all paths joining q_1 to q_2 that consist of two segments – from q_1 to q and from q to q_2. Among all such paths, *the actual light ray can be characterized as that path for which the optical length L takes on an extreme value, that is, for which*

$$\frac{\partial L}{\partial q} = 0.$$

This is (our Gaussian approximation to) the famous Fermat principle of "least time." Let us now check whether this extremum is a maximum or minimum. Let us substitute $p_1 = (n_1/d_1)(q - q_1)$ and $p_2 = (n_2/d_2)(q_2 - q)$ into our formula for L to obtain a third expression for L:

$$L = n_1 d_1 + n_2 d_2 + \tfrac{1}{2}\{(n_1/d_1)(q - q_1)^2 + (n_2/d_2)(q_2 - q)^2 - Pq^2\}.$$

The coefficient of q^2 is $\frac{1}{2}(n_1/d_1 + n_2/d_2 - P)$. Thus the extremum is a *minimum* if

$$(n_1/d_1) + (n_2/d_2) - P > 0$$

and a *maximum* if

$$(n_1/d_1) + (n_2/d_2) - P < 0.$$

If $P > 0$ we see that we get a minimum for small values of d_1 and d_2 but a maximum for large values of d_1 and d_2. The situation is indeterminate (and we cannot, in general solve for q) when

$$n_1/d_1 + n_2/d_2 = P,$$

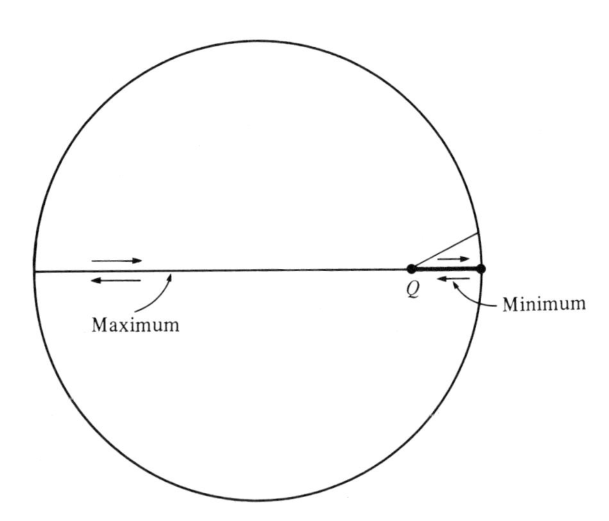

Figure 3.2.

which is precisely the condition that the planes are conjugate. Thus, we get a minimum if the conjugate plane to z_1 does not lie between z_1 and z_2 and a maximum otherwise. The fact that L is minimized only up to the first conjugate point is true in a more general setting and is known as the Morse index theorem. (For more on Morse theory, see Milnor 1963.)

To see an intuitively obvious example of this phenomenon, let us consider light being reflected from a concave spherical mirror. We take a point Q inside the sphere and let the light shine along a diameter so that it bounces back to Q. Then it is clear that the distance to the mirror is a local minimum if Q is closer than the center, and a maximum otherwise (see Figure 3.2).

4. From Gaussian optics to linear optics

What happens if we drop the assumption of rotational symmetry but retain the approximation that all terms higher than the first order in the angles and distances to one can be ignored? First of all, in specifying a ray, we now need four variables: q_x and q_y, which specify where the ray intersects a plane transverse to the z axis, and two angles, θ_x and θ_y, which specify the direction of the ray. A direction in three-dimensional space is specified by a unit vector, $v = (v_x, v_y, v_z)$. If v is close to pointing in the positive z direction, it will have the form $v = (\theta_x, \theta_y, v_z)$, where $v_z \doteq 1 - \frac{1}{2}(\theta_x^2 + \theta_y^2) \doteq 1$, provided θ_x and θ_y are small. Again, we replace the θ variables by p variables, where $p_x = n\theta_x$ and $p_y = n\theta_y$. (If the medium is anisotropic, as is the case in certain kinds of crystals, the relation between the θ variables and the p variables can be more complicated, but we will not concern ourselves with that here.) All of this, of course, is taking place at some fixed plane. If we consider two planes z_1 and z_2, the ray will correspond to vectors

$$u_1 = \begin{pmatrix} q_{x1} \\ q_{y1} \\ p_{x1} \\ p_{y1} \end{pmatrix} \quad \text{and} \quad u_2 = \begin{pmatrix} q_{x2} \\ q_{y2} \\ p_{x2} \\ p_{y2} \end{pmatrix}$$

at the respective planes.

Our problem is to find the form of the relationship between u_1 and u_2. Since we are ignoring all higher-order terms, we know that

$$u_2 = Mu_1,$$

where M is some 4×4 matrix. Our problem is to ascertain what kind of 4×4 matrices can actually arise in linear optics. The most obvious guess is

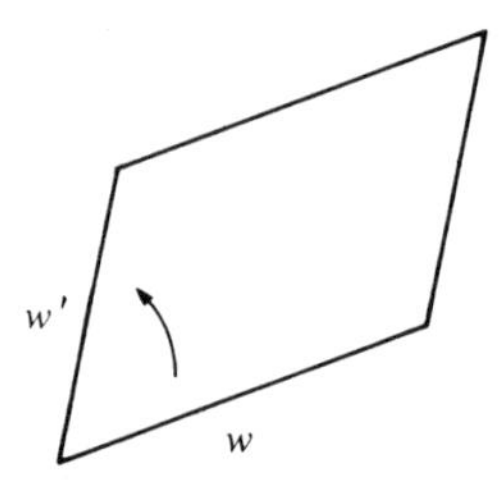

Figure 4.1.

that M must satisfy the requirement $\det M = 1$. This is not the right answer, however. It is true that all optical matrices must have unit determinant, but it is not true that all 4×4 matrices of determinant 1 can actually arise as transformation matrices in linear optics. There is a stronger condition that must be imposed. In order to explain what this stronger condition is, we first go back and reformulate the condition that a 2×2 matrix has determinant 1. We then formulate this condition in four variables. Let

$$w = \begin{pmatrix} q \\ p \end{pmatrix} \qquad \text{and} \qquad w' = \begin{pmatrix} q' \\ p' \end{pmatrix}$$

be two vectors in the plane. We define an antisymmetric "product," $\omega(w, w')$ between these two vectors by the formula

$$\omega(w, w') = qp' - q'p.$$

The geometric meaning of $\omega(w, w')$ is that it represents the oriented area of the parallelogram spanned by the vectors w and w' (see Figure 4.1). It is clear from both the definition and the geometry that ω is antisymmetric:

$$\omega(w, w') = -\omega(w', w).$$

A 2×2 matrix preserves area and orientation if and only if its determinant equals 1. Thus a 2×2 matrix M has determinant 1 if and only if

$$\omega(Mw, Mw') = \omega(w, w')$$

for all w and w'. Now suppose that

$$u = \begin{pmatrix} q_x \\ q_y \\ p_x \\ p_y \end{pmatrix} \qquad \text{and} \qquad u' = \begin{pmatrix} q'_x \\ q'_y \\ p'_x \\ p'_y \end{pmatrix}$$

are two vectors in four-dimensional space. We define

$$\omega(u, u') = q_x p'_x - q'_x p_x + q_y p'_y - q'_y p_y.$$

The product ω is still antisymmetric.

$$\omega(u', u) = -\omega(u, u');$$

but the geometric significance of ω is not so transparent.

It turns out that a 4×4 matrix M can arise as the transformation matrix of a linear optical system if and only if

$$\omega(Mu, Mu') = \omega(u, u'),$$

for all vectors u and u'. These kinds of matrices are called (linear) canonical transformations in the physics literature, and are called (linear) symplectic transformations in the mathematics literature. They (and their higher-dimensional generalizations) play a crucial role in theoretical mechanics and geometry.

After developing some of the basic facts about the group of linear symplectic transformations in four variables, we shall see that our arguments showing that Gaussian optics is equivalent to $Sl(2, \mathbb{R})$ can be used to show that linear optics is equivalent to $Sp(4, \mathbb{R})$, the group of linear symplectic transformations in 4 variables.

In general, let V be any (finite-dimensional, real) vector space. A bilinear form Ω on V is any function Ω: $V \times V \to \mathbb{R}$ that is linear in each variable when the other variable is held fixed; that is, $\Omega(u, v)$ is a linear function of v for each fixed u and a linear function of u for each fixed v. We say that Ω is antisymmetric if $\Omega(u, v) = -\Omega(v, u)$ for all u and v in V. We say that Ω is nondegenerate if the linear function $\Omega(u, \cdot)$ is not identically zero unless u itself is zero. An antisymmetric, nondegenerate bilinear form on V is called a *symplectic form.* A vector space possessing a given symplectic form is called a *symplectic vector space*, or is said to have a *symplectic structure.* If V is a symplectic vector space with symplectic form Ω, and if A is a linear transformation of V into itself, we say that A is a *symplectic transformation* if $\Omega(Au, Av) = \Omega(u, v)$ for all u and v in V. Later on we shall see that every symplectic vector space must be even-dimensional and that every symplectic linear transformation must have determinant 1 and, hence, be invertible. It is clear that the inverse of any symplectic transformation must be symplectic and that the product of any two symplectic transformations must be symplectic. Hence, the collection of all symplectic linear transformations form a group, which is known as the symplectic group (of V), and is denoted by $Sp(V)$.

Now let us assume that $V = \mathbb{R}^n + \mathbb{R}^n$ and write the typical vector in V as

$$u = \begin{pmatrix} q \\ p \end{pmatrix}, \qquad \text{where} \qquad q = \begin{pmatrix} q_1 \\ \vdots \\ q_n \end{pmatrix} \qquad \text{and} \qquad p = \begin{pmatrix} p_1 \\ \vdots \\ p_n \end{pmatrix}.$$

On V there is the symplectic form Ω given by

$$\Omega(u, u') = p \cdot q' - p' \cdot q,$$

where $\cdot$ denotes ordinary scalar product in $\mathbb{R}^n$. In terms of the scalar product $u \cdot u' = q \cdot q' + p \cdot p'$ we can write this as

$$\Omega(u, u') = u' \cdot Ju,$$

where J is the $2n \times 2n$ matrix $\begin{pmatrix} 0 & I \\ -I & 0 \end{pmatrix}$ and I is the $n \times n$ identity matrix. A linear transformation T on V is symplectic if, for all u and u',

$$\Omega(Tu, Tu') = \Omega(u, u').$$

We can write this as

$$T^t J T u \cdot u' = Ju \cdot u',$$

where T^t denotes the transpose of T relative to the scalar product on V. Since this is to hold for all u and u' we must have

$$T^t J T = J.$$

We can write

$$T \begin{pmatrix} q \\ q \end{pmatrix} = \begin{pmatrix} Aq + Bp \\ Cq + Dp \end{pmatrix},$$

where A, B, C, and D are $n \times n$ matrices; that is,

$$T = \begin{pmatrix} A & B \\ C & D \end{pmatrix}.$$

Then

$$T^t = \begin{pmatrix} A^t & C^t \\ B^t & D^t \end{pmatrix},$$

where A^t denotes the n-dimensional transpose of A, etc. The condition $T^t J T = J$ becomes the conditions $A^t C = C^t A$, $B^t D = D^t B$, and $A^t D - C^t B = I$. Notice that T^{-1}, which is also symplectic, is given by

$$T^{-1} = \begin{pmatrix} D^{\mathrm{t}} & -B^{\mathrm{t}} \\ -C^{\mathrm{t}} & A^{\mathrm{t}} \end{pmatrix},$$

and so we also have

$$DC^{\mathrm{t}} = CD^{\mathrm{t}} \qquad \text{and} \qquad BA^{\mathrm{t}} = AB^{\mathrm{t}}.$$

We now turn to the problem of justifying the assertion that the group of linear symplectic transformations (in four dimensions) is precisely the collection of all transformations of linear optics. As in the case of Gaussian optics, the argument can be split into two parts: The first is a physical part showing that (in the linear approximation) the matrix $\begin{pmatrix} I & 0 \\ -P & I \end{pmatrix}$, where $P = P^{\mathrm{t}}$ is a symmetric matrix, corresponds to refraction at a surface between two regions of constant index of refraction (and that every P can arise) and that $\begin{pmatrix} I & dI \\ 0 & I \end{pmatrix}$ corresponds to motion in a medium of constant index of refraction, where d is the optical distance along the axis. The second is a mathematical argument showing that every symplectic matrix can be written as a product of matrices of the above types. Let us begin with the mathematical part, which is valid in $2n$ dimensions. First observe that

$$\begin{pmatrix} I & I \\ 0 & I \end{pmatrix}\begin{pmatrix} I & 0 \\ -I & I \end{pmatrix}\begin{pmatrix} I & I \\ 0 & I \end{pmatrix} = \begin{pmatrix} 0 & I \\ -I & 0 \end{pmatrix},$$

that

$$\begin{pmatrix} 0 & I \\ -I & 0 \end{pmatrix}^3 = \begin{pmatrix} 0 & -I \\ I & 0 \end{pmatrix},$$

and that

$$\begin{pmatrix} 0 & I \\ -I & 0 \end{pmatrix}\begin{pmatrix} I & 0 \\ -P & I \end{pmatrix}\begin{pmatrix} 0 & -I \\ I & 0 \end{pmatrix} = \begin{pmatrix} I & P \\ 0 & I \end{pmatrix}.$$

Thus we can obtain all matrices of the type $\begin{pmatrix} I & P \\ 0 & I \end{pmatrix}$, where P is symmetric.

For P symmetric and nonsingular we have

$$\begin{pmatrix} I & P^{-1} \\ 0 & I \end{pmatrix}\begin{pmatrix} I & 0 \\ -P & I \end{pmatrix}\begin{pmatrix} I & P^{-1} \\ 0 & I \end{pmatrix} = \begin{pmatrix} 0 & P^{-1} \\ -P & 0 \end{pmatrix}$$

and

$$\begin{pmatrix} 0 & -I \\ I & 0 \end{pmatrix}\begin{pmatrix} 0 & P^{-1} \\ -P & 0 \end{pmatrix} = \begin{pmatrix} P & 0 \\ 0 & P^{-1} \end{pmatrix}.$$

Therefore we can get every matrix of the form $\begin{pmatrix} P & 0 \\ 0 & P^{-1} \end{pmatrix}$ with P symmetric. We claim that every nonsingular $n \times n$ matrix A can be written as the product of three symmetric matrices. We prove this fact first for the case $n = 2$ (the case of linear optics). We have a matrix $\begin{pmatrix} a & b \\ c & d \end{pmatrix}$. If $bc \neq 0$, consider the equation

$$\begin{pmatrix} \lambda & 0 \\ 0 & \mu \end{pmatrix}\begin{pmatrix} x & y \\ y & z \end{pmatrix} = \begin{pmatrix} a & b \\ c & d \end{pmatrix}$$

or

$$\lambda x = a; \qquad \lambda y = b; \qquad \mu y = c; \qquad \mu z = d.$$

We may choose $x = a$, $\lambda = 1$, $y = b$, $\mu = c/b$, and $z = d/\mu$. If $bc = 0$, then $ad \neq 0$ since the matrix is nonsingular. Then

$$\begin{pmatrix} a & b \\ c & d \end{pmatrix} = \begin{pmatrix} 0 & 1 \\ 1 & 0 \end{pmatrix}\begin{pmatrix} c & d \\ a & b \end{pmatrix},$$

and we have returned to the previous case, completing the proof for $n = 2$.

We now prove the corresponding fact in n dimensions. (We are indebted to Y. Yelamed and to T. Goodwillie for the following argument.) We can write any nonsingular $n \times n$ matrix as PO, where P is positive definite and O is orthogonal. Since P is symmetric, we must show that we can write every orthogonal matrix as the product of two symmetric ones. Now we can block diagonalize O within the group of orthogonal matrices; that is, we can find an orthogonal matrix R such that $O = RAR^{-1}$, where A consists of blocks of matrices along the diagonal at which each block is either a 2×2 matrix with nonzero off-diagonal entries or is a 1×1 matrix with ± 1 on the diagonal. Applying the result for the 2×2 case (and trivially for the 1×1 case), we see that $A = S_1 S_2$, where S_1 and S_2 are symmetric. Then $O = (RS_1R^{-1})(RS_2R^{-1})$, and the conjugate of a symmetric matrix by an orthogonal one is still symmetric, proving that every orthogonal matrix can be written as the product of two symmetric ones, and, hence, that every nonsingular matrix can be written as the product of three symmetric ones.

Now $P^{-1}Q^{-1}\dots R^{-1} = [(PQ\dots R)^t]^{-1}$ if $P, Q, \dots, R$ are symmetric. We thus can get every symplectic transformation of the form

$$\begin{pmatrix} A & 0 \\ 0 & (A^t)^{-1} \end{pmatrix}.$$

Now consider a symplectic matrix $\begin{pmatrix} A & B \\ C & D \end{pmatrix}$ with A nonsingular. Then

$$\begin{pmatrix} I & 0 \\ -E & I \end{pmatrix}\begin{pmatrix} A & B \\ C & D \end{pmatrix} = \begin{pmatrix} A & B \\ C - EA & D - EB \end{pmatrix}.$$

Choose $E = CA^{-1}$, which is possible since CA^{-1} is symmetric. We then get

$$\begin{pmatrix} A & B \\ 0 & D - EB \end{pmatrix}$$

with $D - EB = (A^t)^{-1}$. Multiplying by $\begin{pmatrix} A^{-1} & 0 \\ 0 & A^t \end{pmatrix}$ gives

$$\begin{pmatrix} I & A^{-1}B \\ 0 & I \end{pmatrix}.$$

We can thus obtain any symplectic matrix with A nonsingular. Now suppose that A is singular. Then we can find invertible $n \times n$ matrices L and M such that

$$LAM = \begin{pmatrix} I_r & 0 \\ 0 & 0 \end{pmatrix},$$

where I_r is the $r \times r$ identity matrix, $r = \operatorname{rank} A$. We may thus pre- and postmultiply by

$$\begin{pmatrix} L & 0 \\ 0 & (L^t)^{-1} \end{pmatrix} \quad \text{and} \quad \begin{pmatrix} M & 0 \\ 0 & M^t \end{pmatrix},$$

respectively, and assume that our matrix $\begin{pmatrix} A & B \\ C & D \end{pmatrix}$ is such that

$$A = \begin{pmatrix} I_r & 0 \\ 0 & 0 \end{pmatrix}.$$

We can write

$$C = \begin{pmatrix} C_1 & C_2 \\ C_3 & C_4 \end{pmatrix},$$

where C_1 is the $r \times r$ component of C, etc. Then

$$A^t C = \begin{pmatrix} C_1 & C_2 \\ 0 & 0 \end{pmatrix},$$

and the condition that $A^t C$ be symmetric implies that $C_2 = 0$ and that C_1 is symmetric. Now

$$\begin{pmatrix} I & E \\ 0 & I \end{pmatrix}\begin{pmatrix} A & B \\ C & D \end{pmatrix} = \begin{pmatrix} A + EC & B + ED \\ C & D \end{pmatrix}.$$

We choose

$$E = \begin{pmatrix} 0 & 0 \\ 0 & I_{n-r} \end{pmatrix}.$$

Then E is symmetric and

$$A + EC = \begin{pmatrix} I_r & 0 \\ C_3 & C_4 \end{pmatrix}.$$

We claim that this matrix is nonsingular. For otherwise C_4 would be singular. But this would mean that there is some vector $v \neq 0$ whose first r components vanish and which is sent into zero by C, since $C_2 = 0$. Thus $Av = 0$ and $Cv = 0$, implying that $(D^t A - B^t C)v = 0$, contradicting the condition that $D^t A - B^t C = I$. Thus $A + EC$ is nonsingular, and we have returned to the preceding case.

We now turn our attention to the physical aspects of the problem. As in Gaussian optics, we describe the incoming light ray by its direction $v = (v_x, v_y, v_z)$ and its intersection with the plane parallel to the x-y plane passing through the point z on the optical axis. Here $\|v\|^2 = v_x^2 + v_y^2 + v_z^2 = 1$. Now

$$v_z = (1 - v_x^2 - v_y^2)^{1/2} = 1 - \tfrac{1}{2}(v_x^2 + v_y^2) + \cdots \doteq 1,$$

since we are ignoring quadratic terms in v_x and v_y, which are assumed small. We set

$$p_x = nv_x, \qquad p_y = nv_y,$$

where n is the index of refraction. Moving a distance t along the optical axis is the same (up to quadratic terms in v_x and v_y) as moving a distance t along the line through v and hence

$$q_{2x} - q_{1x} = tv_x$$

and

$$q_{2y} - q_{1y} = tv_y$$

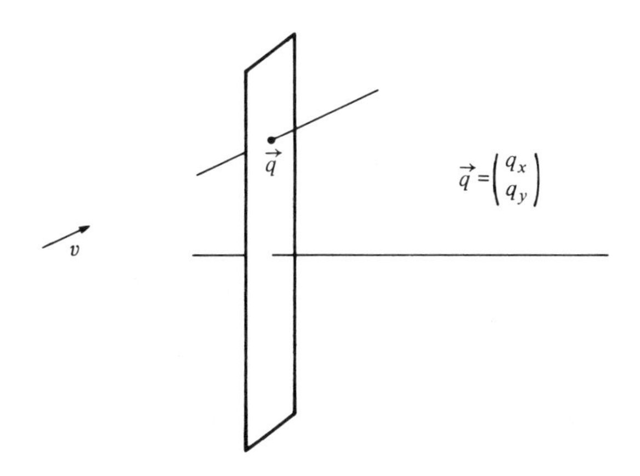

Figure 4.2.

or

$$\begin{pmatrix} \vec{q}_2 \\ \vec{p}_2 \end{pmatrix} = \begin{pmatrix} I & dI \\ 0 & I \end{pmatrix} \begin{pmatrix} \vec{q}_1 \\ \vec{p}_1 \end{pmatrix}$$

where $d = t/n$ (see Figure 4.2).

Now let us turn to refraction. We may assume that our surface is quadratic, and is given by

$$z' - z = \tfrac{1}{2} k\vec{q}\cdot\vec{q},$$

where k is a symmetric (2×2) matrix. The normal to this surface at the point q is given by

$$u = (k\vec{q}, -1).$$

(Up to quadratic terms and higher, u has length 1.) The projection of a vector v onto the tangent plane to the surface at q is given by

$$v - (v\cdot u)u.$$

Writing $v = (v_x, v_y, 1) = (\vec{v}, 1)$ we see that $v\cdot u = k\vec{q}\cdot\vec{v} - 1$ and

$$v - (v\cdot u)u = (\vec{v}, 1) - (k\vec{q}\cdot\vec{v} - 1)(k\vec{q}, -1).$$

Ignoring the quadratic term $k\vec{q}\cdot\vec{v}$ this becomes

$$(\vec{v} + k\vec{q}, 0).$$

Snell's law says that $n_1(v_1 - (v_1\cdot u)u) = n_2(v_2 - (v_2\cdot u)u)$. In the linear approximation, with $\vec{p}_1 = n\vec{v}_1$ and $\vec{p}_2 = n\vec{v}_2$, this becomes

$$\vec{p}_1 - n_1 k\vec{q} = \vec{p}_2 - n_2 k\vec{q}$$

or

$$\vec{p}_2 = \vec{p}_1 - P\vec{q},$$

where

$$P = -(n_1 - n_2)k.$$

We thus get the refraction matrix $\begin{pmatrix} I & 0 \\ -P & I \end{pmatrix}$. This concludes our proof that linear optics is isomorphic to the study of the group $Sp(4, \mathbb{R})$.

There is one more point relating to Gaussian optics that deserves mention. Suppose that our optical system is rotationally invariant; then at each refracting surface, the power matrix P is of the form $P = cI$, where c is a scalar and I is the 2×2 identity matrix. It is clear that the collection of matrices that one can get by multiplying such matrices with $\begin{pmatrix} I & dI \\ 0 & I \end{pmatrix}$ will be of the form $\begin{pmatrix} aI & bI \\ cI & dI \end{pmatrix}$. These form a subgroup of $Sp(4, \mathbb{R})$ that is isomorphic to $Sl(2, \mathbb{R}) = Sp(2, \mathbb{R})$. Note that a matrix of the above form, when acting on

$$\begin{pmatrix} q_x \\ q_y \\ p_x \\ p_y \end{pmatrix},$$

is the same as $\begin{pmatrix} a & b \\ c & d \end{pmatrix}$ acting separately on $\begin{pmatrix} q_x \\ p_x \end{pmatrix}$ and $\begin{pmatrix} q_y \\ p_y \end{pmatrix}$. Thus, in our study of Gaussian optics the restriction to paraxial rays was unnecessary. We could have treated skew rays by simply treating the x and y components separately in the same fashion. This is a consequence of the linear approximation.

The basic formula for the optical length

$$L = L_{\text{axis}} + \tfrac{1}{2}(p_2 \cdot q_2 - p_1 \cdot q_1)$$

(where the p and q components are now vectors) is proved exactly as it was in the Gaussian case, by looking at what happens at each of our basic components. There is no point in repeating the proof.

Two planes are called nonconjugate if, in the optical matrix relating them, the matrix B is nonsingular. Then we can solve the equations

$$q_2 = Aq_1 + Bp_1$$

and

$$p_2 = Cq_1 + Dp_1$$

for p_1 and p_2 as

$$p_1 = -B^{-1}Aq_1 + B^{-1}q_2$$

and

$$p_2 = (C - DB^{-1}A)q_1 + DB^{-1}q_2.$$

We can then write

$$L = L_{\text{axis}} + W(q_1, q_2),$$

where

$$W(q_1, q_2) = \tfrac{1}{2}[DB^{-1}q_2 \cdot q_2 + B^{-1}Aq_1 \cdot q_1 - 2(B^{\text{t}})^{-1}q_1 \cdot q_2].$$

(In proving this formula we make use of the identity

$$-B^{\text{t}-1} = C - DB^{-1}A,$$

which follows for nonsingular B from $A^{\text{t}}D - B^{\text{t}}C = I$.) A direct computation (using the above identity) shows that (in the obvious sense)

$$\frac{\partial L}{\partial q_2} = p_2$$

and

$$\frac{\partial L}{\partial q_1} = -p_1.$$

Thus a knowledge of L allows us to determine p_1 and p_2 in terms of q_1 and q_2.

In the next section we will move from linear optics to geometrical optics. In geometrical optics we will have to be a little more careful about our choice of p variables. In linear optics we have identified θ, $\sin\theta$, and $\tan\theta$, where θ is a small angle. It turns out that $\sin\theta$ is the right variable to use. More precisely, let us assume that the index of refraction is now a smoothly varying function in space (perhaps rapidly changing across an interface) so that the optical rays will now be smooth curves instead of broken line segments. We will still use the z axis to parametrize these rays (we are staying away from large deviations from the optical axis), and so a ray will be given by a pair of functions $x(z)$ and $y(z)$ and we will write $\dot{x}$ for dx/dz.

We set

$$p_x = \frac{\dot{x}n(x, y, z)}{\sqrt{1 + \dot{x}^2 + \dot{y}^2}}, \qquad p_y = \frac{\dot{y}n(x, y, z)}{\sqrt{1 + \dot{x}^2 + \dot{y}^2}};$$
$$q_x = x, \qquad q_y = y$$

as the parameters describing a light ray at the z plane. The basic assertion of geometrical optics is that the transformation from one z plane to another is a symplectic diffeomorphism. Hamilton derived this from Fermat's principle. We shall present the argument in Section 6.

5. Geometrical optics, Hamilton's method, and the theory of geometrical aberrations

Let (V_1, ω_1) and (V_2, ω_2) be two copies of $\mathbb{R}^4$ (or more generally of $\mathbb{R}^{2n}$ or, even more generally, they can be any two symplectic manifolds of the same dimension).

On $V_1 \times V_2$ we can put the symplectic form

$$\Omega = \pi_2^* \omega_2 - \pi_1^* \omega_1,$$

where π_1 and π_2 denote the projections onto the first and second components. By abuse of notation we shall usually drop the π^*'s and simply write this as

$$\Omega = \omega_2 - \omega_1.$$

A smooth map T: $V_1 \to V_2$ is called a symplectic diffeomorphism (or symplectomorphism or canonical transformation) if $T^*\omega_2 = \omega_1$ and if T is a diffeomorphism. (Notice that the condition $T^*\omega_2 = \omega_1$, together with the fact that the ω are nondegenerate, implies that dT_x is injective for each x in V_1; that is, that T is a local diffeomorphism since V_1 and V_2 have the same dimension. Since all the considerations in this section will be purely local, we shall simply work with the condition $T^*\omega_2 = \omega_1$.)

The submanifold graph $T = (v, Tv)$ is a dim V_1 submanifold of $V_1 \times V_2$ and we shall let ι: graph $T \to V_1 \times V_2$ denote its standard injection as a submanifold. Then it is clear that T is symplectic if and only if

$$\iota^*\Omega = 0.$$

(A submanifold Λ of $V_1 \times V_2$ with the property that dim $\Lambda = \frac{1}{2}\dim(V_1 \times V_2) = \dim V_1$ and that satisfies $\iota^*\Omega = 0$ is called a *Lagrangian* (or sometimes a canonical) relation. If π_1 maps Λ diffeomorphically onto V_1, then we can write Λ = graph T, where T is a symplectic diffeomorphism.)

In the case that $V_1 = V_2 = \mathbb{R}^n + \mathbb{R}^n$ we can consider the form

$$\alpha = p_2 \cdot dq_2 - p_1 \cdot dq_1$$

on $V_1 \times V_2$ and observe that $d\alpha = \Omega$. Thus, for any symplectic diffeomorphism T we have $d\iota^*\alpha = 0$.

Let us suppose that the projection $\begin{pmatrix} q_1 \\ p_1 \end{pmatrix}\begin{pmatrix} q_2 \\ p_2 \end{pmatrix} \to \begin{pmatrix} q_1 \\ q_2 \end{pmatrix}$ is a diffeomorphism (locally) when restricted to graph T. Then we can introduce the q_1's and q_2's as coordinates on graph T and write (at least locally)

$$\iota^*\alpha = dL$$

for some function L. (Hamilton called this function the *point characteristic* of the optical system, as it depends on the incoming and outgoing "points" q_1 and q_2.) From L we can reconstruct the submanifold graph T and, hence, the mapping T by the formulas

$$dL = p_2 \cdot dq_2 - p_1 \cdot dq_1 \qquad \text{so} \quad p_2 = \frac{\partial L}{\partial q_2} \qquad \text{and} \qquad p_1 = -\frac{\partial L}{\partial q_1}.$$

In the case that T is a linear transformation, the condition that the projection onto the space of (q_1, q_2) be a local diffeomorphism is equivalent, as we have seen, to the condition that B be nonsingular. The planes are not conjugate to one another. In that case we have seen that we may take L to be the optical length of the unique ray joining q_1 to q_2 and that the formula for L is given by

$$L = L_{\text{axis}} + W,$$

where

$$W(q_1, q_2) = \tfrac{1}{2}[DB^{-1}q_2 \cdot q_2 + B^{-1}Aq_1 \cdot q_1 - 2(B^t)^{-1}q_1 \cdot q_2].$$

In the general case it can be proved that L still coincides with the optical length. But for the moment it is enough to know that we can always find some function L that generates the symplectic transformation T.

Note that the important geometrical information lies in T, or what is the same, in graph T; and the important geometrical condition is that

$$\iota^*\Omega = 0.$$

The fact that we could choose q_1 and q_2 as coordinates and choose α as above was only auxiliary information. For example, suppose we could solve for q_1, q_2 in terms of p_1, p_2, that is, that the projection onto the p_1, p_2 subspace were a diffeomorphism on graph T. (Geometrically this means that we can use the incoming and outgoing angles to uniquely determine the ray.) Then we could consider the form

$$\beta = q_1 \cdot dp_1 - q_2 \cdot dp_2.$$

Notice that

$$d\beta = d\alpha = dp_2 \wedge dq_2 - dp_1 \wedge dq_1 = \Omega$$

everywhere and hence $d\beta = 0$ on graph T. Also notice that

$$\beta = \alpha - d(q_2 \cdot p_2 - q_1 \cdot p_1).$$

Hence, we may set

$$A = L - (q_2 \cdot p_2 - q_1 \cdot p_1)$$

and conclude that

$$dA = \beta = q_1 \cdot dp_1 - q_2 \cdot dp_2$$

on graph T. Hamilton called A the *angle characteristic* and the preceding equation can be written as

$$q_1 = \frac{\partial A}{\partial p_1}, \qquad q_2 = -\frac{\partial A}{\partial p_2}.$$

As above, we can consider A as "generating" the transformation T.

Finally, suppose we let

$$\gamma = -(q_2 \cdot dp_2 + p_1 \cdot dq_1) = \alpha - d(p_2 \cdot q_2).$$

Then again $d\gamma = d\alpha$ everywhere. We can write $\gamma = dM$, where $M = L - p_2 q_2$ on graph T. Suppose that we can solve for p_1 and q_2 in terms of q_1 and p_2 on graph T. Then we get

$$q_2 = -\frac{\partial M}{\partial p_2} \quad \text{and} \quad p_1 = -\frac{\partial M}{\partial q_1}$$

as equations generating the transformation T. Hamilton called the function M the *mixed characteristic* of the system. The "asymmetric" assumption involved in the mixed characteristic may seem a bit strange at first, but let us point out that this is exactly the situation that arises when we want to study what happens near a pair of conjugate planes. Indeed, let us assume that the planes are in perfect focus with each other in the linear approximation; that is, that the matrix $B = 0$. Then the optical matrix between these two focal planes is given by

$$\begin{pmatrix} A & 0 \\ C & A^{-1} \end{pmatrix}$$

with $A \neq 0$. Thus $q_2 = Aq_1$, $p_1 = A(p_2 - Cq_1)$ solves for p_1, q_2 in terms of q_1 and p_2. For a plane a distance t from the plane, the matrix is

$$\begin{pmatrix} 1 & t \\ 0 & 1 \end{pmatrix} \begin{pmatrix} A & 0 \\ C & A^{-1} \end{pmatrix} = \begin{pmatrix} A + tC & tA^{-1} \\ C & A^{-1} \end{pmatrix}$$

and

$$q_2 = Aq_1 + tp_2,$$
$$p_1 = A(p_2 - Cq_1).$$

Thus, up to an additive constant, in the linear approximation

$$M(q_1, p_2, t) = -Ap_2 \cdot q_1 + \tfrac{1}{2}Cq_1 \cdot q_2 - \tfrac{1}{2}tp_2^2,$$

where we have made the dependence on t explicit. Notice that

$$q_2 = \frac{-\partial M}{\partial p_2} = Aq_1 + tp_2,$$

as required. The focal plane is determined by the requirement that the position of the arriving rays depend only on q_1; here $\partial q_2/\partial p_2 = t = 0$, as expected. Suppose that M is a differentiable function that is not necessarily quadratic, but that our system is rotationally symmetric. Then M can only depend on q_1 and p_2 via the functions

$$u = \|q_1\|^2 = q_{x1}^2 + q_{y1}^2,$$
$$v = \|p_2\|^2 = p_{x2}^2 + p_{y2}^2,$$

and

$$w = 2q_1 \cdot p_2.$$

Making the dependence on the initial and final planes explicit, we can write

$$M(q_1, p_2; z_1, z_2) = \hat{W}(u, v, w; z_1, z_2)$$

for some function $\hat{W}$. The linear terms in $\hat{W}$ will give rise to the quadratic expression for M written above (where now A and C are scalar matrices). We see that there are no third-order terms in the expansion of M, and thus the aberrations start with the terms that arise from the quadratic terms in $\hat{W}$, which give rise to fourth-order terms in M, and hence third-order terms in the expression of q_2 and p_1 as functions of q_1 and p_2. These third-order aberrations are known as the *Seidel aberrations.* Before describing them in detail, we will examine the optical significance of expanding in terms of q_1 and p_2.

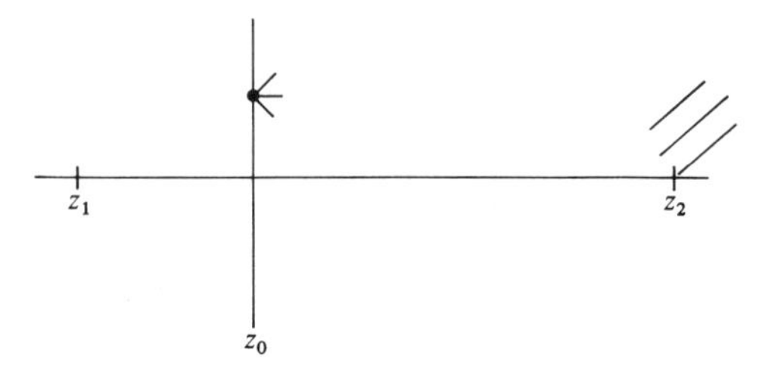

Figure 5.1.

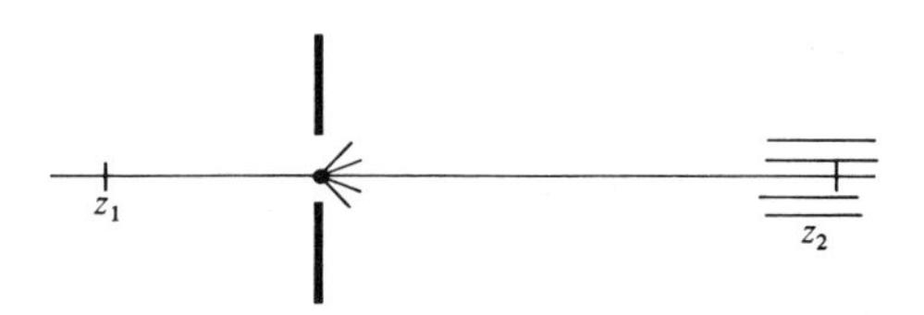

Figure 5.2.

Suppose we have two planes z_1 and z_2 that are conjugate in the Gaussian approximation. Suppose that z_0 represents the first local plane of the system so that all rays through a fixed point at z_0 emerge as parallel rays at z_2 (Figure 5.1). Now suppose that we put a diaphragm at z_0 (Figure 5.2). When the diaphragm is almost shut, all the rays will emerge at z_2 with p_2 close to 0. Thus opening the diaphragm corresponds to increasing the values of $|p_2|$. Hence, expansion in terms of p_2 describes the effects of opening the diaphragm.†

We now describe the Seidel aberrations in detail. We introduce $p_{x_2}, p_{y_2}, q_{x_1}, q_{y_1}$ as coordinates on graph T and write

$$q_{x_2} = -\frac{\partial M}{\partial p_{x_2}}, \qquad q_{y_2} = -\frac{\partial M}{\partial p_{y_2}};$$

$$p_{x_1} = -\frac{\partial M}{\partial q_{x_1}}, \qquad p_{y_1} = -\frac{\partial M}{\partial q_{y_1}};$$

where $M = W_1 + W_2$. The terms from W_1 give the linear approximation to the expansion of the q_2's in terms of the q_1 and p_2. We let Δq_{x_2} denote the aberration in q_{x_2} (i.e., the deviation of q_{x_2} from its value if W were linear).

† The rest of this section can be omitted on first reading.

Thus

$$\Delta q_{x_2} = -\frac{\partial W_2}{\partial p_{x_2}},$$

so

$$\Delta q_{x_2} = -2\left(\frac{\partial W_2}{\partial w} q_{x_1} + \frac{\partial W_2}{\partial v} p_{x_2}\right), \qquad \Delta q_{y_2} = -2\left(\frac{\partial W_2}{\partial w} q_{y_1} + \frac{\partial W_2}{\partial v} p_{y_2}\right);$$

$$\Delta p_{x_1} = -2\left(\frac{\partial W_2}{\partial u} q_{x_1} + \frac{\partial W_2}{\partial w} p_{x_2}\right), \qquad \Delta p_{y_1} = -2\left(\frac{\partial W_2}{\partial u} q_{y_1} + \frac{\partial W_2}{\partial w} p_{y_2}\right).$$

The most general quadratic form in three variables depends on six parameters. With a view to later interpretation let us set

$$W_2 = -\left[\tfrac{1}{4}Fu^2 + \tfrac{1}{4}Av^2 + \tfrac{1}{8}(C-D)w^2 + \tfrac{1}{2}Duv + \tfrac{1}{2}Euw + \tfrac{1}{6}Bvw\right].$$

Then our equations become

$$\Delta q_{x_2} = \left[Eu + \tfrac{1}{3}Bv + \tfrac{1}{2}(C-D)w\right]q_{x_1} + \left[Av + Du + \tfrac{1}{3}Bw\right]p_{x_2};$$

$$\Delta q_{y_2} = \left[Eu + \tfrac{1}{3}Bv + \tfrac{1}{2}(C-D)w\right]q_{y_1} + \left[Av + Du + \tfrac{1}{3}Bw\right]p_{y_2}.$$

To investigate the meaning of these equations we consider incoming rays with

$$q_{y_1} = 0, \qquad q_{x_1} = x$$

and introduce polar coordinates

$$p_{x_2} = \rho \cos \varphi, \qquad p_{y_2} = \rho \sin \varphi,$$

so that $u = x^2, v = \rho^2$, and $w = 2x\rho\cos\varphi$ and we get

$$\Delta q_{x_2} = A\rho^3 \cos \varphi + \tfrac{1}{3}B\rho^2(3\cos^2\varphi + \sin^2\varphi)x + C(\rho\cos\varphi)x^2 + Ex^3.$$

$$\Delta q_{y_2} = A\rho^3 \sin \varphi + \tfrac{2}{3}B\rho^2(\sin\varphi\cos\varphi)x + D(\rho\sin\varphi)x^2$$

so that A, B, C, D, and E are the coefficients of terms involving increasing powers of x. For $x = 0$ only the first term appears. This represents the fact that the point is not perfectly focused, but is imaged on a disk of radius $A\rho^3$, where $\rho^2 = p_{x_2}^2 + p_{y_2}^2$ is determined, as we remarked above, by the opening of the diaphragm. This is spherical aberration.

Suppose we have corrected for spherical aberration (or are willing to ignore it), and let us consider the contribution of the second term. For fixed ρ and x, the curve

$$\Delta_B q_{x_2} = \tfrac{1}{3}B\rho^2 x(3\cos^2\varphi + \sin^2\varphi) = \tfrac{1}{3}B\rho^2 x(2 + \cos 2\varphi),$$

$$\Delta_B q_{q_2} = \tfrac{2}{3}B\rho^2 x \sin\varphi\cos\varphi = \tfrac{1}{3}B\rho^2 x \sin 2\varphi$$

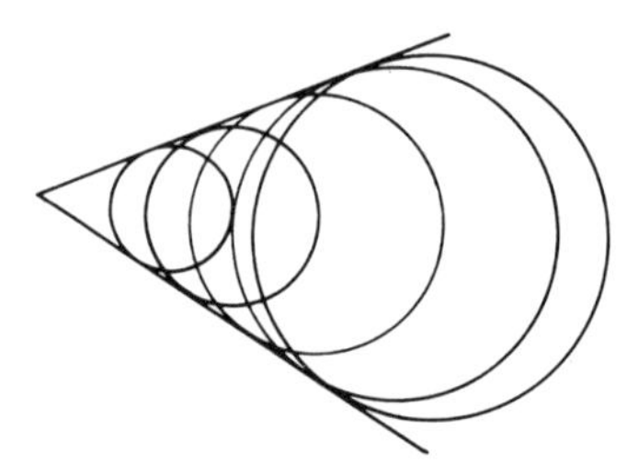

Figure 5.3.

describes the circle of center $= \frac{2}{3}B\rho^2 x$ and radius $= \frac{1}{3}B\rho^2 x$ twice as φ goes from 0 to 2π. If we now put a ring diaphragm (i.e., one that stops all rays except those passing through a circular ring) at the focal plane, then rotational symmetry guarantees that the corresponding rays all have $p_{x_2}^2 + p_{y_2}^2$ constant, and thus they image on one of the above circles. These circles are illustrated in Figure 5.3: The effect is that rays emanating from x form a planelike image reminiscent of a comet; hence, this aberration is known as *coma*.

Suppose the system is corrected for spherical aberration and coma so that $A = B = 0$, and let us consider the effects of C and D:

$$\Delta_{C,D} q_{x_2} = Cx^2 \rho \cos \varphi;$$

$$\Delta_{C,D} q_{y_3} = Dx^2 \rho \sin \varphi.$$

We see that rays passing through the ring diaphragm form an ellipse on the image plane. In order to understand this effect, note what happens if we try to improve the focus by moving the plane $z = z_1$ by an amount s. Thus we wish to consider $M_{z_1, z_2 + s}$. We assume that n_2 is constant. Then the ray given by $(x, 0; p_{x_2}, p_{y_2})$ will intersect the plane $z = z_2 + s$ at $x_2(s)$, $y_2(s)$, where

$$x_2(s) = x_2 + s\frac{p_{x2}}{n_2} = mx + \left(C_x^2 + \frac{s}{n_2}\right)\rho \cos \varphi,$$

$$y_2(s) = y_2 + s\frac{p_{y2}}{n_2} = \left(Dx^2 + \frac{s}{n_2}\right)\rho \sin \varphi,$$

thus describing an ellipse about mx with axes

$$Cx^2 + \frac{s}{n_2} \qquad \text{and} \qquad Dx^2 + \frac{s}{n_2}.$$

We can make the first axis vanish by choosing

$$s_p = -n_2 Cx^2 = -\frac{n_2}{m^2} Cq_{x_2}^2$$

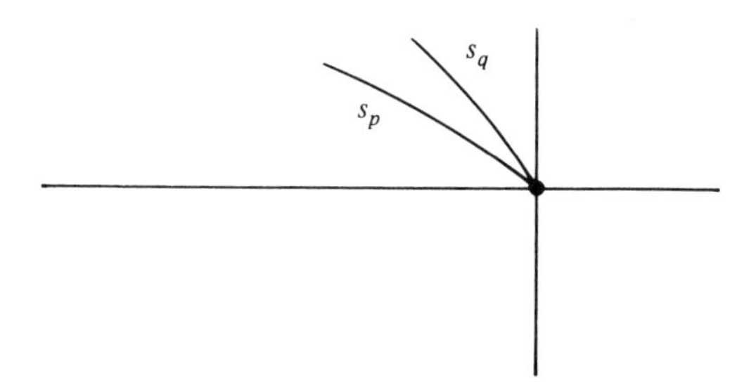

Figure 5.4.

up to higher-order terms and the second axis vanishes by choosing

$$s_q = -n_u D x^2 = -\frac{n_2}{m^2} D q_{x_2}^2 + \cdots .$$

Thus s_p will not coincide with s_q for $x \neq 0$ unless $C = D$ (i.e., the rays emanating from $(x, 0)$ do not focus), but we have two focal lines instead of a focal point. This effect is known as *astigmatism.* The fact that the focusing does not occur on a plane but rather along the surfaces of rotation swept out by the curves s_p or s_q is known as curvature of the field (Figure 5.4). The difference $s_p - s_q = (n_1/m^2)(D - C)q_{x_2}^2$ is called the astigmatism, and the average

$$\tfrac{1}{2}(s_p + s_q) = -\tfrac{1}{2}(n_1/m^2)(D + C)q_{x_2}^2$$

is called the curvature of field.

Suppose that A, B, C, and D all vanish. Then

$$\Delta_E q_{x_2} = Ex^3, \qquad \Delta_E q_{y_2} = 0.$$

In this case (to this order of approximation) all the points q_{x_1}, q_{y_1} are focused perfectly. But the map from $q_{x_1} q_{y_1}$ to $q_{x_2} q_{y_2}$ is not linear. Indeed we have $(x, 0)$ going into $mx + Ex^3$ and so, by symmetry,

$$q_{x_2} = m q_{x_1} + E q_{x_1}(q_{x_1}^2 + q_{y_1}^2),$$
$$q_{y_2} = m q_{y_1} + E q_{y_1}(q_{x_1}^2 + q_{y_1}^2).$$

This effect is known as *distortion.* If E and m have the same sign, then a point is moved further away from the origin depending on its distance, thus the image of a centered square appears as shown in Figure 5.5. This is known as pincushion distortion. If E and m have opposite signs then the image of a centered square is that shown in Figure 5.6. This is known as barrel distortion.

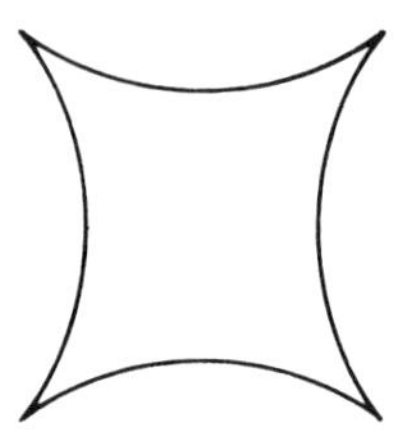

Figure 5.5.

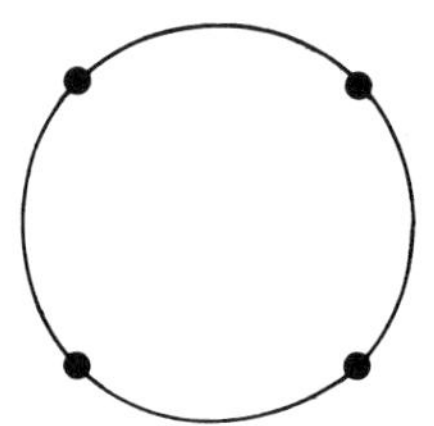

Figure 5.6.

6. Fermat's principle and Hamilton's principle

We have not yet shown why we expect the transformation from one set of ray parameters to another in geometric optics to be symplectic. In this section we show how this is a consequence of Fermat's principle. We assume that the index of refraction is a smoothly varying function n. For any path γ that has nonsingular projection onto the z axis, we can use z as parameter, and so we write

$$\gamma(z) = \begin{pmatrix} x(z) \\ y(z) \\ z \end{pmatrix}.$$

The optical length is defined as

$$\mathbf{J}(\gamma) = \int n(1 + \dot{x}^2 + \dot{y}^2)^{1/2}\, dz$$
$$= \int \mathbf{J}(x, y, \dot{x}, \dot{y}, z)\, dz.$$

When we want to compute the optical length of the path between two planes the integral is computed between the two values z_1 and z_2. Fermat's principle asserts that the light rays are those paths that are extrema for the optical length between any two fixed endpoints. We now describe

Hamilton's method for finding these extrema – that is, Hamilton's version of the Euler–Lagrange equations:

The function $\mathbf{J}$ depends on five variables

$$\mathbf{J} = \mathbf{J}(x, y, \dot{x}, \dot{y}, z) = n(x, y, z)(1 + \dot{x}^2 + \dot{y}^2)^{1/2}.$$

We introduce new variables q_x, q_y, p_x, p_y, z to define a mapping $\phi: \mathbb{R}^5 \to \mathbb{R}^5$ by the formula

$$q_x = x,$$
$$q_y = y,$$
$$p_x = \frac{\partial \mathbf{J}}{\partial \dot{x}} = \frac{n\dot{x}}{(1 + \dot{x}^2 + \dot{y}^2)^{1/2}},$$
$$p_y = \frac{\partial \mathbf{J}}{\partial \dot{y}} = \frac{n\dot{y}}{(1 + \dot{x}^2 + \dot{y}^2)^{1/2}}.$$

Define the function H by

$$H(q_x, q_y, p_x, p_y, z) = p_x\dot{x} + p_y\dot{y} - \mathbf{J},$$

where we regard $\dot{x}$, $\dot{y}$, and $\mathbf{J}$ as functions of q_x, q_y, p_x, p_z, z via the map ϕ^{-1}. That is, we should really write the preceding equation as

$$H(q_x, q_y, p_x, p_y, z) = p_x(\dot{x} \circ \phi^{-1}) + p_y(\dot{y} \circ \phi^{-1}) - \mathbf{J} \circ \phi^{-1}.$$

Note that

$$(p_x \circ \phi)\dot{x} + (p_y \circ \phi)\dot{y} - H = \mathbf{J}.$$

This has the following important consequence: For any smooth curve γ where

$$\gamma(z) = \begin{pmatrix} x(z) \\ y(z) \\ z \end{pmatrix},$$

we get a curve $\dot{\gamma}$ in $\mathbb{R}^5$:

$$\dot{\gamma}(z) = \begin{pmatrix} x(z) \\ y(z) \\ \dot{x}(z) \\ \dot{y}(z) \\ z \end{pmatrix}$$

and, hence, a smooth curve

$$\hat{\gamma} = \phi \circ \dot{\gamma}$$

in $\mathbb{R}^5$.

Now $\hat{\gamma}^* dq_x = \dot{x}\, dz$ and $\hat{\gamma}^* dq_y = \dot{y}\, dz$, so

$$\hat{\gamma}^*(p_x dq_x + p_y dq_y - H\, dz) = \mathbf{J}(\dot{\gamma}(z))\, dz.$$

In other words, if we define the form θ on $\mathbb{R}^5$ by

$$\theta = p_x dq_x + p_y dq_y - H\, dz,$$

we see that

$$\mathbf{J}(\gamma) = \int_{\hat{\gamma}} \theta.$$

Consider the form

$$d\theta = dp_x \wedge dq_x + dp_y \wedge dq_y - dH \wedge dz$$

on $\mathbb{R}^5$. This form is closed and clearly has rank 4. It therefore defines at each point of $\mathbb{R}^5$ a one-dimensional subspace spanned by those vectors η that satisfy

$$i(\eta)\, d\theta = 0.$$

If we choose η of the form $\eta = \partial/\partial z + A(\partial/\partial q_x) + B(\partial/\partial q_y) + C(\partial/\partial p_x) + D(\partial/\partial p_y)$, then

$$\begin{aligned} i(\eta)\, d\theta = {} & \left(C + \frac{\partial H}{\partial q_x}\right) dq_x + \left(D + \frac{\partial H}{\partial q_y}\right) dq_y \\ & - \left(A - \frac{\partial H}{\partial p_x}\right) dp_z - \left(B - \frac{\partial H}{\partial p_y}\right) dp_y \\ & - \left(+ A\frac{\partial H}{\partial q_x} + B\frac{\partial H}{\partial q_y} + C\frac{\partial H}{\partial p_x} + D\frac{\partial H}{\partial p_y}\right) dz. \end{aligned}$$

In other words, a curve v in $\mathbb{R}^5$ given by

$$v(z) = \begin{pmatrix} q_x(z) \\ q_y(z) \\ p_x(z) \\ p_y(z) \\ z \end{pmatrix}$$

will be everywhere tangent to the lines defined by $i(\eta)\,d\theta = 0$ if and only if

$$\frac{dq_x}{dz} = \frac{\partial H}{\partial p_x}, \tag{6.1}$$

$$\frac{dq_y}{dz} = \frac{\partial H}{\partial p_y}, \tag{6.2}$$

$$\frac{dp_x}{dz} = -\frac{\partial H}{\partial q_x}, \tag{6.3}$$

$$\frac{dp_y}{dz} = -\frac{\partial H}{\partial q_y}. \tag{6.4}$$

Now $\partial H/\partial p_x = \dot{x}$, and thus the first of these equations implies that $\phi^{-1}v$ satisfies

$$\frac{dx(\phi^{-1}v)}{dz} = \dot{x}$$

and, similarly,

$$\frac{dy(\phi^{-1}v)}{dz} = \dot{y}.$$

Thus Equations (6.1) and (6.2) imply that there exists a curve γ such that

$$v = \hat{\gamma}.$$

It is a standard theorem in the calculus of variations (Loomis and Sternberg, 1968, p. 535) that γ is an extremal for $J(\gamma)$ if and only if $\hat{\gamma}$ satisfies these equations. Thus

γ is an extremal of optical length if and only if $\hat{\gamma}$ satisfies

$$i(\hat{\gamma}')\,d\theta \equiv 0.$$

Now we can consider the space of all curves satisfying this equation. This space is a four-dimensional manifold M, and we put local coordinates on M by choosing (locally) some four-dimensional surface S transversal to these curves and using local coordinates on S (Figure 6.1). The restriction of $d\theta$ to S is closed and nondegenerate. Let us denote it by ω_S. Suppose we had two such surfaces S_1 and S_2 that meet the same curves. Then we get a map T from S_1 to S_2 by simply following the curves from one surface to the second, as shown in Figure 6.2. We claim that $T^*\omega_{S_2} = \omega_{S_1}$. Indeed, consider any

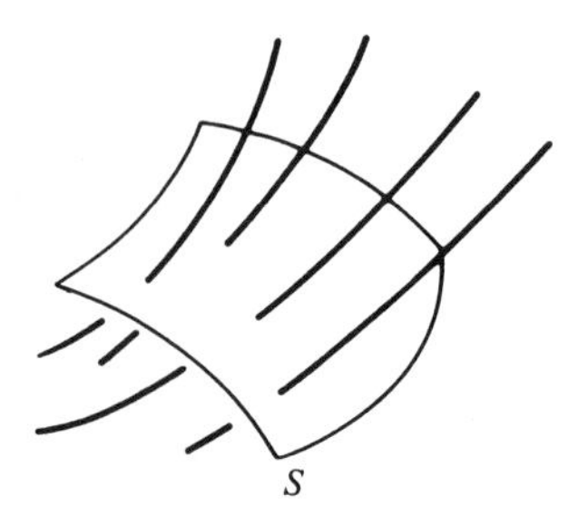

Figure 6.1.

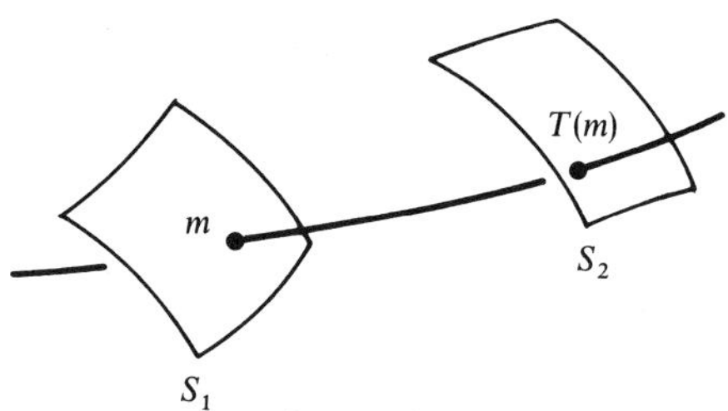

Figure 6.2.

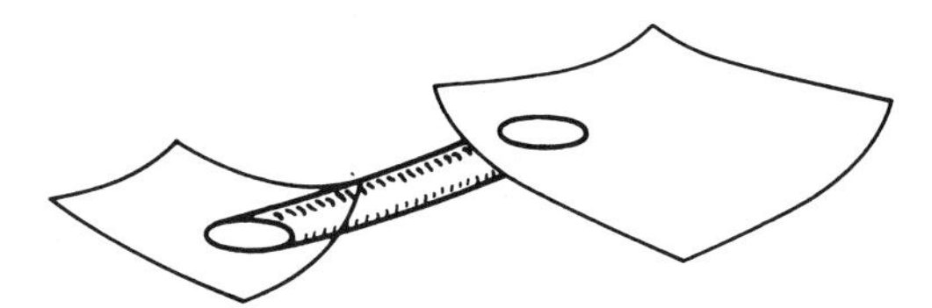

Figure 6.3.

small region R in S_1 and its image, TR on S_2, and let D denote the domain in $\mathbb{R}^5$ swept out by the curves from R to $TR \subset S_2$.

The boundary of D consists of TR, R, and the curves joining ∂R to ∂TR (Figure 6.3). The restriction of $d\theta$ to the portion of the boundary containing the curves must vanish, since these curves are all tangent to $i(\eta)\,d\theta = 0$. Thus

$$\int_D d(d\theta) = \int_{TR} d\theta - \int_R d\theta.$$

But $d(d\theta) = 0$, so

$$\int_{TR} \omega_{S_2} = \int_R \omega_{S_1}$$

for all regions R and hence

$$T^*\omega_{S_2} = \omega_{S_1}.$$

In particular, if we take S_1 to be the surface given by $z = z_1$, then

$$\omega_{S_1} = dp_x \wedge dq_x + dp_y \wedge dq_y$$

on S_1. This proves our assertion that the map T of optics is symplectic.

Notice that the right way of organizing these facts is to say that the manifold M of all light rays carries a natural symplectic structure. The choice of a transversal plane to the optical axis gives us a way of choosing a surface S in $\mathbb{R}^5$ transversal to the curves and gives local coordinates on M relative to which the symplectic form takes the expression

$$\omega = dp_x \wedge dq_x + dp_y \wedge dq_y.$$

The fact that the optical transformation is symplectic is simply the assertion that symplectic forms on M have the above expression in terms of these local coordinates.

7. Interference and diffraction

We have already briefly mentioned the Young interference experiment, performed circa 1800. Light from a distant source, not shown in Figure 7.1, passes through a slit S then through two narrow slits S_1 and S_2 parallel to S and equidistant from it and then falls on a screen. A series of alternating

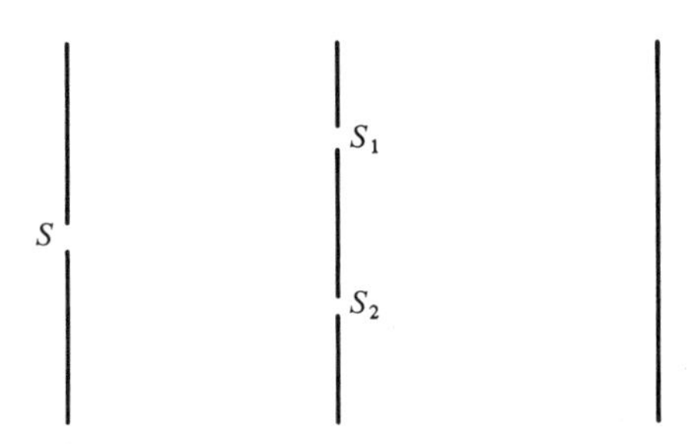

Figure 7.1.

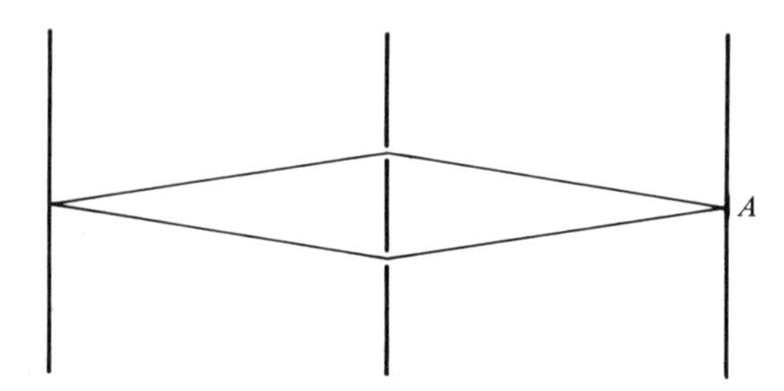

Figure 7.2.

bright and dark bands appears on the screen (Figure 1.2), with a bright band at the center. If either of the slits S_1 or S_2 is covered so that light passes through only one slit, the dark lines disappear and a diffuse broad band replaces the alternating bands. If we use white light, the effect is somewhat fuzzy; but if we use light of a definite color, it becomes much more sharp. For example, we might separate the white light by a prism to the left of S and only allow one component color to pass through S. Then the bright lines are very sharp, and the bright positions on the screen have the following very interesting characterization: For any point A on the central vertical line of the screen, if γ_A is the closed path going from S to S_1 to A to S_2 back to S along straight line segments (Figure 7.2), and we let $L(\gamma_A)$ denote the optical length of this path, then there is a length λ whose value depends on the color of the light such that A is a bright position if and only if $L(\gamma_A) = n\lambda, n \in \mathbb{Z}$. Young realized that this experiment shows that, associated with each color of light, there is an intrinsic length λ – the *wavelength* of the light. Newton had concluded from his famous experiments with prisms in 1666 that color was an intrinsic property of light: Once we separate white light into rays of different colors, each ray retains its color under further refraction or reflection. Newton observed that such rays have their own characteristic angle of refraction in a glass prism (corresponding to the fact that the index of refraction of the glass depends on the color of the light). Thus the subjective notion of color could be correlated with a numerically measurable quantity – an angle of refraction. However, this angle depends also on the material used; experimenters using different kinds of glass would get different angles for the same kind of light. Young's experiment furnished a measurable property of the light itself, independent of the material of the prism. In the *Philosophical Transactions* for 1802 Young refers to his discovery of a "simple and general law." The law is that wherever two portions of the same light arrive at the eye by different routes, either exactly or very nearly in the same direction, the light becomes most intense where the difference of the routes is a multiple of a certain length and least intense in the intermediate state of the interfering portions; and this length is different for light of different colours.

Fresnel suggested that the integrality condition be written in the form

$$\exp[2\pi i(1/\lambda)L(\gamma_A)] = 1$$

and be explained as follows[†]: Associated with each point on the light rays of geometrical optics is a complex number c. As the light propagates along the ray, not only does the amplitude $|c|$ decrease due to the attenuation of the light, but the phase changes. As we move along the ray a distance of optical length l, the change in phase is given by multiplication by the factor $\exp(2\pi il/\lambda)$, where λ is the wavelength of the light. If light arrives at some point Q from various rays we add the c's for each ray to obtain a total value C, and the intensity I of the light is proportional to $|C|^2$. This is Fresnel's explanation of the Young interference experiment: For any point A on the screen, let c_i be the contribution coming from the SS_iA path, $i = 1, 2$. The two complex numbers c_1 and c_2 will have the same phase precisely when the integrality condition is satisfied, and so the two contributions will reinforce each other. In fact, as long as the two optical lengths do not differ by very much, there will not be much difference in the attenuation of the light along the different paths, and the intensity at any point A will be given roughly by

$$I = |c|^2|\{\exp[2\pi i(L(\gamma_A)/\lambda)] + 1\}|^2,$$

which accounts for interference bands.

Let us formulate Fresnel's ideas in more mathematical terms. Suppose that z_1 and z_2 are two nonconjugate planes. Thus, for every pair of points (q_1, q_2) there is a unique light ray that joins them and whose optical length is given by a well-determined function $L(q_1, q_2)$. Suppose that we know the light distribution $c_1(q_1)$ at every point in the z_1 plane. Our problem is to determine the light distribution $c_2(\cdot)$ on the z_2 plane. Our optical data are now functions (instead of ray parameters), and the optical system is determined by a transformation on functions instead of a symplectic transformation as is the case in geometrical optics. Now let q_2 be a point in the

[†] We are, for the moment, oversimplifying Fresnel's explanation in that we are ignoring the phenomenon of polarization. We shall come back to this later on. The polarization effect also has to do with Young's requirement that the light arriving via different routes comes in at "nearly the same direction."

z_2 plane. As the light propagates from q_1 to q_2 it will undergo both a phase change and attenuation, so that when it arrives at q_2 it will contribute the term

$$a(q_1, q_2, \lambda) \exp[2\pi i L(q_1, q_2)/\lambda] c_1(q_1)$$

to the total expression for the light at q_2, where L is the optical length of the path and $a(L)$ is the attenuation factor along the optical path. To find the light at q_2 we must integrate this expression over q_1 to get the total of all the contributions. Of course, to do this we must know the function a.

We now propose to employ Fresnel's ideas, but in the linear approximation. In terms of the diagram of theories of Figure 1.4, we are adding an "intermediate" theory, one which takes account of interference and diffraction, but not geometric aberration (and, temporarily, ignores polarization). For lack of a better name, let us call this approximate theory "Fresnel optics." In Fresnel optics we will be able to determine the value of a (up to a sign) from the knowledge of L. This is true for the following reasons:

(i) We are in a situation where linear optics is valid. This is so because we know, for one reason or another (for instance by the presence of stops or diaphragms in the apparatus), that $c_1(q_1)$ essentially vanishes for large values of q_1 and q_2. We assume that a is a relatively slowly varying function. Hence, we may assume that a is a constant and move it to the left of the integration sign.

Thus our formula for c_2 is

$$c_2(q_2) = a \int c_1(q_1) \exp[2\pi i L(q_1, q_2)/\lambda] \, dq_1,$$

where a is a constant, as yet undetermined but L is a known (quadratic) function of q_1 and q_2. The intensity at the point q_2 is given by

$$I_2(q_2) = |c_2(q_2)|^2$$

and the intensity at q_1 is given by

$$I_1(q_1) = |c_1(q_1)|^2.$$

(ii) We assume that the total intensity of the light on the z_2 plane is the same as the total intensity of the light on the z_1 plane; that is:

$$\int |c_2(q_2)|^2 \, dq_2 = \int |c_1(q_1)|^2 \, dq_1.$$

If this is possible at all (and we shall see that it is), it will determine $|a|$, so that a will be determined up to a phase factor.

(iii) The "constant" a will depend on the wavelength λ. Assume (in partial contradiction to condition (i)) that our diaphragm openings are large in comparison to λ and that we are located at a (vertical) distance of many wavelengths from the boundaries of any of the stops or diaphragms. Then we would expect that the phase of the arriving light should be the same as if the stops were not present. For example, suppose we have no lenses or other apparatus but simply place a half-plane opaque screen to block out the incoming light. We would expect that high above the screen the light should behave as if the screen were not present. This will allow us to determine the phase of a.

In order to work out the details we will need some facts about integrals of quadratic exponentials, so we digress slightly in the next section to develop these basic results.

8. Gaussian integrals

Everything in this section stems from the basic fact that

$$\frac{1}{\sqrt{2\pi}}\int_{-\infty}^{\infty} e^{-x^2/2}\,dx = 1.$$

This is proved by taking the square of the left-hand side:

$$\begin{aligned}\left(\frac{1}{\sqrt{2\pi}}\int_{-\infty}^{\infty} e^{-x^2/2}\,dx\right)^2 &= \frac{1}{2\pi}\int_{-\infty}^{\infty}\int_{-\infty}^{\infty} e^{-(x^2+y^2)/2}\,dx\,dy\\ &= \frac{1}{2\pi}\int_{0}^{2\pi}\int_{-\infty}^{\infty} e^{-r^2/2}\,r\,dr\,d\theta\\ &= \int_{0}^{\infty} e^{-r^2/2}\,r\,dr = 1.\end{aligned}$$

Now

$$\frac{1}{\sqrt{2\pi}}\int_{-\infty}^{\infty} e^{-x^2/2-\eta x}\,dx$$

converges for all complex values of η and uniformly in any compact region. Hence, it defines an analytic function that may be evaluated by taking η to be real and then using analytic continuation. For real η, we simply complete the square and make a change of variables:

$$\frac{1}{\sqrt{2\pi}}\int \exp(-x^2/2 - x\eta)\,dx = \frac{1}{\sqrt{2\pi}}\int \exp[-(x+\eta)^2/2 + \eta^2/2]\,dx$$

$$= \exp(\eta^2/2)\frac{1}{\sqrt{2\pi}}\int \exp[-(x+\eta)^2/2]\,dx$$

$$= e^{\eta^2/2}.$$

This is true for all complex values of η. In particular, taking $\eta = i\zeta$, we get

$$\frac{1}{\sqrt{2\pi}}\int \exp(-x^2/2 - i\zeta x)\,dx = \exp(-\zeta^2/2).$$

In other words, the Fourier transform of $\exp(-x^2/2)$ is $\exp(-\zeta^2/2)$. Now set $x = \sqrt{\lambda}u$, where $\sqrt{\lambda} > 0$ is some positive real number. Then $dx = \sqrt{\lambda}\,du$ and setting $\xi = \sqrt{\lambda}\zeta$ we get

$$\frac{1}{\sqrt{2\pi}}\int e^{-\lambda u^2/2}e^{-i\xi u}\,du = \frac{1}{\sqrt{\lambda}}e^{-\xi^2/2\lambda}.$$

Now this last integral converges for any λ with $\operatorname{Re}\lambda > 0$ and $\lambda \neq 0$. Indeed, for $S > R > 0$

$$\int_R^S e^{-\lambda x^2/2}e^{-i\xi x}\,dx = -\int_R^S \frac{1}{\lambda x}\frac{d}{dx}(e^{-\lambda x^2/2})e^{-i\xi x}\,dx$$

$$= \frac{e^{-\lambda R^2/2}e^{-i\xi R}}{\lambda R} - \frac{e^{-\lambda S^2/2}e^{-i\xi S}}{\lambda S} + \int_R^S e^{-\lambda x^2/2}\frac{d}{dx}\left(\frac{e^{i\xi x}}{\lambda x}\right)dx.$$

In this last integral we can write

$$e^{-\lambda x^2/2} = -\frac{1}{\lambda x}\frac{d}{dx}e^{-\lambda x^2/2}$$

and integrate by parts once again. We thus see that

$$\left|\int_R^S e^{-\lambda x^2/2}e^{-i\xi x}\,dx\right| = 0\left(\frac{1}{|\lambda|R}\right).$$

We thus see that

$$\frac{1}{\sqrt{2\pi}}\int e^{-\lambda x^2/2}e^{-i\xi x}\,dx$$

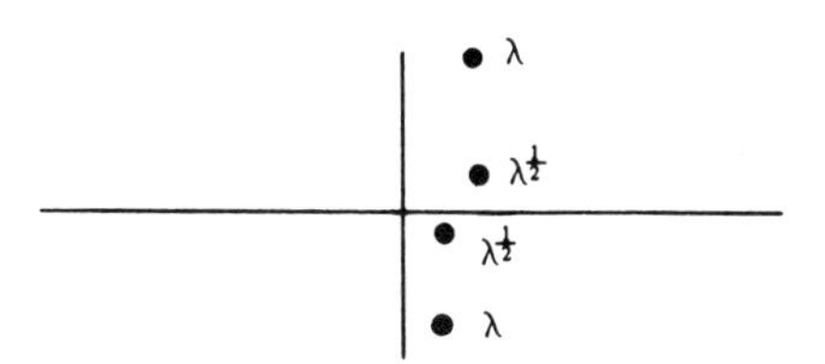

Figure 8.1.

converges uniformly (but not absolutely) for λ in any region of the form

$$\operatorname{Re}\lambda > 0$$
$$|\lambda| > \delta \qquad \text{any } \delta > 0.$$

By analytic continuation, its value is given by the same formula as for λ real and positive, where the square root must now be taken to be the analytic continuation of the positive square root (Figure 8.1). In particular, if we set $\lambda = -ir$, $r > 0$, then

$$\lambda^{1/2} = |\lambda|^{1/2} e^{-\pi i/4}$$

and

$$\frac{1}{\sqrt{2\pi}} \int e^{irx^2/2} e^{-i\xi x}\, dx = e^{\pi i/4} r^{-1/2} e^{-i\xi^2/2r}$$

and, similarly with $\lambda = ir$,

$$\frac{1}{\sqrt{2\pi}} \int e^{-irx^2/2} e^{-i\xi x}\, dx = e^{-\pi i/4} r^{-1/2} e^{i\xi^2/2r}.$$

Now take

$$x = \begin{pmatrix} x_1 \\ \vdots \\ x_n \end{pmatrix}$$

to be a vector variable and let

$$Q = \begin{pmatrix} \pm r_1 & & 0 \\ & \ddots & \\ 0 & & \pm r_n \end{pmatrix}$$

be an $n \times n$ matrix with all the $r_i > 0$ and sgn Q = number of $+$'s minus number of $-$'s. Then $|r_1| \ldots |r_n| = |\det Q|$, and by multiplying the formulas for the one-dimensional integrals, we get

$$\frac{1}{(2\pi)^{n/2}}\int \exp\left(\frac{i}{2}Qx\cdot x\right)\exp(-i\xi\cdot x)\,dx = |\det Q|^{-1/2}\exp\left(\frac{\pi i}{4}\operatorname{sgn} Q\right) \times \exp\left(-\frac{i}{2}Q^{-1}\xi\cdot\xi\right),$$

where now ξ is also a vector and Q^{-1} is the inverse matrix. We have proved this for diagonal Q. But given any nonsingular symmetric matrix Q, we can find an orthogonal matrix O such that OQO^{-1} is diagonal. This proves that the above formula is true for all such Q. This is the formula we need for Fresnel optics.

9. Examples in Fresnel optics

In this section we work out some examples of Fresnel's formula, and propose how to determine the constant a that occurs in it. We will then justify this choice in the succeeding sections. From a mathematical point of view it is a bit easier to work out the integrals first in one dimension; that is, to first do Fresnel optics with linear optics replaced by Gaussian optics. The mathematics is a bit simpler but some care is needed in the physical interpretation. For instance, we will want to talk of a one-dimensional diaphragm that blocks out all light except that arriving at $-h < q < h$, as in Figure 9.1a. Physically, this corresponds to a slit like that shown in Figure 9.1b. In Gauss–Fresnel optics the intensity of light at q_2 on the z_2 "plane" is given by

$$I(q_2) = |c_2(q_2)|^2,$$

where

$$c_2(q_2) = a\int c_1(q_1)\exp\left[2\pi i L\,(q_1, q_2)/\lambda\right]dq_1$$

with

$$L = W + L_{\text{axis}}$$

and

$$W(q_1, q_2) = (1/2B)(Aq_1^2 + Dq_2^2 - 2q_1q_2),$$

where a is an overall constant describing the attenuation of the system, and where $\begin{pmatrix} A & B \\ C & D \end{pmatrix}$ is the optical matrix.

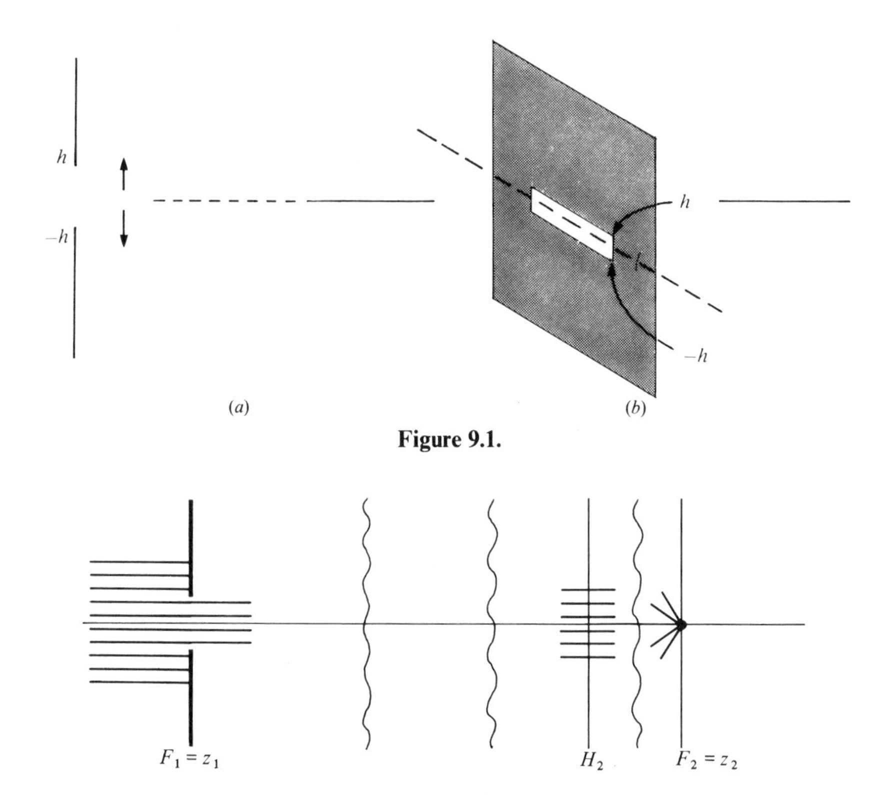

Figure 9.1.

Figure 9.2.

Let us now work out the integral in a number of simple cases. Suppose that z_1 and z_2 are the focal planes of the lens system, so that $A = D = 0$ and $B = f$ is the focal length. (This special case is known as Fraunhofer diffraction.) Let us assume, first of all, that light arrives from a distant source located on the optical axis, so that the light rays are all parallel to the axis, and that the arriving light all has the same c value near the axis. We place an opaque screen at z_1 with a slit of width $2h$ centered on the axis (Figure 9.2), allowing some of the light to pass through. Thus, $c_1(q_1) = c$ for $-h < q_1 < h$, where c is a constant, and $c_1(q_1) = 0$ for $|q_1| > h$. Geometrical optics predicts that all the light will be focused at $q_2 = 0$. Let us see what Fresnel's formula gives. The integral becomes

$$\begin{aligned} c_2(q_2) &= ac \int_{-h}^{h} \exp(-2\pi i q_1 q_2 / f\lambda)\, dq_1 \\ &= 2ach \left[\frac{\sin(2\pi q_2 h / f\lambda)}{2\pi q_2 h / f\lambda} \right] \end{aligned}$$

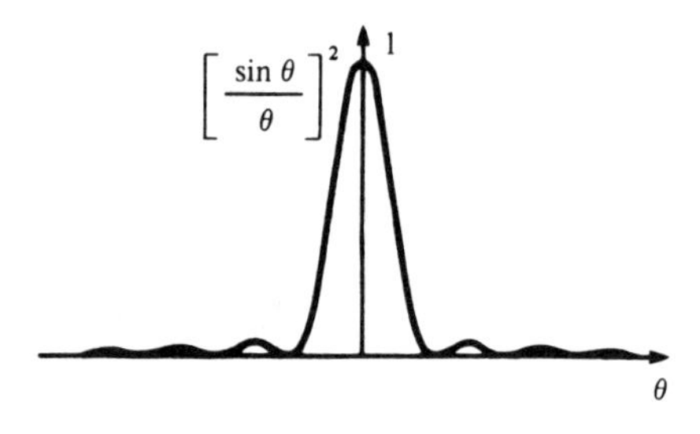

Figure 9.3.

so that

$$I(q_2) = |c_2(q_2)|^2$$
$$= 4a^2h^2|c|^2\left[\frac{\sin(2\pi hq_2/f\lambda)}{2\pi hq_2/f\lambda}\right]^2$$

Now $\lim_{\theta\to 0}[\sin\theta/\theta] = 1$ and $\lim_{\theta\to\infty}[\sin\theta/\theta] = 0$, and, in fact, the graph of the function $[\sin\theta/\theta]^2$ looks something like Figure 9.3. The total intensity of the light arriving at the z_2 plane is given by $\int I(q_2)\,dq_2$. Substituting the above formula for $I(q_2)$ gives, by the change-of-variables formula for an integral.

$$\int I(q_2)\,dq_2 = (4a^2h^2|c|^2|f\lambda|/2\pi h)J,$$

where

$$J = \int_{-\infty}^{\infty}\left[\frac{\sin\theta}{\theta}\right]^2 d\theta = \pi;$$

so the total light intensity arriving at the z_2 plane is $2a^2h|c|^2|f\lambda|$. The total light intensity passing through the screen at z_1 is $2h|c|^2$. We expect that the total light leaving the z_1 plane will arrive somewhere on the z_2 plane, which then tells us that the factor a must have its absolute value given by

$$|a| = |f\lambda|^{-1/2}.$$

(More generally, a similar computation, which will be presented later will show that

$$|a| = |B\lambda|^{-1/2}$$

for the one-dimensional case and

$$|a| = |\det \lambda B|^{-1/2} = \lambda^{-1}|\det B|^{-1/2}$$

Figure 9.4.

for the two-dimensional, linear-optics case, where $\begin{pmatrix} A & B \\ C & D \end{pmatrix}$ is the optical matrix of the system.)

If we substitute this value of $|a|$ into the expression we obtained above for $I(q_2)$ we get

$$I(q_2) = (4h^2/|f\lambda|)|c|^2 \left[\frac{\sin(2\pi h q_2/f\lambda)}{2\pi h q_2/f\lambda}\right]^2.$$

We see that the light is not all focused at the point $q_2 = 0$, but is spread over a spot centered at $q_2 = 0$. The intensity falls off to its first zero at $q_2 = \pm(f\lambda/2h)$. Since the intensity falls off quadratically with q_2, most of the light will be concentrated between the zeros nearest the origin, and we can use $|f\lambda|/h$ as a measure of the lack of focusing due to diffraction. For fixed f and h, the spot becomes narrower as $\lambda \to 0$. In this sense, Gaussian optics is the "short-wavelength limit" of wave optics. Of course, in optics we cannot pass to this limit without going out of the range of visible light. For λ and f fixed, the spot becomes narrower as we enlarge h. If we wish to reduce the effects of diffraction, it is in our interest to make h large. However, for large values of h, the Gaussian approximation will, in general, no longer be valid, and one encounters a lack of focusing due to geometrical aberrations. Thus, in a sense, the aberration effects and the diffraction effects are working against each other. One of the main problems in the design of optical instruments is to try to arrange matters so that the next-order terms beyond the Gaussian approximation, the third-order aberrations, vanish, in which case we can allow h to be large enough to minimize the effects of diffraction.

Now let us see how the above computation changes when we replace the distant on-axis source by a distant source off the optical axis. (We still assume $A = D = 0$, $B = f$, and a centered slit of width $2h$ at the first focal plane.) Assume that the light arrives at $q_1 = 0$ with a value $c = c(0)$. The light arriving at some other value of q_1 will have traveled an additional distance $l = q_1 \sin\theta$, where $\theta = p_1/n_1$ or $l = q_1 p_1/n_1$ (see Figure 9.4). The optical length of the additional path is $n_1 l = q_1 p_1$, and, hence, the light will arrive at the point q_1 on the z_1 plane with the value

$$c(q_1) = \exp(2\pi i q_1 p_1/\lambda)\, c(0)$$
$$= \exp(2\pi i q_1 p_1/\lambda)\, c$$

That means that

$$c_{p_1}(q_2) = c \int \exp(2\pi i q_1 p_1/\lambda) \exp(-2\pi i q_1 q_2/f\lambda)\, dq_1$$
$$= c \int \exp[-2\pi i q_1(q_2 - f p_1)/f\lambda]\, dq_1.$$

If $c_0(q_2)$ denotes the value computed above for $p_1 = 0$, we can write this last equation as

$$c_{p_1}(q_2) = c_0(q_2 - f p_1).$$

The spot of light is now centered at the point fp_1 in the z_2 plane, which is precisely the point where all the incoming lines would be focused according to geometrical optics. Thus the shift in phase of the arriving light at z_1 is translated into a shift of the image point at z_2. Let us now turn to a famous example of a two-dimensional diffraction integral, where we take $|a| = \lambda^{-1}|\det B|^{-1/2}$. We consider the case where there are no lenses at all in the optical system, so that the optical matrix is given by $\begin{pmatrix} I & dI \\ 0 & I \end{pmatrix}$, where d is the distance traversed. Thus

$$c_2(q_{x2}, q_{y2}) = a \int c_1(q_{x1}, q_{y1}) \exp\{2\pi i[(q_{x2} - q_{x1})^2 + (q_{y2} - q_{y1})^2]/2d\lambda\}\, dq_{x1}\, dq_{y1}$$

where

$$|a| = |\lambda d|^{-1}.$$

Let us evaluate this integral for the following situation: Light parallel to the optical axis arrives at a plane at which there is a circular opaque disk perpendicular to the axis, with its center on the axis (Figure 9.5). Thus $c_1 = 0$ inside the circle and $c_1 = c$ is constant outside the circle. We shall take $q_{x2} = q_{y2} = 0$ so that we are positioned on the axis. We write x for

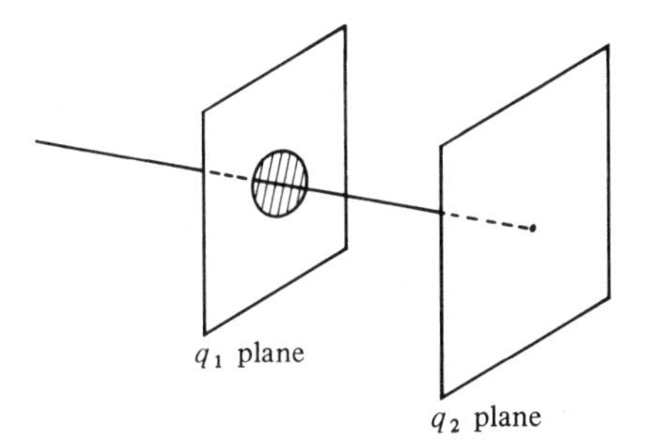

Figure 9.5.

q_{x1} and y for q_{y1} and R for the radius of the disk. The above integral becomes c times

$$
\begin{aligned}
&(\lambda d)^{-1} \int_{x^2+y^2>R^2}^{\infty} \exp[2\pi i(x^2+y^2)/2d\lambda]\, dx\, dy \\
&= \lim_{\varepsilon \to 0} (\lambda d)^{-1} \int_R^{\infty} \int_0^{2\pi} \exp[2\pi(i-\varepsilon)r^2/2d\lambda]\, r\, dr\, d\theta \\
&= \lim_{\varepsilon \to 0} \frac{2\pi}{\lambda d} \int_R^{\infty} \exp[2\pi(i-\varepsilon)r^2/2d\lambda]\, r\, dr \\
&= \lim_{\varepsilon \to 0} \int_b^{\infty} \exp[(i-\varepsilon)s^2/2]\, s\, ds, \qquad b = \left(\frac{2\pi}{d\lambda}\right)^{1/2} R \\
&= -e^{ib^2/2}.
\end{aligned}
$$

Thus

$$I_2(0,0) = |c|^2.$$

The intensity of the light along the axis is the same as if the disk were not present! There is a white spot in the shadow of a circular disk. (Of course this is valid where the linear approximation is valid. So the disk can't be too large in comparison to the distance d.)

Fresnel presented his memoir on diffraction for the prize of the French Academy in 1818. Poisson, Biot, and Laplace were on the committee to judge the paper, and all three were supporters of the corpuscular theory of light. Fresnel had calculated various cases of his formula and compared them with experiment. But Poisson noticed that one could calculate the integral for a circular disk, and come to the conclusion that there is a white spot in the center of the shadow of a circular disk. He raised this as an objection to the whole theory, since it was patently ridiculous. However,

Arago and Fresnel went and performed the experiment, and, sure enough the spot was there, whereupon Fresnel was awarded the prize. This phenomenon is now known as the "Poisson spot."

10. The phase factor

We now turn to the problem of determining the correct phase factor in the constant a, whose absolute value we have already determined. We thus write

$$a = u|a|,$$

where u is some number of absolute value 1 that must be determined, and

$$|a| = \lambda^{-1}|\det B|^{-1/2}$$

in two-dimensional linear optics, and

$$a = (\lambda B)^{-1/2}$$

in the one-dimensional Gaussian situation. The principle that we shall use to determine u is that if we are at a distance of many wavelengths (i.e., multiples of λ) vertically distant from any stops or diaphragms in our apparatus, the light should behave approximately as if these stops are not present. For simplicity we argue in the Gaussian case. Suppose that we are in an optical situation with no lenses at all; simply put an opaque screen at z_1 that blocks out all the light arriving with $q_1 < 0$, and examine the light which arrives at z_2, which is a distance d from z_1 (Figure 10.1). Let us assume that the arriving light comes from a distant on-axis point and so arrives with $p_1 = 0$ and with constant phase. Thus

$$\begin{aligned} c(q_1) &= c \quad \text{for} \quad q_1 \geqslant 0 \\ &= 0 \quad \text{for} \quad q_1 < 0 \end{aligned}$$

and the optical matrix is $\begin{pmatrix} 1 & d \\ 0 & 1 \end{pmatrix}$, so that

$$\begin{aligned} W(q_1, q_2) &= (1/2d)(q_1^2 + q_2^2 - 2q_1q_2) \\ &= (1/2d)(q_1 - q_2)^2. \end{aligned}$$

We get

$$\begin{aligned} c(q_2) &= uc(d\lambda)^{-1/2} \int_0^\infty \exp[\pi i(q_2 - q_1)^2/\lambda d]\, dq_1 \\ &= uc\phi(q_2/(\lambda d)^{1/2}), \end{aligned}$$

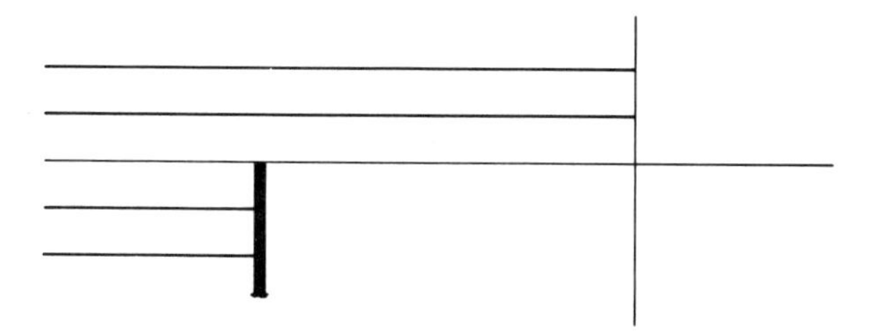

Figure 10.1.

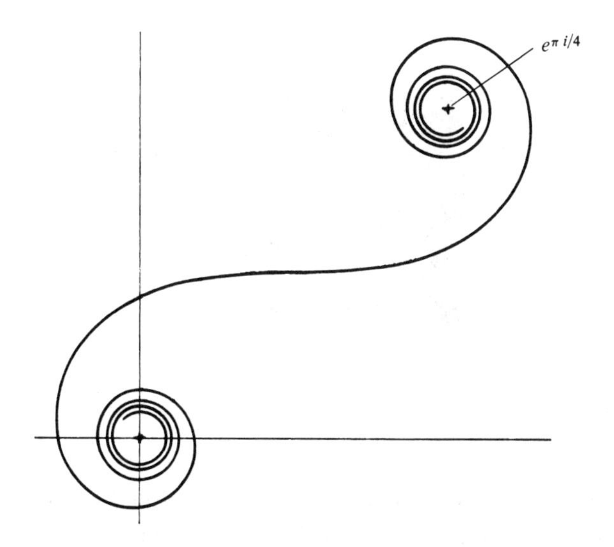

Figure 10.2.

where ϕ is the function defined by the integral

$$\phi(w) = \int_0^\infty \exp[\pi i(w - r)^2]\, dr = \int_{-w}^\infty \exp(\pi i t^2)\, dt.$$

These integrals are known as Fresnel integrals and have been tabulated; ϕ behaves as in Figure 10.2.

We have already proved, in our section on Gaussian integrals, that

$$\lim_{w \to \infty} \phi(w) = e^{\pi i/4}.$$

For large negative values of w the value of the integral is practically zero. As w approaches 0 from $-\infty$, it spirals outward from 0 gradually widening its

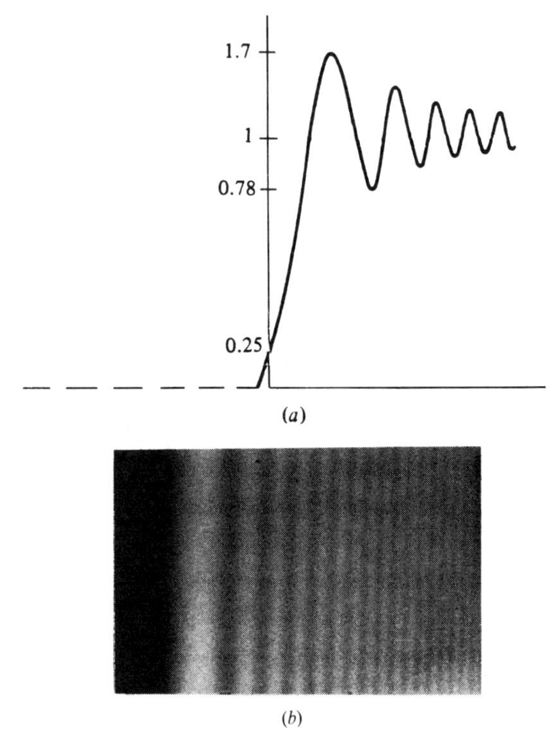

Figure 10.3. Illumination near the geometric shadow of the edge of a screen and the observed diffraction figure.

swing until at $w = 0$ it moves over to swing about the point $e^{\pi i/4}$ in a gradually diminishing spiral as w goes to $+\infty$. (These spirals are known as the Cornu spirals.) We see (Figure 10.3) that there will be some illumination in the geometrical shadow, and for sufficiently large values of q_2 the light arriving at the z_2 plane will have the same intensity as the light passing the z_1 planes; in other words, the effect of the opaque screen can be ignored.

We would expect for large enough q_2 that the phase of the arriving light should also be unaffected by the lower half screen at z_1. If the screen were not present, the light arriving at z_2 would have the value $\exp(2\pi i d/\lambda)c$ for all values of q_2. The evaluation of the Fresnel integral as $e^{\pi i/4}$ for large w gives the value $ue^{\pi i/4}c$ for the light arriving at large values of q_2. How do we alter the integral formula for the diffracted light to make these two values agree? The first obvious step is to use the correct arbitrary constant in the eikonal. The optical length between q_1 at z_1 and q_2 at z_2 is

$$[d^2 + (q_2 - q_1)^2]^{1/2} \doteq d + (1/2d)(q_2 - q_1)^2.$$

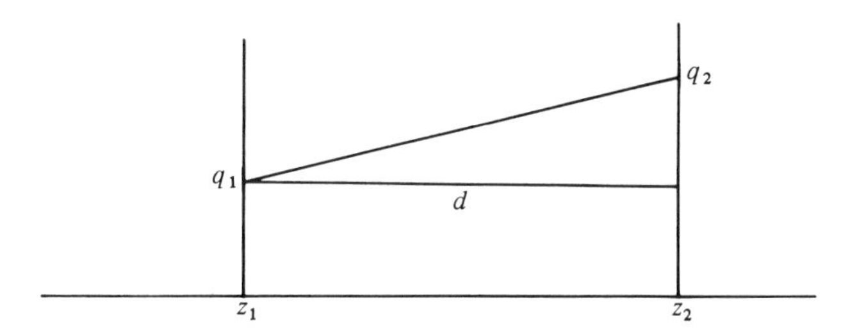

Figure 10.4.

(See Figure 10.4). We should use the formula

$$L(q_1, q_2) = d + (1/2d)(q_2 - q_1)^2$$

instead of our previous expression, which did not have the d in it. Including the d will modify the earlier computation by multiplying it by an overall factor of $e^{2\pi i d/\lambda}$. With this corrected value of L our integral formula now gives $ue^{\pi i/4}e^{2\pi i d/\lambda}c$ for the light which arrives at large values of q_2. We are still off by the factor $e^{\pi i/4}$, which is independent of λ, d, or anything else in the problem. There isn't much we can do about it at this stage other than to compensate for it in our integral formula: Choose $u = e^{-\pi i/4}$. Thus the general formula now reads

$$e^{-\pi i/4}|B\lambda|^{-1/2}\int c(q_1)\exp[2\pi i L(q_1, q_2)/\lambda]\,dq_1,$$

where $L(q_1, q_2)$ is the optical length of the light ray joining q_1 to q_2. The phase factor $e^{-\pi i/4}$ occurring in front of the integral certainly looks ad hoc and mysterious. It caused a great deal of resistance, at first, to the acceptance of Fresnel's ideas. [Some 30 years after Fresnel proposed his formula Helmholtz showed that this factor is a consequence of the wave equation and an integral formula in vector calculus known today as Stokes' theorem (see Guillemin-Sternberg (1977), Chapter I for a modern presentation of Helmholtz's argument).] The formula as written is still not quite right; it is only valid for positive values of B. When B is negative the correct phase factor is $e^{+\pi i/4}$. (We could have expected this had we interchanged the role of z_1 and z_2 in the preceding discussion.) The correct formula for the diffracted light is thus given by

$$\exp(\mp \pi i/4)|B\lambda|^{-1/2}\int c(q_1)\exp[2\pi i L(q_1, q_2)/\lambda]\,dq_1,$$

where $L(q_1, q_2)$ is the optical length of the light ray joining q_1 to q_2 and where $\pm$ refers to the sign of B.

Before giving further illustrations and applications of Fresnel's formula, we should discuss here the nature of this formula and the relation between Fresnel's wave optics and Gaussian optics. In Gaussian optics each optical system is described by a matrix that gives the relation between the incoming and outgoing light rays. In Fresnel optics the same optical system is described by an integral transform relating the entire incoming light distribution to the outgoing light distribution:

Gaussian optics	Wave optics (in the Gaussian approximation)
matrix	integral transform
$M = \begin{pmatrix} A & B \\ C & D \end{pmatrix}$	F
$\begin{pmatrix} q_1 \\ p_1 \end{pmatrix} \to M\begin{pmatrix} q_1 \\ p_1 \end{pmatrix} = \begin{pmatrix} Aq_1 + Bp_1 \\ Cq_1 + Dp_1 \end{pmatrix}$	$c_1 \to c_2 = Fc_1$

$$(Fc_1)(q_2) = \exp(\mp \pi i/4)|B\lambda|^{-1/2} \int c_1(q_1) \times \exp[2\pi i L(q_1, q_2)/\lambda] dq_1,$$

where

$$L(q_1, q_2) = \frac{1}{2B}(Aq_1^2 + Dq_2^2 - 2q_1q_2) + d$$

The explicit formula for F given in the last line is only valid when $B \neq 0$. Although we expect that the operator F from functions to functions be defined for all optical systems, it need not be given by a smooth integral kernel. For example, if the planes z_1 and z_2 coincide so that we are dealing with a "trivial" optical system, then $M = \begin{pmatrix} 1 & 0 \\ 0 & 1 \end{pmatrix}$ is the identity matrix and F is the identity operator. The identity operator is a perfectly nice operator, but it cannot be represented by a smooth integral kernel. (Its integral kernel is $\delta(x - y)$, that is, $f(x) = \int f(y)\delta(x - y)\, dy$.) To see that we are on the right track, let us examine what happens to the transformation formula for free-space transmission (no lenses) as the optical distance approaches 0. Thus for $d > 0$

$$(F_d c)(x) = \exp(-\pi i/4)\, |d\lambda|^{-1/2} \int c(y) \exp[(2\pi i/2d\lambda)(x - y)^2]\, dy.$$

We claim that

$$\lim_{d \to 0+} F_d c = c.$$

To prove this we shall make use of several facts from the theory of Fourier transforms.

(i) Let f be a smooth function that vanishes rapidly at infinity together with its derivatives; that is,

$$|f^{(n)}(x)| \leqslant \frac{c}{1 + |x|^k},$$

for each k and n, where c depends on k, n, and f. Its Fourier transform $\mathscr{F}f$ is defined by

$$(\mathscr{F}f)(\xi) = \left(\frac{1}{2\pi}\right)^{1/2} \int_{-\infty}^{\infty} e^{-i\xi x} f(x)\, dx.$$

The Fourier inversion formula says that

$$f(x) = \left(\frac{1}{2\pi}\right)^{1/2} \int_{-\infty}^{\infty} e^{+i\xi x} (\mathscr{F}f)(\xi)\, d\xi$$

so that

$$\mathscr{F}(\mathscr{F}f)(x) = f(-x)$$

and

$$\mathscr{F}^4 f = f.$$

(ii) The Fourier transform carries convolution into multiplication. Let $\phi * f$ be defined by

$$\phi * f(x) = \left(\frac{1}{2\pi}\right)^{1/2} \int_{\infty}^{\infty} \phi(x - y) f(y)\, dy$$

for functions ϕ and f. Then

$$\mathscr{F}(\phi * f) = (\mathscr{F}\phi)(\mathscr{F}f).$$

We can write this equation as follows: Let c_ϕ denote the operator of convolution by ϕ so that $c_\phi(f) = \phi * f$. Let M_ψ denote the operator of multiplication by ψ so that

$$M_\psi g = \psi g.$$

Then setting $f = \mathscr{F}^{-1} g$ and $\psi = \mathscr{F}\phi$ we see that

$$\mathscr{F} c_\phi \mathscr{F}^{-1} = M_{\mathscr{F}\phi}.$$

(iii) The Fourier transform of

$$\frac{1}{\sqrt{t}} e^{-x^2/2t}$$

is

$$e^{-t\xi^2/2}$$

for any t with $\operatorname{Re} t > 0$. The square root in this formula is chosen to be positive for real positive t and analytically continued for $\operatorname{Re} t \geqslant 0$. In particular, letting t approach the point iP on the positive imaginary axis we see that the Fourier transform of $(1/|t|^{1/2}) \exp(-x^2/2t)$ approaches $\exp(+\pi i/4) \exp(-iP\xi^2/2)$. Thus

$$\mathscr{F}(|d\lambda|^{-1/2} \exp[(2\pi i/d\lambda)x^2/2]) = \exp(\pi i/4) \exp\left(-i\left|\frac{d\lambda}{2\pi}\right|\xi^2/2\right).$$

Now F_d is convolution with

$$\exp(-\pi i/4)\, |d\lambda|^{-1/2} \exp[(2\pi i/d\lambda)x^2/2]$$

and, hence, the Fourier transform of $F_d c$ is given by

$$\mathscr{F}(F_d c)(\xi) = \exp\left(-i\left|\frac{d\lambda}{2\pi}\right|\xi^2/2\right)(\mathscr{F}c)(\xi).$$

It is clear that as $d \to 0$, $\mathscr{F}(F_d c) \to \mathscr{F}c$, and hence, by the Fourier inversion formula,

$$F_d c \to c.$$

Note that the correct phase factor, in particular the $e^{\pm\pi i/4}$, plays a crucial role in this computation.

Having emphasized the importance of the role of the phase factors $e^{\pm\pi i/4}$, we now point out that these very factors show that the preceding description can't be 100 percent correct. Indeed, if F_{21} is the Fresnel transform between the planes z_1 and z_2 and F_{32} the transform between the planes z_2 and z_3, then we want $F_{31} = F_{32} \circ F_{21}$ to be the transform between the planes z_1 and z_3. Now the expression for L in the formula for F has a quadratic term and a constant term. The constant terms come out of the integral and behave nicely under composition:

$$d_{31} = d_{21} + d_{32}$$

d_{31}

d_{21} d_{32}

so

$$\exp(2\pi i d_{31}/\lambda) = \exp(2\pi i d_{21}/\lambda) \exp(2\pi i d_{32}/\lambda).$$

We can ignore the constant term in L if we want to check whether or not

$$F_{32} \circ F_{21} = F_{31}.$$

Thus we may replace L by

$$W(q_1, q_2) = \frac{1}{2B}(Aq_1^2 + Dq_2^2 - 2q_1 q_2)$$

and write U for the corresponding operator.
Now let us consider five points z_1, z_2, z_3, z_4, z_5.

z_1 z_2 z_3 z_4 z_5

Let's assume the same optical matrix $\begin{pmatrix} 0 & 1 \\ -1 & 0 \end{pmatrix}$ applies between each successive pair of points. Then

$$U_{21} = U_{32} = U_{43} = U_{54} = U,$$

where

$$Uc(x) = e^{-\pi i/4}|\lambda|^{-1/2}\int e^{-2\pi ixy/\lambda}c(y)\,dy$$

or, scaling the lengths so that $\lambda = 2\pi$

$$Uc(r) = e^{-\pi i/4}\frac{1}{\sqrt{2\pi}}\int e^{-ixy}c(y)\,dy$$

so that the operator U is just the Fourier transform, multiplied by $e^{-\pi i/4}$

$$U = e^{-\pi i/4}\mathscr{F},$$

where $\mathscr{F}$ denotes the Fourier transform. Now recall that the Fourier inversion formula implies that $\mathscr{F}^4 c = c$, and, indeed,

$$\begin{pmatrix} 0 & 1 \\ -1 & 0 \end{pmatrix}^4 = \begin{pmatrix} 1 & 0 \\ 0 & 1 \end{pmatrix}.$$

On the other hand, because of the phase factor $e^{-\pi i/4}$, $U^4 = -I$, instead of getting $U^4 = I$. Thus the preceding prescription cannot be completely correct and we cannot hope to assign a Fresnel transform to each optical *matrix* so that the identity $F_{31} = F_{32} \circ F_{21}$ holds.

What we *can* do is the following: Let us go back to the original components of our optical systems, with straight line propagation and refraction at a surface. Every optical system can be built up out of a succession of such components. We can write down the Fresnel transform for each such component and then *define* the transform to be the composition of the transform associated with each component. Then, of course, $F_{31} = F_{32} \circ F_{21}$ holds automatically. We shall show that for any Gaussian optical system whose matrix $\begin{pmatrix} A & B \\ C & D \end{pmatrix}$ satisfies $B \neq 0$, the preceding formula is *correct up to sign.*

The situation is very much like extracting a square root. Suppose that we decide to assign the operation of multiplication by $z^{1/2}$ to the complex number z with $|z| = 1$. If we want 1 to be assigned to the identity operator and the operators to depend continuously on z near $z = 1$, we must define the square root to be $(e^{i\theta})^{1/2} = e^{i\theta/2}$. On the other hand, taking $\theta = 2\pi$ we find that $e^{i\theta/2} = -1$. To rectify the situation, we must pass to the "double cover" of the circle. In much the same way, the operators of Fresnel optics (in the Gaussian approximation) form a double cover of the group of 2×2 matrices of determinant 1.

We have already worked out Fresnel's formula for straight-line propagation. If the optical distance is d, then the operator is given by

$$[F_d c](r) = \exp(-\pi i/4)\,|\lambda d|^{-1/2} \int c(y) \exp\left\{2\pi i\left[d + \frac{1}{2d}(x-y)^2\right]\Big/\lambda\right\} dy$$

and, dropping the axial contribution,

$$\left[U_{\left(\begin{smallmatrix} 1 & d \\ 0 & 1 \end{smallmatrix}\right)} c\right](x) = \exp(-\pi i/4)\,|\lambda d|^{-1/2} \int c(y) \exp\left[2\pi i \frac{1}{2d\lambda}(x-y)^2\right] dy.$$

Suppose we have a refracting surface of power P, so that its matrix is $\begin{pmatrix} 1 & 0 \\ -P & 1 \end{pmatrix}$. We have computed the change in optical length as we pass through this surface and found it to be $-\frac{1}{2}Pq^2$ for any ray at height q. Thus Fresnel's prescription gives

$$\left[U_{\left(\begin{smallmatrix} 1 & 0 \\ -P & 1 \end{smallmatrix}\right)}\right] = \text{multiplication by } \exp(-iPx^2/2\lambda);$$

that is,

$$\left[U_{\left(\begin{smallmatrix} 1 & 0 \\ -P & 1 \end{smallmatrix}\right)} c\right](x) = \exp(-iPx^2/2\lambda)c(x).$$

Now

$$U_{\left(\begin{smallmatrix} 1 & P \\ 0 & 1 \end{smallmatrix}\right)}$$

is convolution by $\exp(\pm\pi i/4)\,|P\lambda|^{-1/2} \exp(2\pi i x^2 P/2\lambda)$, where $-$ is used for $P > 0$ and $+$ for $P < 0$. Assertions (ii) and (iii) about the Fourier transform listed on page 65 imply that

$$\mathscr{F}U_{\begin{pmatrix}1&P\\0&1\end{pmatrix}}\mathscr{F}^{-1}=U_{\begin{pmatrix}1&0\\-P&1\end{pmatrix}}.$$

Having defined the operators U for our two basic components, we then define the operator for any system as the composite of the various components. Let us work through some examples. Suppose that we have a single refracting surface of power P with displacements to the right and left of the surface

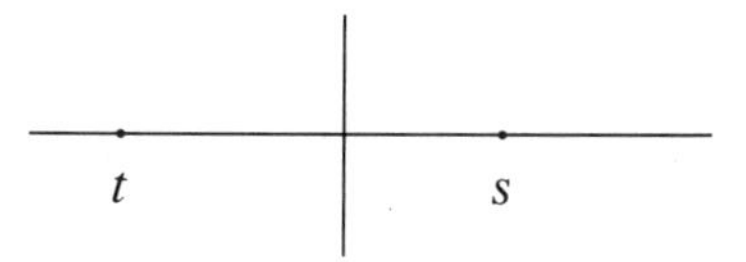

with t, s optical lengths. We thus want to compute

$$U_{\begin{pmatrix}1&s\\0&1\end{pmatrix}}\circ U_{\begin{pmatrix}1&0\\-P&1\end{pmatrix}}\circ U_{\begin{pmatrix}1&t\\0&1\end{pmatrix}}.$$

In Gaussian optics the matrix of this system is obtained by multiplying the matrices to obtain

$$\begin{pmatrix}1&s\\0&1\end{pmatrix}\begin{pmatrix}1&0\\-P&1\end{pmatrix}\begin{pmatrix}1&t\\0&1\end{pmatrix}=\begin{pmatrix}1-sP&t+s-Pst\\-P&1-tP\end{pmatrix}=\begin{pmatrix}A&B\\C&D\end{pmatrix}.$$

We shall assume that $B\neq 0$. Let us choose units so that $\lambda=2\pi$. Then, applied to any function c, the composite of the three operators, evaluated at x is

$$\exp(-\pi i/4)|2\pi s|^{-1/2}\int\exp[i(x-z)^2/2s]$$

$$\times\left[\exp(-iPz^2/2)\exp(-\pi i/4)|2\pi t|^{-1/2}\int\exp[i(z-y)^2/2t]c(y)\,dy\right]dz.$$

So we must evaluate the integral

$$\int\exp[i(x-z)^2/2s]\exp(-iPz^2/2)\exp[i(z-y)^2/2t]\,dz.$$

We can write the exponent as

$$i\left[x^2/2s+y^2/2t+\frac{1}{2}\left(\frac{1}{s}+\frac{1}{t}-P\right)z^2-\left(\frac{x}{s}+\frac{y}{t}\right)z\right].$$

The first two terms do not depend on z so their exponential can be pulled out as a factor. The integral to evaluate thus becomes

$$\int\exp\left[i\left(\frac{1}{s}+\frac{1}{t}-P\right)z^2/2-i\left(\frac{x}{s}+\frac{y}{t}\right)z\right]dz,$$

which is $(2\pi)^{1/2}$ times the Fourier transform of $\exp[i(1/s + 1/t - P)z^2/2]$ evaluated at the point $x/s + y/t$. Now the Fourier transform of $\exp(iQz^2/2)$ is $\exp(\pm\pi i/4)|Q|^{-1/2}\exp(-i\xi^2/2Q)$. The $\pm$ is given by the sign of Q. Substituting $\xi = x/s + y/t$ gives

$$(2\pi)^{1/2}\exp(\pm\pi i/4)\left|\frac{1}{s}+\frac{1}{t}-P\right|^{-1/2}\exp\left\{-i\left(\frac{x}{s}+\frac{y}{t}\right)^2\bigg/\left[2\left(\frac{1}{s}+\frac{1}{t}-P\right)\right]\right\}.$$

Resubstituting all the factors gives

$$(2\pi)^{-1/2}e^{-\pi i/4}e^{-\pi i/4}e^{\pm\pi i/4}|s|^{-1/2}|t|^{-1/2}\left|\frac{1}{s}+\frac{1}{t}-P\right|^{-1/2}$$
$$\times\exp i\left\{\frac{x^2}{2s}+\frac{y^2}{2t}-\left(\frac{x}{s}+\frac{y}{t}\right)^2\bigg/\left[2\left(\frac{1}{s}+\frac{1}{t}-P\right)\right]\right\}.$$

Now

$$B = s + t - Pst = st\left(\frac{1}{s}+\frac{1}{t}-P\right)$$

and

$$\frac{D}{2B} = \frac{1-tP}{2(s+t-Pst)}$$

is the coefficient of x^2 in the brackets of the exponential and similarly, $A/2B$ is the coefficient of y^2. Thus the above expression simplifies to

$$e^{-\pi i/4}e^{-\pi i/4}e^{\pm\pi i/4}|2\pi B|^{-1/2}\exp\left(\frac{i}{2B}[Dx^2 + Ay^2 - 2xy]\right).$$

Up to the phase factors in front, this is our original formula for the transform associated with $\begin{pmatrix} A & B \\ C & D \end{pmatrix}$. Now let us examine these phase factors. If $B > 0$ then the third factor is $e^{+\pi i/4}$ and the product simplifies to $e^{-\pi i/4}$ and we recover our original formula. If $B < 0$, then the third factor is $e^{-\pi i/4}$ and we get $e^{-3\pi i/4} = -e^{+\pi i/4}$ so the answer differs from the preceding one by an overall minus sign. Recall that $1/s + 1/t - P > 0$ if the conjugate point to z does not lie between z_1 and z_2, whereas $1/s + 1/t - P < 0$ if z_2 is beyond the conjugate point.

The preceding discussion shows that we must modify Fresnel's original description: Not only does the phase of light get multiplied by $e^{id/\lambda}$ as we move along a ray. We get an extra factor of $-i = e^{-\pi i/4}\cdot e^{-\pi i/4}$ as the light passes through a conjugate point. As strange as this prescription seems, this fact was experimentally observed by Gouy in 1890. We have described and explained his experiment in Guillemin and Sternberg, 1977, Chapter I.

11. Fresnel's formula

Let us work with the n-dimensional case, $Sp(2n, \mathbb{R})$, since it requires no additional effort. For convenience, we take $\lambda = 2\pi$. (This can be thought of as a choice of units of length, which can be undone at the end of the argument.) We recall that matrices of the form

$$\begin{pmatrix} I & dI \\ 0 & I \end{pmatrix}, \qquad d > 0,$$

and

$$\begin{pmatrix} I & 0 \\ -P & I \end{pmatrix}, \qquad P = P^t,$$

generate $Sp(2n, \mathbb{R})$. We now define the operators U_d and V_{P} by

$$(U_d c)(x) = \exp(-\pi i n/4)\, d^{-n/2}(2\pi)^{-n/2} \int \exp[i(x - y)^2/2d]c(y)\, dy$$

and

$$(V_{\mathrm{P}} c)(x) = \exp(-iPx \cdot x/2)c(x).$$

We define

$$\rho(U_d) = \begin{pmatrix} I & d \\ 0 & I \end{pmatrix}$$

and

$$\rho(V_{\mathrm{P}}) = \begin{pmatrix} I & 0 \\ -P & I \end{pmatrix},$$

and for any product of U's and V's, the product of the corresponding elements in $Sp(2n, \mathbb{R})$. So, for instance:

$$\rho(U_{d_1} V_{\mathrm{P}_1} U_{d_2} V_{\mathrm{P}_2}) = \begin{pmatrix} I & d_1 I \\ 0 & I \end{pmatrix} \begin{pmatrix} I & 0 \\ -\mathrm{P}_1 & 1 \end{pmatrix} \begin{pmatrix} I & d_2 I \\ 0 & I \end{pmatrix} \begin{pmatrix} I & 0 \\ -\mathrm{P}_2 & I \end{pmatrix},$$

etc. We have thus constructed a homomorphism of the group G generated by the U's and V's onto $Sp(2n, \mathbb{R})$.† We claim that *if* $X \in G$ *is such that*

† Strictly speaking, we must show that map ρ is well defined, that is, that it depends on the element $X \in G$ and not on the way that X is written as a product of U's and V's. This follows from the proof of the next assertion.

$\rho(X) = \begin{pmatrix} A & B \\ C & D \end{pmatrix}$ *with B nonsingular, then the operator X is given by*

$$(Xc)(x) = i^{\#} e^{-\pi i n/4} |\det B|^{-1/2} (2\pi)^{-n/2} \int e^{iW(x,y)} c(y)\, dy,$$

where

$$W(x, y) = \tfrac{1}{2}[DB^{-1}x \cdot x + B^{-1}Ay \cdot y - 2B^{t-1}y \cdot x]$$

and

$$\# \text{ is even if } \det B > 0,$$
$$\# \text{ is odd if } \det B < 0.$$

In particular, for such X, the matrix $\rho(X)$ determines X up to sign.

We will prove this fact by induction on the number of factors of X. It is true by definition for $X = U_d$ (with $\# = 0$). If

$$X = U_d V_{\mathrm{P}}$$

then

$$(Xc)(x) = \exp(-\pi i n/4) d^{-n/2} (2\pi)^{-n/2} \times \int \exp[i(x - y)^2/2d - iPy \cdot y/2] c(y)\, dy$$

and

$$\rho(X) = \begin{pmatrix} I - dp & dI \\ -P & I \end{pmatrix}.$$

Since $B^{-1}A = d^{-1}I - P$, we see that the exponential is precisely $iW(x, y)$, as required. So again, the formula is correct with $\# = 0$.

We thus need only multiply on the left – we assume the formula for X – and check that it remains true for $V_{\mathrm{P}}X$ and $U_d X$. Now

$$\rho(V_{\mathrm{P}}X) = \begin{pmatrix} I & 0 \\ -P & I \end{pmatrix} \begin{pmatrix} A & B \\ C & D \end{pmatrix} = \begin{pmatrix} A & B \\ C - PA & D - PB \end{pmatrix}$$

so that B is unchanged and the term DB^{-1} in W is replaced by $DB^{-1} - P$. But this is just the same as multiplying by $\exp(-iPx \cdot x/2)$. Thus we must check what happens when we multiply on the left by U_t.

$$\rho(U_t X) = \begin{pmatrix} I & tI \\ 0 & I \end{pmatrix} \begin{pmatrix} A & B \\ C & D \end{pmatrix} = \begin{pmatrix} A + tC & B + tD \\ C & D \end{pmatrix}$$

and we are assuming that t is such that $B + tD$ is nonsingular. We must compute $U_t X$. It is clearly given by the integral kernel $(U_t X_c)(z) = \int F(z, y)c(y)dy$, where $F(z, y)$ is given by

$$e^{-n\pi i/4} e^{-n\pi i/4} i^{\#} t^{-n/2} |\det B|^{-1/2} (2\pi)^{-n/2}$$

$$\times (2\pi)^{-n/2} \int \exp\left\{\frac{i}{2}[(z-q)^2/t + DB^{-1}q\cdot q + B^{-1}Ay\cdot y - 2B^t y\cdot q]\right\} dq.$$

We can write the $(2\pi)^{-n/2}\int \exp\{\ldots\}\, dq$ as

$$\exp\left\{\frac{i}{2}[t^{-1}z^2 + B^{-1}Ay\cdot y]\right\}(2\pi)^{-n/2}$$

$$\times \int \exp\left[\frac{i}{2}(t^{-1} + DB^{-1})q\cdot q - i(t^{-1}z + B^{t-1}y)\cdot q\right] dq$$

and the integral in this expression is just the Fourier transform of $\exp[i(t^{-1} + DB^{-1})q\cdot q/2]$ evaluated at the point $(t^{-1}z + B^{t-1}y)$. This is a Gaussian integral that we know how to compute. By our general formula, it is given by

$$\exp\left[\frac{\pi i}{4}\operatorname{sgn}(t^{-1}I + DB^{-1})\right]|\det(t^{-1}I + DB^{-1})|^{-1/2}$$

$$\times \exp\left[-\frac{i}{2}(t^{-1} + DB^{-1})^{-1}(t^{-1}z + B^{t-1}y)\cdot(t^{-1}z + B^{t-1}y)\right].$$

We substitute this into the original expression and make the following observations:

$$t^{-1}I + DB^{-1} = t^{-1}(B + tD)B^{-1}$$

and hence

$$|\det(t^{-1}I + DB^{-1})|^{-1/2}\cdot t^{-n/2}\cdot|\det B|^{-1/2} = |\det(B + tD)|^{-1/2}.$$

Also, $t^{-1}I + DB^{-1}$ is a symmetric matrix. For small positive values of t it is positive definite. Also for these values of t, the term $\exp[(\pi i/4)\operatorname{sgn}(t^{-1}I + DB^{-1})]$ exactly cancels the term $e^{-\pi i n/4}$, and also $\det(B + tD)$ has the same sign as $\det B$. For larger values of t, the matrix $t^{-1}I + DB^{-1}$ may have some negative eigenvalues and

$$\exp\left[\frac{\pi i}{4}\operatorname{sgn}(t^{-1}I + DB^{-1})\right]\exp(-\pi i n/4) = i^b,$$

where b is the number of negative eigenvalues. But

$$(-1)^b = \operatorname{sgn}(\det(t^{-1}I + DB^{-1})) = \operatorname{sgn}\det(B + tD)/\operatorname{sgn}\det B,$$

so the phase factor in front is $i^{\# + b}$, where $(-1)^{\# + b} = \operatorname{sgn}\det(B + tD)$. Thus all the constant factors in the formula for $U_t X$ check. We must now examine the exponential terms. For this we simply collect coefficients in z and y in the quadratic expression. The pure z term is $i/2$ times

$$t^{-1}z \cdot z - (t^{-1}I + DB^{-1})^{-1}t^{-1}z \cdot t^{-1}z$$

and, writing $(t^{-1}I + DB^{-1}) = t^{-1}(B + tD)B^{-1}$, this becomes

$$\begin{aligned} t^{-1}z \cdot z - B(B + tD)^{-1}t^{-1}z \cdot z &= \big(t^{-1}(B + tD) - t^{-1}B\big)(B + tD)^{-1}z \cdot z \\ &= D(B + tD)^{-1}z \cdot z, \end{aligned}$$

which is exactly the term in the expression of $W(z, y)$ required for the matrix

$$\rho(U_t X) = \begin{pmatrix} A + tC & B + tD \\ C & D \end{pmatrix}.$$

The coefficient involving y and z is $-i/2$ times

$$B(B + tD)^{-1}tt^{-1}z \cdot B^{t-1}y + B(B + tD)^{-1}tB^{t-1}y \cdot t^{-1}z,$$

which, since $(I + tDB^{-1})^{-1} = [(B + tD)B^{-1}]^{-1} = B(B + tD)^{-1}$ is symmetric, is just

$$2B(B + tD)^{-1}t \cdot t^{-1}z \cdot B^{t-1}y = 2(B + tD)^{-1}z \cdot y,$$

which is the correct expression. Finally, the expression quadratic in y is $\frac{i}{2}$ times

$$\begin{aligned} &B^{-1}Ay \cdot y - B(B + tD)^{-1}tB^{t-1}y \cdot B^{t-1}y \\ &\quad = [B^{-1}Ay - (B + tD)^{-1}tB^{t-1}y] \cdot y. \end{aligned}$$

Now $-B^{t-1} = C - DB^{-1}A$, since our matrix is symmetric, and so

$$\begin{aligned} B^{-1}A - (B + tD)^{-1}tB^{t-1} &= B^{-1}A + (B + tD)^{-1}t(C - DB^{-1}A) \\ &= (B + tD)^{-1}[(B + tD)B^{-1}A + tC - tDB^{-1}A] \\ &= (B + tD)^{-1}(A + tC), \end{aligned}$$

as required.

We have defined a group of operators G on Hilbert space and a homomorphism ρ from G to $Sp(2n, \mathbb{R})$ and have seen that the map ρ is 2 to 1 over those symplectic matrices with B nonsingular. From this it is not hard to prove (we shall give the details in the next section) that ρ is 2 to 1 everywhere; that is, that G is a double covering of the symplectic group. It is called the metaplectic group and is usually denoted by $Mp(2n, \mathbb{R})$. We shall follow this nomenclature from now on. The action of $Mp(2n, \mathbb{R})$ on Hilbert space given above is called the metaplectic representation. Thus Fresnel optics is equivalent to the study of the metaplectic representation of $Mp(4, \mathbb{R})$. We should point out that the metaplectic representation is *unitary*: Every operator U in our collection satisfies

$$\|Uc\| = \|c\|,$$

where

$$\|c\|^2 = (2\pi)^{-n/2} \int |c(q)|^2 \, dq.$$

For a proof we observe that since the product of any number of unitary operators is unitary, we need only verify that each of our basic operators, U_d and V_P is unitary. Now V_P is clearly unitary, since it consists of multiplication by a function of absolution value 1. On the other hand, a basic theorem in the theory of Fourier transforms, Plancherel's theorem, asserts that the Fourier transform F is unitary. Since U_d is conjugate to V_{dI} via the Fourier transform, we conclude that U_d is unitary, and hence every $U \in Mp(2n, \mathbb{R})$ is unitary.

12. Fresnel optics and quantum mechanics

In this section we shall present some applications of Fresnel's formula to quantum mechanics. In particular, we shall see that Fresnel's formula provides an explicit solution to Schrödinger's equation for an interesting class of systems. For this purpose we must describe the relation between Fresnel's optics and quantum mechanics. It is well known that the founders of "wave mechanics," which then developed into "quantum mechanics" were guided by an analogy: They felt that just as there was a wave optics that was a more accurate physical theory than Hamilton's geometrical optics, there should be a wave mechanics standing in the same relation to classical mechanics. The purpose of this section is to show that for *linear* mechanical systems, this analogy can be given a precise mathematical form, and is nothing but the relation between the metaplectic representation and the symplectic group $Sp(n, \mathbb{R})$, as described in Section 11.

We begin with some elementary observations. Notice that the operators of the form V_P all commute with one another and, in fact,

$$V_{P_1} \circ V_{P_2} = V_{P_1 + P_2}$$

and, of course, the image of these operators under ρ, satisfy the corresponding identity

$$\begin{pmatrix} I & 0 \\ -P_1 & I \end{pmatrix}\begin{pmatrix} I & 0 \\ -P_2 & I \end{pmatrix} = \begin{pmatrix} I & 0 \\ -(P_1 + P_2) & I \end{pmatrix}.$$

In particular, for a given P, the set of operators V_{tP} forms a one-parameter group of operators: $V_{(s+t)P} = V_{sP} \circ V_{tP}$. Similarly the operators

$$U_P = \mathscr{F} V_P \mathscr{F}^{-1}$$

form a commutative family with

$$U_{P_1} \circ U_{P_2} = U_{(P_1 + P_2)},$$

since conjugation by $\mathscr{F}$ is a group isomorphism. Recall that the operator U_P is given by the formula

$$(U_P c)(x) = \exp(-\pi i n/4) \exp(\pi i \operatorname{sgn} P/4) |\det P|^{-1/2}$$
$$\times \int \exp[iP^{-1}(x - y)\cdot(x - y)/2] c(y)\, dy.$$

In particular, for fixed P, the operators U_{tP} form a one-parameter group of operators: $U_{(s+t)P} = U_{sP} U_{tP}$.

More generally, any family of operators $U(t) \in Mp(2n, \mathbb{R})$ that depend continuously on t and satisfy $U(t_1 + t_2) = U(t_1)U(t_2)$ is called a one-parameter group of operators. To such a one-parameter group of operators there will be a corresponding one-parameter group of matrices $M(t)$. Of course, $M(t)$ will be close to the identity matrix for small values of $|t|$. For such small $|t|$ we can uniquely recover $U(t)$ from $M(t)$, since only one of the two possible choices of the Fresnel operator corresponding to $M(t)$ will be close to the identity. Of course, once we know $U(t)$ for small $|t|$, we know it for all t since $U(t) = U(t/n)^n$.

Since $M(t)$ is a one-parameter group of symplectic matrices, it is easy to prove that M depends differentiably on t and we can write

$$M'(0) = K \qquad \text{and} \qquad M(t) = e^{tK}.$$

Thus $M(t)$ and hence $U(t)$ are determined by K.

We can also consider the operator $U'(0)$ defined by

$$U'(0)c = \lim_{t \to 0} \frac{U(t)c - c}{t}.$$

Here this limit need not exist for all c, but we do expect it to exist for well-behaved c. It is a standard theorem that $U'(0)$ is a skew adjoint operator, so we can write

$$U'(0) = -iH,$$

where H is a self-adjoint operator, and

$$\frac{dU}{dt} = -iHU.$$

This last equation is known as Schrödinger's equation with Hamiltonian H.

Here are some examples (with $n = 1$ and units chosen so that $\lambda = 2\pi$):

Matrices

K	$M(t)$
$\begin{pmatrix} 0 & 1 \\ 0 & 0 \end{pmatrix}$	$\begin{pmatrix} 1 & t \\ 0 & 1 \end{pmatrix}$
$\begin{pmatrix} 0 & 0 \\ -1 & 0 \end{pmatrix}$	$\begin{pmatrix} 1 & 0 \\ -t & 1 \end{pmatrix}$
$\begin{pmatrix} 0 & 1 \\ -1 & 0 \end{pmatrix}$	$\begin{pmatrix} \cos t & \sin t \\ -\sin t & \cos t \end{pmatrix}$

Operators in Hilbert Space

H	$U(t)$
$-\frac{1}{2}\frac{d^2}{dx^2}$	$[U(t)f](x) = \exp(\pm\pi i/4)\lvert t\rvert^{-1/2}\int e^{i(x-y)^2/2t}f(y)\,dy$
$(Hf)(x) = \frac{1}{2}x^2 f(x)$	$(U(t)f)(x) = e^{-itx^2/2}f(x)$
$(Hf)(x) = -\frac{1}{2}\left(\frac{d^2}{dx^2} - x^2\right)f(x)$	$U(t)f(x) = (-i)^{\#}e^{-\pi i/4}\lvert\sin t\rvert^{-1/2}(2\pi)^{-1/2} \times \int \exp\left\{\frac{i}{2\sin t}[(\cos t)[x^2 + y^2] - 2xy]\right\}f(y)\,dy$

The entry for H in the third row follows from the entry for the first two rows since the mapping from K to H is linear (on any common domain of definition).

In the usual treatments of quantum mechanics one *starts* with the operator H and then seeks to solve the corresponding Schrödinger equation. We have found via Fresnel's formula an explicit solution of

Schrödinger's equation for those H's arising in the left-hand column (for arbitrary values of n). The corresponding one-parameter group $M(t)$ of (q, p) space is called the "classical motion" corresponding to H. For example, looking at the H in the first line of our table, the one-parameter group $M(t)$ is

$$\begin{pmatrix} q \\ p \end{pmatrix} \to \begin{pmatrix} q + tp \\ p \end{pmatrix}.$$

If we interpret q as the position and p as the momentum of a classical particle, then these equations represent the motion of a particle (of mass $m = 1$) moving in the absence of any external force – the so-called free particle, and $-\frac{1}{2}(d^2/dx^2)$ is known as the Hamiltonian of a free (nonrelativistic) particle. In our formulas above, and in most of what follows, we choose units so that $m = 1$ and $\hbar = 1$. If we do not use these units, then the Hamiltonian H for a free particle becomes

$$H = -\frac{\hbar}{2m}\frac{d^2}{dx^2},$$

and we merely have to replace t by $t\hbar/m$ in $U(t)$ to get the solution to this Hamiltonian. Let ψ be some function of x that is square integrable over $\mathbb{R}$, and let us set

$$\psi(x, t) = \left(U\left(\frac{\hbar}{m}t\right)\psi\right)(x).$$

Then $\psi(\cdot, t)$ is called a *wavefunction* at time t and its interpretation is that

$$\rho(x, t) = |\psi(x, t)|^2$$

represents the probability density at time t of finding the particle in some region of space. (Thus instead of light intensity we talk of probability density.) It is usual to take ψ such that $\|\psi\|_2 = 1$, and then by the unitarity of $U(t)$ we know that $\|\psi(\cdot, t)\|_2 = 1$ for all t, so that at any time the total probability of finding the particle somewhere in space equals 1. For example, suppose that we take our initial $\psi = \psi(\cdot, 0)$ to be a Gaussian:

$$\psi(x, 0) = A \exp\left[-\frac{x^2}{2a^2} + ik_0 x\right].$$

The normalization to $\|\psi\|_2 = 1$ requires that we choose A so that

$$|A|^2 = \frac{1}{a\sqrt{\pi}}.$$

We can now substitute into the integral expression for U to obtain a formula for $\psi(x, t)$ for any time t. The integral reduces to a Gaussian integral of the type studied in Section 8, and a straightforward computation shows that

$$\psi(x,t) = \frac{A}{[1 + i(\hbar t/ma^2)]^{1/2}} \exp\left\{\frac{x^2 - 2ia^2k_0x + i(\hbar t/2m)k_0^2a^2}{2a^2[1 + i(\hbar t/ma^2)]}\right\}.$$

The probability density at time t is thus given by

$$\rho(x,t) = |\psi(x,t)|^2 = \frac{|A|^2}{[1 + (\hbar t/ma^2)^2]^2} \exp\left\{-\frac{[x - (\hbar k_0/m)t]^2}{a^2[1 + (\hbar t/ma^2)^2]}\right\}.$$

Notice that this is again a Gaussian density, but now the maximum has moved from $x = 0$ to $x = (\hbar k_0/m)t$. The maximum of the "wave" is moving with velocity $\hbar k_0/m$. At the same time the "wave packet" has broadened from a width measured by a (the denominator in the exponent of the Gaussian at $t = 0$) to the value given by

$$a' = a\left[1 + \left(\frac{\hbar t}{ma^2}\right)^2\right]^{1/2}$$

at time t. This is the typical behavior of a relatively sharply peaked initial wave packet ψ. It moves with a certain velocity while simultaneously spreading out. All of the detailed information is contained, of course, in the explicit expression for the integral kernel for $U(t)$.

The Hamiltonian $H = \frac{1}{2}(-d^2/dx^2 + x^2)$ is known as the Hamiltonian of the harmonic oscillator. (We are back in units where $\hbar = m = 1$ and we are assuming that the frequency of our oscillator is 1). In this case, the classical trajectories are circles: The matrix $M(t)$ is just rotation through angle t. Again, our formula for $U(t)$, a special case of Fresnel's formula, gives the explicit solution. In the physics literature the formula for the solution to the harmonic oscillator Hamiltonian is known as *Mehler's formula*.

For the case of the harmonic oscillator the one-parameter group $U(t)$ is a representation of a compact group, the circle, and, hence, on general principles (the Peter–Weyl theorem), the space $L^2(\mathbb{R})$ decomposes into a direct sum of irreducibles that must be one-dimensional, since the circle is abelian. Thus $\frac{1}{2}(-d^2/dx^2 + x^2)$ has a discrete spectrum. We can actually compute its spectrum as follows:

$$\frac{d}{dx} x^n e^{-x^2/2} = (nx^{n-1} - x^{n+1})e^{-x^2/2}$$

so

$$\frac{d^2}{dx^2} x^n e^{-x^2/2} = \big(n(n-1)x^{n-2} - (2n+1)x^n + x^{n+2}\big)e^{-x^2/2}$$

and hence,

$$\frac{1}{2}\left(-\frac{d^2}{dx^2} + x^2\right)x^n e^{-x^2/2} = \left(n + \frac{1}{2}\right)x^n e^{-x^2/2} - \tfrac{1}{2}n(n-1)x^{n-2}e^{-x^2/2}.$$

Thus, if we let $\mathscr{F}_n$ denote the space of all functions of the form $p(x)\exp(-x^2/2)$ with $p(x)$ a polynomial of degree at most n, we see that

$$H\mathscr{F}_n \subset \mathscr{F}_n$$

and H takes the triangular form

$$\begin{pmatrix} \frac{1}{2} & \cdot & \cdot & \cdot \\ 0 & (1+\frac{1}{2}) & \cdot & \cdot \\ \vdots & 0 & (2+\frac{1}{2}) & \cdot \\ 0 & \cdot & \cdot & (n+\frac{1}{2}) \end{pmatrix}$$

with respect to the monomial basis $\exp(-x^2/2), x\exp(-x^2/2), \ldots, x^n\exp(-x^2/2)$. From this follows that the numbers $(n+\frac{1}{2})$ are all eigenvalues of H and that the corresponding eigenvectors are of the form $H_n(x)\exp(-x^2/2)$, where $H_n(x)$ is a polynomial of degree n (called the nth Hermite polynomial). To show that these are all the eigenvalues it is enough to prove that if $f \subset L^2(\mathbb{R})$ is such that $\big(f, x^n\exp(-x^2/2)\big) = 0$ for all n then $f = 0$. But if $(f, x^n e^{(-x^2/2)}) = 0$ for all n then

$$(f, e^{-(x-a)^2/2}) = \lim_{n\to\infty} e^{-a^2/2}\sum_0^n \left(f, \frac{(ax)^k}{n!}e^{-x^2/2}\right) = 0$$

so

$$\varphi(a) = \sqrt{2\pi}(f * e^{-x^2/2})(a) = \int f(x)e^{-(x-a)^2/2}dx = 0$$

for all a. But, taking the Fourier transform we see that

$$0 = (\mathscr{F}\varphi)(\xi) = (\mathscr{F}f)(\xi)e^{-\xi^2/2} \qquad \forall\,\xi,$$

and since $\exp(-\xi^2/2)$ is positive everywhere we conclude that $\mathscr{F}f$ and hence $f = 0$. Thus the spectrum of $\frac{1}{2}(-(d^2/dx^2) + x^2)$ consists precisely of the numbers $(n+\frac{1}{2})$, and the Hermite functions $H_n(x)\exp(-x^2/2)$ form an orthonormal basis of $L^2(\mathbb{R})$.

We now wish to determine which Hamiltonians actually arise from the metaplectic representation. For this purpose we must first describe which matrices are in the Lie algebra of the symplectic group, that is, which matrices K are such that e^{sK} is symplectic for all s. Differentiating the equation

$$(e^{sK})J(e^{sK})^{\mathrm{t}} = J$$

with respect to s and setting $s = 0$ shows that K must satisfy

$$KJ + JK^{\mathrm{t}} = 0,$$

and, conversely, if K satisfies this identity, then since the derivative of e^{sK} with respect to s is Ke^{sK} we see that the derivative of $(e^{sK})J(e^{sK})^{\mathrm{t}}$ with respect to s vanishes identically. At $s = 0$ its value is J, so we see that $KJ + JK^{\mathrm{t}} = 0$ is a necessary and sufficient condition for e^{sK} to be symplectic for all s. The set of such K's is called the symplectic algebra and will be denoted by $sp(2n)$. If we write K out as

$$K = \begin{pmatrix} L & Q \\ P & M \end{pmatrix}$$

the condition becomes

$$M + L^{\mathrm{t}} = 0, \qquad Q = Q^{\mathrm{t}}, \qquad P = P^{\mathrm{t}}$$

so the most general symplectic matrix looks like

$$\begin{pmatrix} L & Q \\ P & -L^{\mathrm{t}} \end{pmatrix}$$

where L is an arbitrary $n \times n$ matrix and P and Q are symmetric. The space $sp(2n)$ is a Lie algebra: It is closed under the Lie bracket (commutator) operation

$$[k_1, k_2] = k_1 k_2 - k_2 k_1.$$

This can be checked directly, or can be seen to follow from standard facts about groups of matrices. Notice that the set of all matrices of the form

$$\begin{pmatrix} 0 & 0 \\ P & 0 \end{pmatrix}, \qquad P = P^{\mathrm{t}},$$

is a commutative subalgebra:

$$\left[\begin{pmatrix} 0 & 0 \\ P_1 & 0 \end{pmatrix}, \begin{pmatrix} 0 & 0 \\ P_2 & 0 \end{pmatrix}\right] = 0.$$

We claim that these matrices together with $\begin{pmatrix} 0 & I \\ 0 & 0 \end{pmatrix}$ generate $sp(2n)$ as an algebra. Indeed,

$$\left[\begin{pmatrix} 0 & I \\ 0 & 0 \end{pmatrix}, \begin{pmatrix} 0 & 0 \\ P & 0 \end{pmatrix}\right] = \begin{pmatrix} P & 0 \\ 0 & -P \end{pmatrix},$$

and if P_1 and P_2 are symmetric, then

$$\left[\begin{pmatrix} P_1 & 0 \\ 0 & -P_1 \end{pmatrix}, \begin{pmatrix} P_2 & 0 \\ 0 & -P_2 \end{pmatrix}\right] = \begin{pmatrix} P_1 P_2 - P_2 P_1 & 0 \\ 0 & P_1 P_2 - P_2 P_1 \end{pmatrix}$$

and $P_1 P_2 - P_2 P_1$ is antisymmetric. Thus we can get all matrices of the form

$$\begin{pmatrix} L & 0 \\ 0 & -L^t \end{pmatrix}$$

as combinations of successive brackets of $\begin{pmatrix} 0 & 0 \\ P & 0 \end{pmatrix}$'s and $\begin{pmatrix} 0 & I \\ 0 & 0 \end{pmatrix}$. Finally

$$\left[\begin{pmatrix} L & 0 \\ 0 & -L^t \end{pmatrix}, \begin{pmatrix} 0 & I \\ 0 & 0 \end{pmatrix}\right] = \begin{pmatrix} 0 & L + L^t \\ 0 & 0 \end{pmatrix}$$

so we get all of $sp(2n)$. For any symmetric matrix P, the one-parameter group of matrices

$$\begin{pmatrix} I & 0 \\ tP & I \end{pmatrix}$$

corresponds to the one parameter group V_{-tP} of operators, so

$$\left.\frac{dV_{-tP}}{dt}\right|_{t=0} \equiv \text{multiplication by } iPx \cdot x/2$$

and similarly $\begin{pmatrix} 0 & I \\ 0 & 0 \end{pmatrix}$ corresponds to the operator

$$\frac{i}{2}\left(\frac{\partial^2}{\partial x_1^2} + \cdots + \frac{\partial^2}{\partial x_n^2}\right).$$

Now, as operators,

$$\begin{aligned} \frac{i}{2}\frac{\partial^2}{\partial x_j^2}\left(\frac{i}{2}\sum P_{kl} x_k x_l\right) &= -\frac{1}{4}\frac{\partial}{\partial x_j}\left(P_{jl} x_l + P_{kj} x_k + \sum P_{kl} x_k x_l \frac{\partial}{\partial x_j}\right) \\ &= -\frac{1}{4}\left(2\delta_{jk}\delta_{jl} P_{jj} + 4\sum_l P_{jl} x_l \frac{\partial}{\partial x_j} + \sum P_{kl} x_k x_l \frac{\partial^2}{\partial x_j^2}\right) \end{aligned}$$

since P is symmetric, so

$$\left[\frac{i}{2}\left(\frac{\partial^2}{\partial x_1^2}+\cdots+\frac{\partial^2}{\partial x_n^2}\right), \frac{i}{2}\sum P_{kl}x_k x_l\right] = -\left(\sum P_{jk}x_k\frac{\partial}{\partial x_j}+\frac{1}{2}\operatorname{tr} P\right)$$

and the commutator of two such operators L_P and $L_{P'}$ is easily computed to be

$$-\sum A_{ij}x_j\frac{\partial}{\partial x_i},$$

where $A = PP' - P'P$ is antisymmetric, and so, for a general $\begin{pmatrix} L & 0 \\ 0 & -L^t \end{pmatrix}$, we expect the corresponding operator to be

$$-\left[\sum L_{ij}x_j\partial/\partial x_i + \tfrac{1}{2}\operatorname{tr} L\right]$$

(where the second term means multiplication by $\frac{1}{2}\operatorname{tr} L$).

Finally,

$$\left(x_k\frac{\partial}{\partial x_l}\right)\frac{\partial^2}{\partial x_j^2} - \frac{\partial^2}{\partial x_j^2}x_k\frac{\partial}{\partial x_l} = -2\delta_{kj}\frac{\partial}{\partial x_l}\frac{\partial}{\partial x_j},$$

so

$$\left[-\left(\sum L_{ij}x_i\frac{\partial}{\partial x_j}+\frac{1}{2}\operatorname{tr} L\right), \frac{i}{2}\sum\frac{\partial^2}{\partial x_y^2}\right] = \frac{i}{2}\sum\left(L_{jk}+L_{kj}\right)\frac{\partial}{\partial x_j}\frac{\partial}{\partial x_k},$$

showing that the operator corresponding to $\begin{pmatrix} 0 & Q \\ 0 & 0 \end{pmatrix}$ should be

$$\frac{i}{2}\sum Q_{jk}\frac{\partial}{\partial x_j}\frac{\partial}{\partial x_k}.$$

Now for any two matrices or operators A and B, we have

$$\left.\frac{d}{dt}e^{sB}e^{tA}e^{-sB}\right|_{t=0} = e^{sB}Ae^{-sB}$$

and

$$\frac{d}{ds}e^{sB}Ae^{-sB} = BA - AB.$$

Since we know that the operators U_t and V_P generate $G = Mp(2n)$ and we know that the map ρ from $Mp(2n)$ into $Sp(2n)$ is a homomorphism, we know (and can also check directly) that the assignments

$$\begin{pmatrix} 0 & 0 \\ P & 0 \end{pmatrix} \to \frac{i}{2}\sum P_{kl}x_k x_l,$$

$$\begin{pmatrix} L & 0 \\ 0 & -L^t \end{pmatrix} \to -\left(\sum L_{jk}x_k \frac{\partial}{\partial x_j} + \frac{1}{2}\operatorname{tr} L\right),$$

and

$$\begin{pmatrix} 0 & Q \\ 0 & 0 \end{pmatrix} \to \frac{i}{2}\sum Q_{jk}\frac{\partial}{\partial x_j}\frac{\partial}{\partial x_k}$$

is an isomorphism of Lie algebras. These operators (or rather $-i$ times them) are the Hamiltonians that arise from Fresnel optics.

The operators so obtained are all skew adjoint and have a common dense domain of definition, the Schwartz space of rapidly decreasing smooth functions. By a standard theorem in the theory of Lie groups this implies that there is a corresponding unitary representation U of the universal covering group $\widetilde{Sp}(2n, \mathbb{R})$ on the Hilbert space of functions inducing the given representation of Lie algebras. Now it is easy to see that the fundamental group of $Sp(2n, \mathbb{R})$ is $\mathbb{Z}$; indeed, a standard theorem says that the fundamental group of a Lie group is the same as the fundamental group of its maximal compact subgroup. It is easy to check that any maximal compact subgroup is (conjugate to) $U(n)$ and that $SU(n)$ is simply connected, so that the fundamental group of $U(n)$ is $\mathbb{Z}$. So $Sp(2n, \mathbb{R})$ has a unique double cover; call it $Mp(2n, \mathbb{R})$. We have a tower of coverings

$$\begin{array}{ccl} \widetilde{Sp}(2n, \mathbb{R}) & \xrightarrow{\tilde{U}} & \text{unitary operators on } H \\ \downarrow \bar{\pi} & \nearrow & \\ Mp(2n, \mathbb{R}) & & \\ \downarrow & & \\ Sp(2n, \mathbb{R}) & & \end{array} \qquad (\pi : \widetilde{Sp}(2n, \mathbb{R}) \to Sp(2n, \mathbb{R}))$$

and have checked that if $\pi T = \begin{pmatrix} A & B \\ C & D \end{pmatrix}$ with B nonsingular, then $\tilde{U}(T)$ can be only one of two operators and so can depend only on $\bar{\pi} T \in Mp(2n, \mathbb{R})$. Since the set of $\begin{pmatrix} A & B \\ C & D \end{pmatrix}$ with B nonsingular is dense on $Sp(2n, \mathbb{R})$, we conclude that $\tilde{U}(T)$ depends only on $\bar{\pi} T$ in general. Thus we really have a unitary representation, U, of $Mp(2n, \mathbb{R})$ whose infinitesimal representation on the Lie algebra is the given one. Thus, in terms of the notation of the preceding section, we see that G is really the double cover of $Sp(2n, \mathbb{R})$, as was promised.

13. Holography

Before continuing with a theoretical analysis of the relations between classical and quantum mechanics, we pause to consider holography, a technological application of Fresnel's formula.

We begin by recalling the formula for the free-space transmission of light. If c denotes the light amplitude on a given plane z_0, then the light distribution at optical length d units to the right is given by $c_d = U_d c$ (where we are ignoring the overall constant phase factor coming from the optical length of the z axis).

In terms of the Fourier transforms

$$\tilde{c} = \mathscr{F}c \qquad \text{and} \qquad \tilde{c}_d = \mathscr{F}c_d,$$

Fresnel's formula is

$$\tilde{c}_d(\xi) = \exp(-i\lambda\, d\xi^2/2)\tilde{c}(\xi).$$

We have not yet said anything about where the plane z_0 is or how the light distribution c got there. We wish to compare two possible sets of circumstances.

(i) There is nothing at all at z_0. The light arriving at z_0 comes from some object source lying to the left of z_0 (Figure 13.1). An observer situated d units to the right of z_0 simply sees the object. Of course, it is the light c_d arriving at the observer that determines what he sees, and this is completely determined by c via the above formulas.

(ii) The plane z_0 is immediately to the right of some translucent film. The film absorbs some portion of the light passing through it and transmits the rest. The fraction transmitted at height x above the axis is $t(x)$. We shine a monochromatic plane wave (of wavelength λ) parallel to the z axis through

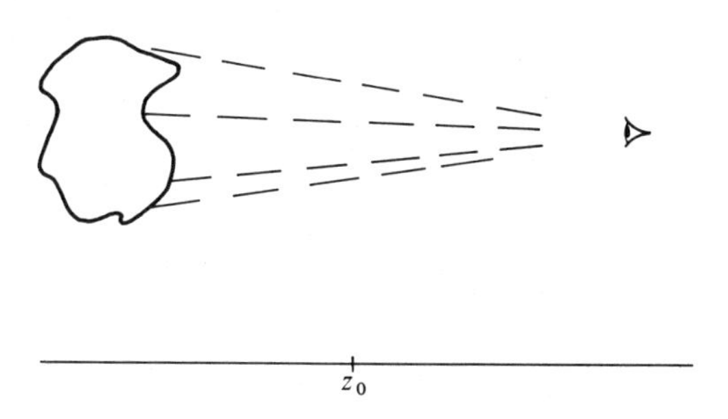

Figure 13.1.

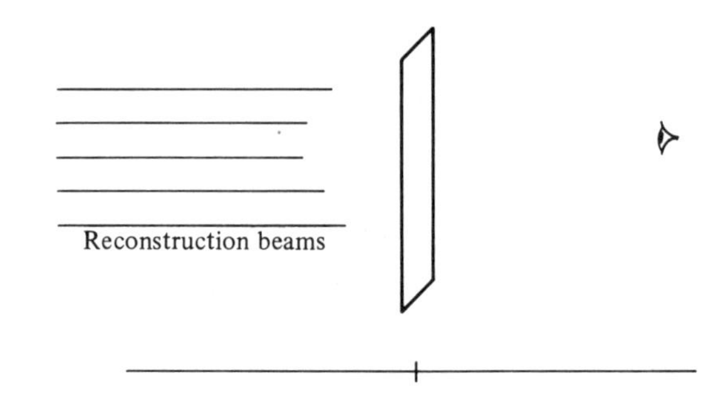

Figure 13.2.

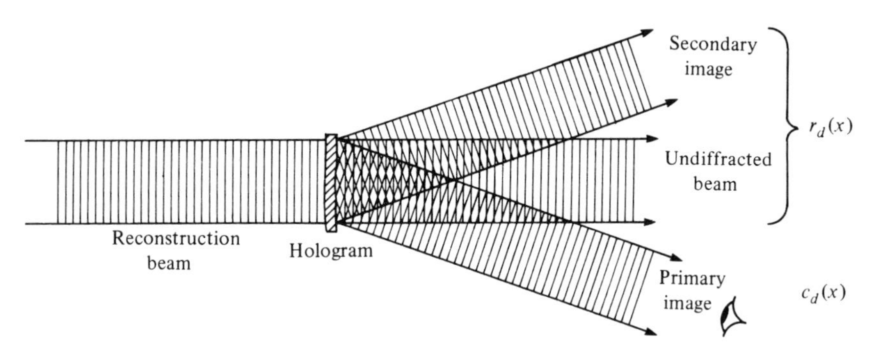

Figure 13.3.

the film (Figure 13.2). Thus the light arriving at z_0 is $kt(x)$, where k is some (complex) constant.

The goal of holography is to arrange things so that $kt(x) = c(x)$. Then an observer situated to the right would not be able to distinguish between case (i) and case (ii). We may not quite be able to arrange that $kt(x) = c(x)$. But suppose that $\tilde{c}(\xi)$ has its support in a neighborhood of some $\xi_0 \neq 0$, so that approximately all the light is coming to z_0 from some fixed direction. Now suppose that

$$\mathscr{F}(kt)(\xi) = k\tilde{t}(\xi) = \tilde{c}(\xi) + \tilde{r}(\xi),$$

where $\operatorname{supp} \tilde{r}(\xi) \cap \operatorname{supp} \tilde{c}(\xi) = \varnothing$. Thus the light corresponding to $\tilde{c}$ and $\tilde{r}$ are traveling in different directions. Then at some distance to the right, the light coming from c will be spatially separated from the light coming from r. The total light distribution on the $z_0 + d$ plane will not equal c_d, but near $\operatorname{supp} c_d$ the two light distributions will be the same. Thus, an observer situated at $\operatorname{supp} c_d$ will not be able to distinguish case (ii) from case (i). He will just see the object (Figure 13.3). Of course, if $kt = c + r + s$, where c

and r are as above and s is very small in comparison with c, then an observer situated at supp c_d will see the object with a certain amount of blurring or other distortion.

The task is to prepare the film. This can be done photographically. Within a certain range (avoiding under or over exposure) a photographic emulsion responds (darkens) proportionally to the *intensity* of the total arriving radiation. The (negative) film so obtained thus has a transmission function

$$t(x) = 1 - \gamma I(x),$$

where γ is a proportionality factor depending on the film, and $I(x)$ is the intensity of the radiation arriving at x. Suppose we shine a plane wave parallel to the axis onto the plane and superimposed it upon the light arriving from the object. We assume that the light illuminating the object is coherent with the plane wave source. Then the *amplitude* of the light arriving at x is

$$a + c(x) = a + \frac{1}{\sqrt{2\pi}} \int \tilde{c}(\xi) e^{i\xi x}\, d\xi,$$

where a is a complex constant. The intensity at x is

$$I(x) = |a + c(x)|^2 = |a|^2 + \frac{1}{\sqrt{2\pi}} \int \bar{a}\tilde{c}(\xi) e^{i\xi x}\, dx + \frac{1}{\sqrt{2\pi}} \int a\bar{\tilde{c}}(\xi) e^{-i\xi x}\, d\xi + |c(x)|^2.$$

Now suppose that we use a very intense plane wave so that $|a|$ is much larger than $|c(x)|$ while γ, the sensitivity of the film, is small. Then $\gamma|c(x)|^2$ can be ignored relative to the remaining three terms in $\gamma I(x)$ and

$$t(x) = 1 - \gamma I(x) = -\gamma a c(x) + 1 - \gamma |a|^2 + \frac{1}{\sqrt{2\pi}} \int a\bar{\tilde{c}}(\xi) e^{-i\xi x}\, d\xi.$$

The first term is proportional to $c(x)$. The second term is just a plane wave passing through. We can write the third term as $(1/\sqrt{2\pi}) \int a\bar{\tilde{c}}(-\xi) e^{i\xi x}\, d\xi$. If c is supported near $\xi_0 \neq 0$ then the integrand in this last integral is supported near $-\xi_0$. In fact, it is not hard to see that the third term is proportional to the light that would arrive from an (imaginary) object situated at the mirror image of the real object in the plane.

14. Poisson brackets

In Section 12 we had occasion to study the "infinitesimal generators" of various one-parameter groups of linear symplectic transformations. In this section we wish to study these "infinitesimal symplectic transformations" in more detail, and also consider the nonlinear case.

Let $X = \mathbb{R}^{2n} = \mathbb{R}^n + \mathbb{R}^n$ and ω its symplectic form

$$\omega = -\sum dp_i \wedge dq_i. \tag{14.1}$$

(More generally, for the purpose of the next chapter, we may let X be an arbitrary symplectic manifold and ω its symplectic form. We will attempt to formulate most of the proofs so that they apply in the more general setting. A general theorem asserts that *locally* we can always introduce coordinates so that ω has the form (14.1).)

Let $\phi_t\colon \mathrm{X} \to X$ be a one-parameter group of symplectic diffeomorphisms; in other words

$$\phi_t^* \omega = \omega \tag{14.2}$$

for all t. Let ξ be the vector field on X which is the infinitesimal generator of ϕ_t. By a standard formula in differential geometry (Loomis and Sternberg, 1968, p. 456, or for a more general formula, see Section 22), we know that, whether or not (14.2) holds,

$$\frac{d}{dt}\phi_t^* \omega = i(\xi)\, d\omega + di(\xi)\omega$$

where $i(\xi)$ denotes interior product by ξ. In the case at hand $d\omega = 0$ so (14.2) implies

$$di(\xi)\omega = 0. \tag{14.3}$$

A vector field satisfying (14.3) is called an infinitesimal symplectic transformation (an infinitesimal canonical transformation in the physics literature). Locally on X (and globally if $X = \mathbb{R}^{2n}$ or, more generally if $H^1(X) = 0$), (14.3) is equivalent to the existence of a smooth function H determined up to a (local) constant such that

$$i(\xi)\omega = dH. \tag{14.4}$$

To emphasize the relation between ξ and H we shall write ξ_H for ξ when (14.4) holds. Thus, for ω given by (14.1), the vector field ξ_H is given by

$$\xi_H = \sum \frac{\partial H}{\partial p_i}\frac{\partial}{\partial q_i} - \frac{\partial H}{\partial q_i}\frac{\partial}{\partial p_i}. \tag{14.5}$$

The trajectories of (14.5) are thus the solution curves of the system of differential equations:

$$\frac{dq_i}{dt} = \frac{\partial H}{\partial p_i}; \qquad \frac{dp_i}{dt} = -\frac{\partial H}{\partial q_i}, \tag{14.6}$$

which are known as Hamilton's equations in the physics literature. The function H is called the (classical) Hamiltonian. If H_1 and H_2 are two functions, we define their Poisson bracket by

$$\{H_1, H_2\} = -\xi_{H_1} H_2. \tag{14.7}$$

By (14.4)

$$\{H_1, H_2\} = -i(\xi_{H_1}) dH_2 = -i(\xi_{H_1}) i(\xi_{H_2})\omega$$

so

$$\{H_1, H_2\} = -\{H_2, H_1\}. \tag{14.8}$$

It follows from (14.5) and (14.7) that

$$\{H_1, H_2\} = \sum \frac{\partial H_1}{\partial q_i}\frac{\partial H_2}{\partial p_i} - \frac{\partial H_2}{\partial q_i}\frac{\partial H_1}{\partial p_i} \tag{14.9}$$

is the local expression for the Poisson bracket and from (14.9) it follows that

$$\{H_1\{H_2, H_3\}\} = \{\{H_1, H_2\}, H_3\} + \{H_2, \{H_1, H_3\}\}. \tag{14.10}$$

More generally, we can let A be any commutative ring with elements f_1, f_2, etc. A Poisson bracket on A is a binary relation $\{\,,\}$: $A \times A \to A$ sending $(f_1, f_2) \to \{f_1, f_2\}$ and satisfying the relations

antisymmetry: $\{f_1, f_2\} = -\{f_2, f_1\}$

derivation property: $\{f_1, f_2 f_3\} = \{f_1, f_2\} f_3 + f_2\{f_1, f_3\}$

Jacobi's identity: $\{f_1\{f_2, f_3\}\} = \{\{f_1, f_2\}, f_3\} + \{f_2, \{f_1, f_3\}\}$.

In other words, the Poisson bracket makes A into a Lie algebra with the additional requirement that the bracket acts as a derivation of the commutative multiplication. We shall encounter a number of examples of such *Poisson algebras* in what follows.

For the moment, however, we restrict our attention to the example at hand. Let H_1 and H_2 be smooth functions on X. Then,

$$\begin{aligned} D_{\xi_{H_1}}(i(\xi_{H_2})\omega) &= i(D_{\xi_{H_1}}\xi_{H_2})\omega + i(\xi_{H_2}) D_{\xi_{H_1}}\omega \\ &= i(D_{\xi_{H_1}}\xi_{H_2})\omega \end{aligned}$$

since

$$D_{\xi_{H_1}}\omega = 0.$$

But

$$i(\xi_{H_2})\omega = dH_2 \qquad \text{and} \qquad D_{\xi_{H_1}}(dH_2) = -d\{H_1, H_2\}$$

so

$$\xi_{\{H_1,H_2\}} = -D_{\xi_{H_1}}\xi_{H_2}. \tag{14.11}$$

Now the group, Diff X, of all diffeomorphisms of X, acts on the space of all functions on X where $\phi \in$ Diff X sends the function f into

$$\rho_{\mathscr{F}}(\phi)f = \phi^{*-1}f = f \circ \phi^{-1}.$$

If ϕ_t is a one-parameter group of diffeomorphisms with infinitesimal generator ξ and f a smooth function, then

$$\frac{d}{dt}\rho_{\mathscr{F}}(\phi_t)f\bigg|_{t=0} = -D_\xi f$$

and so we get an action of the Lie algebra $D(X)$ of all vector fields, on the space of all functions defined by

$$\dot{\rho}_{\mathscr{F}}(\xi)f = -D_\xi f = -i(\xi)\,df = -\xi f. \tag{14.12}$$

Let $\mathscr{F}(X)$ denote the space of smooth functions on X and let $\operatorname{Ham}(X) \subset D(X)$ denote the space of all vector fields of the form ξ_H. We have a linear map from $\mathscr{F}(X)$ to $\operatorname{Ham}(X)$ sending each function H into the corresponding vector field ξ_H. If X is connected, the kernel of this map consists of the constants. We can express them by saying that the sequence

$$0 \to \mathbb{R} \to \mathscr{F}(X) \to \operatorname{Ham}(X) \to 0 \tag{14.13}$$

is an exact sequence of maps, in the sense that the image of each map is the kernel of the next. (Here the map $\mathbb{R} \to \mathscr{F}(X)$ is the injection of $\mathbb{R}$ as the constant functions.)

Equations (14.7) and (14.12) say that the action of H_1 on $\mathscr{F}(X)$ given by (left) Poisson bracket is consistent with the action, $\dot{\rho}_{\mathscr{F}}(\xi_{H_1})$, of ξ_{H_1} on $\mathscr{F}(X)$.

The group Diff(X) acts on vector fields by the representation ρ_D given by

$$\rho_D(\phi)\eta = \phi^{*-1}\eta, \qquad \text{for any } \eta \in D(X).$$

The corresponding infinitesimal action is given by

$$\dot{\rho}_D(\xi)\eta = -D_\xi\eta.$$

Equation (14.11) says that (left) Poisson bracket by H_1 is consistent with the action $\dot{\rho}_D(\xi_{H_1})$ on Ham (X). Finally, the map of $D(X) \times D(X) \to D(X)$ sending $(\xi, \eta) \to \dot{\rho}_D(\xi)\eta$ makes $D(X)$ into a Lie algebra with

$$[\xi, \eta] = \dot{\rho}_D(\xi)\eta.$$

Equation (14.11) can also be regarded as asserting that the maps in (14.13) are homomorphisms of Lie algebras. Here $\mathbb{R}$ is thought of as the trivial Lie algebra. From (14.9) it is clear that the constants (and only the constants) have Poisson bracket 0 with all functions. Thus the image of $\mathbb{R}$ is the center of $\mathscr{F}(X)$. We say that (14.13) describes $\mathscr{F}(X)$ as a central extension of Ham (X).

Let us work out some examples. Take $X = \mathbb{R}^2$. Taking $H(q,p) = \frac{1}{2}p^2$ in (14.5), we see that

$$\xi_{p^2/2} = p\frac{\partial}{\partial q}$$

and (14.6) becomes

$$\frac{d}{dt}\begin{pmatrix} q \\ p \end{pmatrix} = \begin{pmatrix} 0 & 1 \\ 0 & 0 \end{pmatrix}\begin{pmatrix} q \\ p \end{pmatrix}.$$

Thus the function $\frac{1}{2}p^2$ corresponds to the linear vector field $p(\partial/\partial q)$ which can also be thought of as the matrix $\begin{pmatrix} 0 & 1 \\ 0 & 0 \end{pmatrix}$ belonging to the Lie algebra $sp(2)$ of the symplectic group. A similar computation yields the table

Quadratic polynomial	Vector field	Matrix
$\frac{1}{2}p^2$	$p\frac{\partial}{\partial q}$	$\begin{pmatrix} 0 & 1 \\ 0 & 0 \end{pmatrix}$
$\frac{1}{2}q^2$	$-q\frac{\partial}{\partial p}$	$\begin{pmatrix} 0 & 0 \\ -1 & 0 \end{pmatrix}$
pq	$q\frac{\partial}{\partial q} - p\frac{\partial}{\partial p}$	$\begin{pmatrix} 1 & 0 \\ 0 & -1 \end{pmatrix}$

and we have an isomorphism between the algebra of quadratic homogeneous polynomials (in two variables) and the symplectic algebra. Thus, for example, the harmonic-oscillator Hamiltonian $\frac{1}{2}(p^2 + q^2)$ corresponds to the infinitesimal rotation $\begin{pmatrix} 0 & 1 \\ -1 & 0 \end{pmatrix}$. The same result clearly holds in $2n$ dimensions: A matrix $\begin{pmatrix} X & Y \\ Z & W \end{pmatrix}$ belongs to the Lie algebra $sp(2n, \mathbb{R})$ if and only if

$$\begin{pmatrix} X & Y \\ Z & W \end{pmatrix}\begin{pmatrix} 0 & I \\ -I & 0 \end{pmatrix} + \begin{pmatrix} 0 & I \\ -I & 0 \end{pmatrix}\begin{pmatrix} X & Y \\ Z & W \end{pmatrix}^{\mathrm{t}} = 0$$

or

$$\begin{aligned} X^{\mathrm{t}} &= -W, \\ Y &= Y^{\mathrm{t}} \\ Z &= Z^{\mathrm{t}}. \end{aligned} \tag{14.14}$$

It follows from (14.5) that the quadratic form

$$Q_T(q, p) = \tfrac{1}{2}(Yp\cdot p - Zq\cdot q + 2Xq\cdot p)$$

corresponds to the matrix

$$T = \begin{pmatrix} X & Y \\ Z & -X^{\mathrm{t}} \end{pmatrix}.$$

We can also write this as

$$Q_T(u) = \tfrac{1}{2}(Tu, u). \tag{14.15}$$

We can replace the discussion of Section 12 as follows, using the isomorphism between the algebra of quadratic homogeneous polynomials under Poisson bracket and the Lie algebra of the symplectic group: We have associated to each homogeneous quadratic polynomial H_{cl} a self-adjoint operator H_{qu} on the Hilbert space L^2 in such a way that the map from H_{cl} to $-iH_{\mathrm{qu}}$ carries Poisson brackets into commutators; that is, we have a representation of the Lie algebra of homogeneous quadratic polynomials. We have also "integrated" this Lie algebra representation so as to obtain a representation of the metaplectic group.

15. The Heisenberg group and representation

The space of homogeneous quadratic polynomials forms a subalgebra of the algebra, $\mathscr{F}(X)$ of all functions on $X = \mathbb{R}^{2n}$ under Poisson bracket. But so does the algebra of inhomogeneous quadratic polynomials. The purpose of this section is to extend the results of the preceding sections from the algebra of homogeneous polynomials to the algebra of inhomogeneous polynomials. There are a number of subtle points here that require some care.

First of all, we observe that whereas the space of homogeneous quadratic polynomials forms a subalgebra, the space of homogeneous linear polynomials does not:

$$\{q, p\} = 1$$

for $n = 1$, or, more generally, for arbitrary n

$$\{q_k, p_l\} = \delta_{kl} = \begin{matrix} 1 & \text{if } k = l, \\ 0 & \text{if } k \neq l. \end{matrix} \qquad (15.1)$$

The vector field corresponding to p_i is $\partial/\partial q_i$, which represents infinitesimal translation in the q_i direction and similarly, the vector field corresponding to q_i is $-\partial/\partial p_i$. Thus the image of the space of homogeneous linear polynomials is the space of all constant vector fields. These *do* form a Lie algebra, the commutative Lie algebra of all infinitesimal translations, which we may denote by $\mathbb{R}^{2n}$. Thus $\mathbb{R}^{2n}$ is a commutative subalgebra of Ham (X). The homomorphism from $\mathscr{F}(X)$ to Ham (X) has the constants as kernel, and a constant Poisson commutes with every function. So replacing the q_i by $q_i + c_i$ and the p_j by $p_j + d_j$, where the c's and d's are constants, will not affect relation (15.1). Thus there is *no* subalgebra of $\mathscr{F}(\mathrm{X})$ that projects isomorphically onto $\mathbb{R}^{2n}$. Thus, if we take $g = \mathbb{R}^{2n}$ we have a homomorphism (actually an isomorphism) i of $g \to \text{Ham}(X)$, but there is no way of lifting this to a homomorphism $\lambda: g \to \mathscr{F}(X)$ so as to make the diagram

$$\begin{array}{ccccccccc} 0 & \to & \mathbb{R} & \to & \mathscr{F}(X) & \to & \text{Ham}(X) & \to & 0 \\ & & & & & \nwarrow^{\lambda} & \uparrow i & & \\ & & & & & & g & & \end{array}$$

commute. This is in contrast to the situation in the preceding section, where for $g = sp(2n)$ we were able to find λ (which identified $sp(2n)$ with the space of homogeneous quadratic polynomials). In the next chapter we will discuss conditions on a general Lie algebra g and homomorphism i that determine when such a λ exists and to what extent it is unique.

In any event the algebra of (inhomogeneous) linear functions does form a Lie subalgebra of $\mathscr{F}(\mathbb{R}^{2n})$ that projects homomorphically onto the commutative algebra of infinitesimal translations $\mathbb{R}^{2n}$. Now we can also consider the *group* of all translations of $\mathbb{R}^{2n}$ and look for a *group* $\mathscr{H}(\mathbb{R}^{2n})$, that projects homomorphically such that this projection is the group-theoretical version of the above homomorphism.

Let V be a symplectic vector space with symplectic form ω, and let S denote the unit circle in the complex plane. The Heisenberg group $\mathscr{H}(V)$ is the set

$$\mathscr{H}(V) = V \times S$$

endowed with the multiplication

$$(v_1, z_1)(v_2, z_2) = \left(v_1 + v_2, z_1 z_2 \exp\left(\frac{i}{2}\omega(v_1, v_2)\right)\right). \tag{15.2}$$

Readers should check the axioms for a group are satisfied by this law of multiplication. In particular, check that the inverse is given by

$$(v, z)^{-1} = (-v, z^{-1}) \tag{15.3}$$

and that the set $(tv, e^{it\theta})$ is a (and in fact the most general) one-parameter group. Both of these facts follow from the antisymmetry of ω, in particular, from the fact that $\omega(v, v) = 0$. Thus we may identify the Lie algebra of $\mathscr{H}(V)$ as a vector space with

$$h(V) = V \times i\mathbb{R}.$$

Conjugation in the group is given by

$$(v_1, z_1)\cdot(v_2, z_2)\cdot(v_1, z_1)^{-1} = (v_2, z_2 \exp[i\omega(v_1, v_2)]), \tag{15.4}$$

again by the antisymmetry of ω. In particular, the adjoint representation of $\mathscr{H}(V)$ on $h(V)$ is given by

$$\begin{aligned} \mathrm{Ad}_{(v_1, z_1)}(u, i\theta) &\stackrel{\text{def}}{=} \left.\frac{d}{dt}\right|_{t=0} (v_1, z_1)(tu, e^{it\theta})(v_1, z_1)^{-1} \\ &= \left.\frac{d}{dt}\right|_{t=0} (tu, e^{it(\theta + \omega(v_1, u))}) \\ &= (u, i(\theta + \omega(v, u))). \end{aligned}$$

The Lie bracket, by definition, is given by

$$\begin{aligned} [(u_1, i\theta_1), (u_2, i\theta_2)] &= \left.\frac{d}{dt}\right|_{t=0} \mathrm{Ad}_{(tu_1, \exp(it\theta_1))}(u_2, i\theta_2) \\ &= (0, i\omega(u_1, u_2)) \end{aligned}$$

or, since the i is completely superfluous in this formula, we may identify $h(V)$ with $V \times \mathbb{R}$ and write the Lie bracket as

$$[(u_1, \theta_1), (u_2, \theta_2)] = (0, \omega(u_1, u_2)). \tag{15.5}$$

The Lie algebra $h(V)$ is known as the Heisenberg algebra. We may identify the (homogeneous) linear functions on V with V using the form ω. (This is precisely the identification of linear functions with constant vector fields – they are thought of as vectors in V.) Then $V \times \mathbb{R}$ can be identified with the inhomogeneous linear functions on V and it is easy to check that this is an isomorphism of Lie algebras.

Let us illustrate this identification when $V = \mathbb{R}^2$ where ω is given by

$$\omega\left(\begin{pmatrix} q \\ p \end{pmatrix}, \begin{pmatrix} q' \\ p' \end{pmatrix}\right) = qp' - q'p.$$

The vector $\begin{pmatrix} 1 \\ 0 \end{pmatrix}$ is identified with the linear function that sends $\begin{pmatrix} q \\ p \end{pmatrix}$ into the number

$$\omega\left(\begin{pmatrix} 1 \\ 0 \end{pmatrix}, \begin{pmatrix} q \\ p \end{pmatrix}\right) = p.$$

Thus, the element $\left(\begin{pmatrix} 1 \\ 0 \end{pmatrix}, 0\right)$ of the Heisenberg algebra is identified with the function p. Similarly, the element $\left(\begin{pmatrix} 0 \\ 1 \end{pmatrix}, 0\right)$ of the Heisenberg algebra is identified with the function $-q$. Finally, the element $\left(\begin{pmatrix} 0 \\ 0 \end{pmatrix}, 1\right)$ of the Heisenberg algebra is identified with the constant function, 1. In the Heisenberg algebra we have

$$\left[\left(\begin{pmatrix} 1 \\ 0 \end{pmatrix}, 0\right), \left(\begin{pmatrix} 0 \\ 1 \end{pmatrix}, 0\right)\right] = \left(\begin{pmatrix} 0 \\ 0 \end{pmatrix}, 1\right),$$

while

$$\{p, -q\} = 1$$

This exhibits the isomorphism between the Heisenberg algebra and the algebra of the linear inhomogeneous functions under Poisson bracket.

For the case of $V = \mathbb{R}^{2n} = \mathbb{R}^n + \mathbb{R}^n$, the elements $\left(\begin{pmatrix} \vec{q} \\ 0 \end{pmatrix}, 1\right)$ and $\left(\begin{pmatrix} 0 \\ \vec{p} \end{pmatrix}, 1\right)$, where $\vec{q}$, $\vec{p} \in \mathbb{R}^n$, generate the group $\mathscr{H}(V)$ and we have the multiplication rules

$$\left(\begin{pmatrix} \vec{q}_1 \\ 0 \end{pmatrix}, 1\right)\left(\begin{pmatrix} \vec{q}_2 \\ 0 \end{pmatrix}, 1\right) = \left(\begin{pmatrix} \vec{q}_1 + \vec{q}_2 \\ 0 \end{pmatrix}, 1\right),$$

$$\left(\begin{pmatrix} 0 \\ \vec{p}_1 \end{pmatrix}, 1\right)\left(\begin{pmatrix} 0 \\ \vec{p}_2 \end{pmatrix}, 1\right) = \left(\begin{pmatrix} 0 \\ \vec{p}_1 + \vec{p}_2 \end{pmatrix}, 1\right),$$

$$\left(\begin{pmatrix} 0 \\ \vec{p} \end{pmatrix}, 1\right)\left(\begin{pmatrix} \vec{q} \\ 0 \end{pmatrix}, 1\right) = \left(\begin{pmatrix} \vec{q} \\ \vec{p} \end{pmatrix}, e^{-i\vec{p}\cdot\vec{q}/2}\right).$$

We shall now construct a unitary representation of the group $\mathscr{H}(\mathbb{R}^{2n})$ on the space $L^2(\mathbb{R}^n)$. In fact, we can give an "optical" interpretation of the action of the generators: The vector $\begin{pmatrix} \vec{q} \\ 0 \end{pmatrix}$ corresponds to translation in the vertical direction, that is, to moving the optical axis to $\begin{pmatrix} \vec{q} \\ 0 \end{pmatrix}$ from $\begin{pmatrix} 0 \\ 0 \end{pmatrix}$. We therefore define

$$\tau\left(\left(\begin{pmatrix} \vec{q} \\ 0 \end{pmatrix}, 1\right)\right)c = c_{\vec{q}},$$

where

$$c_{\vec{q}}(x) = c(x - \vec{q}).$$

The vector $\begin{pmatrix} 0 \\ \vec{p} \end{pmatrix}$ corresponds to tilting the axis through "angle" p, which has the effect of changing the phase (see our discussion of interference in Section 9). We define

$$\tau\left(\left(\begin{pmatrix} 0 \\ \vec{p} \end{pmatrix}, 1\right)\right)c = e^{i\vec{p}\cdot x}c.$$

Notice that

$$\tau\left(\begin{pmatrix} 0 \\ \vec{p} \end{pmatrix}, 1\right)\tau\left(\begin{pmatrix} \vec{q} \\ 0 \end{pmatrix}, 1\right)c = e^{i\vec{p}\cdot x}c_{\vec{q}}$$

and multiplication in the Heisenberg group gives

$$\left(\begin{pmatrix} 0 \\ \vec{p} \end{pmatrix}, 1\right)\left(\begin{pmatrix} \vec{q} \\ 0 \end{pmatrix}, 1\right) = \left(\begin{pmatrix} \vec{q} \\ \vec{p} \end{pmatrix}, e^{-i\vec{p}\cdot\vec{q}/2}\right).$$

$$\tau\left(\left(\begin{pmatrix} \vec{q} \\ \vec{p} \end{pmatrix}, e^{-i\vec{p}\cdot\vec{q}/2}\right)\right)c = \tau\left(\left(\begin{pmatrix} 0 \\ 0 \end{pmatrix}, e^{-i\vec{p}\cdot\vec{q}/2}\right)\right)\tau\left(\left(\begin{pmatrix} \vec{q} \\ \vec{p} \end{pmatrix}, 1\right)\right)c = e^{i\vec{p}\cdot x}c_q.$$

To determine the representation, we must choose the value of $\tau\left(\begin{pmatrix} 0 \\ 0 \end{pmatrix}, z\right)$. If we expect the representation to be irreducible (and it turns out that this will be the case) then, since $\left(\begin{pmatrix} 0 \\ 0 \end{pmatrix}, z\right)$ is in the center of Heisenberg group, $\tau\left(\begin{pmatrix} 0 \\ 0 \end{pmatrix}, z\right)$ must be multiplication by a scalar, and for τ to be a representation, this scalar must be z^n for some integer n. The standard physics convention, and the one that matches with the formulas of the preceding sections, is to take $n = -1$. Thus we define

$$\left[\tau\left(\left(\begin{pmatrix}\vec{q}\\ \vec{p}\end{pmatrix}, z\right)\right)c\right](x) = z^{-1}e^{-i\vec{p}\cdot\vec{q}/2}e^{i\vec{p}\cdot x}c(x - \vec{q}). \tag{15.6}$$

It now follows that

$$\begin{aligned}
&\left(\tau\left(\begin{pmatrix}\vec{q}'\\ \vec{p}'\end{pmatrix}, z'\right)\tau\left(\begin{pmatrix}\vec{q}\\ \vec{p}\end{pmatrix}, z\right)c\right)(x)\\
&\quad = z'^{-1}z^{-1}e^{-i\vec{p}'\cdot\vec{q}'/2}e^{-i\vec{p}\cdot\vec{q}/2}e^{i\vec{p}'\cdot x}e^{i\vec{p}(x-\vec{q}')}c(x - \vec{q} - \vec{q}')\\
&\quad = (z'ze^{i(\vec{q}'\circ\vec{p} - \vec{p}'\cdot\vec{q})/2})^{-1}e^{-[i(\vec{p}+\vec{p}')/2\cdot(\vec{q}+\vec{q}')}e^{i(\vec{p}+\vec{p}')\circ x}c(x - (\vec{q} + \vec{q}'))
\end{aligned}$$

proving that

$$\tau(v', z')\tau(v, z) = \tau(v + v', zz'e^{i\omega(v',v)/2}),$$

and we see that τ is a representation of $\mathscr{H}(V)$.

The group $Sp(V)$ preserves the linear and symplectic structure of V and hence acts as a group of automorphisms of $\mathscr{H}(V)$. Explicitly, for $T \in Sp(V)$, we define

$$T(v, z) = (Tv, z).$$

The group $Mp(V)$ is the double cover of $Sp(V)$ and therefore also acts as automorphisms of $\mathscr{H}(V)$: If $X \in Mp(V)$ and $\bar{X}$ denotes its image in $Sp(V)$, we define

$$X(v, z) = (\bar{X}v, z).$$

We have the metaplectic representation U of $Mp(V)$ and the Heisenberg representation τ of $\mathscr{H}(V)$. We claim that they are related by the basic formula

$$U(X)\tau(a)U(X^{-1}) = \tau(Xa). \tag{15.7}$$

To prove this, it is enough to prove it for generators; so we may assume that

$$U(X) = U_d, \qquad \bar{X} = \begin{pmatrix} I & dI\\ 0 & I\end{pmatrix}$$

or

$$U(X) = V_P, \qquad \bar{X} = \begin{pmatrix} I & 0\\ -P & I\end{pmatrix}$$

in the notation of Section 4, and that

$$a = \left(\begin{pmatrix}\vec{q}\\ 0\end{pmatrix}, 1\right)$$

or

$$a = \left(\begin{pmatrix} 0 \\ \vec{p} \end{pmatrix}, 1\right),$$

so there are four cases to check. Actually, two of them are trivial: If $a = \left(\begin{pmatrix} 0 \\ \vec{p} \end{pmatrix}, 1\right)$ and $\bar{X} = \begin{pmatrix} I & 0 \\ -P & I \end{pmatrix}$, then both $\tau(a)$ and $U(X)$ are multiplication operators and hence commute, while

$$\begin{pmatrix} I & 0 \\ -P & I \end{pmatrix}\begin{pmatrix} 0 \\ \vec{p} \end{pmatrix} = \begin{pmatrix} 0 \\ \vec{p} \end{pmatrix}.$$

The case is similar for $a = \left(\begin{pmatrix} \vec{q} \\ 0 \end{pmatrix}, 1\right)$ and $\bar{X} = \begin{pmatrix} I & dI \\ 0 & I \end{pmatrix}$.

If $U(X) = V_P$ and $a = \left(\begin{pmatrix} \vec{q} \\ 0 \end{pmatrix}, 1\right)$, then

$$\bar{X}a = \left(\begin{pmatrix} \vec{q} \\ -P\vec{q} \end{pmatrix}, 1\right)$$

so

$$\left(\tau(\bar{X}a)c\right)(x) = e^{iP\vec{q}\cdot\vec{q}/2}e^{-iP\vec{q}\cdot x}c(x - \vec{q})$$

and

$$\begin{aligned} (V_{-P}c) &= e^{iPx\cdot x/2}c(x) \\ [(\tau(a)V_{-P})c](x) &= \exp[iP(x - \vec{q})\cdot(x - \vec{q})/2]c(x - \vec{q}) \\ &= \exp(iPx\cdot x/2 + iP\vec{q}\cdot\vec{q}/2 - iP\vec{q}\cdot x)c(x - \vec{q}) \end{aligned}$$

so

$$\left(V_P\tau(a)V_{-P}c\right)(x) = e^{iPq\cdot q/2}e^{-iPq\cdot x}c(x - q),$$

as required. The remaining case we leave to the reader as an exercise in Gaussian integrals (or they can be reduced to the preceding case by the Fourier transform).

We can reformulate the conclusion of our basic formula (15.7) using the notion of the semidirect product of two groups whose definition we now recall: Let H and N be groups and let H act as automorphisms of N. We shall denote the action of $X \in H$ on $a \in N$ by Xa, and the group multiplication in H and in N by $\cdot$. We can then make the set $H \times N$ into a group by the law

$$(X, a)\cdot(Y, b) = (X\cdot Y, a\cdot(Xb)).$$

It is easy to check that this satisfies all the conditions for a group multiplication. The resulting group is called the semidirect product of H and N. Suppose that ρ and τ are representations of H and N on the same space V. Suppose furthermore that

$$\rho(X)\tau(a)\rho(X^{-1}) = \tau(Xa).$$

Then we can define a representation σ of the semidirect product of H and N by

$$\sigma(X, a) = \tau(a)\rho(X).$$

Then

$$\begin{aligned}\sigma(X, a)\sigma(Y, b) &= \tau(a)\rho(X)\tau(b)\rho(Y)\\ &= \tau(a)\rho(X)\tau(b)\rho(X^{-1})\rho(X)\rho(Y)\\ &= \tau(a \cdot Xb)\rho(X \cdot Y)\\ &= \sigma\big((X, a)(Y, b)\big)\end{aligned}$$

so that σ is indeed a representation.

In our case, we take $H = Mp(V)$ (where $V = \mathbb{R}^n + \mathbb{R}^n$) and $N = \mathscr{H}(V)$. It is easy to check that the Lie algebra of the semidirect product in this case is precisely the space of inhomogeneous quadratic polynomials. We have constructed a representation of the semidirect product $Mp(V)$Ⓢ $\mathscr{H}(V)$ and thus a "quantization procedure" for all inhomogeneous quadratic polynomials, which was one of our goals.

The question naturally now arises as to how far the quantization procedure extends beyond the quadratic polynomials. After all, in our original justification of the (linear) Fresnel formula in Section 8, we made the approximating and simplifying assumption that the attenuation factor a was a constant. There is no reason to believe this to be true in a nonlinear theory. The answer to our question is, at least as far as polynomials are concerned, *not at all*! Before going to the demonstration of this assertion in the next section, we pause to summarize some relevant facts about the Heisenberg and metaplectic representation that will make the results of the next section even more striking.

It is not difficult to check that the representation τ of the Heisenberg group is irreducible. A theorem of Stone and von Neumann (see Mackey, 1968, pp. 49–53 for a proof) says that it is the unique irreducible representation of $\mathscr{H}(V)$ with the property that

$$\tau(0, z) = z^{-1}I,$$

where I is the identity operator. That is, any other irreducible representation of $\mathscr{H}(V)$ with this property must be unitarily equivalent to τ, and (by the irreducibility of τ) the unitary operator effecting the equivalence is determined up to a scalar. In particular, for $M \in Sp(V)$ the representation τ_M defined by

$$\tau_M(a) = \tau(Ma)$$

must be unitarily equivalent to τ, so there exists a unitary operator U_M such that

$$\tau_M = U_M \tau U_M.$$

As U_M is determined up to a constant multiple, it follows that

$$U_{M_1} U_{M_2} = c(M_1, M_2) U_{M_1 M_2},$$

where $c(M_1, M_2)$ is some complex number of absolute value 1. In other words, the mapping $M \to U_M$ is a *projective representation* of the symplectic group. From general principles we know that this corresponds to an ordinary (unitary) representation of the universal covering group of the symplectic group. We know from the preceding discussion that it is, in fact, a representation of the double cover,

$$U_{\tilde{X}} = \rho(X).$$

We are thus in the following situation: If we believe in the primacy of the Heisenberg commutation relations (i.e., in the Heisenberg group and its unique irreducible representation), then almost all of the preceding discussion is superfluous as far as quantum mechanics is concerned. The Heisenberg representation is unique; it determines a unique projective representation of the symplectic group from general principles of representation theory. This determines a unique representation of the universal cover of $Sp(V)$ and hence, in particular, a unique method of quantizing quadratic polynomials. All of this without any computation at all! It requires some work (but only a modest amount) to verify that it is a representation of the double cover and, perhaps, the explicit formulas are of some use. But starting with the Heisenberg group we could dispense with all the preceding discussion and proceed completely abstractly. The rub comes from the results of the next section. Having committed ourselves to the Heisenberg (canonical) commutation relations, it turns out that we cannot get beyond quadratic polynomials.

16. The Groenwald–van Hove theorem

The representation of the space of quadratic polynomials that we have been studying is given (for $n = 1$) by the table

Polynomial	Corresponding operator
1	$-i$
q	$-ix$
p	$-\dfrac{\partial}{\partial x}$
q^2	$-ix^2$
pq	$-\left[x\dfrac{\partial}{\partial x} + \frac{1}{2}\right]$
p^2	$i\dfrac{\partial^2}{\partial x^2}$

(where on the right, ix means the operator of multiplication by ix, etc.). As a warm-up to reading this section, the reader should check directly that the table does indeed give a representation of Lie algebras, that is, that the Poisson bracket of any two functions on the left corresponds to the commutator of the corresponding operators on the right.

We shall prove the following facts (we follow the presentation of Chernoff (1981)):

(i) *The algebra of quadratic polynomials is a maximal subalgebra (under Poisson bracket) of the algebra of all polynomials.*

Let $\dot{\sigma}$ denote our representation so that $\dot{\sigma}(q) = ix$, etc. Let

$$A(f) = i\dot{\sigma}(f)$$

so that

$$[A(f), A(g)] = \frac{1}{i}A(\{f, g\}) \tag{16.1}$$

for all quadratic polynomials. Write

$$Q = A(q) \qquad \text{and} \qquad P = A(p)$$

and observe from the table that

$$A(q^2) = Q^2 \qquad \text{and} \qquad A(p^2) = P^2. \tag{16.2}$$

Then (we shall prove):

(ii) *If* A *is any linear map from polynomials to an associative algebra satisfying (16.1) and (16.2), then we must have*

$$A(q^3) = Q^3, \qquad A(p^3) = P^3 \tag{16.3}$$

and

$$A(q^2p) = \tfrac{1}{2}(Q^2P + PQ^2), \qquad A(qp^2) = \tfrac{1}{2}(P^2Q + QP^2). \tag{16.4}$$

A direct computation shows that

$$\tfrac{1}{9}\{q^3, p^3\} = q^2p^2 = \tfrac{1}{3}\{pq^2, p^2q\}, \tag{16.5}$$

but a more tedious computation shows that

$$\frac{1}{9i}[Q^3, P^3] = Q^2P^2 - 2iQP - \frac{2}{3} \tag{16.6}$$

and, further,

$$\frac{1}{3i}\left[\frac{1}{2}(PQ^2 + Q^2P), \frac{1}{2}(P^2Q + QP^2)\right] = Q^2P^2 - 2iQP - \frac{1}{3}. \tag{16.7}$$

Putting these facts together, we arrive at the theorem of Groenwald and van Hove – there is no way of extending the metaplectic representation to include any nonquadratic polynomial. Indeed, by (i), the subalgebra generated by this polynomial together with the quadratic ones must be the algebra of all polynomials. But then (16.3), (16.5), and (16.6), on the one hand, and (16.4), (16.5), and (16.7), on the other, assign two distinct values to $A(q^2p^2)$, yielding a contradiction.

For the proof of (i) we first observe that the space S_l of homogeneous polynomials of degree l is irreducible under $sp(\mathbb{R}^2)$. Indeed,

$$\{pq, p^jq^k\} = (j-k)p^jq^k$$

and the values $j - k$ are all distinct as j and k range over all nonnegative integers with $j + k = l$. It follows that any invariant subspace must contain a monomial. But

$$\{\tfrac{1}{2}q^2, p^jq^k\} = jp^{j-1}q^{k+1},$$

so repeated bracketing by $\frac{1}{2}q^2$ shows that any invariant subspace must contain q^l and that repeated bracketing with $\frac{1}{2}p^2$ gives all monomials. (A similar argument works in n dimensions, showing the space S_l of homogeneous polynomials of degree l in the p_i's and q_j's is irreducible under $Sp(2n)$.)

If f is any nonquadratic polynomial in the p's and q's, repeated bracketing by q or p would yield a cubic polynomial with nonvanishing homogeneous cubic component. Since quadratic polynomials already belong to our set, we can subtract off the quadratic terms and get a nonzero element of S_3. Thus the subalgebra generated by f and the quadratic polynomials must contain all of S_3. But

$$\{q^3, p^3\} = 9q^2p^2$$

is a nonzero element of S_4 and hence, by the irreducibility of S_4, our subalgebra must contain all of S_4. Then

$$\{q^3, p^4\} = 12q^2p^3,$$

and so we get all of S_5, etc. This proves (i).

We now prove (16.3). First observe that

$$\{q^2, p^2\} = 4pq$$

and so

$$4A(pq) = \frac{1}{i}[Q^2, P^2] = \frac{1}{2}[QP + PQ].$$

Write $A(q^3) = X$. Now

$$\{q^3, q\} = 0 \qquad \text{and} \qquad \{q^3, p\} = 3q^2,$$

and so we must have

$$[X, Q] = 0 \qquad \text{and} \qquad [X, P] = 3iQ^2.$$

But

$$[Q^3, Q] = 0 \qquad \text{and} \qquad [Q^3, P] = 3iQ^2,$$

so if we set

$$X - Q^3 = Y,$$

then

$$[Y, Q] = [Y, P] = 0.$$

But

$$\{q^3, pq\} = 3q^3,$$

which implies that

$$3iX = [X, \tfrac{1}{2}(PQ + QP)].$$

But

$$[Y, PQ] = [Y, P]Q + P[Y, Q] = 0$$

so

$$[X, \tfrac{1}{2}(PQ + QP)] = [Q^3, \tfrac{1}{2}(PQ + QP)]$$

so

$$3iX = [Q^3, \tfrac{1}{2}(PQ + QP)] = 3iQ^3$$

so

$$X = Q^3,$$

proving (16.3) for q (with a similar argument for p). Now

$$\{q^3, p^2\} = 6q^2 p,$$

hence

$$\begin{aligned} A(q^2, p) &= \frac{1}{6i}[Q^3, P^2] \\ &= \frac{1}{6i}(P[Q^3, P] + [Q^3, P]P) \\ &= \frac{1}{6i}(3PQ^2 + 3Q^2P), \end{aligned}$$

proving (16.4) for q^2p (and a similar argument gives the result for p^2q). The proof of (16.5), (16.6), and (16.7) are left to the reader.

17. Other quantizations

It is not clear how to evaluate the Groenwald–van Hove theorem. It is true that the algebra of inhomogeneous quadratic polynomials is a maximal subalgebra of the space of all polynomials under Poisson bracket, and hence we do not expect to be able to "quantize" anything beyond quadratic polynomials in a way consistent with our metaplectic quantization. Yet there will exist interesting groups that are not contained in the inhomogeneous metaplectic group, which act as symplectic transformations and have a concomitant unitary representation, and so will have their own "quantization rules."

To illustrate what we have in mind, let us return to rotationally symmetric optics, and consider the "trivial" case where the index of refraction is constant. In the geometric-optics approximation, the light rays are just straight lines. In the wave-optics approximation (still ignoring polarization effects) we would be considering solutions of the reduced wave equation

$$\left(\frac{\partial^2 u}{dx^2} + \frac{\partial^2 u}{dy^2} + \lambda^2 u\right) = 0,$$

where λ is the wavelength of the light. In both approximations the group of Euclidean motions of the plane $E(2)$ acts as a group of symmetries. Let us carry out the details of this and related examples.

Let V be a real vector space equipped with a nondegenerate symmetric scalar product $(\,,)$. We let $O(V)$ denote the corresponding orthogonal group; that is, the group of linear transformations A that satisfy

$$(Au, Av) = (u, v) \qquad \forall\, u, v \in V.$$

We let $E(V)$ denote the group of Euclidean motions, so an $a \in E(V)$ is a transformation of the form

$$x \to Ax + b \qquad A \in O(V),\, b \in V$$

for any $x \in V$. We can describe this transformation in matrix form by writing

$$a = \begin{pmatrix} A & b \\ 0 & 1 \end{pmatrix}$$

and x as $\begin{pmatrix} x \\ 1 \end{pmatrix}$ so

$$\begin{pmatrix} A & b \\ 0 & 1 \end{pmatrix}\begin{pmatrix} x \\ 1 \end{pmatrix} = \begin{pmatrix} Ax + b \\ 1 \end{pmatrix}.$$

The product of two Euclidean motions can be computed by "matrix multiplication"

$$a_1 a_2 = \begin{pmatrix} A_1 & b_1 \\ 0 & 1 \end{pmatrix}\begin{pmatrix} A_2 & b_2 \\ 0 & 1 \end{pmatrix} = \begin{pmatrix} A_1 A_2 & A_1 b_2 + b_1 \\ 0 & 1 \end{pmatrix}.$$

The Lie algebra of $O(V)$, denoted by $o(V)$, consists of those linear transformations η that satisfy

$$(\eta u, v) + (u, \eta v) = 0.$$

The dimension of $o(V)$ is clearly given by $\frac{1}{2}d(d-1)$, where $d = \dim V$, and this is the same as $\dim \bigwedge^2(V)$. Given $e \wedge f \in \bigwedge^2(V)$, we can define a linear transformation $l_{e \wedge f} \in \mathrm{Hom}(V, V)$ by

$$l_{e \wedge f} u = -(f, u)e + (e, u)f;$$

and then

$$(l_{e \wedge f} u, v) = -(f, u)(e, v) + (e, u)(f, v)$$

is clearly antisymmetric in u and v. Thus $l_{e \wedge f} \in o(V)$. Now $l_{e \wedge f}$ is clearly linear in e and f. Thus we have defined a linear map from $\wedge^2 V \to o(V)$ and it is easy to check that this map is an isomorphism. Henceforth we shall drop the symbol l and simply write

$$(e \wedge f)u = -(f, u)e + (e, u)f.$$

This identification of $\wedge^2(V)$ with $o(V)$ depends only on the fact that $(\ ,\ )$ is nondegenerate, it need not be positive definite.

The Lie algebra of $E(V)$, denoted by $e(V)$, can be described as the set of all matrices of the form

$$\begin{pmatrix} \eta & \xi \\ 0 & 0 \end{pmatrix} \qquad \eta \in o(V), \xi \in V.$$

The Lie bracket is just the matrix commutator so

$$\left[\begin{pmatrix} \eta_1 & \xi_1 \\ 0 & 0 \end{pmatrix}, \begin{pmatrix} \eta_2 & \xi_2 \\ 0 & 0 \end{pmatrix}\right] = \begin{pmatrix} \eta_1\eta_2 - \eta_2\eta_1 & \eta_1\xi_2 - \eta_2\xi_1 \\ 0 & 0 \end{pmatrix}.$$

Since the zeros in the bottom row are irrelevant to this computation, we shall frequently drop them and write a typical element of $e(V)$ as (η, ξ) and the above bracket relations are written as

$$[(\eta_1, \xi_1), (\eta_1, \xi_2)] = (\eta_1\eta_2 - \eta_2\eta_1, \eta_1\xi_2 - \eta_2\xi_1).$$

The three examples that we shall be primarily interested in are $V = \mathbb{R}^2$ or $\mathbb{R}^3$, with the standard Euclidean metric, and $V = \mathbb{R}^4$, with the Minkowski metric. For the case $V = \mathbb{R}^n$, with the standard Euclidean metric, we shall write $E(n)$ instead of $E(\mathbb{R}^n)$, etc.

Thus, the Lie algebra $e(2)$ is three-dimensional. If e_1, e_2 is an orthonormal basis of $\mathbb{R}^2$, then a basis of $e(2)$ is given by

$(0, e_1)$ which we shall write as e_1,

$(0, e_2)$ which we shall write as e_2,

and

$(e_1 \wedge e_2, 0)$ which we shall write as $e_1 \wedge e_2$.

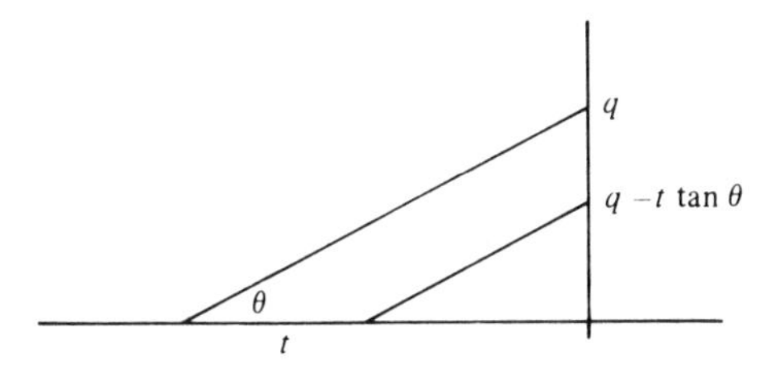

Figure 17.1.

The bracket relations are

$$[e_1, e_2] = 0,$$

$$[e_1 \wedge e_2, e_1] = e_2,$$

$$[e_1 \wedge e_2, e_2] = -e_1.$$

Now the group of Euclidean motions acts on the space of straight lines in the plane. Recall that q denotes the y intercept (see Figure 17.1). Hence the one-parameter group of translations in the e_2 direction is

$$\begin{pmatrix} q \\ p \end{pmatrix} \to \begin{pmatrix} q + t \\ p \end{pmatrix}.$$

The corresponding infinitesimal transformation is clearly given by Poisson bracket with the function p.

Translation in the e_1 direction is

$$\begin{pmatrix} q \\ p \end{pmatrix} \to \begin{pmatrix} q - t \tan\theta \\ p \end{pmatrix}, \qquad \sin\theta = p$$

or

$$\begin{pmatrix} q \\ p \end{pmatrix} \to \begin{pmatrix} q - t \dfrac{p}{\sqrt{1 - p^2}} \\ p \end{pmatrix},$$

and the infinitesimal generator is Poisson bracket by the function $(1 - p^2)^{1/2}$. Finally, it is easy to check that

$$\{-q\sqrt{1 - p^2}, p\} = -\sqrt{1 - p^2}$$

and

$$\{-q\sqrt{1 - p^2}, \sqrt{1 - p^2}\} = p$$

so $-q(1-p^2)^{1/2}$ corresponds to the infinitesimal rotation $e_1 \wedge e$. We thus have an isomorphism of $e(2)$ with an algebra of functions (under Poisson bracket) given by

$$e_2 \to p,$$
$$e_1 \to \sqrt{1-p^2},$$
$$e_1 \wedge e_2 \to -q\sqrt{1-p^2}.$$

Note that the image algebra is *not* contained in the algebra of inhomogeneous quadratic polynomials. We shall soon describe a family of unitary irreducible representations of $E(2)$ that will give us a quantization procedure for this image algebra; that is, a consistent quantization scheme for the three Hamiltonians p, $(1-p^2)^{1/2}$, $q(1-p^2)^{1/2}$. It is just that this scheme *cannot* be made consistent with the metaplectic representation. Note also, that in the Gaussian limit, where we ignore expressions cubic or higher, our Euclidean algebra "tends to" the four-dimensional algebra spanned by the functions 1, q, p, and $\frac{1}{2}p^2$.

Let us spend some time now analyzing this limit process, which is known as a "contraction," that goes from the three-dimensional Euclidean algebra $e(2)$ to the algebra of the "free system" in Gaussian optics. We shall present the formal theory of this procedure in Chapter V.

Writing out the complete 3×3 matrix form of the generators of $e(2)$ we have

$$e_1 = \begin{pmatrix} 0 & 0 & 1 \\ 0 & 0 & 0 \\ 0 & 0 & 0 \end{pmatrix},$$

$$e_2 = \begin{pmatrix} 0 & 0 & 0 \\ 0 & 0 & 1 \\ 0 & 0 & 0 \end{pmatrix},$$

$$e_1 \wedge e_2 = \begin{pmatrix} 0 & 1 & 0 \\ -1 & 0 & 0 \\ 0 & 0 & 0 \end{pmatrix}.$$

In the Gaussian approximation, we are assuming that the off-axis distances are all very small. This suggests that we should introduce units of length in the vertical (perpendicular to axis) direction that are much smaller than the units we will use to measure along the axis. In other words, we should conjugate all the above matrices by

$$M_c = \begin{pmatrix} 1 & 0 & 0 \\ 0 & c & 0 \\ 0 & 0 & 1 \end{pmatrix},$$

where c is some large number.

We obtain the subalgebra $M_c e(2) M_{c^{-1}}$ conjugate to $e(2)$ in $gl(3)$ with generators

$$M_c e_1 M_{c^{-1}} = \begin{pmatrix} 0 & 0 & 1 \\ 0 & 0 & 0 \\ 0 & 0 & 0 \end{pmatrix},$$

$$M_c e_2 M_{c^{-1}} = \begin{pmatrix} 0 & 0 & 0 \\ 0 & 0 & c \\ 0 & 0 & 0 \end{pmatrix},$$

$$M_c(e_1 \wedge e_2) M_c^{-1} = \begin{pmatrix} 0 & c^{-1} & 0 \\ -c & 0 & 0 \\ 0 & 0 & 0 \end{pmatrix}.$$

We are concerned here with small vertical displacements and angles. So we will replace $M_c e_2 M_{c^{-1}}$ by

$$c^{-1} M_c e_2 M_{c^{-1}} = \begin{pmatrix} 0 & 0 & 0 \\ 0 & 0 & 1 \\ 0 & 0 & 0 \end{pmatrix}$$

and $M_c(e_1 \wedge e_2) M_c^{-1}$ by

$$c^{-1} M_c(e_1 \wedge e_2) M_{c^{-1}} = \begin{pmatrix} 0 & c^{-2} & 0 \\ -1 & 0 & 0 \\ 0 & 0 & 0 \end{pmatrix}.$$

Let us set $\epsilon = c^{-2}$. We have thus introduced the basis

$$\begin{pmatrix} 0 & 0 & 1 \\ 0 & 0 & 0 \\ 0 & 0 & 0 \end{pmatrix}$$

$$\begin{pmatrix} 0 & 0 & 0 \\ 0 & 0 & 1 \\ 0 & 0 & 0 \end{pmatrix}$$

$$\begin{pmatrix} 0 & \epsilon & 0 \\ -1 & 0 & 0 \\ 0 & 0 & 0 \end{pmatrix}$$

for the subalgebra $M_c e(2) M_{c^{-1}}$. We can say this a little differently: Let g be an abstract three-dimensional vector space with basis, X, Y, J, and define the linear map

$$f_\epsilon : g \to gl(3)$$

by

$$f_\epsilon(X) = \begin{pmatrix} 0 & 0 & 1 \\ 0 & 0 & 0 \\ 0 & 0 & 0 \end{pmatrix},$$

$$f_\epsilon(Y) = \begin{pmatrix} 0 & 0 & 0 \\ 0 & 0 & 1 \\ 0 & 0 & 0 \end{pmatrix},$$

$$f_\epsilon(J) = \begin{pmatrix} 0 & \epsilon & 0 \\ -1 & 0 & 0 \\ 0 & 0 & 0 \end{pmatrix}.$$

The image of f_ϵ is a subalgebra of $gl(3)$ and so f_ϵ induces a bracket, $[\,,]_\epsilon$ on g. Explicitly

$$[X, Y]_\epsilon = 0,$$

$$[J, X]_\epsilon = -Y,$$

$$[J, Y]_\epsilon = \epsilon X.$$

For $\epsilon > 0$, all of these algebra structures on g are isomorphic to $e(2)$. For $\epsilon = 0$, we get an algebra isomorphic to the algebra of vector fields (in two dimensions) spanned by $p(\partial/\partial q)$, $\partial/\partial q$, and $\partial/\partial p$. Note that this is the three-dimensional algebra of vector fields corresponding to the four-dimensional algebra of functions spanned by $\frac{1}{2}p^2$, p, q, 1 (since the constant function 1 goes into the 0 vector field). In fact, we can keep track of the parameter ϵ in terms of Poisson brackets and vector fields in p and q all along:

Generator in g	Function	Vector field
X	$\epsilon^{-2}(1 - \epsilon^2 p^2)^{1/2}$	$\dfrac{p}{(1 - \epsilon^2 p^2)^{1/2}} \dfrac{\partial}{\partial q}$
Y	p	$\dfrac{\partial}{\partial q}$
J	$q(1 - \epsilon^2 p^2)^{1/2}$	$\dfrac{-\epsilon^2 pq}{(1 - \epsilon^2 p^2)^{1/2}} \dfrac{\partial}{\partial q} - (1 - \epsilon^2 p^2)^{1/2} \dfrac{\partial}{\partial p}$

Notice that there is some trouble with the first row in the middle column at $\epsilon = 0$ – the image of X blows up. We can express this in a mathematically more precise way. Let ϕ_ϵ be the linear map from g to vector fields in p, q given on generators by the rule that assigns the element in the third column to the element of the first. Thus

$$\phi_\epsilon(X) = \frac{p}{(1 - \epsilon^2 p^2)^{1/2}} \frac{\partial}{\partial q},$$

etc. Then ϕ_ϵ is an isomorphism of g, with bracket $[\,,]_\epsilon$ into the Hamiltonian vector fields in the plane. For $\epsilon > 0$, there is a unique subalgebra of functions that projects isomorphically onto the image of ϕ_ϵ. Thus there is an isomorphism ψ_ϵ from $e(2)$ into $\mathscr{F}(\mathbb{R}^2)$ so that the diagram

$$\begin{array}{ccccccccc} 0 & \to & \mathbb{R} & \to & \mathscr{F}(\mathbb{R}^2) & \to & \mathrm{Ham}(\mathbb{R}^2) & \to & 0 \\ & & & & \nwarrow \psi_\epsilon & & \uparrow \phi_\epsilon & & \\ & & & & & & g & & \end{array}$$

commutes. On generators, the map ψ_ϵ assigns to elements of the first column the corresponding element of the second column in the above table. For $\epsilon = 0$, there is no isomorphic lifting of $\phi_\epsilon(g)$ into $\mathscr{F}(\mathbb{R}^2)$, and so we must pass to the four-dimensional algebra $\{1, q, p, \frac{1}{2}p^2\}$, which projects onto the three-dimensional algebra $\phi_0(g)$. It is in this sense that the family of three-dimensional algebras $\{\epsilon^{-2}(1 - \epsilon^2 p^2)^{1/2}, p, q(1 - \epsilon^2 p^2)^{1/2}\}$ tends to the four-dimensional algebra $\{1, \frac{1}{2}p^2, p, q\}$. We shall return to the general theory of this contraction process in Chapter V.

We can also follow the contraction at the level of unitary representations; that is, in quantum mechanics. We have already observed that in the wave theory, we should be considering the space of solutions of the reduced wave equation

$$\Delta u + \lambda^2 u = 0,$$

with the Euclidean group acting in the standard fashion:

$$[\rho(a)u](\vec{x}) = u(a^{-1}\vec{x})$$

for any $a \in E(2)$ and $\vec{x} \in \mathbb{R}^2$. We can represent the most general solution of the reduced wave equation as

$$u(x, y) = \int \exp\{i[x\lambda \cos\theta + y\lambda \sin\theta]\} f(\theta)\, d\theta.$$

(In other words, the Fourier transform of u is supported on the circle of radius λ.) We have:

$$u(x - x_0, y) = \int \exp\{i[x\lambda \cos\theta + y\lambda \sin\theta]\} \exp(-ix_0\lambda \cos\theta) f(\theta)\, d\theta,$$

so we define

$$\left[\hat{\rho}\left(\begin{pmatrix} 1 & 0 & x_0 \\ 0 & 1 & 0 \\ 0 & 0 & 1 \end{pmatrix}\right) f\right](\theta) = \exp(-ix_0\lambda \cos\theta) f(\theta),$$

and similarly

$$\left[\hat{\rho}\left(\begin{pmatrix} 1 & 0 & 0 \\ 0 & 1 & y_0 \\ 0 & 0 & 1 \end{pmatrix}\right) \hat{f}\right](\theta) = \exp(-iy_0\lambda \sin\theta) f(\theta)$$

and

$$\left[\hat{\rho}\left(\begin{pmatrix} \cos\theta_0 & \sin\theta_0 & 0 \\ -\sin\theta_0 & \cos\theta_0 & 0 \\ 0 & 0 & 1 \end{pmatrix}\right) \hat{f}\right](\theta) = f(\theta - \theta_0).$$

If we let f range over $L^2[0, 2\pi]$, this gives the unitary representation of $E(2)$ corresponding to the reduced wave equation with parameter λ. The corresponding infinitesimal representation of the Lie algebra $e(2) = g_1$ is given by

$$\dot{\hat{\rho}}_\lambda(X) = i\lambda \cos\theta, \qquad \dot{\hat{\rho}}_\lambda(Y) = i\lambda \sin\theta, \qquad \text{and} \qquad \dot{\hat{\rho}}_\lambda(J) = \frac{\partial}{\partial\theta}.$$

This, of course, corresponds to the value of $\epsilon = 1$. For $\epsilon > 0$, we get a representation of g_ϵ, obtained by setting

$$\dot{\hat{\rho}}_{\lambda,\epsilon}(X) = i\lambda \cos \epsilon^{1/2}\theta, \qquad \dot{\hat{\rho}}_{\lambda,\epsilon}(Y) = \epsilon^{1/2} i\lambda \sin \epsilon^{1/2}\theta,$$

and

$$\dot{\hat{\rho}}_{\lambda,\epsilon}(J) = \frac{\partial}{\partial\theta}.$$

If we simply let $\epsilon \to 0$ in these representations, keeping λ fixed, the limiting representation that we get is not very interesting. However, something interesting happens when we take $\lambda = \epsilon^{-1}$. (More generally, we could let $\lambda \to \infty$ in such a way that $\epsilon\lambda \to h$, where h is some constant.) Then

$$\dot{\hat{\rho}}_{\epsilon^{-1},\epsilon}(X) = i\epsilon^{-1} \cos \epsilon^{1/2}\theta, \qquad \dot{\hat{\rho}}_{\epsilon^{-1},\epsilon}(Y) = i\epsilon^{-1/2} \sin \epsilon^{1/2}\theta,$$

and

$$\dot{\hat{\rho}}_{\epsilon^{-1},\epsilon}(J) = \frac{\partial}{\partial\theta}.$$

As before, there is some trouble with the first term, which is tending to infinity. If we were allowed to "renormalize" by subtracting off the scalar operator $i\epsilon^{-1}$ and then let $\epsilon \to 0$, we would obtain the limiting operators

$$\dot{\hat{\sigma}}(X) = -i\tfrac{1}{2}\theta^2, \qquad \dot{\hat{\sigma}}(Y) = i\theta, \qquad \text{and} \qquad \dot{\hat{\sigma}}(J) = \frac{\partial}{\partial\theta}.$$

The three operators involving θ generate a four-dimensional Lie algebra isomorphic to the algebra spanned by 1, q, p, $\frac{1}{2}p^2$, and it is easy to see that the representation we obtain is precisely the metaplectic representation (where we have applied the Fourier transform to pass from x to θ). Thus we must justify and/or explain the process of subtracting off the scalar operator. For this we need to recall the definition of a projective representation, a notion that we have already briefly alluded to: If H is a vector space, we let $Gl(H)$ denote the group of all invertible linear transformations on H, and $Pl(H)$ denotes the quotient group, $Pl(H) = Gl(H)/cI$, where cI denotes the subgroup consisting of all nonzero scalar operators. A projective representation of a group G on H is a homomorphism of G into $Pl(H)$. Similarly, we can let $gl(H)$ denote the Lie algebra of all linear transformations of H. The set of scalar operators, $\mathbf{c}$ is the center of $gl(H)$ and hence we can form the quotient algebra $pl(H) = gl(H)/\mathbf{c}$. A projective representation of a Lie algebra g on a vector space H is a homomorphism of g into $pl(H)$. Any ordinary representation of g on H (i.e., a homomorphism of g into $gl(H)$), determines a projective representation by simply passing to the quotient. But not every projective representation of g need come from an ordinary representation. We can think of a projective representation as assigning to every element of g an "operator up to a scalar operator." We can now formulate our limiting process for the $\rho_{\epsilon^{-1},\epsilon}$ by saying that the *projective* representations associated to the $\rho_{\epsilon^{-1},\epsilon}$ tend to a limiting projective representation as $\epsilon \to 0$, and that this limiting projective representation does not come from an ordinary representation of g_0 but does come from the restriction of the metaplectic representation to the four-dimensional extended algebra. Note the similarity between the classical and quantum mechanical situations. In the classical case, we had a family of homomorphisms of g into Ham(X) that could be lifted to $\mathscr{F}(X)$ for $\epsilon > 0$, but not for $\epsilon = 0$, where we had to pass to a central extension of g_0. Similarly, in the quantum mechanical case, we have a family of homomorphisms of g into $pl(H)$ that can be lifted to $gl(H)$ for $\epsilon > 0$, but not for $\epsilon = 0$, where we must pass to the same central extension. The reason for this analogy can best be understood in the framework of geometric quantization, cf. Kostant (1970), Simms and Woodhouse (1976), Guillemin and Sternberg (1977), Snyaticki (1980), and Woodhouse (1980).

We should also point out that the limit process that we have described, the passage by contraction from the Euclidean algebra $e(2)$ to the algebra of "free Gaussian optics," is the counterpart of a very familiar limiting process – the passage from the Poincaré group to the Galilean group: We need only replace the optical axis by the time axis and the off-axial direction by 3-space. Here are the details: Special relativity assumes that space-time consists of a four-dimensional space $\mathbb{R}^{1,3}$ equipped with a scalar product of signature $+ - - -$. The corresponding Euclidean group, called the Poincaré group, is the group of symmetries of special relativity and its Lie algebra consists of all 5×5 matrices of the form

$$\begin{pmatrix} B & b \\ 0 & 0 \end{pmatrix} \qquad B \in o(1,3),\ b \in \mathbb{R}^{1,3}.$$

Our own individual psychological space and time gives a space-time splitting, so that we can write a point in $\mathbb{R}^{1,3}$ as a column vector (t, x_1, x_2, x_3), which we may write as $\begin{pmatrix} t \\ x \end{pmatrix}$. Thus we can write the most general element of the Poincaré algebra as a 5×5 matrix of the form

$$\begin{pmatrix} 0 & v_1 & v_2 & v_3 & t \\ v_1 & 0 & a_{12} & a_{13} & x_1 \\ v_2 & -a_{12} & 0 & a_{23} & x_2 \\ v_3 & -a_{13} & -a_{23} & 0 & x_3 \\ 0 & 0 & 0 & 0 & 0 \end{pmatrix},$$

which we will write more simply as

$$\begin{pmatrix} 0 & v^{\mathrm{t}} & t \\ v & A & x \\ 0 & 0 & 0 \end{pmatrix}, \qquad x \in \mathbb{R}^3,\ v \in \mathbb{R}^3,\ t \in \mathbb{R},\ A \in o(3).$$

This is the form of the typical element of the Poincaré algebra relative to a coordinate system in which the speed of light is set equal to 1. Our psychological perception tends to exaggerate spatial distances relative to time units in such a coordinate system, so that the speed of light takes on a value c. To find the form of the Poincaré algebra in a coordinate system where the speed of light is c, we must expand the spatial coordinates by a factor of c. This is the same as conjugating the above matrices by the matrix

$$M_c = \begin{pmatrix} 1 & 0 & 0 & 0 & 0 \\ 0 & c & 0 & 0 & 0 \\ 0 & 0 & c & 0 & 0 \\ 0 & 0 & 0 & c & 0 \\ 0 & 0 & 0 & 0 & 1 \end{pmatrix}$$

so as to obtain matrices of the form

$$\begin{pmatrix} 0 & c^{-1}v^{t} & t \\ cv & A & cx \\ 0 & 0 & 0 \end{pmatrix}.$$

For each positive value of c, we obtain a different subalgebra of 5×5 matrices. They are all isomorphic (and indeed conjugate) to the Poincaré algebra. Within the framework of our distended coordinates, the vector cv gives a "velocity" (or "boost") transformation and the vector cx gives a spatial translation. If we are accustomed to dealing with velocities and displacements which are small relative to the velocity of light, it makes sense to introduce $\bar{v} = cv$ and $\bar{x} = cx$ and so parametrize the elements of one such algebra (corresponding to a fixed c) as

$$\begin{pmatrix} 0 & c^{-2}\bar{v}^{t} & t \\ \bar{v} & A & \bar{x} \\ 0 & 0 & 0 \end{pmatrix}.$$

Let us now set, as before, $\epsilon = c^{-2}$ and drop the bars over the v's and x's. We thus obtain, for each $\epsilon > 0$, a map of the ten-dimensional space spanned by the A's, v's, x's, and t's into the space of 5×5 matrices. Explicitly, this map is

$$(A, v, x, t) \rightsquigarrow \begin{pmatrix} 0 & \epsilon v^{t} & t \\ v & A & x \\ 0 & 0 & 0 \end{pmatrix}.$$

For each ϵ, we get an induced Lie bracket structure $[\,,]_\epsilon$ on g. Explicitly, if

$$\xi_1 = (A_1, v_1, x_1, t_1) \qquad \text{and} \qquad \xi_2 = (A_2, v_2, x_2, t_2)$$

then

$$[\xi_1, \xi_2]_\epsilon = \big([A_1, A_2] + \epsilon(v_1 \otimes v_2^{t} - v_2 \otimes v_1^{t}),\quad A_1 v_2 - A_2 v_1,\\ A_1 x_2 + t_2 v_1 - A_2 x_1 - t_1 v_2,\quad \epsilon(v_1 \cdot x_2 - v_2 \cdot x_1)\big).$$

For $\epsilon = 0$, the image algebra is the Galilean algebra (corresponding to an infinite speed of light). We can also find an analog for the algebras of functions under Poisson brackets that we introduced in the optical situation, where we now look at a six-dimensional space with coordinates $q_x, q_y, q_z, p_x, p_y, p_z$. Letting $p^2 = p_x^2 + p_y^2 + p_z^2$, we can, for $\epsilon > 0$ consider the ten-dimensional algebra spanned by

$$p_x, \quad p_y, \quad p_z, \quad \epsilon^{-2}(1 + \epsilon^2 p^2)^{1/2}, \quad q_x(1 + \epsilon^2 p^2)^{1/2},\\ q_y(1 + \epsilon^2 p^2)^{1/2}, \quad q_z(1 + \epsilon^2 p^2)^{1/2}, \quad q_x p_y - q_y p_x,\\ q_x p_z - q_z p_x, \quad \text{and} \quad q_y p_z - q_z p_y.$$

It is easy to check that this algebra is isomorphic to g_ϵ for $\epsilon > 0$, and is thus the corresponding algebra of Hamiltonian vector fields. For $\epsilon = 0$, the limiting algebra of Hamiltonian vector fields exists and is isomorphic to the Galilean algebra. Similar results hold on the quantum mechanical level. Of course, this particular homomorphism of the algebra g_ϵ into functions is not the only one we might want to examine and deform. The general theory of such homomorphism will be presented in Chapter V.

18. Polarization of light

In the preceding section we constructed a one-parameter family of representations of the Euclidean group $E(2)$ that depend on the parameter λ, which corresponds to the wavelength of light. Indeed, throughout most of the previous discussion, the parameter λ was present, although on occasion, we found it convenient to choose our unit of length so that $\lambda = 2\pi$ "in order to make the λ disappear from the final formulas." But both in the Euclidean and Fresnel situations, the parameter λ, describing different unitary representations, corresponds to a physically observable parameter – the wavelength. We could also introduce the parameter λ into the geometrical-optics approximation by observing that the symplectic form $\lambda dp \wedge dq$ is preserved by our symplectic diffeomorphisms. This looks very artificial at the moment, although we will see that it lies behind the deformation procedure of the preceding section.

For a true theory of light we should pass from two dimensions to three, and thus look at the representations of the three-dimensional Euclidean group. Similarly, we should look at various possible homomorphisms of the group $E(3)$ into the group of symplectic diffeomorphisms of various spaces. We might then hope to give some physical interpretation to some of these representations or actions. In order to give the physical interpretation we first must recall and describe a physical phenomenon that we have been avoiding until now, namely, the polarization of light.

Light of a fixed wavelength can be obtained by passing white light through a prism and selecting the light emerging at a definite angle. We can think of the original white light as being a superposition of light of various wavelengths: The physical operation performed by the prism can be thought of as "projecting" onto a simple component of the original white light. There are infinitely many components, corresponding to the possible values of the wavelength. If we start with light of a fixed wavelength and pass it through a Polaroid filter, it turns out that the emerging light is also, in some sense that we will explain below, simpler than the light we started with. Arago and Fresnel conducted a whole series of clever experiments that investigated the interference and diffraction properties of polarized light. They found, for example, that two beams of light parallel to one another will not interfere if one beam has passed through a Polaroid filter and the other through a second Polaroid filter oriented at a right angle to the first. (Of course they did not have Polaroid filters at their disposal, but used other polarizing devices.) If the beams are at right angles to each other (and come from some coherent source), then, again, they will or will not interfere according to how they are polarized. By a careful analysis of the underlying Euclidean geometry of the experiments, they came to the conclusion that all the effects could be explained if the following modifications are made in the description of the nature of light that we gave in Section 7:

Light at any point is described by a complex vector c rather than a complex number. Thus $c = (c_x, c_y, c_z)$, where each of the components is a complex number. The intensity of the light is given by

$$I = \|c\|^2 = |c_x|^2 + |c_y|^2 + |c_z|^2.$$

The light vector at any point is obtained by adding the light vector contributions associated with each of the light rays arriving at that point. The light vector associated with any light ray is always perpendicular to the ray. Thus, for example, if the ray is parallel to the z axis, we have $c_z = 0$ for any light vector associated with the ray. As light of a fixed wavelength λ propagates along a ray, it is multiplied by the phase factor $e^{2\pi i l/\lambda}$, where l is the optical length of the path. It is also multiplied by a factor accounting for the attenuation of the light and for the fact that the plane perpendicular to the ray is changing. In the Gaussian approximation, where we take $\cos(p/n) \doteq 1$, we may ignore the change in direction of this plane and use the same value for a as given in the preceding section. In fact the same diffraction formula can be used for the Gaussian approximation with the understanding that c is now a vector.

Let us now fix attention on the light vector associated with a position on the fixed ray. We examine what is happening at some distance from refracting surfaces. We may thus assume that the ray is a straight line, say the z axis. Then $c_z = 0$ and the light vector is specified by the two-dimensional vector $\begin{pmatrix} c_x \\ c_y \end{pmatrix}$. The light is said to be *linearly polarized* if the two complex numbers c_x and c_y have the same phase; thus c is linearly polarized if

$$\begin{pmatrix} c_x \\ c_y \end{pmatrix} = e^{2\pi i\phi} \begin{pmatrix} r_x \\ r_y \end{pmatrix}$$

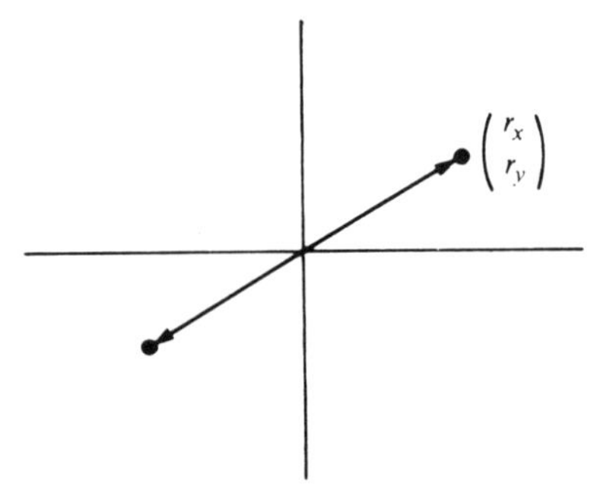

Figure 18.1.

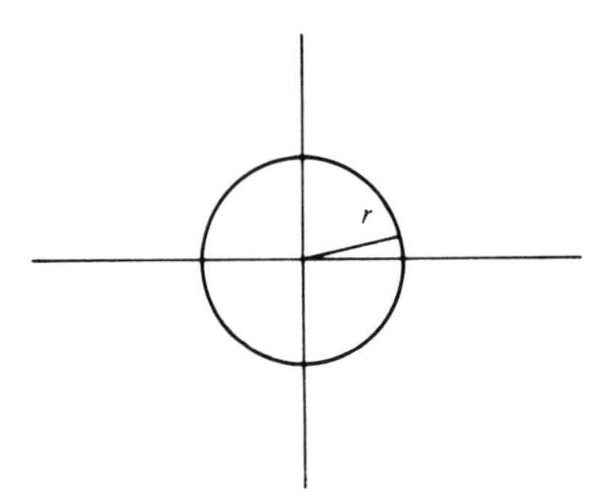

Figure 18.2.

for real numbers r_x and r_y. Notice that if c is linearly polarized, then so is wc for any complex number w. In particular, $e^{2\pi il/\lambda}c$ is again linearly polarized, so that the light remains linearly polarized as it propagates along the ray. The reason for this name is the following: Suppose that we consider the real part of the vector $e^{2\pi il/\lambda}c$ for varying values of l. Thus

$$\mathrm{Re}\,[e^{2\pi il/\lambda}c] = \begin{pmatrix} r_x \cos t \\ r_y \cos t \end{pmatrix}, \qquad t = 2\pi[l/\lambda + \phi],$$

moves back and forth on the line through the vector $\begin{pmatrix} r_x \\ r_y \end{pmatrix}$ as l varies (Figures 18.1 and 18.2). Light is said to be *circularly polarized* if c has the form $c = e^{2\pi i\phi}\begin{pmatrix} 1 \\ \pm i \end{pmatrix}$. Then

$$\mathrm{Re}\,[e^{2\pi il/\lambda}c] = r\begin{pmatrix} \cos t \\ \pm \sin t \end{pmatrix}$$

describes a circle in the counterclockwise direction for $+$ (in which case we talk of right-handed circularly polarized light) and in the clockwise direction for $-$ (left-handed circular polarization). In the general case, $\mathrm{Re}[e^{2\pi il/\lambda}c]$ describes an ellipse, and we say that the light is *elliptically polarized.*

A linear polarizer is an apparatus (such as a sheet of Polaroid filter) that converts the light passing through it into linearly polarized light. Let $\begin{pmatrix} r_x \\ r_y \end{pmatrix}$ be the vector describing the direction of polarization of the emerging polarized light. We say that the apparatus is an ideal linear polarizer if all light that enters the apparatus in a state of linear polarization parallel to $\begin{pmatrix} r_x \\ r_y \end{pmatrix}$ is 100 percent transmitted. In other words, if we place two ideal linear polarizers one behind the other and they are both oriented parallel to each other, then the effect is the same as if the second polarizer were absent. On the other hand, if the two linear polarizers are oriented at right angles to each other, no light is transmitted. In fact, by considering various possible relative orientations of two linear polarizers placed in sequence, we conclude the following:

Let L be an ideal linear polarizer whose emerging light vector is in the direction of the real unit vector $u = \begin{pmatrix} u_x \\ u_y \end{pmatrix}$, so $u_x^2 + u_y^2 = 1$. Then L acts on any light vector $c = \begin{pmatrix} c_x \\ c_y \end{pmatrix}$ by converting it into the light vector $L_u c$, where L_u is the 2×2 matrix giving the projection onto the line through u; thus

$$L_u = \begin{pmatrix} u_x u_x & u_x u_y \\ u_y u_x & u_y u_y \end{pmatrix}.$$

For example, if the light emerging from the linear polarizer is all parallel to the x axis, it will act by multiplying the light vector by $\begin{pmatrix} 1 & 0 \\ 0 & 0 \end{pmatrix}$. If the light emerges linearly polarized in the y direction the matrix is the projection matrix $\begin{pmatrix} 0 & 0 \\ 0 & 1 \end{pmatrix}$. Another device that is of frequent use in polarization optics is known as a linear retarder. Relative to the appropriate axes, it has the effect of multiplying each of the two coordinates of the light vector by (different) phase factors. Thus, if the axes are the x and y axes, the effect of such a device is the same as multiplication by a diagonal matrix

$$\begin{pmatrix} e^{i\delta_1} & 0 \\ 0 & e^{i\delta_2} \end{pmatrix} = e^{i\delta_1} \begin{pmatrix} 1 & 0 \\ 0 & e^{-i\Delta} \end{pmatrix}, \qquad \text{where } \Delta = \delta_1 - \delta_2.$$

Since we are not usually interested in the overall phase factor $e^{i\delta_1}$, it is the *angle of retardation* Δ that is the important parameter in the device. If $\Delta = \pi/2$, for example, the device is called a *quarter wave plate* (with fast axis along the x axis). More generally, let $R(\theta)$ denote the rotation matrix

$$R(\theta) = \begin{pmatrix} \cos\theta & -\sin\theta \\ \sin\theta & \cos\theta \end{pmatrix}.$$

Then the most general linear retarder, with fast axis at angle θ to the x axis, can obviously be obtained by conjugating the preceding matrix by the rotation matrix $R(\theta)$; that is

$$e^{i\delta_1} R(\theta) \begin{pmatrix} 1 & 0 \\ 0 & e^{-i\Delta} \end{pmatrix} R(-\theta) = R(\theta) \begin{pmatrix} e^{i\delta_1} & 0 \\ 0 & e^{i\delta_2} \end{pmatrix} R(-\theta)$$

represents the most general linear retarder. For example, suppose we want to describe a quarter wave plate with fast axis at 45° relative to the x axis. Then $\theta = \pi/4$, and $\Delta = \pi/2$. If we choose the overall phases with $\delta_1 = \pi/4$, then matrix multiplication gives

$$(1/\sqrt{2}) \begin{pmatrix} 1 & i \\ i & 1 \end{pmatrix}$$

as the matrix representing this device. (Note that this device has the effect of changing light that is linearly polarized along the x axis into right-handed circularly polarized light.) In general, any succession of linear polarizers and retarders can now be analyzed in terms of the multiplication of their corresponding 2×2 complex matrices. In this way, the study of the most general polarizer is reduced to the study of the collection of complex 2×2 matrices under multiplication. This is not surprising since the set of all possible light vectors on a given ray is just a two-dimensional complex vector space.

From an intuitive physical point of view, linearly polarized light with its preferred direction of oscillation in space is perhaps the easiest form of light to comprehend. But from the group-theoretical viewpoint, especially that of the Euclidean group, circularly polarized light is the simplest for the following reason: Suppose we are in a situation where Euclidean geometry applies – in empty space with a constant index of refraction so that the light rays are all straight lines. The group $E(3)$ acts transitively on the (four-dimensional) space of straight lines; that is, we can find a Euclidean motion carrying any given straight line into any other. Let us focus our attention on one particular straight line, say the z axis. The subgroup $E(3)_z$, which fixes this straight line, consists of all possible rotations about the z axis and translations along the z axis (we are assuming that we are dealing with the proper Euclidean group, excluding reflections). In other words $E(3)_z$ is the direct product $E(3) = T \times \mathbb{R}$, where T consists of all rotations about the z axis and $\mathbb{R}$ consists of all translations along the z axis. We can use (θ, l) to denote the element of $E(3)_z$ consisting of rotation through angle θ about the z axis and translation through a distance l along the z axis. Rotation through angle θ is given by the matrix

$$\begin{pmatrix} \cos\theta & -\sin\theta \\ \sin\theta & \cos\theta \end{pmatrix}$$

relative to the basis spanned by the unit vectors in the x and y directions. These form a basis for the vector space perpendicular to the z axis. But remember that we are considering our light vector to lie in the complexification of this vector space. In this complexified vector space, the vectors $\begin{pmatrix} 1 \\ i \end{pmatrix}$ and $\begin{pmatrix} 1 \\ -i \end{pmatrix}$ are eigenvectors for the above rotation matrix with eigenvalues $e^{-i\theta}$ and $e^{i\theta}$. If we ignore the attenuation factor, then translation a distance l has the effect of multiplying the light vector by the factor $e^{2\pi i l/\lambda}$. If we define the functions $\chi_{s,p}$ by

$$\chi_{s,p}(\theta, l) = e^{i(s\theta + pl)},$$

then (for integral values of s and arbitrary real values of p) the functions $\chi_{s,p}$ are the *characters* of the group $E(3)_z$. We see that circularly polarized light corresponds to the characters with values

$$s = \pm 1, \qquad p = 2\pi/\lambda,$$

where λ is the wavelength of the light and $+1$ corresponds to left-handed circularly polarized light and -1 to right-handed circularly polarized light.

The geometrical significance of the identification of circularly polarized light with characters of $E(3)_z$ depends on the following facts, whose proof we defer to the next chapter: Let G be a Lie group and g its Lie algebra, so that G acts on g via the adjoint representation. (For example, if G is given as a group of matrices, then g consists of those matrices, X such that $\exp tX \in G$ for all t, and $a \in G$ acts on $X \in g$ by sending it into aXa^{-1}.) Let g^* denote the dual space of g; that is, the space of linear functions on g and let $\langle \alpha, \xi \rangle$ be the value of the linear function $\alpha \in g^*$ on $\xi \in g$. Then G has a linear representation on g^* given by

$$\langle a\alpha, \xi \rangle = \langle \alpha, \mathrm{Ad}_{a^{-1}}\xi \rangle.$$

In the case where we have identified G as a group of matrices, so that ξ becomes identified with some matrix X, this last equation becomes

$$\langle a\alpha, X\rangle = \langle \alpha, a^{-1}Xa\rangle.$$

For each $\alpha \in g^*$, let $G \cdot \alpha$ denote the G orbit through α so that

$$G \cdot \alpha = \{a\alpha, a \in G\}.$$

Then $G \cdot \alpha$ carries a natural G-invariant symplectic form ω, which can be described as follows: Since G acts on $G \cdot \alpha$, every $\xi \in g$ determines a vector field on $G \cdot \alpha$ and, in particular, a tangent vector ξ_β at each $\beta \in G \cdot \alpha$. Since G acts transitively on $G \cdot \alpha$, the vectors of the form ξ_β fill out the whole tangent space to $G \cdot \alpha$ at β as ξ ranges over all of g. If ξ_β and η_β are two such tangent vectors, then the value of the symplectic form ω on them is given by

$$\omega_\beta(\xi_\beta, \eta_\beta) = -\langle \beta, [\xi, \eta]\rangle.$$

Of course, we must prove that this formula is well defined (that the right-hand side is independent of the ξ or η we choose to represent the ξ_β and η_β) and that the form ω so defined is nondegenerate and closed. We will prove these facts in Chapter II. We have thus produced a large family of symplectic manifolds on which the group G acts transitively. We shall also prove that if the group G satisfies a certain mild condition (and all the Euclidean groups in three or more dimensions do), then these are essentially all the transitive symplectic G spaces. More precisely, any symplectic manifold on which G acts transitively as a group of symplectic diffeomorphisms is a covering space of some such orbit by a covering map that is locally a symplectic diffeomorphism.

So we must compute the orbits of $E(3)$ acting on the dual of its Lie algebra. We shall do this in the next section. We state the results of the computation here: There is a two parameter family of four-dimensional orbits, a one-parameter family of two-dimensional orbits, and a single zero-dimensional orbit (the origin). Each of the four-dimensional orbits is equivalent, as a G space, to the space of straight lines in three-dimensional space; they differ in their symplectic structures. In other words, there is a two-parameter family of $E(3)$-invariant symplectic structures on the space of straight lines. Let us describe the parameterization of these symplectic structures. Let us fix a straight line, say the z axis. The subgroup fixing the z axis is just our subgroup $E(3)_z$. We may thus identify the space of straight lines as the homogenous space $E(3)/E(3)_z$. The claim is that there is a two-parameter family of elements $\alpha \in e(3)^*$ with the property that the subgroup fixing α is exactly $E(3)_z$. Now rotations about the z axis have a nontrivial action on infinitesimal translations along the x or y axis and on infinitesimal rotations about the x or y axes. So if α is to be left-invariant by $E(3)_z$, it must vanish when evaluated on infinitesimal translations along or rotations about the x or y axes. Thus α is completely determined by its values on infinitesimal translations along and rotations about the z axis. Let r denote infinitesimal rotation about the z axis chosen so that $\exp 2\pi r = id$ and let t denote some nonzero infinitesimal translation along the z axis. (Thus a choice of t is essentially the same as a choice of a unit of length.) We can then write

$$\langle \alpha, r \rangle = s \qquad \text{and} \qquad \langle \alpha, t \rangle = p,$$

and we see that α is completely determined by the two real parameters s and p. We also see that p has the units of inverse length. We shall see in the next section that $p = 0$ corresponds to a two-dimensional orbit, so that the four-dimensional orbits are precisely those with s arbitrary but $p > 0$ so we may write

$$p = 2\pi/\lambda,$$

where λ is some length. The most general element of the Lie algebra $e(3)_z$ can be written as

$$\theta r + lt,$$

and, using the exponential map from the Lie algebra $e(3)_z$ to the Lie group $E(3)_z$, we can write the general element of $E(3)_z$ as

$$(\theta, l) = \exp(\theta r + lt).$$

We can then try to define the function χ_α on $E(3)_z$ by

$$\chi_\alpha(\exp(X)) = e^{i\langle \alpha, X \rangle},$$

or, more explicitly,

$$\chi_\alpha((\theta, l)) = e^{i(s\theta + pl)}.$$

The problem with this "definition" is that for $\theta = 2\pi$, $\exp(2\pi r) = id$, whereas $e^{is \cdot 2\pi} = 1$ if and only if s is an integer. Thus we can only define χ_α when s is an integer, in which case it coincides with the character $\chi_{s,p}$ that we introduced earlier. Thus we can think of the general α as an "infinitesimal character" that can be exponentiated to an honest character if and only if the integrality (or "quantum") condition – that s be an integer, is satisfied.

We shall see later that this is a typical situation for the general Lie group case. We shall also see that if we use the universal covering group of the Euclidean group (which essentially involves replacing the group $SO(3)$ by $SU(2)$) then the correct quantum condition becomes that s be half-integral. In sum, we see that purely group-theoretical notions – the problem of classifying transitive symplectic $E(3)$ spaces – give rise to physical parameters that we can identify with wavelength and polarization.

19. The coadjoint orbit structure of a semidirect product

In this section we shall compute the action of $E(3)$ on the dual of its Lie algebra and determine the orbits for this action. The principles involved in this computation are valid in a more general setting, that of the semidirect product of any Lie group with a vector space. Let H be a Lie group and suppose that we are given a representation of H on some real vector space. We can then form the semidirect product of H and V. Let us denote the semidirect product by G. Thus, for example, if $H = SO(3)$ and $V = \mathbb{R}^3$, the semidirect product is just the group of Euclidean motions. As in the case of $E(3)$ we can write the elements of G as $(\dim V + 1) \times (\dim V + 1)$ matrices of the form $\begin{pmatrix} a & v \\ 0 & 1 \end{pmatrix}$, where $a \in H \subset Gl(V)$ and $v \in V$ and we can write elements of the Lie algebra g as $\begin{pmatrix} A & x \\ 0 & 0 \end{pmatrix}$, where $A \in h \subset gl(v)$ and $x \in V$. Thus h is identified with the set of $\begin{pmatrix} A & 0 \\ 0 & 0 \end{pmatrix}$ and V with the set of $\begin{pmatrix} 0 & x \\ 0 & 0 \end{pmatrix}$. Now

$$\begin{aligned} \mathrm{Ad}_{\left(\begin{smallmatrix} a & v \\ 0 & 1 \end{smallmatrix}\right)} \begin{pmatrix} A & x \\ 0 & 0 \end{pmatrix} &= \begin{pmatrix} a & v \\ 0 & 1 \end{pmatrix} \begin{pmatrix} A & x \\ 0 & 0 \end{pmatrix} \begin{pmatrix} a & v \\ 0 & 1 \end{pmatrix}^{-1} \\ &= \begin{pmatrix} a & v \\ 0 & 1 \end{pmatrix} \begin{pmatrix} A & x \\ 0 & 0 \end{pmatrix} \begin{pmatrix} a^{-1} & -a^{-1}v \\ 0 & 1 \end{pmatrix} \\ &= \begin{pmatrix} aAa^{-1} & ax - aAa^{-1}v \\ 0 & 0 \end{pmatrix}. \end{aligned}$$

Let us write the most general element of g^* as (α, p), where $\alpha \in h^*$ and $p \in V^*$ so that

$$\left\langle (\alpha, p), \begin{pmatrix} A & x \\ 0 & 0 \end{pmatrix} \right\rangle = \langle \alpha, A \rangle + \langle p, x \rangle.$$

Then

$$\left\langle [\mathrm{Ad}^{\#}_{\left(\begin{smallmatrix} a & v \\ 0 & 1 \end{smallmatrix}\right)}(\alpha, p)], \begin{pmatrix} A & x \\ 0 & 0 \end{pmatrix} \right\rangle = \left\langle (\alpha, p), \mathrm{Ad}_{\left(\begin{smallmatrix} a^{-1} & -a^{-1}v \\ 0 & 1 \end{smallmatrix}\right)} \begin{pmatrix} A & x \\ 0 & 0 \end{pmatrix} \right\rangle$$

$$= \left\langle (\alpha, p), \begin{pmatrix} a^{-1}Aa & a^{-1}x + a^{-1}Av \\ 0 & 0 \end{pmatrix} \right\rangle$$

$$= \langle \alpha, a^{-1}Aa \rangle + \langle p, a^{-1}Av \rangle + \langle p, a^{-1}x \rangle.$$

For $p \in V^*$ and $v \in V$, define $p \odot v \in h^*$ by

$$\langle (p \odot v), A \rangle = \langle p, Av \rangle. \tag{19.1}$$

Let $\mathrm{Ad}^{\# h}$ denote the coadjoint action of H on h^*. Then we have proved that

$$\mathrm{Ad}^{\#}_{\left(\begin{smallmatrix} a & v \\ 0 & 1 \end{smallmatrix}\right)}(\alpha, p) = (\mathrm{Ad}^{\# h}_a \alpha + a^{*-1} p \odot v, a^{*-1} p). \tag{19.2}$$

The G orbits in g^* are thus fibered over the H orbits in V^*. Suppose we pick an H orbit N in V^* and want to describe the various G orbits in g^* "sitting over" N. Pick some point $p \in N$, let $H_p \subset H$ be the isotropy group of p, and let $h_p \subset h$ be the Lie algebra of H_p. Since h_p is a subalgebra of h, we get a projection, $\pi_p : h^* \to h_p^*$, where $\pi_p \alpha$ is simply the restriction of α, which is a linear function on all of h to the subspace h_p. We claim that

$$\alpha - \alpha^0 = p \odot v \qquad \text{for some} \quad v \in V$$

if and only if

$$\pi_p \alpha = \pi_p \alpha^0.$$

Indeed, $\langle p \odot v, \eta \rangle = \langle \eta p, v \rangle$ for any $\eta \in h$. If $\eta \in h_p$ so that $\eta_p = 0$, this implies $\langle p \odot v, \eta \rangle = 0$, so $\pi_p(\alpha - \alpha') = 0$. Conversely, the set of all elements of $gl(V) = V \otimes V^*$ that are orthogonal to $p \otimes v \subset V^* \otimes V$, is $(n^2 - n) =$ dimensional and contains the $(n^2 - n)$-dimensional space of all B satisfying $B^* p = 0$ (by the preceding result applied to $h = gl(V)$). Thus these two spaces coincide. The image of $p \otimes V$ in h^* under the map $V^* \otimes V \to h^*$ dual to the map $h \to gl(V) = V^*$ is $p \odot V$. From this it follows that the orthogonal space of $p \odot V$ in h is just the inverse image in h of $gl(V)_p$; that is h_p.

Thus, once we have picked an orbit N of H on V^*, and a point $p \in N$, the classification of the orbits above N reduce to the classification of the H_p orbits in h_p^*. Let us see what these results say when V is a vector space equipped with a nondegenerate scalar product and $H = SO(V)$ is the connected component of the orthogonal group for this scalar product. Thus, if $V = \mathbb{R}^3$ with the positive definite scalar product, $H = SO(3)$ and

$G = E(3)$ is the Euclidean group in three dimension. If $V = \mathbb{R}^{1,3}$ is Minkowski space, then $H = L$ is the Lorentz group and G is the Poincaré group. We may identify V with V^* using the scalar product, and also identify h with h^*. We may also identify h with $\bigwedge^2 V$, where the decomposable 2-vector $u \wedge w$ is identified with the operator $A_{u \wedge w}$:

$$A_{u \wedge w} v = (w, v)u - (v, u)w,$$

where $(\,,)$ denotes the scalar product on V. Let us examine the meaning of the operator $\odot$ given by (19.1) in this case. We have

$$\begin{aligned} \langle p, A_{u \wedge w} v \rangle &= (p, (w, v)u - (v, u)w) \\ &= (p, u)(w, v) - (p, w)(v, u) \\ &= \det \begin{pmatrix} (p, u) & (p, w) \\ (v, u) & (v, w) \end{pmatrix}. \end{aligned}$$

But this determinant is precisely the scalar product of the elements $p \wedge v$ and $u \wedge w$ in $\bigwedge^2(V)$ with respect to the scalar product induced from the scalar product on V. If we denote this by $(\,,)$ as well, then (19.1) becomes

$$\langle p \odot v, A_{u \wedge w} \rangle = \langle p \wedge v, u \wedge w \rangle$$

or, in other words,

$$p \odot v = p \wedge v.$$

There is also some slight simplification in (19.2). Under the identification of $\bigwedge^2 V$ with $o(V)^*$, the coadjoint action of $SO(V)$ on $o(V)^*$ becomes identified with the usual action of $SO(V)$ on $\bigwedge^2 V$. Also, under the identification of V with V^*, a^{*-1} becomes simply a. Writing $\begin{pmatrix} a & v \\ 0 & 1 \end{pmatrix} (\beta, p)$ instead of

$$\mathrm{Ad}^{\#}_{\left(\begin{smallmatrix} a & v \\ 0 & 1 \end{smallmatrix}\right)} (\beta, p),$$

(19.2) can be written as

$$\begin{pmatrix} a & v \\ 0 & 1 \end{pmatrix} (\beta, p) = (a\beta + ap \wedge v, ap). \tag{19.3}$$

In other words, the rotation a acts on (β, p) by simply rotating β and p. The translation through v does not change p but does replace β by $\beta + p \wedge v$. In three dimensions this transformation law for (β, p) is familiar from elementary mechanics – we can think of p as linear momentum and β as angular momentum. Both p and β are rotated arbitrarily; the linear momentum is independent of the origin of coordinates, but a translation does change the origin and modifies β as above. We will return to this point shortly.

The orbits of $SO(V)$ acting on V are the connected components of the $(\dim V - 1)$ dimensional "spheres" $\|p\|^2 =$ const., $p \neq 0$, together with the zero-dimensional orbit, consisting of the origin $\{0\}$.

For the case of $E(3)$ these orbits are just two-dimensional spheres. So, for example, we may apply a suitable rotation to the element (β, p) to bring it to the form (γ, ke_3), where e_3 denotes the unit vector in the z direction and $k = \|p\|$. There are now two alternatives, $k = 0$ or $k \neq 0$. If $k = 0$, so that $p = 0$, then the action of the translations in (19.3) is trivial and the rotation a sends $(\beta, 0)$ into $(a\beta, 0)$. Again this subdivides into two cases: If $\beta = 0$ then we get the zero-dimensional coadjoint orbit $\{(0,0)\}$. If $\beta \neq 0$ then the coadjoint orbit through β is the two-dimensional sphere $\{(a\beta, 0)\}$. These are the zero- and two-dimensional coadjoint orbits of $E(3)$. (The physical significance of the two-dimensional orbits will be discussed later.) Now let us turn to the case where $p \neq 0$. If $v = ae_1 + be_2 + ce_3$, then, with $p = ke_3$,

$$p \wedge v = kae_3 \wedge e_1 + kbe_3 \wedge e_2.$$

Writing

$$\gamma = xe_1 \wedge e_3 + ye_2 \wedge e_3 + se_1 \wedge e_2,$$

we can set

$$\gamma + p \wedge v = se_1 \wedge e_2.$$

In other words, on every orbit with $p \neq 0$, there is a unique point of the form

$$\alpha = (se_1 \wedge e_2, ke_3), \qquad k = \|p\|.$$

It is clear that the subgroup of $E(3)$ that leaves α fixed is precisely the group $E(3)_z$ of those Euclidean motions that fix the z axis. Thus, as $E(3)$ spaces, these orbits look like the space of straight lines in $\mathbb{R}^3$ and the different symplectic structures are parameterized by the real number s and the positive number k, as predicted in the preceding section. Notice that if $\alpha = (se_1 \wedge e_2, ke_3)$, then

$$(se_1 \wedge e_2) \wedge ke_3 = (sk)e_1 \wedge e_2 \wedge e_3.$$

For general (β, p) we have

$$\beta \wedge p = (a\beta + ap \wedge v) \wedge ap,$$

since $ap \wedge v \wedge ap = 0$ and $a\beta \wedge ap = (\det a)\, \beta \wedge p$ and $\det a = 1$. So the functions $k(\alpha) = \|p\|$, defined for all $\alpha = (\beta, p)$, and $s(\alpha)$, defined for α with

$p \neq 0$ by $\beta \wedge p = s(\alpha)k(\alpha)e_1 \wedge e_2 \wedge e_3$, are invariant under the action of $E(3)$ and the four-dimensional orbits are described by the different possible values of the functions k and s.

Now let us return to the analogy between (19.3) and the transformation law for linear and angular momentum and formulate this analogy in a more mathematically precise manner. Suppose that we had a system of N classical particles. The phase space of this system consists of all points of the form

$$(q_1, \ldots, q_N; p_1, \ldots, p_N) = z \in \mathbb{R}^{6N},$$

where $q_i \in \mathbb{R}^3$ represents the position of the ith particle and $p_i \in \mathbb{R}^3$ represents its momentum. The group $E(3)$ acts on this phase space: A Euclidean motion applies to all the position vectors simultaneously, translation has no effect on momentum but rotations act on all the momentum vectors simultaneously so that the action is given by

$$\begin{pmatrix} a & v \\ 0 & 1 \end{pmatrix}(q_1, \ldots, q_N; p_1, \ldots, p_N) = (aq_1 + v, \ldots, aq_N + v; ap_1, \ldots, ap_N). \tag{19.4}$$

The total linear momentum is the function p defined on phase space by

$$p(z) = p_1 + \cdots + p_N \tag{19.5}$$

and the total angular momentum is the function given by

$$\beta(z) = p_1 \wedge q_1 + \cdots + p_N \wedge q_N. \tag{19.6}$$

We can now define the map Φ from the phase space to $e(3)^*$ given by

$$\Phi(z) = (\beta(z), p(z)).$$

Combining equations (19.4)–(19.6) and comparing them with (19.3), we see that for any $b = \begin{pmatrix} a & v \\ 0 & 1 \end{pmatrix} \in E(3)$ we have

$$\Phi(bz) = b\Phi(z). \tag{19.7}$$

As we shall see in the next chapter, the map Φ and Equation (19.7) have generalizations to arbitrary groups G and certain of their symplectic actions. The study of this "generalized moment map," its geometry, and its physical applications will be one of our major concerns.

For now we concentrate our attention on $E(3)$ and examine what happens for a *single* particle ($N = 1$). The phase space is now six-dimensional. The group $E(3)$ does not act transitively on all of phase space since $\|p\|$ is preserved. The orbit of $E(3)$ through a point (q, p) with $p \neq 0$ is clearly five-dimensional, consisting of all (q', p') with $\|p'\| = \|p\|$. The map Φ must carry this five-dimensional orbit onto a single orbit of $E(3)$ acting on $e(3)^*$ because of (19.7). Since $\Phi(q, p) = (p \wedge q, p)$, we see that the image of Φ is precisely the orbit with $s = 0$ and $k = \|p\|$. Two points (q, p) and (q', p') have the same image under Φ if and only if $p' = p$ and $p \wedge q' = p \wedge q$. For $p \neq 0$ this last condition can hold if and only if $q - q'$ is some multiple of p. Thus

$$\Phi^{-1}(\Phi(q, p)) = \{(q + tp, p)\}$$

is a straight line, which we may think of as the trajectory of the particle with momentum p. In this sense we see that the four-dimensional symplectic orbits of $E(3)$ with $s = 0$ correspond to the space of trajectories of classical particles; the straight lines of the coadjoint orbit are the trajectories of the particle. We know $s = 0$ because a standard classical particle has no "intrinsic angular momentum." The existence of other coadjoint orbits with $s \neq 0$ suggests that there should exist particles with an intrinsic angular momentum or "spin." In fact, in an experiment done in the 1930s by Beck at the suggestion of Einstein, the photon was shown to have just such an intrinsic angular momentum measured mechanically.

We now come to another important point. In the preceding section we treated k as an object having units of inverse length. Indeed, this is the natural choice of units from the group-theoretical point of view, since in $e(3)^*$ p is something dual to translations. In the present section we have been discussing k or p in terms of momentum. What is the relation between the two? Let us try to give an answer from a purely Euclidean point of view: Suppose that mechanics had developed before the invention of clocks. So we could observe the trajectories of particles, their collisions and deflections, but not their velocities. For instance we might be able to observe tracks in a bubble chamber or on a photographic plate. (In the case of light, all of the work described above was done before there was any accurate measurement of the velocity of light.) The configuration space of a single particle is just the three-dimensional Euclidean space $\mathbb{R}^3$, the corresponding phase space is six-dimensional with coordinates (q, p), where q and p are 3-vectors. As we have seen, each free classical particle; that is, each of the five-dimensional orbits of $E(3)$ acting on the six-dimensional phase space is parametrized by its value of $k = \|p\|$. In the absence of clocks we cannot measure velocity so we cannot distinguish between a "light particle moving fast" or a "heavy particle moving slowly." (Of course, from the scattering experiments themselves, we would be led to discover new conserved quantities such as energy, and thus be led to enlarge the group. But we do not want to go into this point.) Without some way of relating momentum to length, we would introduce "independent units" of momentum, perhaps by combining particles in various ways and performing collision experiments. But we know that the "natural units" should be inverse length. A single experiment, the photoelectric effect, involving an interaction between light and one of our "particles" would then give us the conversion factor, and allow us to write $\|p\| = h/\lambda$. Thus, from this group-theoretical point of view, Planck's constant h is a conversion factor from the "independent" units of momentum to the "natural" units of inverse length. Of course, the story did not develop that way. The "conversion factor" was first found between "energy" and "inverse time"; but to explain this would involve us in large groups such as the Galilean or Poincaré groups. We shall study the symplectic homogeneous manifolds for these groups in Chapter V.

20. Electromagnetism and the determination of symplectic structures

In this section we give a rapid review of the basic facts in electromagnetic theory, with an emphasis on the intimate relations between the objects of the theory and the geometry of space and of space-time. We shall present the theory in two versions: The nonrelativistic, pre-Maxwellian version and then in the full setting of Maxwell's equations. In the pre-Maxwellian version, we shall see that the geometry of space is intimately related to the dielectric properties of the vacuum, while the electric field and charge determine the "Hamiltonian" and the magnetic field and electric charge determine the symplectic structure on the six-dimensional phase space of a particle. In Maxwell's version, we shall see that the dielectric properties of the vacuum determine the conformal structure of space-time, and the electromagnetic field (and charge) determines the symplectic structure of the eight-dimensional phase space. More precisely, we wish to establish the following points:

(A) In electrostatics, the dielectric properties of the vacuum determine Euclidean geometry.

(B) In magnetostatics the magnetic field determines the symplectic structure on phase space.

(C) In Maxwell's electrodynamics the constitutive properties of the vacuum determine the conformal geometry of space-time.

(D) The electromagnetic field determines the symplectic structure of the relativistic particle.

Our treatment is (essentially) an extract from Sternberg (1978).

A. Electrostatics: The dielectric properties of the vacuum determine Euclidean geometry

We begin by formulating in geometrical terms the fundamental objects of electrostatics. The electric field strength E is a linear differential form, which, when integrated along any path, gives the voltage drop across that path; thus the units of E will be voltage/length. Since voltage has units energy/charge and force has units energy/length, we can also write the units of E as force/charge. We emphasize that E, as a geometrical object, is a linear differential form on three-dimensional space, because, by its physical definition, it is something that assigns voltage differences to paths by integration. The second fundamental object in electrostatics is the dielectric displacement. It is a 2-form, D, which, when integrated over the boundary surface of any region, gives the total charge contained in that region. Thus D is a 2-form that satisfies the equation

$$dD = 4\pi\rho \, dx \wedge dy \wedge dz,$$

where the 3-form $\rho \, dx \wedge dy \wedge dz$ represents the density of charge in any three-dimensional region. By its definition, D assigns charge to surfaces by a process of integration. Therefore, by its defining properties, it is given as a 2-form on three-dimensional space (and not as a vector field as prescribed in the standard treatments). There are two fundamental equations in electrostatics. The first of these postulates a relationship between E and D determined by the medium. The second asserts that $dE = 0$. Now E is a 1-form and D is a 2-form. We cannot have functional relationships between a 1-form and a 2-form on a three-dimensional space without imposing severe geometrical restrictions on the space. More specifically, the relationship between E and D in what is known as a homogeneous isotropic medium is as follows: One postulates that there is a preferred rectangular (i.e., Euclidean) coordinate system x, y, and z and, therefore, a well-defined $*$ operator given by

$$*dx = dy \wedge dz,$$
$$*dy = -dx \wedge dz,$$
$$*dz = dx \wedge dy,$$

which relates 1-forms to 2-forms and that

$$D = \epsilon * E,$$

where ϵ is a function of the medium, known as its dielectric constant. *In vacuo* the function ϵ is a constant, ϵ_0, known as the dielectric constant of the vacuum. Now in three dimensions the * operator associated to a Riemann metric completely determines the metric. Giving the * operator is the same as giving the metric. Therefore, the law of electrostatics $D = \epsilon_0 * E$ completely determines the Riemannian geometry of space, which is asserted by the experimental laws of electrostatics to be Euclidean. We refer the reader to Loomis and Sternberg (1968) or Sternberg (1983, p. 23) or any modern book on advanced calculus for the basic properties of the * operator.

The statement that it is the dielectric properties of the vacuum that determine Euclidean geometry is not merely a mathematical sophistry. In fact, the forces between charged bodies in any medium are determined by the dielectric properties of that medium. Since the forces that bind together macroscopic bodies as we know them are principally electrostatic in nature, it is the dielectric property of the vacuum that fixes our rigid bodies. We use rigid bodies as measuring rods to determine the geometry of space. It is in this very real sense that the dielectric properties of the vacuum determine Euclidean geometry.

We should emphasize once again, in view of what is going to follow, that it is not the field itself, the E or the D, that determines the Euclidean geometry, but rather the response of the vacuum to the presence of the field *in potentio*; the fixing of the relationship between the E and the D that determines Euclidean geometry. Euclidean geometry in turn, determines the equations of motion of a free, uncharged particle in space. Giving a Riemannian geometry determines a scalar product on the tangent space at any point and, therefore, on the cotangent space as well. The equations of motion of a free particle of mass m are determined by the Hamiltonian function $\mathscr{H}_m$ on $T^*\mathbb{R}^3$, where

$$\mathscr{H}_m(q, p) = \frac{1}{2m} \|p\|^2,$$

where $q = (x, y, z)$ is the position and $p = (p_x, p_y, p_z)$ is the momentum of the free particle.

The effect of an actual electrostatic field is to modify the equations of motion of a charged particle. We consider a particle whose charge e is sufficiently small so that its effect on the electromagnetic field can be ignored. So we are dealing with the passive equations of a test particle of small charge in the presence of a given electrostatic field. The field equation $dE = 0$ is locally equivalent to the existence of a function ϕ such that $E = -d\phi$. Let us assume that we are dealing with a region of space that is simply connected, so that we will take ϕ to be globally defined in our region. Then the equations of motion of a charged particle are given by a modified Hamiltonian $\mathscr{H}_{m,e,\phi}$, where

$$\mathscr{H}_{m,e,\phi}(q, p) = \frac{1}{2m} \| p \|^2 + e\phi(q).$$

In all of the above discussion we have been regarding charge as an independent unit. Therefore, strictly speaking, we should consider the electric field strength E not as a numerical-valued linear differential form, but rather as vector-valued linear differential form with values in a one-dimensional space dual to the charges. Then the choice of units of charge would amount to the choice of a basis in this one-dimensional vector space. Actually, as we shall see later on, the correct formulation will be to consider the field strength as a vector-bundle-valued differential form and the precise geometrical character of this vector bundle will be elucidated.

B. Magnetostatics: The magnetic field determines the symplectic structure on phase space

In the preceding section we wrote down the Hamiltonian for a charged test particle in the presence of an electric field. In writing such a Hamiltonian, we implicitly took for granted that one would derive Hamilton's equations from this Hamiltonian by the standard procedure. That is, we took for granted that there existed a symplectic form ω on the phase space $T^*\mathbb{R}^3$ and that the vector field describing the differential equations of motion of the particle was derived from the Hamiltonian by the standard procedure:

$$i(\xi_{\mathscr{H}})\omega = d\mathscr{H}.$$

We also took for granted that the symplectic form ω was the canonical symplectic form carried by $T^*\mathbb{R}^3$ in virtue of its being a cotangent bundle. That is, we took

$$\omega = dq_x \wedge dp_x + dq_y \wedge dp_y + dq_z \wedge dp_z.$$

In magnetostatics, there are also two fundamental quantities, the *magnetic flux*, B, and the *magnetic loop tension.* Faraday's law of induction says that associated with any system of magnets there is a certain flux. If γ is a closed circuit and S is a surface whose boundary is γ, then the change of flux through S (by, for example, moving the magnets) induces an electromotive force around γ. Put more mathematically, let

$$B = B_x\, dy \wedge dz - B_y\, dx \wedge dz + B_z\, dx \wedge dy.$$

Let S be a surface with boundary γ, then Faraday's law of induction says that

$$\frac{d}{dt}\int_S B = -\int_\gamma E.$$

From its definition as a flux we see that we must regard B as a 2-form on three-dimensional space.

Now, the presence of a magnetic flux changes the motion of a small test particle in its presence. The change in the equations of motion can be described as follows. We are given B as a 2-form on $\mathbb{R}^3$. By the standard projection π of $T^*\mathbb{R}^3$ onto $\mathbb{R}^3$, assigning to each point in phase space the corresponding point in configuration space, we can regard B equally well as being defined on $T^*\mathbb{R}^3$. With this new identification we would write

$$B = B_x\, dq_y \wedge dq_z - B_y\, dq_x \wedge dq_z + B_z\, dq_x \wedge dq_y.$$

Here we write q_x for $x \circ \pi$, etc.

It is a fundamental law of electromagnetism that the integral of B around any closed surface in space vanishes ("There is no true magnetism," to quote Hertz), thus

$$dB = 0.$$

This means that the form

$$\omega_{e,B} = \omega + eB$$

is a closed 2-form defined on $T^*\mathbb{R}^3$. We claim that $\omega_{e,B}$ is also nondegenerate. Indeed, let us examine the equations

$$i(\xi_{\mathscr{H}})\omega_{e,B} = d\mathscr{H}.$$

Let us write

$$\begin{aligned}\xi_{\mathscr{H}} &= a\frac{\partial}{\partial q} + b\frac{\partial}{\partial p}\\ &= a_x\frac{\partial}{\partial q_x} + a_y\frac{\partial}{\partial q_y} + a_z\frac{\partial}{\partial q_z} + b_x\frac{\partial}{\partial p_x} + b_y\frac{\partial}{\partial p_y} + b_z\frac{\partial}{\partial p_z}.\end{aligned}$$

Then

$$i(\xi_{\mathscr{H}})\omega_{e,B} = a_x(dp_x - eB_y\, dq_z + eB_z\, dq_y) - b_x\, dq_x + \text{two similar terms.}$$

On the other hand,

$$\begin{aligned}d\mathscr{H} &= \frac{\partial\mathscr{H}}{\partial q}dq + \frac{\partial\mathscr{H}}{\partial p}dp\\ &= \frac{\partial\mathscr{H}}{\partial q_x}dq_x + \frac{\partial\mathscr{H}}{\partial p_x}dp_x + \text{two similar terms.}\end{aligned}$$

Comparing the coefficient of dp_x shows that

$$a_x = \frac{\partial \mathscr{H}}{\partial p_x}$$

and comparing the coefficient of dq_x shows that

$$b_x = -\frac{\partial \mathscr{H}}{\partial q_x} + e\left(\frac{\partial \mathscr{H}}{\partial p_y} B_z - \frac{\partial \mathscr{H}}{\partial p_z} B_y\right)$$

plus similar equations for a_y, a_z, b_y, and b_z. In particular, we see that the a's and b's are completely determined and, hence, that the form $\omega_{e,B}$ is nondegenerate. If we take the Hamiltonian $\mathscr{H}$ to be $\mathscr{H}_{m,e,\varphi}$ as described in the preceding section, then

$$\frac{\partial \mathscr{H}}{\partial p_x} = \frac{1}{m} p_x \qquad \text{so} \qquad \frac{dq_x}{dt} = \frac{1}{m} p_x, \text{ etc.,}$$

$$\frac{\partial \mathscr{H}}{\partial q_x} = e\frac{\partial \phi}{\partial x} = -eE_x \qquad \text{so} \qquad \frac{dp_x}{dt} = e\left(E_x + \frac{p_y}{m} B_z - \frac{p_z}{m} B_y\right),$$

and the differential equations that we obtain are precisely the classical differential equations for a charged particle in the presence of a given external electric and magnetic field. Thus, the presence of the magnetic field B modifies the equations of motion of a charged particle by modifying the symplectic structure on the cotangent bundle.

The magnetic flux also makes its presence felt by affecting the equations of motion of a magnet considered as a spinning electrical particle. To describe these equations of motion, we must replace the six-dimensional phase space $T^*\mathbb{R}^3$ by the eight-dimensional space given as the direct product $T^*\mathbb{R}^3 \times S^2$, where S^2 is the standard two-dimensional sphere. Let Ω denote the standard volume form on the unit sphere S^2. On the product space $T^*\mathbb{R}^3 \times S^2$ we can put the symplectic structure given by

$$\omega_{e,B} + s\Omega,$$

where the parameter s is called the spin of the particle. Now, we can think of a vector $u \in S^2$ as determining a vector $su \in \mathbb{R}^3$ of length s. In turn, we can regard, in view of a fixed orientation on space, the vector su as determining a bivector in $\mathbb{R}^3$:

$$S = *(su) \in \wedge^2\mathbb{R}^3,$$

and we can then, at each point of $\mathbb{R}^3$, take the scalar product of S with the

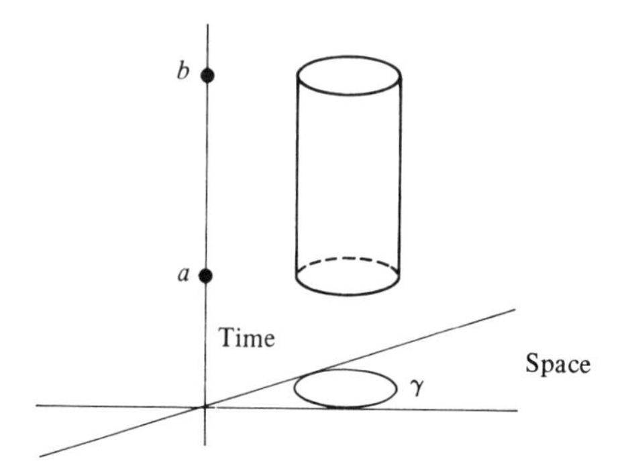

Figure 20.1.

magnetic flux at that point, also regarded as an exterior 2-vector at the point in question. Thus $S \cdot B$ is a well-defined function on $T^*\mathbb{R}^3 \times S^2$. Let us now introduce the modified Hamiltonian $\mathscr{H}_{m,e,\phi,\mu,B}$ given by

$$\mathscr{H}_{m,e,\phi,\mu,B}(q, p, u) = \frac{1}{2m} \| p \|^2 + \phi(q) + \mu S \cdot B.$$

Here the parameter μ is called the magnetic moment of the particle and we will leave it as an exercise to the reader to verify that the equations of motion associated with this Hamiltonian relative to the symplectic structure described above gives precisely what is required for the equations of a charged spinning particle with magnetic moment μ in the presence of an external electric and magnetic field.

C. Maxwell's equations: The constitutive properties of the vacuum determine the conformal geometry of space-time

Electrostatics and magnetostatics are only approximately correct. They must be replaced by Maxwell's theory, which we now quickly review.

We begin by rewriting Faraday's law of induction

$$\frac{d}{dt} \int_S B = - \int_\gamma E$$

in a form that is more accessible to the space-time approach. Consider an interval $[a, b]$ in time and the three-dimensional cylinder $S \times [a, b]$ whose boundary is the two-dimensional cylinder $\gamma \times [a, b]$ together with the top and the bottom of this two-dimensional cylinder. (See Figure 20.1.) Integrating Faraday's law of induction with respect to t from a to b gives the equation

$$\int_{S\times\{b\}} B - \int_{S\times\{a\}} B + \int_{\gamma\times[a,b]} E \wedge dt = 0.$$

Let us set

$$F = B + E \wedge dt$$

so that F is a 2-form defined on four-dimensional space. Let C denote the three-dimensional cylinder, $C = S \times [a,b]$ so that

$$\partial C = S \times \{b\} - S \times \{a\} + \gamma \times [a,b].$$

Now, B is a 2-form involving just the spatial differentials and, therefore, must vanish when restricted to the side $\gamma \times [a,b]$ of the cylinder, while dt and, hence, $E \wedge dt$ must vanish on the top and the bottom. Thus, we can write Faraday's law of induction as

$$\int_{\partial C} F = 0.$$

We can also consider a three-dimensional region C lying entirely in space at one fixed constant time. In this case, ∂C will be a surface on which $dt = 0$ so that

$$\int_{\partial C} F = \int_{\partial C} B$$

and by the absence of true magnetism, this surface integral must vanish. Thus, $\int_{\partial C} F = 0$ for all three-dimensional cubes whose sides are parallel to any three of the four coordinate axes. This is enough to imply that $dF = 0$, where now, of course, d stands for the exterior derivative in 4-space. Since

$$F = B_x\, dy \wedge dz - B_y\, dx \wedge dz + B_z\, dz \wedge dy + E_x\, dy \wedge dt$$
$$+ E_y\, dy \wedge E_z\, dz \wedge dt,$$

the equation $dF = 0$ is equivalent to the four equations

$$\frac{\partial B_x}{\partial x} + \frac{\partial B_y}{\partial y} + \frac{\partial B_z}{\partial x} = 0, \qquad \frac{\partial B_y}{\partial t} - \frac{\partial E_z}{\partial x} + \frac{\partial E_x}{\partial z} = 0,$$

$$\frac{\partial B_x}{\partial t} - \frac{\partial E_y}{\partial z} + \frac{\partial E_z}{\partial y} = 0, \qquad \frac{\partial B_z}{\partial t} - \frac{\partial E_x}{\partial y} + \frac{\partial E_y}{\partial x} = 0.$$

We will use Faraday's law to define B so that the units of B are

$$\frac{\text{voltage} \cdot \text{time}}{\text{area}} = \frac{\text{energy} \cdot \text{time}}{\text{charge} \cdot (\text{length})^2}.$$

Ampère's law relates current to magnetism. It says that the electric current flux through a surface S whose boundary is γ equals the magnetic loop tension around γ. According to Maxwell's great discovery, we must write the electric current flux as the sum of two terms $\partial D/\partial t + 4\pi j$, where D is the dielectric displacement and j is the current density of moving charges. (For slowly varying fields, the first term is negligible in comparison to the second and did not appear in Ampère's original formulation.) The magnetic loop tension is obtained by integrating a linear differential form H, called the "magnetic field strength" around γ. Thus Ampère's law says

$$\int_S \left(\frac{\partial D}{\partial t} + 4\pi j \right) = \int_\gamma H.$$

Consider the three-dimensional cylinder $C = S \times [a, b]$ as before and set

$$G = D - H \wedge dt.$$

We integrate Ampère's law from a to b with respect to t and get

$$\int_{\partial C} G = -4\pi \int_C j \wedge dt$$

If we consider a three-dimensional region R at constant time, then dt vanishes on ∂R and the integral of G over ∂R is the same as the integral of D.

This equals $4\pi \times$ (the total charge in R) which is $4\pi \int \rho \, dx \wedge dy \wedge dz$. Thus, for regions which lie in constant time, we have

$$\int_{\partial R} G = 4\pi \int \rho \, dx \wedge dy \wedge dz.$$

Let us set

$$J = \rho \, dx \wedge dy \wedge dz - j \wedge dt$$

and we see that

$$\int_{\partial C} G = 4\pi \int_C J$$

for any three-dimensional cube whose sides are parallel to the coordinate axes. Thus

$$dG = 4\pi J,$$

from which it follows that $dJ = 0$.

We summarize:

Set	then Maxwell's equations say
$F = B + E \wedge dt$	$dF = 0$
$G = D - H \wedge dt$	$dG = 4\pi J.$

Note that Maxwell's equations are invariant under smooth, orientation-preserving changes of coordinates.

We will use Ampère's law to define H. D has units charge/area so $\partial D/\partial t$ has units charge/(area · time) and j has units current/area = charge/(area · time). Thus H has units charge/(time · length). *In vacuo* we have the constitutive relations

$$D = \epsilon_0 * E \qquad \text{and} \qquad B = \mu_0 * H.$$

ϵ_0 has units:

$$\frac{\text{charge}}{\text{area}} \times \frac{\text{length}}{\text{voltage}} = \frac{(\text{charge})^2}{\text{energy} \cdot \text{length}}.$$

μ_0 has units

$$\frac{\text{energy} \cdot \text{time}}{\text{charge} \cdot (\text{length})^2} \times \frac{\text{time} \cdot \text{length}}{\text{charge}} = \frac{\text{energy} \cdot (\text{time})^2}{(\text{charge})^2 \cdot \text{length}}.$$

Thus $1/\epsilon_0\mu_0$ has units $(\text{length})^2/(\text{time})^2 = (\text{velocity})^2$. Thus the theory of electromagnetism has a fundamental velocity built into it. It was Maxwell's great discovery that this velocity is exactly c – the velocity of light! So introduce $c\,dt$ instead of dt and the four-dimensional $*$ operator becomes

$$\begin{aligned}
*(dx \wedge dy) &= -c\,dz \wedge dt,\\
*(dy \wedge dz) &= c\,dy \wedge dt,\\
*(dy \wedge dz) &= -c\,dx \wedge dt,\\
*(dx \wedge c\,dt) &= dy \wedge dz,\\
*(dy \wedge c\,dt) &= -dx \wedge dz,\\
*(dz \wedge c\,dt) &= dx \wedge dy.
\end{aligned}$$

Then

$$\begin{aligned}
*F = &-c(B_z\,dz \wedge dt + B_y\,dy \wedge dt + B_z\,dx \wedge dt)\\
&+ (1/c)(E_x\,dy \wedge dz - E_y\,dx \wedge dz + E_z\,dx \wedge dy).
\end{aligned}$$

The constitutive equations can be written as

$$G = \sqrt{\epsilon_0/\mu_0} \times * F.$$

Now in four dimensions, the $*$ operator from $\bigwedge^2 \to \bigwedge^2$ does *not* determine the metric. It does determine it up to scalar multiple at each point. Thus the constitutive properties of the vacuum determine the conformal Lorentzian structure of space-time.

From now on we shall use coordinates in which $c = 1$. Thus x_0, x_1, x_2, x_3 will be coordinates on space-time with

$$ds^2 = dx_0^2 - dx_1^2 - dx_2^2 - dx_3^2,$$

the Lorentz metric, and $* dx_1 \wedge dx_2 = dx_0 \wedge dx_3$, etc.

D. The Lorentz force: The electromagnetic field determines the symplectic structure on the phase space of space-time

We denote our four-dimensional space-time by M so that T^*M denotes its cotangent bundle, and ω the canonical 2-form on T^*M. Suppose we are given an electromagnetic field F, which is a 2-form on M. Via the projection $\pi\colon T^*M \to M$ we can pull F back to T^*M to obtain a 2-form on T^*M, which we shall continue to denote by F. We define

$$\omega_{e,F} = \omega + eF,$$

where e is an electric charge. One half of Maxwell's equations assert that $dF = 0$, so that $d\omega_{e,F} = 0$. Since F involves only dq's (i.e., differentials coming from M), it is easy to check, using the same argument as in the preceding subsection, that $\omega_{e,F}$ is nondegenerate. Thus e and F determine a symplectic structure on T^*M. Let $\mathscr{H}$ be a function on T^*M and suppose that the Hamiltonian equations corresponding to $\mathscr{H}$ and to the canonical symplectic form ω describe the equations of motion of a free uncharged particle. Then the Hamiltonian equations corresponding to H relative to the symplectic form $\omega_{e,F}$ describe the equations of motion of a charged particle of charge e in the presence of the external electromagnetic field F. As before, we are assuming that the charge e is sufficiently small that we can neglect the influence of the particle on the electromagnetic field. If we take $M = \mathbb{R}^{1,3}$ to be Minkowski space with the standard Lorentz metric, and we take

$$\mathscr{H}(p, q) = \tfrac{1}{2}\|p\|^2 = \tfrac{1}{2}(p_0^2 - p_1^2 - p_2^2 - p_3^2),$$

then it is easy to check, using the same methods as in subsection C, that this gives the standard Lorentz equations. We shall redo this calculation in a more general setting later on. We have introduced the effect of the electromagnetic field by keeping the same Hamiltonian $\mathscr{H}$ but modifying the symplectic structure. In the standard physics literature, there is another procedure, called "minimal coupling," for obtaining the Lorentz equation that keeps the original symplectic structure but modifies the Hamiltonian. Let us pause to show that the two procedures are formally equivalent. The minimal coupling prescription is as follows: We can, at least locally, find a 1-form A on M called a 4-potential, satisfying $dA = F$. We can think of A as a section of T^*M and introduce a modified Hamiltonian $\mathscr{H}_{e,A}$, where

$$\mathscr{H}_{e,A}(q, p) = \mathscr{H}\big(q, p - eA(\pi(q))\big).$$

The second-order differential equations determined by the Hamiltonian equations of $\mathscr{H}_{e,A}$ relative to the canonical form ω are again the Lorentz equations. To see why the two procedures are the same, let us introduce the map $\varphi_{e,A}$ of T^*M into itself defined by

$$\varphi_{e,A}(q, p) = \big(q, p + eA(q)\big).$$

Let $\theta = p \cdot dq$ be the canonical (action) form on T^*M so that $\omega = -d\theta$. Then

$$\varphi_{e,A}{}^* \theta = \theta + e\pi^* A.$$

Here A is a 1-form on M so π^*A is a 1-form on T^*M and $d\pi^*A = \pi^*\, dA = \pi^*F$. We have been writing F for π^*F so, applying d to the preceding equation, we get

$$\varphi_{e,A}^*\omega = \omega + eF = \omega_{e,F}.$$

On the other hand, $\varphi_{e,A}^{-1}(q, p) = \big(q, p - eA(q)\big)$ so

$$(\varphi_{e,A}^{-1})^* \mathscr{H} = \mathscr{H}_{e,A}.$$

It is now clear that the solution curves of $\mathscr{H} = \varphi_{e,A}{}^* \mathscr{H}_{e,A}$ relative to $\omega_{e,A} = \varphi_{e,A}{}^*\omega$ are the images, under $\varphi_{e,A}$ of the solution curves of $\mathscr{H}_{e,A}$ relative to ω. Since $\pi \circ \varphi_{e,A} = \pi$, we obtain the same trajectories on M from one system as from the other.

There are some important differences between the space-time treatment described in this section, and the discussion in Subsections A and B. First of all, in contrast to the electrostatic case, the Riemannian (Lorentzian) metric of space-time is not completely determined by the electromagnetic constitutive properties of the vacuum, and hence the Hamiltonian $\mathscr{H}$ must be specified. The special theory of relativity assumes that $M = \mathbb{R}^{1,3}$ with its flat metric, and $\mathscr{H} = \frac{1}{2}\|p\|^2$, as above. (The constitutive properties of the vacuum would allow any metric conformally equivalent to the flat metric and hence any Hamiltonian of the form $\mathscr{H}(q, p) = \frac{1}{2}\lambda(q)\|p\|^2$, where λ is any function on $\mathbb{R}^{1,3}$.) The general theory of relativity supposes the existence of a Lorentzian metric g determined from Einstein's equation that provides a theory of the gravitational force in addition to electromagnetism. Then Maxwell's equations are replaced by the equations $dF = 0$ and $d * F = J$, where now the $*$ operator is that determined by g. (To the extent that g is not conformally equivalent to a flat metric, this implies a modification in Maxwell's equations.) Then $\mathscr{H}$ is taken to be $\mathscr{H}(q, p) = \frac{1}{2}\|p\|^2$, where now $\|p\|^2$ means the square length of the covector p relative to g. The trajectories of the Hamiltonian vector field corresponding to $\mathscr{H}$ relative to the canonical form ω – of the "free" Hamiltonian equations – project onto the geodesics of g. The presence of an electromagnetic field F modifies these equations for a charged particle as indicated above.

A second difference between the relativistic formulation of this subsection and the nonrelativistic formulation in Subsections A and B lies in the concept of mass. In Subsection A we introduced the mass as a parameter in the Hamiltonian on a six-dimensional phase space. Here T^*M is eight-dimensional and the mass is introduced by considering the seven-dimensional manifold $\mathscr{H} = \frac{1}{2}m^2$. The vector field $\xi_{\mathscr{H}}$ is tangent to this submanifold, where $\xi_{\mathscr{H}}$ denotes the Hamiltonian vector field associated to $\mathscr{H}$ by the form $\omega_{e,F}$. Let $\omega_{e,F,m}$ denote the restriction of $\omega_{e,F}$ to the submanifold $\mathscr{H} = \frac{1}{2}m^2$. Then $\omega_{e,F,m}$ is a closed 2-form of rank 6 on a seven-dimensional manifold and so has a one-dimensional null foliation spanned by $\xi_{\mathscr{H}}$. The projection onto M of the integral curves of this null foliation are the world lines of the particles of mass m.†

Let us now examine the relativistic equations for a spinning particle. In three dimensions, we identified the spin of a particle at a point in space as an antisymmetric tensor; that is, as an element of $\wedge^2\mathbb{R}^3$. Under the identification of $\wedge^2\mathbb{R}^3$ with $\mathbb{R}^3$ we could regard the set of all spin vectors of constant length as a sphere, and hence were able to use the symplectic structure of the sphere. In four dimensions, we should regard the spin as an element of $\wedge^2(\mathbb{R}^{1,3})$ at each point. We must describe the analog of the sphere and the corresponding symplectic structure. To this end, we recall that for any vector space V carrying a nondegenerate scalar product we may identify $o(V)^*$ with $\wedge^2(V)$. Here $O(V)$ denotes the orthogonal group of V and $o(V)$ the Lie algebra of $O(V)$ and $o(V)^*$ the dual space of $o(V)$. The identification of $o(V)$ with $\wedge^2(V)$ is just the usual identification of infinitesimal orthogonal transformations with antisymmetric tensors of

† The rest of this should be omitted on a first reading.

degree 2. The identification of $o(V)$ with $o(V)^*$ comes from the scalar product on V (or if you like, from the Killing form of $o(V)$). Now if G is any Lie group and g its Lie algebra, then G acts on g via the adjoint representation and on g^*, the dual space of g by the contragredient to the adjoint representation, known as the coadjoint representation.

We shall prove in the next chapter that all of the orbits of G acting on g^* are symplectic manifolds; that is, if $l \in g^*$ and $\mathcal{O} = G \cdot l$, then $\mathcal{O}$ carries a canonical symplectic structure that is invariant under the action of G. Let us denote the symplectic form on $\mathcal{O}$ by $\Omega_{\mathcal{O}}$. If M is a (pseudo-) Riemannian manifold, let B_M denote the bundle of orthonormal frames of M. If $O(V)$ is the orthogonal group of the corresponding metric, then we can form the associated bundle $B_M \times \wedge^2 V/O(V)$, which may be identified with the bundle of antisymmetric 2-tensors $\wedge^2 TM$. With each orbit $\mathcal{O}$, we may form the associated bundle $B_M \times \mathcal{O}/O(V)$ that will be a submanifold of $\wedge^2 TM$, which we shall denote by $\wedge^2 TM_{\mathcal{O}}$. In the case of special relativity, where we may choose a global flat frame, we may use this frame to identify $\wedge^2 TM_{\mathcal{O}}$ with $M \times \mathcal{O}$. A different choice of frame modifies this identification by the action of an element of $O(V)$ on $\mathcal{O}$. Since $\Omega_{\mathcal{O}}$ is invariant under the action of $O(V)$, we see that $\Omega_{\mathcal{O}}$ is well defined as a 2-form on $\wedge^2 TM_{\mathcal{O}}$.

We can use the projection π: $T^*M \to M$ to pull the bundle $\wedge^2 TM_{\mathcal{O}}$ back to T^*M; in other words, we can consider it as a bundle over T^*M. Again, in the case of special relativity, where we can choose a globally defined flat frame on M, we may, using this frame, identify the pulled-back bundle with $T^*M \times \mathcal{O}$ and observe that the form $\Omega_{\mathcal{O}}$ is well defined on it. Finally, we may define the form

$$\omega_{e,F,\mathcal{O}} = \omega_{e,F} + \Omega_{\mathcal{O}}$$

on $T^*M \times \mathcal{O}$. Since $\omega_{e,F}$ and $\Omega_{\mathcal{O}}$ are each closed and nondegenerate, and since they involve different variables, it is clear that $\omega_{e,F,\mathcal{O}}$ is a well-defined symplectic structure on the pulled-back bundle $\pi^{\#} \wedge^2 TM_{\mathcal{O}}$. (Our construction made use of the global trivialization available in special relativity. Later on we shall see how to define the symplectic structure in general and shall see that the Riemannian curvature will enter.) In case $M = \mathbb{R}^3$, the orbits $\mathcal{O}$ are the spheres of radius s (and the origin corresponding to $s = 0$) and the construction reduces to the construction in Subsection B. For $M = \mathbb{R}^{1,3}$ we must examine the orbits of the Lorentz group acting on $\wedge^2(\mathbb{R}^{1,3})$. If $S \in \wedge^2(\mathbb{R}^{1,3})$ then $S \wedge S \in \wedge^4(\mathbb{R}^{1,3})$, which, up to choice of orientation, we may identify with $\mathbb{R}$. Thus $|S \wedge S|$ is a function on $\wedge^2(\mathbb{R}^{1,3})$ invariant under the action of the Lorentz group. Also, the scalar product on $\mathbb{R}^{1,3}$ induces a scalar product on $\wedge^2(\mathbb{R}^{1,3})$ and so $\|S\|^2$ is a second invariant function. It is easy to see that these invariant functions are independent and that, in fact, all the orbits are four-dimensional (except for $\{0\}$, which is zero-dimensional). Thus, for nonzero orbits, the symplectic manifold $\pi^{\#} \wedge^2 TM_{\mathcal{O}}$ is twelve-dimensional. Since, in the nonrelativistic limit, the symplectic manifold corresponding to a spinning particle was eight-dimensional, we must describe a procedure for cutting down four dimensions. Actually, as in the case of a particle without spin, we shall describe a presymplectic manifold whose dimension is one more than the dimension of the nonrelativistic limiting manifold, and whose null foliations project onto the world lines of the spinning particle. We thus want to get to a nine-dimensional manifold; that is, cut down three dimensions. This is done as follows: For any $p \in \mathbb{R}^{1,3}$ consider the constraint

$$*S \wedge p = 0.$$

For example, if $p = re_0$ this implies that S is a linear combination of $e_1 \wedge e_2, e_1 \wedge e_3$, and $e_2 \wedge e_3$. (Here e_0, e_1, e_2, e_3 is an orthonormal basis of $\mathbb{R}^{1,3}$.) In other words, "S is spinning in space in the rest frame of p." By readjusting the basis e_1, e_2, e_3 of space we can arrange that

$$S = se_2 \wedge e_3$$

so that $S \wedge S = 0$ and $\|S\|^2 = s^2$. Thus, for $\|p\|^2 > 0$, the condition $*S \wedge p = 0$ requires us to restrict attention to orbits satisfying $S \wedge S = 0$. We assume that $\|S\|^2 = s^2 > 0$.

Now S is an antisymmetric 2-tensor, as is F. We can therefore form their scalar product $S \cdot F$, which is a scalar-valued function. Let $f > 0$ be any differentiable function of a real variable and consider the submanifold $\mathscr{V}_9$ defined by the equations

$$*S \wedge p = 0 \qquad \text{and} \qquad \mathscr{H}(q, p) = f(eS * F).$$

We claim that $\mathscr{V}_9$ is nine-dimensional. Indeed, since for fixed q, these equations are invariant under the action of the Lorentz group acting on p and S, it is enough to show that the fiber of $\mathscr{V}_9$ over q is five-dimensional, and for this it suffices to show that the isotropy group of some point (q, p, S) of $\mathscr{V}_9$ is a one-dimensional subgroup of the Lorentz group. Since $f > 0$ we can choose $p = re_0$ and hence $S = se_2 \wedge e_3$. It is clear that the isotropy group consists of those orthogonal transformations of space which fix the e_1 axis, and so is one-dimensional. For generic f the restriction of $\omega_{e,F,\mathcal{O}}$ to $\mathscr{V}_9$ has rank 8, and so its null foliation is one-dimensional. The corresponding integral curves describe the motion of a spinning particle.

Notice that the notion of a Lagrangian has been entirely eliminated and the notion of "force" has also been almost completely removed from the theory. It only makes its appearance in the form of inertial forces in writing down $\mathcal{H}$. As we shall see in Chapter III, this last vestige can also be removed, and the entire classical theory of particle motion can be reduced to the construction of presymplectic manifolds.

Epilogue: Why symplectic geometry?

We hope that in the course of this long chapter we have convinced the reader that symplectic geometry plays a crucial role in the formulation of many branches of classical and quantum physics. The next chapter will be highly mathematical – we will spend a lot of effort in developing the mathematical interactions of group theory and symplectic geometry. So it might not be a bad idea to pause here with a more speculative and philosophical section in which we try to present several (probably not all) of the more "fundamental" physical principles that give rise to symplectic geometry. Of course, in any such discussion, one must have some prejudice about what is regarded as a fundamental principle in physics.

Three principles, each quite different from the others, all giving rise to the same systems of Hamilton's equations, are (i) the variational principle (ii) the method of high-frequency approximation, and (iii) the principle of general covariance. Let us illustrate what we mean by each of these principles for the case of the motion of a particle in a Riemannian (or pseudo-Riemannian) manifold. The equations of motion in this case are the Hamiltonian equations corresponding to the function $\mathcal{H}(q, p) = \frac{1}{2}\|p\|^2$ on T^*X, which X is a (pseudo-)Riemannian manifold. The solution curves are geodesics of M (traversed so as to maintain constant (kinetic) energy). The question is: Why these equations – Why geodesics?

Our first "fundamental" principle, (i), says that what is at issue is a variational problem; the Hamilton equations are the Euler–Lagrange equations for the integral $\int L(q, \dot{q})\, ds$ along curves, where $\dot{q}$ denotes the tangent vector to the curve and $L(q, \dot{q}) = \frac{1}{2}\|\dot{q}\|^2$. As the Euler–Lagrange equations are exactly the same as Hamilton's equations, it is difficult to see why the variational principle, taken alone, should be regarded as more basic than the symplectic formulation. Should we regard Fermat's principle, or de Maupertuis's principle as being more fundamental than the basic laws of refraction or the basic equations of mechanics. One currently popular reason for believing so is the existence of a "more exact theory," in which the integral we are varying, for example $I(\gamma) = \int_\gamma L(q, \dot{q})\, ds$, plays an important role. For example, in the Feynman path integral formulation of quantum mechanics, one wants to consider the "integral" of the expression $e^{iI(\cdot)}$ over "all paths." The mathematical difficulties in giving precise meaning to this integral and its generalizations, with "paths" replaced by "fields" are well known. Nevertheless, as of this writing, the path integral method seems to be the universally accepted approach – the problem of finding the basic laws of physics being formulated as the problem of discovering the "true Lagrangian." We refer the reader to the end of Chapter IV of our book (Guillemin and Sternberg 1977) for a study of the relations between the variational calculus and symplectic geometry.

(ii) In the high-frequency approximation, the Hamiltonian equations described above are the bicharacteristic equations for the wave equation on $M \times \mathbb{R}$ (ibid.) Similarly, as indicated in Figure 1.3, if we apply the high-frequency approximation to Maxwell's equations, we obtain geometrical optics as the zeroth-order approximation (the bicharacteristic equation) and a certain portion of the wave phenomena including polarization as the first-order high-frequency approximation (the transport equation and conditions). We have explained this in great detail in Guillemin and Sternberg, 1977, so there is no need to go over the material here. In this context, the equations of motion of symplectic geometry are justified, as we indicated in our introductory paragraphs, as an approximate theory to a more precise theory. Geometrical optics is an approximation to Maxwell's equations. But, of course, this opens the question of why Maxwell's equations? What are the fundamental physical principles that give rise to the more precise theory?

(iii) A third way of getting the equations of geodesics was proposed by Einstein, Infeld, and Hoffman (1938) and extended by Souriau (1974) to incorporate the equations for the Lorentz force. We gave a general formulation of this principle in Guillemin and Sternberg (1978). In Chapter III we will provide a precise mathematical formulation and show how it gives rise to the equations of motion of a classical particle in the presence of a Yang–Mills field as well as to other general applications. We will present a sketch of the general formulation of this principle of general covariance here and explain what it means for the case of general relativity.

Let G be a group acting on a space X. In what follows, we shall pretend that G is a Lie group acting on a manifold X. (In the applications that we have in mind, however, G and X will be infinite-dimensional; though we do not want to go into the mathematical subtleties of defining their differential structure.) For example, let M be a finite-dimensional (say, four-dimensional) manifold, and let X be the space of all pseudo-Riemannian metrics (of signature $+ - - -$) on M. Let $G = \mathrm{Diff}_0(M)$ be the group of all diffeomorphisms of M of compact support. (A diffeomorphism has compact support if it equals the identity outside some compact subset.) The action of G on X in this case is the obvious one: A diffeomorphism φ sends the metric γ into $\varphi^{-1*}\gamma$.

In general, by a covariant theory associated to the action of G on X we shall mean a smooth function F, defined on X, that is invariant under the action of G. Thus

$$F(ax) = F(x) \qquad \forall a \in G.$$

Let x be a point of X and let $B = G \cdot x$ be the G orbit through x. Let μ be an element of T^*X_x so that μ is a linear function on the tangent space, TX_x to X at x. If

$$\mu = dF_x$$

then it is clear that μ vanishes when evaluated on all vectors tangent to B; that is,

$$\mu \in TB_x^0. \tag{I}$$

This innocent-looking equation contains within it the equations for geodesics! Let us explain how. In the case where X is the space of Riemann metrics or pseudo-Riemannian metrics, the tangent space TX_x can be identified with the space of smooth symmetric tensor fields of second order. Let Z denote the space of such smooth symmetric tensor fields. (Notice, and this will be important shortly, that we have identified TX_x with this fixed space Z, which is independent of x – we have "trivialized" the tangent bundle TX.) When we are dealing with infinite-dimensional spaces, we might want to allow our μ to be defined only on some subspace of TX_x. Here, for example, we will consider the subspace Z_0 of smooth tensor fields of compact support. A (continuous) linear function on Z is a "generalized" of "distributional" tensor field. We say that $\mu = \mu_T$ is smooth if there is a smooth symmetric tensor field T such that

$$\langle \mu, v \rangle = \int_M T \cdot v \, d(\mathrm{vol})_x \qquad \forall v \in Z_0,$$

where $T \cdot v$ denotes the contraction of the symmetric tensor fields T and v and where $d(\mathrm{vol})_x$ is the volume measure on M associated to the metric x. For such smooth μ_T, it turns out that condition (I) is equivalent to the condition that the covariant divergence of T vanish. We say that μ is (smoothly) concentrated along a curve if there is a curve C and if there is a (smooth) tensor field T, defined on the curve C, such that

$$\langle \mu, v \rangle = \int_C T \cdot v \, ds \qquad \forall v \in Z_0.$$

It turns out that for μ concentrated along a curve C, condition (I) implies that C must be a geodesic! Thus the equations for geodesics are a consequence of (I). We shall provide the details of this remarkable result in Chapter III.

Let us explore a little further the significance of (I). Let us suppose that the tangent bundle TX has a given global trivialization. (As we have seen, this is the case for general relativity and will be the case for many of the examples to be considered later on.) We thus assume that we are given a definite identification of each of the tangent spaces TX_x with a fixed vector space Z and that this yields a diffeomorphism of the tangent bundle, TX with $X \times Z$. Then, given $\mu \in Z^*$, we can try to find a point x in X such that

$$dF_x = \mu. \tag{S}$$

As we have remarked, since we are dealing with infinite-dimensional spaces, we may want to consider situations where μ will not be defined as a continuous functional on all of Z but only on some subspace Z_0. Similarly, it may be that F will not be defined on all of X, yet (S) makes good sense when both sides are regarded as linear functions on Z_0. For example, in the case of general relativity, where the points of X are pseudo-Riemann metrics $x = g_{ij}$, Einstein chose as his function F the function given up to constant by

$$F(x) = \int_M g_{ij} R_{ij}(x) d(\mathrm{vol})_x, \tag{E}$$

where $R_{ij}(x)$ is the Ricci curvature of the metric x. (We will discuss similar computations in detail in Chapter III, so the reader need not follow these computations in detail here or even know what the Ricci curvature is in order to follow the general discussion.) There is no reason to expect that the integral written in (E) will converge for all metrics x and so, strictly speaking, F is not defined on all of X. On the other hand, since F is given by an integral of a local expression in x, the formal expression for $dF_x(v)$ will also be given by the integral of a local expression involving x and v; in fact, a certain amount of computation will show that

$$dF_x(v) = \int_M \tfrac{1}{2} T \cdot v \, d(\mathrm{vol})_x,$$

where

$$T_{ij}(x) = R_{ij}(x) - \tfrac{1}{2}R(x)g_{ij},$$

where $R(x)$ is the scalar curvature of the metric $x = g_{ij}$. If $v \in Z_0$, the integral in the expression for $dF_x(v)$ converges, since v, and hence the integrand, has compact support. Thus $dF_x(v)$ makes sense for all x and all $v \in Z_0$, even though $F(x)$ may not be defined. If we take μ to be a smooth element of Z_0^*, then (S) reduces in this case to

$$R_{ij} - \tfrac{1}{2}Rg_{ij} = T_{ij},$$

which are exactly the Einstein field equations if we regard T_{ij} as the Einstein energy-momentum tensor. So, if we regard μ as representing the distribution of matter, then (I) and hence the equations for geodesics, emerges as a necessary condition for the solution of (S). This was the original idea of Einstein, Infeld, and Hoffman. (Of course, since their paper appeared before the advent of distributions or generalized functions, their actual formulation was a little more cumbersome.)

Let us pause to examine some of the virtues and faults of the Einstein–Infeld–Hoffman approach. One problem is, of course, the question of whether we expect to be able to solve (S), which is a nonlinear partial differential equation, with distributional data. Let us put this nontrivial mathematical problem aside. A second, more serious problem is that equation (S) says that we are using μ to determine the metric, x; that is, regarding μ as the source for x. In the applications described above, we assumed that x was given and fixed: We asked for the motion of a particle in the presence of a given gravitational or electromagnetic field. In effect, what we see is that the applications we developed were an approximation – a "passive" approximation to the true physical laws: In reality, a moving electron produces its own electromagnetic field; a moving massive particle produces its own gravitational field. The passive approximation assumes that we can ignore the field produced by the particle, and treat (I) as a condition at a given gravitational field x.

One of the principal virtues of condition (I) is that it does not involve the function F. It is the same for all choices of F. Thus, for example, Einstein's function F in general relativity is, in some sense, the simplest local invariant function of the metric that one can write down. There are, of course, infinitely many more complicated local invariant expressions, involving higher covariant derivatives of the curvature, for example. Einstein himself was unsure whether or not the "cosmological constant" should be included in F. The principle of general covariance does not tell us which function to choose (or that the function need be local). But no matter which function we choose, and hence no matter how we modify the field equations, the necessary conditions (I) remain the same. In this sense, the passive equations of physics have a more general validity than the source equations.

A serious objection to the whole procedure is that it is too perfect. It gives exactly the classical equations of motion of a point particle from the simplest of geometrical principles. Yet we know that the classical equations are wrong – they are only an approximation to the quantum mechanical equations. The situation becomes even more complicated when we try to combine methods (ii) and (iii). Thus, for example, we would regard Maxwell's equations as an approximation to (and/or part of) a source equation (S), valid over a flat region of space-time. We would then get the classical equation of the photon (including polarization and some wave-optics effects) as the high-frequency approximation to a linear approximation to (part of) (S) while the classical equations of motion of a charged particle would be given by the passive approximation (I). (Similarly, the classical equations for the "graviton" and/or the "gluon" would be high-frequency approximations to linearized approximations to (S) while the particle motion of the corresponding "matter fields" would come from (I).) It is somewhat amazing that all of this leads to a symplectic formulation. But what is clearly needed is some extension of the principle of general covariance that includes quantum phenomena. We shall make some attempts in that direction in Chapter III.

II
The geometry of the moment map

This chapter is primarily devoted to the study of the moment map, the generalization to arbitrary Lie groups of the classical notions of total linear or angular momentum. We first review some general facts concerning symplectic geometry. We begin by showing that all finite-dimensional symplectic vector spaces of the same finite dimension look alike, and then we show that, locally, all finite-dimensional symplectic manifolds of the same dimension look alike. We then give some important examples of symplectic manifolds.

21. Normal forms

This section is devoted to the study of the properties of an antisymmetric form b on a vector space V. Unless otherwise specified, all vector spaces will be finite-dimensional and over the real numbers. In most of what follows the ground field is irrelevant, so that most of the results hold true for finite-dimensional vector spaces over any field. Also, for the most part, the results extend to various infinite-dimensional (topological) vector spaces, when appropriate care is taken to formulate the correct notions involving the continuity of the bilinear forms and closure of the various associated subspaces.

Let V be a finite-dimensional real vector space and V^* its dual space. Let $\wedge^2(V^*)$ denote the space of exterior forms (antisymmetric covariant tensors). Thus elements of $\wedge^2(V^*)$ are linear combinations of elements of the form $l \wedge m$, where $l \in V^*$ and $m \in V^*$. Each such $l \wedge m$ defines a bilinear form on V by $(l \wedge m)(v_1, v_2) = l(v_1)m(v_2) - l(v_2)m(v_1)$ and this bilinear form is antisymmetric. Since each element of $\wedge^2(V^*)$ is a linear combination of such $l \wedge m$, and since a linear combination of antisymmetric bilinear forms is again an antisymmetric bilinear form, each element of $\wedge^2(V^*)$ defines an antisymmetric bilinear form on V. It is a theorem in linear algebra that these represent the most general antisymmetric bilinear forms. (This is an example of a theorem that is *not* true in infinite dimensions.) We may thus regard the most general antisymmetric bilinear form, b as an element of $\wedge^2(V^*)$.

Let $e_1,\ldots,e_n$ be a basis of V. We now established a normal form for antisymmetric bilinear form analogous to the (diagonal) canonical form for symmetric quadratic forms. Let $e^{1*},\ldots, e^{n*}$ be the dual basis of V^*. It is a theorem of linear algebra that the $\{e^{i*} \wedge e^{j*}\}_{i<j}$ form a basis of $\wedge^2(V^*)$. Thus we may write the most general antisymmetric bilinear form b as

$$b = \sum_{i<j} a_{ij} e^{i*} \wedge e^{j*}$$

with suitable coefficients a_{ij}.

Theorem 21.1. *For any* $\mathrm{b} \in \wedge^2(\mathrm{V}^*)$ *there exists a basis* $\{\mathrm{f}^{1*},\ldots,\mathrm{f}^{n*}\}$ *of* V^* *so that*

$$b^* = f^{1*} \wedge f^{2*} + \cdots + f^{2(p-1)*} \wedge f^{2p*}. \tag{21.1}$$

This integer p *depends only on* b.

Proof. Suppose that b is given by

$$b = \sum_{i<j} a_{ij} e^{i*} \wedge e^{j*}$$

in terms of the basis $\{e^{i*}\}$. If $b \neq 0$ then we can (by rearranging the basis if necessary) assume that $a_{12} \neq 0$. Then

$$b = \left(e^{1*} - \frac{a_{23}}{a_{12}} e^{3*} - \cdots - \frac{a_{2n}}{a_{12}} e^{n*} \right)$$
$$\wedge (a_{12}e^{2*} + a_{13}e^{3*} + \cdots + a_{1n}e^{n*}) + b^1,$$

where b^1 does not contain any expression involving e^{1*} or e^{2*}. If we set

$$f^{1*} = e^{1*} - \frac{a_{23}}{a_{1n}} e^{3*} - \cdots - \frac{a_{2n}}{a_{12}} e^{n*}$$

and

$$f^{2*} = a_{12}e^{2*} + a_{13}e^{3*} + \cdots + a_{1n}e^{n*},$$

then $f^{1*}, f^{2*}, e^{3*}, \ldots, e^{n*}$ are linearly independent and

$$b = f^{1*} \wedge f^{2*} + b^1.$$

If $b^1 = 0$, the proof is complete. Otherwise we continue the process until we achieve the desired normal form.

The integer p can be described solely in terms of b as follows: The form b defines a linear map of $V \to V^*$ sending $v \in V$ into the linear function $b(v, \cdot)$, where $b(v, \cdot)w = b(v, w)$ for all $w \in V$. Let $E(b) \subset V^*$ denote the range of this linear map of $V \to V^*$. If $g_1, \ldots, g_n$ is any basis of V, then the linear forms $b(g_1, \cdot), \ldots, b(g_n, \cdot)$ span $E(b)$. In particular, if $f_1, \ldots, f_n$ denotes the basis of V dual to the basis $f^{1*}, \ldots, f^{n*}$ of V^* constructed above, then it is clear that

$$b(f_1) = f^{2*}, \qquad b(f_2) = -f^{1*}, \ldots, b(f_{2p-1}) = f^{2p*}, \qquad b(f_{2p}) = -f^{2p-1*},$$

while

$$b(f_k) = 0 \qquad \text{for} \qquad k > 2p.$$

Thus $f^{1*}, \ldots, f^{2p*}$ form a basis of $E(b)$ and

$$2p = \dim\big(E(b)\big) \tag{21.2}$$

shows that p is defined purely in terms of b and is independent of the specific choice of basis. The integer $2p$ is called the *rank* of b. The theorem shows that there are only two invariants of an antisymmetric bilinear form b on a vector space V, the integers $n = \dim V$ and $2p = \operatorname{rank} b$. The form b is called a *symplectic bilinear form* if it is nondegenerate; that is, if the map of $V \to V^*$ is an isomorphism, or, what amounts to the same thing,

$$b(v, w) = 0 \quad \forall w \Rightarrow v = 0.$$

Then $2p = n$. A pair (V, b) consisting of a vector space V and a symplectic form b is called a symplectic vector space (or a vector space with a symplectic structure). We sometimes refer to V as a symplectic vector space when b is understood. As a corollary of Theorem 21.1 we get

Theorem 21.2. *Let* (V, b) *be a symplectic vector space. Then dim* V *is even, say dim* $\mathrm{V} = 2\mathrm{m}$, *and we can find a basis* $\mathrm{u}_1, \ldots, \mathrm{u}_\mathrm{m}, \mathrm{v}_1, \ldots, \mathrm{v}_\mathrm{m}$ *of* V *such that*

$$b(u_i, u_j) = b(v_i, v_j) = 0 \qquad \text{and} \qquad b(u_1, v_j) = \delta_{ij}. \tag{21.3}$$

In particular, all symplectic vector spaces of the same dimension "look alike."

Let us return to the study of a general antisymmetric bilinear form b on a vector space V. For any subspace W we let $W^\perp$ denote the orthogonal subspace of W relative to b so that

$$W^{\perp} = \{u | b(u, w) = 0 \quad \forall w \in W\}.$$

In particular, $V^{\perp}$ is the kernel of the associated map of V into V^*. For any W we have

$$\dim W^{\perp} = \dim V - \dim b(W, \cdot),$$

where $b(W, \cdot)$ is the image of W in V^* under the associated map of $V \to V^*$. Now

$$\dim b(W, \cdot) = \dim W - \dim W \cap V^{\perp},$$

since $W \cap V^{\perp}$ is the kernel of the map of $W \to V^*$. So

$$\dim W + \dim W^{\perp} = \dim V + \dim W \cap V^{\perp}. \tag{21.4}$$

A subspace is called

$$\begin{array}{ll} \text{isotropic} & \text{if } W \subset W^{\perp}, \\ \text{coisotropic} & \text{if } W^{\perp} \subset W, \\ \text{Lagrangian} & \text{if } W = W^{\perp}. \end{array}$$

If follows from (21.4) that

$$\text{if } W \text{ is isotropic, then } \dim W \leqslant \tfrac{1}{2}(\dim V + \dim V^{\perp});$$

$$\text{if } W \text{ is coisotropic, then } \dim W \geqslant \tfrac{1}{2}(\dim V + \dim V^{\perp});$$

$$\text{if } W \text{ is Lagrangian, then } \dim W = \tfrac{1}{2}(\dim V + \dim V^{\perp}).$$

For example, suppose that (V, b) is symplectic and we choose a basis as in Theorem 21.2. Then any subspace spanned by some of the u's (but none of the v's) is isotropic, as is any subspace spanned by only v's (but none of the u's). Any subspace spanned by *all* the u's and some v's (or *all* the v's and some u's) is coisotropic. The subspace spanned by *all* the u's but no v's (or *all* the v's but no u's) is Lagrangian.

Returning to the general case, suppose that W is an isotropic subspace that is not Lagrangian so that $W \subset W^{\perp}$ is a strict inclusion. Choose some $u \in W^{\perp} - W$. Since b is antisymmetric, $u \in u^{\perp}$. On the other hand, $u \in W^{\perp}$, so $W \subset u^{\perp}$. Thus the space $W + \{u\}$ is again isotropic. Proceeding in this way, we obtain a Lagrangian subspace. We have thus proved

Theorem 21.3. *Every isotropic subspace is contained in a Lagrangian subspace. In particular, starting with $\{0\}$, we notice that there exist Lagrangian subspaces. An isotropic subspace is Lagrangian if and only if it is a maximal isotropic subspace.*

As corollaries to the above arguments we see that the following also characterize Lagrangian subspaces:

(i) *An isotropic subspace is Lagrangian if and only if it is coisotropic;*

(ii) *An isotropic subspace* W *is Lagrangian if and only if dim* $W = \frac{1}{2}$ *(dim* V + *dim* $V^{\perp}$).

We need not add that b is symplectic if and only if $V^{\perp} = \{0\}$, in which case dim $V^{\perp}$ disappears from all the preceding formulas.

Let W be a coisotropic subspace. Then b defines a nondegenerate antisymmetric bilinear form on $W/W^{\perp}$ so that each coisotropic subspace W has an associated symplectic vector space $W/W^{\perp}$. In particular, this applies to V itself. For this reason a vector space with an antisymmetric bilinear form is sometimes called a presymplectic vector space. This process of passing from a coisotropic subspace of a symplectic vector space to produce a new, quotient symplectic vector space will be of great importance to us. We will see how it works at the level of symplectic manifolds, where it is called "reduction" and produces new symplectic manifolds from old ones.

22. The Darboux–Weinstein theorem

In this section we prove that locally all symplectic manifolds of the same finite dimension look alike. This fact was first proved by Darboux using induction on the dimension (see Sternberg 1983 for a presentation of Darboux's proof). Here we present a proof due to Weinstein (which is valid in many infinite-dimensional cases). The proof given here also yields an equivariant version of the theorem relative to a compact group action. The first part of this section is an almost verbatim transcription from a previous work of ours (1977, Chap. IV, Sect. 1).

Let X be a manifold and Y an embedded submanifold. If σ is a differential form on X, we shall let σ_{Y} denote the restriction of σ to $\wedge (TX)_{|Y}$. (Thus $\sigma_{|Y}$ can be evaluated on vectors that are not necessarily tangent to Y.)

Theorem 22.1 (Darboux–Weinstein). *Let* Y *be a submanifold of* X *and let* ω_0 *and* ω_1 *be two nonsingular closed 2-forms on* X *such that* $\omega_{0|Y} = \omega_{1|Y}$. *Then there exists a neighborhood* U *of* Y *and a diffeomorphism* $f: U \to X$ *such that*

(i) $f(y) = y$ *for all* $y \in Y$,

(ii) $f^* \omega_1 = \omega_0$.

If G *is a compact group acting on* X *with* ω_0 *and* ω_1 *both invariant under the action of* G *and if the submanifold* Y *is invariant under the action of* G, *then the map* f *can be chosen to commute with the action of* G.

If we take Y to be a point, the theorem asserts that if the two forms agree on the tangent space at a point, then up to a diffeomorphism they agree in some neighborhood of the point. We can use the exponential map of some Riemann metric to give a local diffeomorphism of some neighborhood of the origin in the tangent space at a point p with some neighborhood of p in X. By the results of the preceding section, we know that any two symplectic bilinear forms on the tangent space at p are equivalent by a linear transformation. Combining this with Theorem 22.1, we see that given any two symplectic forms ω_0 and ω_1 defined near p, there is a diffeomorphism g defined near p with $g(p) = p$ and $g^*\omega_1 = \omega_0$. Thus all symplectic forms look alike locally. If a compact group acts on X, and p is a fixed point, we can choose the Riemann metric to be invariant under G by averaging over the group, and also choose the linear equivalence to commute with G. Then Theorem 22.1 implies the following equivariant Darboux's theorem:

Theorem 22.2. *Let* G *be a compact group acting on* X *and let* $x \in X$ *be a fixed point of* G. *Let* ω_0 *and* ω_1 *be* G*-invariant symplectic forms on* X. *Then there exists a* G*-invariant neighborhood* U *of* x *and a* G*-equivariant diffeomorphism* f *of* U *into* X *with* $f(x) = x$ *and* $f^*\omega_1 = \omega_0$.

For the proof of Theorem 22.1 we need a basic formula of differential calculus, which we now recall. Let W and Z be differentiable manifolds and let $\varphi_t\colon W \to Z$ be a smooth one-parameter family of maps of W into Z. In other words, the map $\varphi\colon W \times I \to Z$ given by $\varphi_t(w) = \varphi(w, t)$ is smooth. Then we let ξ_t denote the tangent field along φ_t; that is, $\xi_t\colon W \to TZ$ is defined by letting $\xi_t(w)$ be the tangent vector to the curve $\varphi(w, \cdot)$ at t. If σ is a differential $k + 1$-form on Z, then $\varphi_t^* i(\xi)\sigma$ is a well-defined differential k form on W given by

$$\varphi_t^* i(\xi)\sigma\,(\eta_1, \ldots, \eta_k) = \big(i(\xi_t(w)\sigma)\,(d\varphi_t\,\eta_1, \ldots, d\varphi_t\eta_k).$$

(Note that since ξ_t is not a vector field on Z the expression $i(\xi_t)\sigma$ does not define a differential form on Z.)

Let σ_t be a smooth one-parameter family of forms on Z. Then $\varphi_t^*\sigma_t$ is a smooth family of forms on W and the basic formula of the differential calculus of forms asserts that

$$\frac{d}{dt}\varphi_t^*\sigma_t = \varphi_t^*\frac{d\sigma_t}{dt} + \varphi_t^*\big(i(\xi_t)\,d\sigma_t\big) + d\varphi_t^* i(\xi_t)\sigma_t. \tag{22.1}$$

For the sake of completeness we shall present a proof of this formula at the end of this section.

Let $Y \subset X$ be an embedded submanifold and suppose that there exists a smooth retraction φ_t of X onto Y. Thus we assume that φ_t is a smooth family of maps of $X \to X$ such that

$$\varphi_0 : X \to Y \qquad \text{and} \qquad \varphi_1 = \text{id},$$

and

$$\varphi_t y = y \quad \forall y \in Yt.$$

Notice that if X were a vector bundle and Y were the zero section, then multiplication by t would provide such a retraction (also if X were a convex open neighborhood of the zero section). By choosing a Riemann metric and using the exponential map on the normal bundle of Y, we can thus arrange that some neighborhood of Y has a differentiable retraction onto Y. Then, for any form σ on X we have (in some neighborhood of Y)

$$\sigma - \varphi_0^* \sigma = \int_0^1 \frac{d}{dt}(\varphi_t^* \sigma)\, dt = \int_0^1 (\varphi_t^*(i(\xi_t)\, d\sigma))\, dt + d \int_0^1 (\varphi_t^*(i(\xi_t)\sigma))\, dt$$

$$= I\, d\sigma + dI\sigma, \tag{22.2}$$

where we have set

$$I\beta = \int_0^1 [\varphi_t^*(i(\xi_t)\sigma)]\, dt$$

for any form β on X. In other words $I: \wedge^k (X) \to \wedge^{k-1} (X)$ and

$$\sigma - \varphi_0^* \sigma = dI\sigma + I\, d\sigma.$$

If a compact group G acts on X and leaves Y invariant, we can choose an invariant Riemann metric on X, and then the exponential map and hence the retraction φ_t will commute with the action of G and hence so will I. In other words,

$$Ia^*\beta = a^*I\beta \qquad \forall\, a \in G.$$

Proof of the Darboux–Weinstein theorem. Set

$$\omega_t = (1 - t)\omega_0 + t\omega_1 = \omega_0 + t\sigma,$$

where $\sigma = \omega_1 - \omega_0$. Notice that $\sigma_{|Y} = 0$, so that, in particular,

$$\varphi_0^* \sigma = 0 \qquad \text{and} \qquad d\sigma = 0.$$

Hence, by (22.2), $\sigma = d\beta$, where $\beta = I\sigma$. Again, we see that $\beta_{|Y} = 0$. Now

$$\omega_{t|Y} = \omega_{0|Y} = \omega_{1|Y},$$

and so $\omega_{t|Y}$ is nondegenerate for all $0 \leqslant t \leqslant 1$. We can therefore find some neighborhood of Y on which ω_t is nondegenerate for all $0 \leqslant t \leqslant 1$. We can therefore find a vector field $\boldsymbol{\eta}_t$ such that

$$i(\boldsymbol{\eta}_t)\omega_t = -\beta \tag{22.3}$$

In the case of a compact group action with ω_0 and ω_1 G-invariant, then σ is G-invariant and we can choose the operator I to commute with the action of G and hence arrange that β is G-invariant. Then the unique $\boldsymbol{\eta}_t$ satisfying (22.3) is G-invariant. We can integrate the time-dependent vector field $\boldsymbol{\eta}_t$ to obtain a one-parameter family of maps f_t with $f_0 = \mathrm{id}$, whose tangent vector is $\boldsymbol{\eta}_t$. In the group case, the maps f_t all commute with the action of G. Notice that $f_{t|Y} = \mathrm{id}$. By restricting to a smaller neighborhood of Y we may assume that f_t is also defined for all $0 \leqslant t \leqslant 1$. (Strictly speaking, in proving this fact, we may want $\boldsymbol{\eta}_t$, etc., to be defined for some range of $t > 1$.) Then $f_0 = \mathrm{id}$ and, by (22.1) and the fact that

$$\frac{d}{dt}\,\omega_t = \sigma,$$

we see that

$$f_1^*\omega_1 - \omega_0 = \int_0^1 \frac{d}{dt}(f_t^*\omega_t)\,dt = \int_0^1 f_t^*(\sigma + d(i(\boldsymbol{\eta}_t)\omega_t))\,dt = 0$$

since $d\omega_t = 0$. Thus f_1 provides the desired diffeomorphism, proving the theorem.

Let us now give a proof of (22.1). We first prove the formula in the special case where $W = Z = M \times I$ and φ_t is the map ψ_t: $M \times I \to M \times I$ given by

$$\psi_t(x, s) = (x, s + t).$$

The most general differential form on $M \times I$ can be written as

$$ds \wedge a + b,$$

where a and b are forms on M that may depend on t and s. (In terms of local coordinates, $s, x^1, \ldots, x^n$, these forms are sums of terms that look like

$$c\,dx^{i_1} \wedge \cdots \wedge dx^{i_k},$$

where c is a function of t, s, and x.) To show the dependence on x and s we shall rewrite the above expression as

$$\sigma_t = ds \wedge a(x, s, t)\,dx + b(x, s, t)\,dx.$$

With this notation it is clear that

$$\psi_t^*\sigma_t = ds \wedge a(x, s + t, t)\,dx + b(x, s + t, t)\,dx$$

and therefore

$$\frac{d\psi_t^*\sigma_t}{dt} = ds \wedge \frac{\partial a}{\partial s}(x, s + t, t)\, dx + \frac{\partial b}{\partial s}(x, s + t, t)\, dx$$

$$+ ds \wedge \frac{\partial a}{\partial t}(x, s + t, t)\, dx + \frac{\partial b}{\partial t}(x, s + t, t)\, dx,$$

so that

$$\frac{d\psi_t^*\sigma_t}{dt} - \psi_t^*\left(\frac{d\sigma_t}{dt}\right) = ds \wedge \frac{\partial a}{\partial s}(x, s + t, t)\, dx + \frac{\partial b}{\partial s}(s, s + t, t)\, dx. \quad (22.4a)$$

It is also clear that in this case the tangent to $\psi_t(x, s)$ is $\partial/\partial s$ evaluated at $(x, s + t)$.

In this case $\partial/\partial s$ is a vector field and

$$i\left(\frac{\partial}{\partial s}\right)\sigma_t = a\, dx$$

so

$$\psi_t^*\left(i\left(\frac{\partial}{\partial s}\right)\sigma_t\right) = a(x, s + t, t)\, dx$$

and therefore

$$d\psi_t^*\left(i\left(\frac{\partial}{\partial s}\right)\sigma_t\right) = \frac{\partial a}{\partial s}(x, s + t, t)\, ds \wedge dx + d_x a(x, s + t, t)\, dx \quad (22.4b)$$

(where d_x denotes the exterior derivative of the form $a(x, s + t, t)\, dx$ on the manifold M, holding s fixed). Similarly,

$$d\sigma_t = -ds \wedge d_x a\, dx + \frac{\partial b}{\partial s}\, ds \wedge dx + d_x b\, dx$$

so

$$i\left(\frac{\partial}{\partial s}\right) d\sigma_t = -d_x a\, dx + \frac{\partial b}{\partial s}\, dx$$

and

$$\psi_t^* i\left(\frac{\partial}{\partial s}\right) d\sigma_t = -d_x a(x, s + t, t)\, dx + \frac{\partial b}{\partial s}(x, s + t, t)\, dx. \quad (22.4c)$$

Adding equations (22.4a–c) proves (22.1) for ψ_t.

Now let $\varphi\colon W \times I \to Z$ be given by $\varphi(w, s) = \varphi_s(w)$. Then the image under φ of the lines parallel to I through w in $W \times I$ are just the curves $\varphi_s(w)$

in Z. In other words

$$d\varphi\left(\frac{\partial}{\partial s}\right)_{(w,t)} = \xi_t(w).$$

If we let $\iota\colon W \to W \times I$ be given by $\iota(w) = (w, 0)$, then we can write the map φ_t as $\varphi \circ \psi_t \circ \iota$. Thus $\varphi_t^* \sigma_t = \iota^* \psi_t^* \varphi^* \sigma_t$ and, since ι and φ do not vary with t,

$$\frac{d}{dt}\,\varphi_t^* \sigma_t = \iota^* \frac{d}{dt}\,\psi_t^*(\varphi^* \sigma_t).$$

At the point w, t of $W \times I$, we have

$$i\left(\frac{\partial}{\partial s}\right)\varphi^* \sigma_t = (d\varphi)^* \left\{\left(i\left(d\varphi\,\frac{\partial}{\partial s}\right)\sigma_t\right)\right\} = (d\varphi)^* i(\xi_t)\sigma_t$$

and thus

$$\iota^* \psi_t^* \left(i\left(\frac{\partial}{\partial s}\right)\varphi^* \sigma_t\right) = \iota^* \psi_t^* \varphi^*(i(\xi_t)\sigma_t) = \varphi_t^*(i(\xi_t)\sigma_t).$$

Substituting into the formula for $d/dt[\psi_t^* \varphi^* \sigma_t]$ yields (22.1).

23. Kaehler manifolds†

Although Darboux's theorem says that all symplectic manifolds are locally alike, in actual practice a symplectic manifold usually comes equipped with some additional geometric structure. The most frequently encountered types of symplectic manifolds are cotangent bundles, coadjoint orbits, or Kaehler manifolds. We shall explore each of these types in the course of the book. Here we focus on Kaehler manifolds.

Let V be a $2n$-dimensional vector space over the real numbers. A *complex structure* on V means a real linear transformation, $J\colon V \to V$ such that $J^2 = -I$. This makes V into an n-dimensional complex vector space by declaring

$$(x + iy)v = xv + yJv, \qquad x, y \in \mathbb{R}, v \in V.$$

Conversely, if V is an n-dimensional complex vector space, it can be considered as a $2n$-dimensional real vector space and we take J to be the (real) linear transformation consisting of multiplication by i. Then $J^2 = -I$.

Definition. *A Kaehlerian vector space is a symplectic space* V *with symplectic form ω and a complex structure* J *such that* J $\in$ Sp(V).

† This section can be omitted without loss of continuity.

On the Kaehlerian vector space (V, ω, J), define the real bilinear form b by

$$b(v, w) = \omega(v, Jw). \tag{23.1}$$

We note that

$$\begin{aligned} b(w, v) &= \omega(w, Jv) \\ &= \omega(Jw, J^2 v) && \text{since } J \in Sp(V) \\ &= -\omega(Jw, v) && \text{since } J^2 = -I \\ &= \omega(v, Jv) && \text{since } \omega \text{ is antisymmetric} \\ &= b(v, w) \end{aligned}$$

so *b is symmetric*. We also claim that *b is nonsingular*. Indeed, if $b(v, w) = 0$ for all v, then

$\omega(v, Jw) = 0 \,\forall v$ so $Jw = 0$ since ω is nonsingular, and, hence,

$$w = -J^2 w = 0.$$

Also notice that b is J-invariant; that is,

$$b(Jv, Jw) = b(v, w). \tag{23.2}$$

Indeed,

$$b(Jv, Jw) = \omega(Jv, J^2 w) = \omega(v, Jw) = b(v, w).$$

Finally, notice that

$$\omega(v, w) = b(Jv, w) \tag{23.3}$$

since $J \in Sp(V)$. Conversely, it is clear that if we start with a complex structure on V and a nonsingular symmetric form b invariant with J (i.e., satisfying (23.2)), then defining ω by (23.3) gives us a Kaehler structure.

Definition. *A Kaehler structure on* V *is called positive if the corresponding symmetric form* b *is positive definite.*

Recall that a manifold X is called a *complex manifold* if it possesses an atlas $\{(U, \phi_U)\}$ where the U's are open sets covering X, where ϕ_U is a diffeomorphism of U onto an open subset of $\mathbb{C}^n$, and where the transition maps $\phi_{UV} = \phi_V \phi_U^{-1} : \phi_U(U \cap V) \to \phi_V(U \cap V)$ are holomorphic. If $p \in U \cap V$, then

$$(d\phi_U)_p : TX_p \to \mathbb{C}^n \qquad \text{and} \qquad (d\phi_V)_p : TX_p \to \mathbb{C}^n$$

with

$$(d\phi_V)_p \circ (d\phi_U)_p^{-1} \in Gl(n, \mathbb{C})$$

so that TX_p has a complex structure at every p. (A manifold with a (smoothly varying) complex structure on each tangent space is called an *almost complex manifold*. The condition that X be a complex manifold is much more stringent.)

Definition. *Let* X *be a complex manifold and let* ω *be a symplectic form on* X. *Then* X *is called a Kaehler manifold if at every* $p \in X$ *the complex structure* J_p *on* TX_p *and the antisymmetric form* ω_p *define a Kaehler structure on* TX_p. *The manifold* X *is called a positive Kaehler manifold if the corresponding Kaehler structure is positive at all* p.

Suppose we start with a complex manifold X and a Riemann metric b on it. Then at each $p \in X$ we have J_p and b_p and hence ω_p. The form ω is nonsingular, the question is whether or not $d\omega = 0$. We now give a criterion due to Mumford (1976, p. 87) that is useful in practice to obtain a sufficient condition for ω to be closed. Suppose we are given a group G of diffeomorphisms of X that preserves the complex structure (i.e., acts as holomorphic transformations) and that also preserves the metric b. For each $x \in X$, let G_x denote the isotropy subgroup of x, that is, the group of all $a \in G$ satisfying $ax = x$. Then $da: TX_x \to TX_x$ for each $a \in G_x$. We thus get a representation

$$\rho_x : G_x \to \mathrm{Aut}_{\mathbb{C}}(TX_x).$$

Proposition 23.1. *If* $J_x \in \rho_x(G_x)$ *for all* x, *then* $d\omega = 0$.
(Here, of course, J_x denotes the complex structure, i.e., the operation of multiplication by i, on TX_x.) For simplicity, if $u \in TX_x$ and $a \in G_x$ we shall write au instead of $\rho_x(a)u$.

Proof. Since G preserves the complex structure and b, it also preserves ω and hence $d\omega$. So for any $a \in G_x$ and $u, v, w \in TX_x$,

$$d\omega_x(au, av, aw) = d\omega_x(u, v, w).$$

Taking $a = J_x$ and applying this equation twice gives

$$d\omega_x(u, v, w) = d\omega_x(J_x u, J_x v, J_x w) = d\omega_x(J_x^2 u, J_x^2 v, J_x^2 w).$$

But $J_x = -1$ so $d\omega_x(u, v, w) = -d\omega_x(u, v, w) = 0$. Q.E.D.

As an illustration of the use of this proposition, let V be a finite-dimensional complex Hilbert space and $U(V)$ the corresponding unitary group. Let $\mathbb{P}(V)$ be the corresponding projective space. Then $\mathbb{P}(V)$ is a complex manifold, and the group $Gl(V,\mathbb{C})$ of all complex linear transformations of V acts as complex holomorphic diffeomorphisms of $\mathbb{P}(V)$ so, in particular, $U(V) \subset Gl(V,\mathbb{C})$ does as well. Now the group $U(V)$ is compact, and so, starting with any Riemann metric b_0 on $\mathbb{P}(V)$, we can, by the process of averaging over the group $U(V)$, obtain an invariant Riemann metric b. Furthermore, since $U(V)$ acts transitively on $\mathbb{P}(V)$ and $U(V)_x$ acts irreducibly on $T\mathbb{P}(V)_x$ for each $x \in \mathbb{P}(V)$, we know that this Riemann metric is unique up to positive multiple. (We shall write a formula for it and for the corresponding form ω in terms of the scalar product on V in just a moment.) We claim that the hypotheses of the proposition are satisfied.

Indeed, let

$$\pi : V - \{0\} \to \mathbb{P}(V)$$

be the projection sending $v \in V - \{0\}$ into the line through v. Let $W = v^{\perp}$ be the orthogonal complement of v in V. Under the identification of V with TV_v, we may regard W as a subspace of TV_v, and as such it is clear that $d\pi$ maps W bijectively onto $T\mathbb{P}(V)_v$. Now

$$U(V)_{\pi v} = U(W) \times T,$$

where T is the group of complex numbers of absolute value 1 and under the above identification $\rho(U(V)_{\pi v}) = U(W) =$ the unitary group of the subspace W. As multiplication by i is certainly a unitary transformation, we see that the hypotheses of the proposition are satisfied and, hence, that ω does define a positive Kaehler structure on $\mathbb{P}(V)$.

Let us now write down an explicit formula for the Riemann metric b and hence the symplectic form ω on $\mathbb{P}(V)$. For this purpose, it is convenient to observe that π maps $S(V)$ surjectively onto $\mathbb{P}(V)$, where $S(V)$ denotes the unit sphere in V:

$$S(V) = \{v \mid \|v\| = 1\}.$$

We can now define a map $\psi : \mathbb{P}(V) \to SA(V)$, where $SA(V)$ denotes the space of self-adjoint operators on V, given by

$$\psi(\pi v) = v \otimes v^*,$$

where $v \in S(V)$. In other words, ψ assigns to each line in V the orthogonal projection operator onto this line. The unitary group $U(V)$ acts (by conjugation) on the space $SA(V)$ and preserves the (real) scalar product

$$q(S_1, S_2) = \tfrac{1}{2} \operatorname{tr} S_1 S_2 .$$

Also the map ψ is clearly $U(V)$-invariant. Thus

$$b = \psi^* q,$$

the pull-back of the Euclidean metric on $SA(V)$ to $\mathbb{P}(V)$ is the (unique up to positive multiple) Riemann metric invariant under $U(V)$. Let A and B be linear transformations on V, thought of as elements of the Lie algebra of $Gl(V)$. For each $z \in \mathbb{P}(V)$ we get corresponding tangent vectors A_z and B_z on $T\mathbb{P}(V)_z$. We can evaluate their scalar product $b_z(A_z, B_z)$ as follows: Suppose that $z = \pi v$ with $v \in S(V)$. Then

$$d\psi(A_z) = Av \otimes v^* + v \otimes A^*v^*$$

so

$$\begin{aligned} b(A_z, B_z) &= \tfrac{1}{2}\operatorname{tr}(Av \otimes v^* + v \otimes A^*v^*)(Bv \otimes v^* + v \otimes B^*v^*) \\ &= \operatorname{Re}(\langle Av, v\rangle\langle v, Bv\rangle + \langle Av, Bv\rangle), \end{aligned} \tag{23.4}$$

where $\langle\,,\rangle$ denotes the scalar product on V, and, hence,

$$\omega(A_z, B_z) = \operatorname{Re}(\langle iAv, v\rangle\langle v, Bv\rangle + \langle iAv, Bv\rangle). \tag{23.5}$$

Notice that if A and B belong to $U(V)$ (i.e., are skew adjoint, then $\langle iAv, v\rangle$ is real, while $\langle v, Bv\rangle$ is imaginary, and, hence, the first term above vanishes and

$$\omega(A_z, B_z) = -\operatorname{Im}\langle Av, Bv\rangle = \tfrac{1}{2}\langle[A, B]v, v\rangle \qquad \text{for} \quad A, B \in U(V). \tag{23.6}$$

An elementary but important property of Kaehler manifolds is the following:

Proposition 23.2. *Let* X *be a positive definite Kaehler manifold with symplectic form* ω. *Let* $\iota: \mathrm{Y} \to \mathrm{X}$ *be a complex immersion of the complex manifold* Y *into* X. *Then* Y *is a positive Kaehler manifold with form* $\iota^*\omega$.

Proof. To say that ι is a complex immersion means that $d\iota$ at each point commutes with multiplication by J on the tangent space. Now $d\iota^*\omega = \iota^*d\omega = 0$ so $\iota^*\omega$ is closed. On the other hand, since b is positive definite, ι^*b is a positive definite Riemann metric on Y. Finally, at any $y \in Y$, $\iota^*\omega(w, Jv) = \omega(d\iota_y w, d\iota_y Jv) = \omega(d\iota_y w, J d\iota_y v) = \iota^*b(w, v)$ so $\iota^*\omega$ is nonsingular and similarly is J-invariant. Q.E.D.

For example, the above proposition implies that every nonsingular algebraic subvariety of complex projective space is a Kaehler manifold. Let us explain a group theoretical implication of this observation that is of importance in quantum mechanics. We begin with a special case. Let W be a finite-dimensional Hilbert space and let $V = \wedge^k W$. (In quantum mechanical language we think of W as the one-particle state space of a fermion, and V the k-particle state space.) The group $U(W)$ is represented on V unitarily, and hence acts on the projective space $P(V)$ so as to preserve the Kaehler structure. If we pick some point $z \in P(V)$ and consider its orbit, $U(W)\cdot z$ under $U(W)$, it will not, in general, be a Kaehler submanifold. There is, however, one $U(W)$ orbit (and, it turns out to be the only one) that *is* a Kaehler submanifold, and that is the orbit through those z of the form

$$z = \pi(w_1 \wedge \cdots \wedge w_k).$$

in other words, z is a line through a decomposable vector. The set Y of all such z is a single $U(W)$ orbit, since we can assume in the above expression for z that the w's form part of an orthonormal basis of W (remember that z is the line through $w_1 \wedge \cdots \wedge w_k$ so the length of $w_1 \wedge \cdots \wedge w_k$ is irrelevant) and $U(W)$ acts transitively on the set of all orthonormal bases. But we claim that Y is in fact a complex submanifold. This follows directly, as Y can be characterized by a set of algebraic equations (Y is an algebraic subvariety). It can also be verified group-theoretically by observing that the full complex group $Gl(W)$ acts on V, and hence on $P(V)$ as automorphisms of $P(V)$, as a complex manifold (not preserving the Kaehler structure) and Y is clearly a $Gl(W)$ orbit and hence a complex submanifold. Hence, by the Proposition 23.2 it is a Kaehler submanifold. In the physics and chemistry literature, the space Y is known as the space of "Slater determinants." Its importance lies in the so-called Hartree–Fock approximation, which we pause to briefly describe.

Suppose we are given a Hamiltonian (i.e., a self-adjoint operator) $\mathscr{H}$ on V. The basic problem is then to find the eigenvectors and eigenvalues – the stationary states and energy levels of $\mathscr{H}$. The linear operator $\mathscr{H}$ defines the function Q on $P(V)$ given by

$$Q(z) = (\mathscr{H}v, v)/(v, v) \qquad \text{if } z = \pi v$$

and the problem of locating the eigenvectors (and hence the eigenvalues) of $\mathscr{H}$ is equivalent to locating the critical points of Q on $P(V)$. In particular, the problem of finding the lowest energy level is the same as finding the minimum point of Q on $P(V)$. If V is large (perhaps infinite-dimensional), this can be quite hard, so one replaces $P(V)$ by some smaller space X and hopes that the minimum of Q on X is a good approximation to the minimum of Q on $P(V)$. Sometimes, as in quantum chemistry, one chooses $X = P(U)$, where U is some small Hilbert subspace of V. For example, the molecular orbital and the valence bond "theories" consist of these types of approximations. In the Hartree–Fock approximation one takes $X = Y$ as described above for $V = \wedge^k W$. In the "time-dependent Hartree–Fock approximation," one is given a time-dependent Hamiltonian $\mathscr{H}(t)$ and hence a time-dependent function $Q(t)$. One then "approximates" the solution of the time-dependent Schrödinger equation by the time-dependent Hamiltonian system obtained by restricting $Q(t)$ to Y, and considering the corresponding Hamiltonian system relative to the symplectic structure on Y given by $\imath^*\omega$.

This example has the following group-theoretical generalization: Replace $U(W)$ by an arbitrary compact semisimple Lie group and V by an irreducible representation. So let G be an arbitrary semisimple compact connected Lie group, and suppose that we are given an irreducible representation of G on V. Then G acts on $P(V)$, preserving its Kaehler structure. It turns out that there is exactly one orbit $G \cdot z$ that is a complex submanifold, and this is the orbit through a point z of the form $z = \pi v$, where v is a maximal weight vector (relative to some choice of maximal torus in G). This result is (part of) the celebrated Borel–Weil theorem. There are other orbits that are symplectic – the restriction of ω to them is nonsingular – but are not Kaehler, while the typical orbit, in general will not be symplectic. The description of the symplectic orbits, to be presented below, is joint work of Kostant and Sternberg (1982).

To establish the above-mentioned results, we will need some Lie group technology and some results from Section 26, so the reader may prefer to defer or skip the remainder of this section.

It will follow from results to be proved later (Section 26) on homogeneous symplectic group actions, that if $G \cdot z$ is to be a symplectic manifold, then z must be of the form $z = \pi v$, where v is a weight vector for some maximal torus. Indeed, to say that z has this form means that z is fixed by some maximal torus T. So we must prove that if $G \cdot z$ is to be symplectic, then G_z, the stabilizer of z, must contain some maximal torus. Now we shall prove in Sections 26 and 52 that for a large class of groups, including all semisimple ones, all symplectic homogenous spaces are (covering spaces of) orbits of G acting on the dual of g via the coadjoint representation. If G is semisimple, we may use the Killing form on g to identify g with g^*, and hence regard the orbits as lying in g. If G is compact, the connected component of the stabilizer group of any point in g contains a maximal torus. Hence, if $G \cdot z$ is to be symplectic, the stabilizer group of z must contain a maximal torus.

So let us fix some maximal torus T in G. (Recall that they are all conjugate.) We let t denote the Lie algebra of T and g the Lie algebra of G, with $t_{\mathbb{C}}$ and $g_{\mathbb{C}}$ denoting their complexifications. We have

$$g_{\mathbb{C}} = t_{\mathbb{C}} + \oplus \mathbb{C}E_{\alpha} + \oplus \mathbb{C}E_{-\alpha},$$

where the sum ranges over some system of positive roots, Correspondingly,

$$g = t + \oplus \mathbb{R}(E_{\alpha} - E_{-\alpha}) + \oplus \mathbb{R}i(E_{\alpha} + E_{-\alpha})$$

gives a decomposition of g. Now suppose that $z = \pi v$, where v is a weight vector corresponding to weight λ. Then tangent space to the orbit $G \cdot z$ is spanned by all vectors of the form $d\pi_v(Xv)$ as X ranges over the Lie algebra g. Now for $X \in t$, Xv is some multiple of v, and hence $d\pi_v(Xv) = 0$. The vector $E_{\alpha}v$ is either 0 or is a weight vector of weight $\lambda + \alpha$. In either event, it is perpendicular to v and to any $E_{\beta}v$ for $\beta \neq \alpha$. So the spaces $\mathscr{C}_{\alpha}$ spanned by $(E_{\alpha} - E_{-\alpha})v$ and $i(E_{\alpha} + E_{-\alpha})$ are mutually perpendicular to one another and their image under $d\pi_v$ spans the tangent space to the orbit $G \cdot z$ at z. Now suppose that λ is the maximal weight (relative to our choice of positive roots). Then $\lambda + \alpha$ is not a weight so $E_{\alpha}v = 0$ for all positive α. The space $\mathscr{C}_{\alpha}$ is spanned by $E_{-\alpha}v$ and $iE_{-\alpha}v$ and is clearly stable under multiplication by i. Hence, each of the spaces $d\pi_v(\mathscr{C}_{\alpha})$ is stable under multiplication by J in $TP(V)_z$ and so $G \cdot z$ is a complex submanifold. Conversely, since $i\mathscr{C}_{\alpha}$ is perpendicular to $\mathscr{C}_{\beta}$ and to $i\mathscr{C}_{\beta}$ for any $\beta \neq \alpha$, and $i\mathscr{C}_{\alpha}$ is perpendicular to all multiples of v, the only way that $G \cdot z$ can be a complex submanifold is for $i\mathscr{C}_{\alpha} \subset \mathscr{C}_{\alpha}$ for every α. But $E_{\alpha}v$ is perpendicular to $E_{-\alpha}v$ and if $i\mathscr{C}_{\alpha} \subset \mathscr{C}_{\alpha}$, then each of the four vectors $E_{\alpha}v$, $iE_{\alpha}v$, $E_{-\alpha}v$, and $iE_{-\alpha}v$ belongs to $\mathscr{C}_{\alpha}$, where $\dim_{\mathbb{R}} \mathscr{C}_{\alpha} \leq 2$. The only way that this can happen is for either $E_{\alpha}v = 0$ or $E_{-\alpha}v = 0$. Thus we have proved that if $G \cdot z$ is complex, then for every root α, either $E_{\alpha}v = 0$ or $E_{-\alpha}v = 0$. Let P denote the set of roots such that $E_{\alpha}v = 0$. Since $[E_{\alpha}, E_{\beta}]$ is a nonzero multiple of $E_{\alpha+\beta}$ if α, β and $\alpha + \beta$ are roots, we see that if $\alpha \in P$ and $\beta \in P$ and $\alpha + \beta$ is a root, then $\alpha + \beta \in P$. We can then let $\alpha_1, \ldots, \alpha_k$ be those roots in P that cannot be written as a sum of roots in P, and it follows from the above that every root can be written as $\pm \Sigma m_i \alpha_i$, with the m_i nonnegative integers. In other words, $k = l$ and the $\alpha_1, \ldots, \alpha_l$ form a system of simple roots, and thus λ is a maximal weight relative to this system of simple roots. Q.E.D.

Now let us investigate which orbits are symplectic. By formula (23.6) and the fact that $[E_{\pm\alpha}, E_{\pm\beta}]v$ is a weight vector of weight $\lambda \pm \alpha \pm \beta$ and hence orthogonal to v, we see that each of the spaces $d\pi_v \mathscr{C}_{\alpha}$ are mutually orthogonal to one another with respect to ω for different positive α. So to check that $\iota^*\omega$ is symplectic, we need to know, by another application of (23.6), that if

$$[E_{\alpha}, E_{-\alpha}]v = 0,$$

then

$$E_\alpha v = E_{-\alpha} v = 0.$$

Now $[E_\alpha, E_{-\alpha}] = \mathscr{H}_\alpha \in t$ and $\mathscr{H}_\alpha v = (\lambda \cdot \alpha)v$, where $\lambda \cdot \alpha = \lambda(\mathscr{H}_\alpha)$ denotes the value of the linear function λ on the element $\mathscr{H}_\alpha$. So $G \cdot z$ is symplectic if and only if

$$\lambda \cdot \alpha = 0 \text{ implies that } E_\alpha v = E_{-\alpha} v = 0 \text{ for any root } \alpha. \tag{23.7}$$

To summarize, we have proved:

Proposition 23.3. *Let* G *be a compact semisimple Lie group with an irreducible representation on* V. *Under the action of* G *on* P(V) *the orbit through a point z is symplectic relative to the restriction of the form* ω *of* P(V) *if and only if* $z = \pi v$, *where* v *is a weight vector of some maximal torus of* G *that satisfies condition (23.7). There is only one orbit that is Kaehler, and that is the orbit through* πv *where* v *is a maximal weight vector.*

For example, suppose we take $G = SU(2)$ and V its five-dimensional (spin-2) irreducible representation space. Then $P(V)$ is four-complex and hence eight-real-dimensional, while $SU(2)$ has dimension 3. A typical orbit in $P(V)$ will be three-dimensional, and hence certainly not symplectic. The orbits through lines of weight vectors will be two-dimensional spheres. There are five weights characterized by the values 2, 1, 0, -1, -2. Since 2 and -2 and 1 and -1 give lines on the same orbit, we need only consider 2, 1, and 0. Since 2 is the maximal weight, we know that the sphere corresponding to it is a Kaehler submanifold. There is only one root, and 1 does not vanish on it, so the sphere corresponding to 1 satisfies (23.7) and so is symplectic. The sphere corresponding to the weight 0 does not satisfy (23.7), and the restriction of ω to it is identically 0. Although the criterion (23.7) is easy to check in practice, its geometrical significance is not transparent. An argument with roots and weights will show that it is equivalent to

the stabilizer group of λ, G_λ, *is the same as the stabilizer group,* $G_{[v]}$, *of the ray through* v. (23.8)

In Section 26 we shall explain this criterion for a G orbit to be symplectic in terms of the moment map, and see that it is valid in much greater generality – for an arbitrary "Hamiltonian" G action.

The importance of the symplectic, but not complex orbits is not clear at present. But the unique complex orbit plays a crucial role both in physics and in mathematics. In physics, the points z belonging to the complex orbits are called the "coherent states." The mathematical importance of the complex orbit lies in the fact that we can recover the entire representation from the action of G on it. Indeed, as we shall see later, the vector space V can be identified with the space of holomorphic sections of a canonical holomorphic line bundle sitting over the complex orbit, and the representation of G on V is just the induced action of G on this space of sections. In this way, information about representation theory is coded into the Kaehler geometry.

24. Left-invariant forms and Lie algebra cohomology

Let G be a Lie group. For any $a \in G$ we let l_a denote left multiplication by a, so

$$l_a c = ac.$$

We let r_b denote right multiplication by b^{-1} for any $b \in G$ so that

$$r_b c = cb^{-1}.$$

A differential form ω on G is called left-invariant if

$$l_a^* \omega = \omega$$

for all $a \in G$. Similarly, a vector field ξ on G is called left-invariant if it satisfies

$$l_a^* \xi = \xi$$

for all $a \in G$. We may identify the set of left-invariant vector fields with the Lie algebra g of G. (Indeed, this is one possible definition of the Lie algebra.)[†] Each left-invariant vector field generates a one-parameter group of right multiplications, since it is the right multiplications that commute with all left multiplications. Any subalgebra h of g defines a foliation on G whose leaves consist of cosets aH, where H is the connected subgroup generated by h. Each left-invariant differential form is completely determined by its values on the left-invariant vector fields; we may thus identify the space of left-invariant q-forms with $\bigwedge^q(g^*)$. If ω is left-invariant, so is $d\omega$ and hence d induces a linear map from $\bigwedge^q(g^*)$ to

[†] The natural identification would be with right-invariant vector fields since they generate left multiplications. It will be convenient to use the left-invariant vector fields but this requires an adjustment of the sign in the bracket. We define the Lie bracket for two elements of the Lie algebra as $[\xi, \eta] = D_\xi \eta$.

$\bigwedge^{q+1}(g^*)$, which we shall denote by δ. The formula for δ is standard, and can be deduced inductively from (22.1): the fundamental formula for the Lie derivative

$$D_\xi \omega = i(\xi)\, d\omega + di(\xi)\omega.$$

For example, suppose that ω is a linear differential form. Then $i(\xi)\omega$ is a constant and so the second term in the above expression vanishes. If η is a second left-invariant vector field, then $\omega(\eta) = i(\eta)\omega$ is again a constant so

$$\begin{aligned} 0 = D_\xi\big(\omega(\eta)\big) &= (D_\xi \omega)(\eta) + \omega(D_\xi \eta) \\ &= \big(i(\xi)\, d\omega\big)(\eta) + \omega([\xi, \eta]) \\ &= d\omega(\xi \wedge \eta) + \omega([\xi, \eta]), \end{aligned}$$

and so we obtain

$$\delta\omega(\xi \wedge \eta) = -\omega([\xi, \eta]) \qquad \text{for } \omega \in \textstyle\bigwedge^1(g^*). \tag{24.1}$$

If $\omega \in \bigwedge^2(g^*)$ and ξ, η, and ζ are in g, we can apply the same argument to $0 = D_\xi\big(\omega(\eta \wedge \zeta)\big)$ and use (24.1) to conclude that

$$\delta\omega(\xi \wedge \eta \wedge \zeta) = -\big(\omega([\xi, \eta] \wedge \zeta) + \omega([\zeta, \xi] \wedge \eta) + \omega([\eta, \zeta] \wedge \xi)\big) \\ \text{for } \omega \in \textstyle\bigwedge^2(g^*), \tag{24.2}$$

and so on.

The formulas for δ involve only the Lie algebra structure of g (and hence can be defined on any Lie algebra, where it can be verified directly that $\delta^2 = 0$). We define, as usual

$$B^k(g) = \delta\big(\textstyle\bigwedge^{k-1}(g^*)\big)$$

and

$$Z^k(g) = \ker \delta \subset \textstyle\bigwedge^k(g^*),$$

so that $B^k \subset Z^k$ and

$$H^k(g) = Z^k(g)/B^k(g)$$

is called the kth cohomology of the Lie algebra g. (If the group G is compact, then standard arguments in topology show that averaging over the group allows us to replace arbitrary forms by left-invariant ones in computing the DeRham cohomology of G, and so $H^k(g)$ is the kth cohomology (over the reals) of G. If G is not compact, there need be no such relation. For instance, if G, and hence g, is Abelian, then $\delta \equiv 0$ so $H^k(g) = \bigwedge^k(g^*)$, but G might be $\mathbb{R}^n$, which has no homology.)

In our study of symplectic geometry we will be particularly interested in $H^1(g)$ and $H^2(g)$ for the following reason: Suppose that we are given an exact sequence of Lie algebras

$$0 \to \mathbb{R} \to f \xrightarrow{\rho} h \to 0, \qquad [\mathbb{R}, f] = 0; \tag{24.3}$$

in other words, f is a central extension of h by $\mathbb{R}$. Of course, the case we will be interested in is where $h = \mathrm{Ham}(X)$ and $f = F(X)$ for some symplectic manifold X, as in (14.13). Suppose we are given a homomorphism $\kappa\colon g \to h$. The question is (cf. Section 15): Can we find a homomorphism $\lambda\colon g \to f$ such that $\rho\lambda = \kappa$, and, if so, how many such λ are there?

$$\begin{array}{ccccccccc} 0 & \to & \mathbb{R} & \to & f & \xrightarrow{\rho} & h & \to & 0 \\ & & & & \nwarrow ? & & \uparrow \kappa & & \\ & & & & & & g & & \end{array}$$

Well, since g is a finite-dimensional vector space and since f is a vector space, we can always find a *linear* map $\mu\colon g \to f$ such that $\rho \circ \mu = \kappa$. Since κ is a Lie algebra homomorphism, it follows that for any ξ and η in g

$$\rho(\mu([\xi, \eta]) - [\mu(\xi), \mu(\eta)]) = 0.$$

Thus

$$c(\xi \wedge \eta) \stackrel{\text{def}}{=} \mu([\xi, \eta]) - [\mu(\xi), \mu(\eta)]$$

defines an element c of $\wedge^2(g^*)$. Since both g and f are Lie algebras, we can apply Jacobi's identity on each, to conclude, using (24.2) that $\delta c = 0$. Suppose we make a different choice of μ, which amounts to replacing μ by $\mu + b$, where b is some linear function from g to $\mathbb{R}$; that is, $b \in \wedge^1(g^*) = g^*$. Then, since $b(\xi) \in \mathbb{R}$ commutes with all elements of f, we see that this has the effect of adding the term $b([\xi, \eta])$ to the expression for $c(\xi \wedge \eta)$, in other words, of replacing c by $c - \delta b$. In particular, the cohomology class $[c]$ of c is independent of the choice of μ. If $[c] \neq 0$, there is no way of finding the desired homomorphism λ. If $[c] = 0$, then we can find a b such that the c corresponding to $\mu + b$ vanishes; $\lambda = \mu + b$ is the desired homomorphism. If we have one homomorphism λ, any other one can differ from it by adding a b with $\delta b = 0$; thus the set of such b in other words, $H^1(g)$, measures the indeterminacy in λ. To summarize we have proved:

Proposition 24.1. *Given a central extension (24.3) and a homomorphism* $\kappa\colon \mathrm{g} \to \mathrm{h}$ *there is a well-defined cohomology class* $[\mathrm{c}]$ *that measures the obstruction to the possibility of finding a homomorphism* $\lambda\colon \mathrm{g} \to \mathrm{f}$ *covering* κ.

Such a λ exists if and only if $[\mathrm{c}] = 0$. *If* $[\mathrm{c}] = 0$, *the set of all possible λ is parametrized by* $\mathrm{H}^1(\mathrm{g})$. *In particular, if* $\mathrm{H}^2(\mathrm{g}) = \{0\}$ *then λ always exists, and if* $\mathrm{H}^1(\mathrm{g}) = \{0\}$ *the λ is unique.*

It is a well-known fact, and we shall present the proof in Section 52, that for semisimple Lie algebras, $H^1(g) = H^2(g) = 0$. This is also true for the Euclidean algebra in three or more dimensions (but not in two dimensions) for the Poincaré algebra and others, cf. Section 53. For the Galilean algebra, $\dim H^1(g) = \dim H^2(g) = 1$ and, as we shall see, the cohomology class $[c]$ associated to a symplectic action of the Galilean group is intimately related to the notion of mass. We shall explain this more fully in Sections 52 and 53.

25. Symplectic group actions

We continue with the notation of the beginning of the preceding section. Suppose that G is a Lie group and that ω is a left-invariant form on G. Since right and left multiplication commute, $r_b^*\omega$ is again left-invariant and

$$r_b^*\omega = r_b^* l_b^* \omega.$$

Now $r_b l_b(c) = bcb^{-1}$ and the corresponding action on left-invariant vector fields is usually denoted by Ad_b. Hence $b \rightsquigarrow (r_b{}^{-1})^*$ defines a representation of G on each $\bigwedge^q(g^*)$, which we denote by $\mathrm{Ad}^\#$. Thus

$$\mathrm{Ad}_a^\# \omega = (r_a^{-1})^* \omega.$$

Let M be a differentiable manifold and let $G \times M \to M$ be a left action of G on M. Thus for each $a \in G$ we get a diffeomorphism $\varphi_a : M \to M$:

$$\varphi_a(m) = am.$$

Similarly, for each point m of M we get a map $\psi_m : G \to M$ given by

$$\psi_m(a) = am.$$

Since $b(am) = (ba)m$ we see that

$$\varphi_b \psi_m = \psi_m l_b \tag{25.1}$$

and

$$\psi_{am} = \psi_m r_a^{-1}. \tag{25.2}$$

Now let Ω be an invariant q-form on M, so that

$$\varphi_a^* \Omega = \Omega$$

for all $a \in G$. For each point m, the form $\psi_m^*\Omega$ on G is left-invariant. Indeed, by (25.1)

$$l_b^*\psi_m^*\Omega = \psi_m^*\varphi_b^*\Omega = \psi_m^*\Omega.$$

Furthermore, by (25.2),

$$\psi_{am}^*\Omega = r_{a^{-1}}^*\psi_m^*\Omega = \mathrm{Ad}_a^{\#}\psi_m^*\Omega.$$

We have thus proved:

Proposition 25.1. *An invariant* q*-form* Ω *on* M *defines a map* $\Psi: \mathrm{M} \to \bigwedge^q(\mathrm{g}^*)$ *given by*

$$\Psi(m) = \psi_m^*\Omega.$$

This map is a G *morphism; that is,*

$$\Psi(am) = \mathrm{Ad}_a^{\#}\ \Psi(m).$$

Each $\xi \in g$ defines a one-parameter subgroup of G and hence a one-parameter group of diffeomorphisms of M and hence a vector field on M, which we shall denote by ξ_M. Letting $\xi_M(m)$ denote the value of the vector field ξ_M at the point m, we see that

$$\xi_M(m) = d(\psi_m)_e\xi,$$

where, on the right-hand side of this equation we consider ξ as a tangent vector at the identity, e of the group and $d(\psi_m)_e: TG_e \to TM_m$. Letting $\langle\,,\rangle$ denote the pairing between $\bigwedge^q(g^*)$ and $\bigwedge^q(g)$, it follows that

$$\langle\Psi, \xi_1 \wedge \cdots \wedge \xi_q\rangle = \Omega(\xi_{1M} \wedge \cdots \wedge \xi_{qM}), \qquad \xi_1, \ldots, \xi_q \in g, \quad (25.3)$$

as functions on M.

If Ω is closed, then so is $\Psi(m) = \psi_m^*\Omega$ for each $m \in M$. In particular, if M is a symplectic manifold and Ω is the symplectic form, then

$$\Psi: M \to Z^2(g).$$

We have thus proved:

Theorem 25.1. *Any symplectic action of a Lie group* G *on a symplectic manifold* M *defines a* G *morphism,* $\Psi: \mathrm{M} \to \mathrm{Z}^2(\mathrm{g})$. *Since the map* Ψ *is a* G *morphism,* $\Psi(\mathrm{M})$ *is a union of* G *orbits in* $\mathrm{Z}^2(\mathrm{g})$. *In particular, if the action of* G *on* M *is transitive, then the image of* Ψ *consists of a single* G *orbit in* $\mathrm{Z}^2(\mathrm{g})$.

Suppose that we start with an orbit of G in $Z^2(g)$. Is it the image under some Ψ of a symplectic manifold M? To see what is involved in this question, let ω be some element of $Z^2(g)$, thought of as a closed 2-form on G. Define

$$h_\omega = \{\xi \in g | i(\xi)\omega = 0\}. \tag{25.4}$$

Notice that $h_\omega \subset g_\omega$, the isotropy algebra of ω, since g_ω consists of those ξ with $D_\xi\omega = 0$, and $D_\xi\omega = d(i(\xi)\omega)$ since $d\omega = 0$. If $\xi \in h_\omega$ and $\eta \in g_\omega$, then

$$0 = D_\eta(i(\xi)\omega) = i([\eta, \xi])\omega + i(\xi)D_\eta\omega = i([\eta, \xi])\omega$$

so $[\eta, \xi] \in h_\omega$. Thus h_ω is an ideal in g_ω, and, in particular, h_ω is a subalgebra of g.

The set of left-invariant vector fields on G that lie in h_ω is thus closed under Lie bracket. We can consider the differential system determined by these vector fields, that is, the rule which assigns to each point a of G the subspace of T_aG given by the values of these vector fields at a. (Here we are using the notation of Sternberg, 1983, p. 130.) The fact that h_ω is a subalgebra means that the corresponding differential system is integrable in the sense of Frobenius; in other words, through each point of G there passes a (unique maximal connected) submanifold whose tangent space at every point is the given differential system (cf. Sternberg, 1983, Theorem 5.1, p. 132). These submanifolds are called the leaves of the foliation determined by the differential system. In fact, this integrability is an illustration of a general statement (Theorem 25.2 below) that the null differential system of a closed form on any manifold is integrable in the sense of Frobenius. In the case at hand, we can describe these leaves explicitly: Let H_ω denote the connected subgroup generated by the subalgebra h_ω. Since left-invariant vector fields generate one-parameter groups of right multiplication, it is clear that the leaf through a is exactly $a \cdot H_\omega$, the right coset of H_ω through a.

If H_ω were closed, we could form the quotient space $M = G/H_\omega$ and we would have the projection $\rho: G \to M$ whose fibers are the leaves of our foliation. The following theorem would then imply that M carries a symplectic form with $\bar{\omega} = \rho^*\omega$. In fact, the following theorem, which embodies the principle of *reduction* will be central to much of what follows.

Theorem 25.2. *Let* X *be a differentiable manifold and* ω *a closed form on* X. *Then the set of all vector fields on* X *satisfying*

$$i(\xi)\omega = 0 \tag{25.5}$$

is closed under Lie bracket. In particular, if the dimension of the space of tangent vectors satisfying (25.5) at each point is constant, ω *defines an integrable foliation. Suppose that this foliation is fibrating; that is, that there* *exists a manifold* M *and a submersion* $\rho: \mathrm{X} \to \mathrm{M}$ *such that the leaves of the foliation are all of the form* $\rho^{-1}(\mathrm{m})$ *for* $\mathrm{m} \in \mathrm{M}$. *Then there exists a unique symplectic form* $\bar{\omega}$ *on* M *with* $\omega = \rho^*\bar{\omega}$.

Proof. If ξ and η both satisfy (25.5) then $D_\xi \omega = i(\xi)\, d\omega + di(\xi)\omega = 0$ by (25.5) and the fact that $d\omega = 0$. But then

$$0 = D_\xi(i(\eta)\omega) = i(D_\xi \eta)\omega + i(\eta) D_\xi \omega = i(D_\xi \eta)\omega$$

so

$$i([\xi, \eta])\omega = 0,$$

proving the first assertion. Now suppose that the dimensions are constant, so we get an integrable distribution – that is, a foliation. By the Frobenius theorem (see Sternberg, 1983, p. 132), we can introduce coordinates $x_1, \ldots, x_k$ about any point of X so that the leaves of the foliation are given by

$$x_1 = \text{const.}, \ldots, x_k = \text{const.},$$

and so the tangent space to the foliation is spanned by $\partial/\partial x_{k+1}, \ldots, \partial/\partial x_n$ at each point in the coordinate neighborhood. If we now write

$$\omega = \sum a_{ij} dx_i \wedge dx_j,$$

where the a_{ij} are functions, the condition that $i(\partial/\partial x_{k+1})\omega = 0, \ldots,$ $i(\partial/\partial x_n)\omega = 0$ implies that $a_{ij} = 0$ if i or j is greater than k. The condition that $d\omega = 0$ then implies that the functions a_{ij} cannot depend on the coordinates $x_{k+1}, \ldots, x_n$. If the foliation is fibrating, we can use the coordinates $x_1, \ldots, x_k$ as local coordinates on M, and so, locally, $\omega = \rho^*\bar{\omega}$, where $\rho(x_1, \ldots, x_n) = (x_1, \ldots, x_k)$ in local coordinates, and $\bar{\omega}$ has the same expression as ω in terms of these local coordinates. Since $\rho^{-1}(m)$ is a leaf of the foliation, which by definition is connected, we see that any two points on the same inverse image $\rho^{-1}(m)$ will give the same form $\bar{\omega}_m$ at m and hence $\bar{\omega}$ is well defined and uniquely determined by the condition $\rho^*(\bar{\omega}) = \omega$. The local expression for ω, together with the fact that the null directions (those satisfying (25.5)) are all spanned by the $\partial/\partial x_{k+1}, \ldots, \partial/\partial x_n$, shows that $\bar{\omega}$ is nondegenerate and closed, that is, symplectic.

A situation in which we will want to apply Theorem 25.2 is one where Y is a symplectic manifold with form Ω and $\iota: X \to Y$ is a smooth map, usually an embedding or an immersion, and where $\omega = \iota^*\Omega$. If the dimension of the null foliation (25.5) is constant, we then have the map ρ given by the

theorem, and hence the diagram

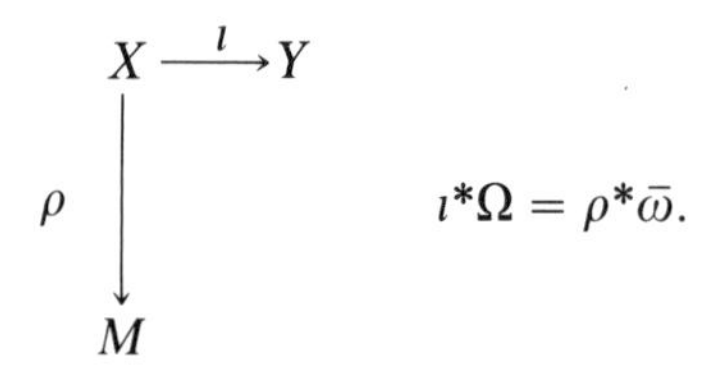

$$\iota^*\Omega = \rho^*\bar{\omega}.$$

Suppose that f_1 and f_2 are two functions on Y such that ι^*f_1 and ι^*f_2 are constant along the fibers of the projection ρ; that is, suppose that there are functions F_1 and F_2 defined on M with

$$\iota^*f_1 = \rho^*F_1 \qquad \text{and} \qquad \iota^*f_2 = \rho^*F_2.$$

We can then inquire about the relation between $\{F_1, F_2\}_M$, the Poisson bracket computed relative to $\bar{\omega}$ on M, and $\{f_1, f_2\}_Y$, the Poisson bracket computed relative to Ω on Y. In general, there need not be any simple relation. For example, if Y is a symplectic vector space, and X is a linear subspace, we could take f_1 and f_2 to be linear functions, given by scalar product with vectors v_1 and v_2 relative to the symplectic form:

$$f_1(y) = (v_1, y), \qquad f_2(y) = (v_2, y),$$

where (,) denotes the symplectic bilinear form on Y. If v_1 and v_2 both belong to $X^\perp$, they induce the identically zero functions on X; that is, $\iota^*f_1 = \iota^*f_2 = 0$. But the Poisson bracket, $\{f_1, f_2\}_Y$ is just the constant function given by the symplectic scalar product of v_1 and v_2 so $\{f_1, f_2\} = (v_1, v_2)$, and there is no reason, in general, why we should expect this to vanish. If, however, the subspace X is coisotropic (if $X^\perp \subset X$), then $(v_1, v_2) = 0$.

In general, the situation is very nice when $\iota: X \to Y$ is a coisotropic immersion, so that $d\iota_x(TX_x)$ is a coisotropic subspace of $TY_{\iota(x)}$ at each point x of X. Let us examine this case. Let f be a function on Y and ξ_f its corresponding vector field, as in Section 14. Let $(\, , \,)_y$ denote the symplectic scalar product on TY_y given by Ω. Then for any tangent vector $v \in TX_x$

$$\langle df_{\iota(x)}, d\iota_x v\rangle = (\xi_f(\iota(x)), d\iota_x v)_{\iota(x)}.$$

If we take v to be such that $d\iota_x v$ lies in $(d\iota_x TX_x)^\perp$, we see that

$d(\iota^*f)_x$ vanishes on the null space of $\iota^*\Omega$, that is, on the space of all vectors satisfying (25.5), if and only if $\xi_f(y) \in d\iota_x TX_x$,

where $y = \iota(x)$.

Or, put more succinctly, ι^*f is constant along the null foliation of X if and only if ξ_f is tangent to the image of X in Y. (Similarly, ι^*f is, locally, a constant on X if and only if ξ_f is everywhere tangent to the image of the null foliation of X.) If f_1 and f_2 are two functions with ξ_{f_1} and ξ_{f_2} both tangent to $\iota(X)$, then

$$\{f_1, f_2\}(y) = (\xi_{f_1}(y), \xi_{f_2}(y))_y = \iota^*\Omega(\xi_{f_1}(y), \xi_{f_2}(y))$$
$$= \rho^*\bar{\omega}(\xi_{f1}(y), \xi_{f2}(y)).$$

If $\iota^*f = \rho^*F$, where F is a function on M and if v is any element of TX_x, then we have

$$\langle df(y), d\iota_x v\rangle = \langle dF, d\rho_x v\rangle = (\xi_F(m), d\rho_x v)_m, \qquad \text{where } m = \rho(x),$$

where $(\ ,\ )_m$ denotes the symplectic scalar product on TM_m, and where ξ_F denotes the vector field on M associated to the function F by the symplectic form of M. But $\langle df(y), d\iota v\rangle = (\xi_f(y), d\iota v)_y = (d\rho\xi_f(y), d\rho v)_m$ so

$$(d\rho\xi_f(y), d\rho v)_m = (\xi_F(m), d\rho v)_m.$$

Since vectors of the form $d\rho_x v$ span all of TM_m we see that $d\rho\xi_f(y) = \xi_F(m)$. Hence

$$\{f_1, f_2\}(y) = (\xi_{f_1}(y), \xi_{f_2}(y))_y = (d\rho\xi_{f_1}(y), d\rho\xi_{f_2}(y))_m$$
$$= (\xi_{F1}(m), \xi_{F2}(m))_m = \{F_1, F_2\}(m)$$

proving

Theorem 25.3. *If $\iota: X \to Y$ is a coisotropic immersion whose null foliation is fibrating with projection map $\rho: X \to M$, and if f_1 and f_2 are functions on Y satisfying $\iota^*f_1 = \rho^*F_1$ and $\iota^*f_2 = \rho^*F_2$ then*

$$\iota^*\{f_1, f_2\}_Y = \rho^*\{F_1, F_2\}_M.$$

Let us now return to our group-theoretical considerations. Let us suppose that the subgroup H_ω is closed. Then $M_\omega = G/H_\omega$ is the symplectic quotient space and the map $\rho: G \to G/H_\omega$ sends any $a \in G$ into its right H_ω coset; $\rho(a) = aH_\omega$. If we let m denote the point H_ω (the identity coset) we see that $\rho(a) = am = \psi_m(a)$. Thus

$$\rho = \psi_m$$

and hence

$$\omega = \rho_m^*(\bar{\omega}).$$

Thus, from the definition of the map Ψ we see that

$$\Psi(m) = \omega.$$

In other words, the orbit in $Z^2(g)$ associated with G/H_ω is precisely the G orbit through ω.

Suppose we start with a symplectic manifold M of the form $M = G/H$, and with symplectic form Ω. Since $TM_H = g/h$ and since G acts transitively on M, the set of vector fields of the form ξ_M fill out the tangent space at each point. Since Ω is symplectic, the only vector field satisfying $i(\eta)\Omega = 0$ is zero. So it follows from (25.3) that h_ω consists precisely of these elements ξ whose corresponding vector field ξ_M vanishes at H (i.e., $\xi \in h$, where h is the Lie algebra of H). In other words, $h = h_\omega$ and hence H_ω is the connected component of H, or, finally, M_ω is a covering space M. Thus, up to covering spaces, the set of transitive symplectic manifolds for G is parametrized by the set of G orbits through elements ω in $Z^2(g)$ whose corresponding group H_ω is closed. We would like therefore to have some criteria that guarantee that an element τ in $Z^2(g)$ has its associated group H_τ closed.

Proposition 25.2 (Kostant–Souriau). *If* $\tau = -\mathrm{d}\beta$ *then* H_τ *is closed and is, in fact, the connected component of the group*

$$\mathrm{G}_\beta = \{\mathrm{c} \in \mathrm{G} | \mathrm{Ad}_\mathrm{c}^\# \ \beta = \beta\}.$$

Thus $\mathrm{G}/\mathrm{H}_\tau$ *is a covering space of the orbit* $\mathrm{G}\cdot\beta$. *In particular, each orbit* $\mathrm{G}\cdot\beta$ *is a symplectic manifold with symplectic form* ω *given at the point* β *by*

$$\omega(\xi_\beta, \eta_\beta) = \beta([\xi, \eta]) \tag{25.6}$$

where ξ_β *is the value at* β *of the vector field corresponding to* $\xi \in \mathrm{g}$ *on the orbit* $\mathrm{G}\cdot\beta$, *and similarly for* η_β.

Proof. By (22.1)

$$D_\xi\beta = i(\xi)d\beta$$

since $i(\xi)\beta$ is a constant, as ξ and β are both left-invariant and hence $d(i(\xi)\beta) = 0$. Thus $\xi \in h_\tau$ if and only if $D_\xi\beta = 0$. But $(D_\xi\beta)(\eta) = -\beta([\xi, \eta])$ since $D_\xi(\beta(\eta)) = 0$ because $\beta(\eta)$ is a constant. Thus H_τ is the connected component of G_β. Now the vector, ξ_β, corresponding to ξ at β is just the image of ξ under the projection ρ of G onto $G/H = G/G_\beta$. Thus the left-hand side of (25.6) is just the value of τ when evaluated on $\xi \wedge \eta$. But $\tau = -d\beta$ so (25.6) follows from (24.1).

If $H^2(g) = \{0\}$ then every element of $Z^2(g)$ is of the form $d\beta$ and if $H^1(g) = \{0\}$ this β is unique; so we get

Theorem 25.4 (Kostant–Souriau). *If* $\mathrm{H}^1(\mathrm{g}) = \mathrm{H}^2(\mathrm{g}) = \{0\}$ *then, up to coverings, every homogeneous symplectic manifold for* G *is an orbit of* G *acting on* g^*.
To prove that H_τ is closed for all τ, it is enough to know that $H^1(g) = \{0\}$:

Proposition 25.3 (Gabber). *Suppose that* $\mathrm{H}^1(\mathrm{g}) = \{0\}$, *so* G *has the property that* $\mathrm{d}\beta = 0$ *implies* $\beta = 0$ *for any left-invariant 1-form* β. *Then* H_τ *is closed for any* τ *and, in fact,* H_τ *is the connected component of* $\mathrm{G}_\tau = \{\mathrm{a} \in \mathrm{G} | \mathrm{Ad}_\mathrm{a}^{\#} \tau = \tau\}$.

Proof. By (22.1)

$$\begin{aligned} D_\xi \tau &= i(\xi)\, d\tau + di(\xi)\tau \\ &= di(\xi)\tau, \end{aligned}$$

so by the hypothesis of the proposition

$$D_\xi \tau = 0 \qquad \text{if and only if } \xi \in \mathrm{h}_\tau.$$

This proves the proposition.

Proposition 25.4 (Chu). *Suppose that* G *is simply connected. Then every* H_τ *is closed.*

Proof. Let us define the affine transformation, $B(\xi)$ of g^* into itself by

$$B_\xi \beta = D_\xi \beta + i(\xi)\tau, \qquad \beta \in g^*.$$

Then

$$D_\xi B_\eta \beta = D_\xi D_\eta \beta + i([\xi, \eta])\tau + i(\eta) D_\xi \tau.$$

Therefore

$$D_\xi B_\eta - D_\eta B_\xi - B_{[\xi,\eta]} = [i(\eta)\, di(\xi) - i(\xi)\, di(\eta) + i[\xi, \eta]]\tau.$$

Now $[(i(\eta)\, di(\xi))\tau](\zeta) = -\tau(\xi, [\eta, \zeta])$ so the right-hand side of the above equation when evaluated on some ζ is given by

$$\tau(\xi, [\eta, \zeta]) - \tau(\eta, [\xi, \zeta]) - \tau([\xi, \eta], \zeta) = 0$$

by (24.2) since $\delta\tau = 0$. Thus the map $\xi \to B_\xi$ is a homomorphism of g into the Lie algebra of all infinitesimal affine transformations of g^*. If G is simply connected, we can find a corresponding homomorphism of G into the group of all affine transformations of g^*. It is then clear that h is the Lie algebra of the subgroup G_0 keeping the origin fixed in this action. So H_τ is the connected component of a closed subgroup and hence closed.

The past few propositions and theorems dealt with the case where the map Ψ was (locally) a diffeomorphism. We now want to discuss the general situation. Many of the theorems we will present take on a simpler form if we make the assumption that $H^1(g) = H^2(g) = 0$, or, more generally, that we are given a Hamiltonian group action, to be defined in the next section. We present the more general formulation for symplectic group actions here for the sake of completeness. The reader may therefore prefer to skip to the next section. First, however, we wish to remind the reader of the notion of "clean intersection," which we will use below and also quite frequently in what follows.

Let $f: X \to Y$ be a smooth map between two differentiable manifolds, and suppose that W is an embedded submanifold of Y. We say that the map f intersects W *cleanly*, if

(i) $f^{-1}(W)$ is a submanifold of X

and

(ii) at each $x \in f^{-1}(W)$, $T(f^{-1}(W))_x = df_x^{-1}(TW_{f(x)})$.

For example, if f is transversal to W, so that for each $x \in f^{-1}(W)$

$$df_x(TX_x) + TW_{f(x)} = TY_{f(x)},$$

then it follows from the implicit function theorem (see Loomis and Sternberg, 1968) that properties (i) and (ii) hold. But f need not be transversal to W for the intersection to be clean. For example, if W lies in a submanifold Z of Y and f maps X into Z and is transversal to W inside Z then (i) and (ii) will hold, even though f is not transversal to W in the larger manifold Y. More generally, let $f: X \to Y$ and g: $W \to Y$ be two differentiable maps. We say that they intersect each other cleanly if the map (f, g): $X \times W \to Y \times Y$ intersects the diagonal in $Y \times Y$ cleanly.

With the notion of clean intersection in hand, let us now return to the study of a symplectic action of a group G on a symplectic manifold M with the associated map Ψ: $M \to Z^2(g)$.

Let $\mathscr{P}$ be a G orbit in $Z^2(g)$ and suppose that Ψ intersects $\mathscr{P}$ cleanly, so that $\Psi^{-1}(\mathscr{P})$ is a submanifold of M, and, at each $m \in \Psi^{-1}(\mathscr{P})$, the tangent space

$$V = T(\Psi^{-1}(\mathscr{P}))_m$$

is given by

$$V = d\Psi_m^{-1}(T\mathscr{P}_{\Psi(m)}).$$

Let $g_M(m)$ denote the subspace of TM_m consisting of all vectors $\xi_M(m)$ as ξ ranges over g. We claim that

$$V = g_M(m) + [g, g]_M(m)^\perp, \tag{25.7}$$

where the $\perp$ is relative to the symplectic form on TM_m. Indeed, since Ψ is equivariant, it is clear that the $\xi_M(m)$ all lie in V and that their image under $d\Psi_m$ span $T\mathscr{P}_{\Psi(m)}$. We thus have

$$V = g_M(m) + \ker d\Psi_m.$$

Now for any pair $\tilde{v}$, $\tilde{w}$ of symplectic vector fields on M we have

$$d(i(\tilde{v})\Omega) = 0$$

so

$$\begin{aligned} -D_{\tilde{w}}(i(\tilde{v})\Omega) &= i([\tilde{w}, \tilde{v}])\Omega \\ &= i(\tilde{w})d(i(\tilde{v})\Omega) + d(i(\tilde{w})i(\tilde{v})\Omega) \\ &= d(\Omega(\tilde{v} \wedge \tilde{w})). \end{aligned}$$

Applied to vector fields of the form ζ_M and η_M we conclude that

$$i([\eta_M, \zeta_M])\Omega = d\langle \Psi, \eta \wedge \zeta \rangle$$

so, for any vector v in TM_m we have

$$\langle d\Psi_m(v), \eta \wedge \zeta \rangle = (v, [\eta, \zeta]_M(m))_m, \tag{25.8}$$

where $(\,,)_m$ denotes the symplectic form, Ω_m, on TM_m. This shows that

$$\ker d\Psi_m = [g, g]_M(m)^\perp$$

and proves (25.7). Notice that (25.7) implies that V is coisotropic with

$$V^\perp = [g, g]_M(m) \cap g_M(m)^\perp.$$

But for any $\xi \in g$, we have $(\xi_M, \eta_M)_m = \langle \Psi(m), \xi \wedge \eta \rangle$ so

$$\xi_M(m) \in g_M(m)^\perp \qquad \text{iff} \qquad \xi \in h_{\Psi(m)}. \tag{25.9}$$

For any $\omega \in Z^2(g)$ let

$$t_\omega = h_\omega \cap [g, g]. \tag{25.10}$$

Then t_ω is a subalgebra of g, and we let T_ω denote the connected subgroup it generates. Notice that T_ω is contained in the isotropy subgroup of ω. It follows from the above discussion that

$$V^\perp = (t_{\Psi(m)})_M(m),$$

which is just the tangent space to the orbit of the group $T_{\Psi(m)}$ through the point m. To summarize, we have proved

Theorem 25.5. *Let* $G \times M \to M$ *be a symplectic* G *action whose associated map* $\Psi: M \to Z^2(g)$ *intersects cleanly with a* G *orbit* $\mathscr{P} \subset Z^2(g)$. *Then* $\Psi^{-1}(\mathscr{P})$ *is a coisotropic submanifold of* M. *The null foliation passing through any point* m *in this submanifold is the orbit through* m *of the group* $T_{\Psi(m)}$.

By a slight modification of the preceding argument we can prove the following:

Theorem 25.6. *Let* $G \times M \to M$ *and* $G \times N \to N$ *be two symplectic actions whose associated maps* $\Psi_M: M \to Z^2(g)$ *and* $\Psi_N: N \to Z^2(g)$ *intersect cleanly. Then* $\Psi_M^{-1}(\Psi_N(N))$ *is a coisotropic submanifold of* M. *Let* m *be a point of this submanifold and* n *a point of* N *with* $\Psi_M(m) = \Psi_N(n)$. *Let* g_n *denote the isotropy subalgebra of* n let $t_n = g_n \cap [g, g]$ *and let* T_n *denote the connected subgroup generated by the subalgebra* t_n. *Then the leaf of the null foliation through* m *is the orbit of* m *under the subgroup* T_n.

Proof. The fact that $\Psi_M^{-1}(\Psi_N(N))$ is coisotropic follows from Theorem 25.5, since it is a union of $\Psi^{-1}(\mathscr{P})$, which are coisotropic, provided that the relevant intersections are clean. But we can also see this directly by a double application of (25.8). Let W denote the tangent space to $\Psi_M^{-1}(\Psi_N(N))$ at m. Then W consists of all $v \in TM_m$ such that there exists a $w \in TN_n$ with

$$(v, [\xi, \eta]_M)_m = (w, [\xi, \eta]_N)_n$$

for all ξ, η in g. Taking $w = 0$ in this equation shows that $W \supset [g, g]_M(m)^\perp$ and clearly $W \supset g_M(m)$ so $W \supset V$, where V is given by (25.7), and hence W is coisotropic. Also, $W^\perp \subset [g, g]_M(m)$, so we must determine which elements of $[g, g]$ give rise to an element of $W^\perp$. Let γ be an element of $\wedge^2 g$ and let (γ) denote its image in g under the Lie bracket mapping $\wedge^2 g \to g$. Then (25.8) says that

$$(v, (\gamma)_M)_m = \langle d\Psi_M(v), \gamma \rangle$$

and v lies in W if and only if $d\Psi_M(v)$ lies in $d\Psi_M(TM_m) \cap d\Psi_N(TN_n)$ so $(\gamma)_M(m)$ lies in $W^\perp$ if and only if γ is orthogonal to the above intersection; that is, if and only if

$$\gamma \in (d\Psi_M(TM_m))^0 + (d\Psi_N(TN_n))^0.$$

Decompose $\gamma = \gamma_1 + \gamma_2$ by the above, where $\gamma_1 \in (d\Psi_M(TM_m))^0$ and $\gamma_2 \in (d\Psi_N(TN_n))^0$. Then another two applications of (25.8) show that

$$(\gamma_1)_M(m) = 0 \qquad \text{and} \qquad (\gamma_2)_N(n) = 0$$

since $(\ ,\ )_m$ and $(\ ,\ )_n$ are nondegenerate. Thus $(\gamma)_M(m) = (\gamma_2)_M(m)$ with $(\gamma_2)_N(n) = 0$; that is, $(\gamma_2) \in t_n$. This completes the proof of Theorem 25.6.

26. The moment map and some of its properties

Suppose that we are given a symplectic action of a Lie group G on a manifold M, with the associated map $\Psi: M \to Z^2(g)$. If $H^1(g) = H^2(g) = \{0\}$, then for each $m \in M$ there is a unique element $\Phi(m)$ of g^*, with $\delta(\Phi(m)) = \Psi(m)$. In other words, there is a unique map $\Phi: M \to g^*$ with

$$\delta\Phi = \Psi. \tag{26.1}$$

We can, however, be more explicit about the map Φ. The action of G on M gives a homomorphism of g into the set of vector fields on M, sending $\xi \in g$ into the vector field ξ_M on m, and ξ_M infinitesimally preserves ω, the symplectic form on M:

$$D_{\xi_M}\omega = 0.$$

But $D_{\xi_M}\omega = di(\xi_M)\omega + i(\xi_M)\, d\omega = di(\xi_M)\omega$ since $d\omega = 0$. If η is a second element of g with corresponding vector field η_M, so that

$$di(\eta_M)\omega = 0,$$

then

$$\begin{aligned} D_{\xi_M}(i(\eta_M)\omega) &= i(D_{\xi_M}\eta_M)\omega + i(\eta_M)D_{\xi_M}\omega \\ &= i([\eta_M, \xi_M])\omega \qquad \text{since} \quad D_{\xi_M}\omega = 0 \\ &= i([\eta_M, \xi_M])\omega \end{aligned}$$

and also

$$D_{\xi_M} i(\eta_M)\omega = di(\xi_M)i(\eta_M)\omega \qquad \text{since} \quad di(\eta_M)\omega = 0.$$

Thus, if $\zeta = [\eta, \xi]$, then $i(\zeta_M)\omega = df$, where f is the function $f = i(\xi_M)i(\eta_M)\omega$. Thus ζ_M is a Hamiltonian vector field corresponding to the function f. Now an examination of (24.1) shows that to say that $H^1(g) = \{0\}$ is the same as saying that $[g, g] = g$ – that every element of g is a linear combination of elements of the form $[\eta, \xi]$. Thus the assertion that $H^1(g) = \{0\}$ implies that we are given a homomorphism from g into the Lie algebra of Hamiltonian vector fields on M.

$$\begin{array}{c} 0 \to \mathbb{R} \to F(M) \xrightarrow{\rho} \mathrm{Ham}(M) \to 0 \\ \nwarrow_{\lambda} \quad \uparrow \\ g \end{array}$$

By Proposition 24.1 we know that this implies that there is a unique homomorphism $\lambda: g \to F(M)$ sending ξ into f_ξ for each $\xi \in g$, where

$$i(\xi_M)\omega = df_\xi.$$

The function f_ξ depends linearly on ξ and so, for each $m \in M$, we can consider the element $\Phi(m) \in g^*$ given by

$$\langle \Phi(m), \xi \rangle = f_\xi(m),$$

where $\langle\, , \rangle$ denotes the pairing between g^* and g. In other words,

$$d\langle \Phi, \xi \rangle = i(\xi_M)\omega \qquad \forall \xi \in g,$$

where we think of Φ as a map from M to g^*. (We will see in a moment that it is the unique map satisfying (26.1).) In the last equation, ξ is a constant element of g, so we can write the left-hand side as $\langle d\Phi, \xi \rangle$, where $d\Phi$ denotes the differential of the g^*-valued function Φ, or, if we like, $d\Phi_m$ gives a linear map of $TM_m \to g^*$ for each $m \in M$. Let $(\, ,)_m$ denote the symplectic bilinear form on TM_m given by ω at m. Then we can write the preceding equation as

$$\langle d\Phi_m(v), \xi \rangle = (\xi_M(m), v)_m \qquad \forall v \in TM_m. \tag{26.2}$$

Put another way, the action of G on M gives a linear map, $\xi \to \xi_M(m)$, of $g \to TM_m$ for each $m \in M$; the transpose of this map would be a linear map from $TM_m^* \to g^*$; but the symplectic form $(\, ,)_m$ allows us to identify TM_m^* with TM_m, and (26.2) can be reformulated as

$d\Phi_m$ *is the transpose of the evaluation map from* g *to* TM_m. (26.2)

Thus, for example, it follows that

$$\ker d\Phi_m = g_M(m)^\perp, \tag{26.3}$$

where $g_M(m)$ denotes the subspace of TM_m consisting of all vectors of the form $\xi_M(m)$. From (26.3) it follows that if G acts transitively on M, so that $g_M(m) = TM_m$ for all m, then $\ker d\Phi = 0$; that is, Φ is an immersion. It also follows from (26.2) that

$$(\mathrm{im}\ d\Phi_m)^0 = \{\xi \in g \text{ with } \xi_M(m) = 0\}, \tag{26.4}$$

where $(\text{im } d\Phi_m)^0 \subset g$ denotes the space of vectors in g that vanishes when evaluated on any element of im $d\Phi_m$. Notice that the right-hand side of (26.4) is the Lie algebra of the group G_m of elements which fix m; it is the stabilizer group of m. Thus

$$d\Phi_m \textit{ is surjective if and only if the stabilizer group of } m \textit{ is discrete.} \tag{26.5}$$

The group G acts on g via the adjoint representation and on vector fields on M due to its action on M. It follows that the mapping $\xi \to \xi_M$ is a G morphism (i.e., commutes with the action of G). Since the map is uniquely determined by the relation between ξ and ξ_M, it follows that the map Φ is a G morphism; that is,

$$\Phi(am) = a\Phi(m) \tag{26.6}$$

for all $a \in G$ and $m \in M$. The action used on the right-hand side of (26.6) is the coadjoint representation of G on g^*:

$$\langle al, \xi \rangle = \langle l, \text{Ad}_{a^{-1}}\xi \rangle, \qquad l \in g^* \text{ and } \xi \in g.$$

If we take $a = \exp t\eta$ to be a one-parameter group, we can differentiate the above equation at $t = 0$, to obtain the expression for the infinitesimal action of g on g^*: The infinitesimal action of η on g^* will be a vector field η_{g*} (which we may identify with a g^*-valued function on g^*), while

$$\frac{d}{dt}\text{Ad}_{a^{-1}}\xi_{t=0} = -[\eta, \xi],$$

and so we get

$$\langle \eta_{g*}(l), \xi \rangle = \langle l, [\xi, \eta] \rangle, \tag{26.7}$$

where $\eta_{g*}(l)$ denotes the value of the vector field η_{g*} at the point l. If we now take $a = \exp t\eta$ in (26.6) and differentiate at $t = 0$ we get

$$\langle d\Phi_m(\eta_M(m)), \xi \rangle = -\langle \Phi(m), [\eta, \xi] \rangle. \tag{26.8}$$

On the other hand, $\langle d\Phi, \xi \rangle = i(\xi_M)\omega$ by the defining property (26.2) of the map Φ. If we compare (26.8) with (25.3) we see that Φ is indeed the unique map satisfying (26.1). Note that Φ has all the properties we described for the total linear and angular momentum map of the group $E(3)$ in Section 19. So we call the map Φ the *moment map.* Let us summarize what we have proved so far:

Theorem 26.1. *Let* G *be a connected Lie group with* $\text{H}^1(\text{g}) = \text{H}^2(\text{g}) = \{0\}$. *Suppose that we are given a symplectic group action of* G *on a symplectic* *manifold* M *with symplectic form* ω. *The corresponding homomorphism of* g *into vector fields on* M *will be denoted by writing* ξ_M *for the vector field corresponding to* $\xi \in$ g. *Then there is a unique homomorphism* $\lambda : \mathrm{g} \to \mathrm{F(M)}$, *where* F(M) *denotes the space of smooth functions on* M, *which is a Lie algebra under Poisson bracket, which satisfies*

$$i(\xi_M)\omega = df_\xi, \tag{26.9}$$

where we have written f_ξ *for* $\lambda(\xi)$. *The homomorphism* λ *is a* G *morphism and allows us to define the map* $\Phi : \mathrm{M} \to \mathrm{g}^*$, *called the moment map, by*

$$\langle \Phi(\mathrm{m}), \xi \rangle = \mathrm{f}_\xi(\mathrm{m}). \tag{26.10}$$

The map Φ *satisfies*

$$\delta\Phi = \Psi, \tag{26.1}$$

$$\langle \mathrm{d}\Phi_{\mathrm{m}}(\mathrm{v}), \xi \rangle = (\xi_{\mathrm{M}}(\mathrm{m}), \mathrm{v})_{\mathrm{m}}, \tag{26.2}$$

and

$$\Phi(\mathrm{am}) = \mathrm{a}\Phi(\mathrm{m}) \tag{26.6}$$

and, as a consequence, (26.3), (26.4), and (26.5).

If $H^2(g) \neq \{0\}$ we know from Section 22 that there may be an obstruction, $[c] \in H^2(g)$ to finding the homomorphism λ. We also know that if $[c] = 0$, λ will exist but will not be unique unless $H^1(g) = \{0\}$. If the group G is not connected, then having chosen a homomorphism λ when possible, there is no guarantee that it is a G morphism. So for a general group G, we must postulate the existence and desired properties of λ: We say that a symplectic action of G on M is *Hamiltonian* if we are given a homomorphism $\lambda : g \to F(M)$, which is a G morphism and satisfies (26.9). If we are given a Hamiltonian group action, we define its moment map by (26.10). It then has all the properties described above.

Theorem 26.1 gives sufficient criteria for a symplectic group action to be Hamiltonian. These criteria are homological criteria on g. In addition, we can give sufficient criteria, involving the topology of M, for the existence of a Hamiltonian action associated to a given symplectic action: If $H^1(M, \mathbb{R}) = 0$, then every closed 1-form is exact. Therefore every symplectic vector field ξ is Hamiltonian – that is, every symplectic vector field is of the form ξ_f, where f is determined up to arbitrary constant. If M is compact, we can fix the constant by requiring that

$$\int f\omega^n = 0, \qquad \text{where } \dim M = 2n.$$

Notice that

$$\{f_1, f_2\}\omega^n = (D_{\xi_{f_2}} f_1)\omega^n = D_{\xi_{f_2}}(f_1\omega^n) \qquad \text{since } D_{\xi_{f_2}}\omega = 0$$
$$= d\big(i(\xi_{f_2})f_1\omega^n\big) \quad \text{since } d(f_1\omega^n) = 0.$$

Thus, by Stokes' theorem

$$\int \{f_1, f_2\}\omega^n = 0$$

for any pair of functions. In particular, the space of functions satisfying $\int f\omega^n = 0$ is a subalgebra, and is clearly G-invariant. We can therefore always use the condition $\int f\omega^n = 0$ to define a G-equivariant algebra homomorphism form g to $F(M)$. We have thus proved

Addendum to Theorem 26.1. *If we are given a symplectic action of any Lie group,* G, *on a compact manifold* M *satisfying* $\mathrm{H}^1(\mathrm{M}, \mathbb{R}) = 0$, *then we can get an associated Hamiltonian action by using the condition* $\int \mathrm{f}\omega^{\mathrm{n}} = 0$ *to fix the Hamiltonian function* f_ξ.

The rest of this section and much of the succeeding ones will be devoted to the study of further properties of the moment map. A reader who is more inclined to see some immediate physical applications may prefer to go directly to Section 28, and to read the rest of this section and the next one as needed.

Let $\mathcal{O}$ be an orbit of G in g^* and suppose that $\Phi: M \to g^*$ intersects $\mathcal{O}$ cleanly. Let m be a point of $\Phi^{-1}(\mathcal{O})$ and let W be the tangent space to $\Phi^{-1}(\mathcal{O})$. By the clean-intersection hypothesis,

$$W = d\Phi_m^{-1}(T\mathcal{O}_{\Phi(m)}).$$

Now G acts transitively on $\mathcal{O}$, and hence the tangent space $T\mathcal{O}_{\Phi(m)}$ is spanned by the vectors $\xi_{g*}(\Phi(m))$.† By the equivariance of Φ, we know that

$$d\Phi_m(\xi_M(m)) = \xi_{g*}(\Phi(m))$$

and so

$$W = g_M(m) + \ker d\Phi_m,$$

† We recall that if we are given an action of G on a manifold N; and if ξ is an element of the Lie algebra g than ξ_N is the vector field corresponding to ξ on N. The representation $\mathrm{Ad}^{\#}$ gives an action of G on g^*. Thus ξ_{g*} denotes the vector field on g^* corresponding to $\xi \in g$. If $\alpha \in g^*$ then $\xi_{g*}(\alpha)$ denotes the value of this vector field at α.

using the notation of (26.3). But, by (26.3) we have $\ker d\Phi_m = g_M(m)^\perp$ and so

$$W = g_M(m) + g_M(m)^\perp, \tag{26.11}$$

and therefore W is coisotropic with $W^\perp \subset W$ given by

$$W^\perp = g_M(m) \cap g_M(m)^\perp. \tag{26.12}$$

Now if $\xi \in g$ is such that $\xi_M(m) \in g_M(m)^\perp$ then by (26.3), $\xi_{g*}(\Phi(m)) = d\Phi_m(\xi_M(m)) = 0$. In other words, $\xi \in g_{\Phi(m)}$, the isotropy algebra of $\Phi(m)$. We have thus proved:

Theorem 26.2 (Kazhdan, Kostant, and Sternberg, 1978). *If* $\Phi\colon \mathrm{M} \to \mathrm{g}^*$ *intersects an orbit* $\mathcal{O}$ *cleanly, then* $\Phi^{-1}(\mathcal{O})$ *is coisotropic and the null foliation through a point* m *in* $\Phi^{-1}(\mathcal{O})$ *is the orbit of* M *under* $\mathrm{G}^0_{\Phi(\mathrm{m})}$, *the connected component of the isotropy subgroup of* $\Phi(\mathrm{m})$.

More generally, suppose that we are given two Hamiltonian G actions with moment maps

$$\Phi_M\colon M \to g^* \qquad \text{and} \qquad \Phi_N\colon N \to g^*,$$

which intersect cleanly. Let m be a point of $\Phi_M^{-1}(\Phi_N(N))$ and W the tangent space to this submanifold at m. Let n be a point of N with $\Phi_M(m) = \Phi_N(n)$. Then by (26.2) W consists of all $v \in TM_m$ such that there is a $w \in TN_n$ with

$$(v, \xi_M(m))_m = (w, \xi_N(n))_n$$

for all $\xi \in g$. Taking $w = 0$ shows that $W \supset g_M(m)^\perp$ and clearly $W \supset g_M(m)$, so that W is coisotropic. Also, $W^\perp \subset g_M(m)^{\perp\perp} = g_M(m)$, so our problem is to determine which $\xi \in g$ are such that $\xi_M(m) \in W^\perp$. Now $v \in W$ if and only if $d\Phi_m(v) \in d\Phi_M(TM_m) \cap d\Phi_N(TN_n)$, and thus, by (26.2), $\xi_M(m) \in W^\perp$ if and only if

$$\xi \in (d\Phi_M(TM_m) \cap d\Phi_N(TN_n))^0 = (d\Phi_M(TM_m))^0 + (d\Phi_N(TN_n))^0.$$

By (26.4), $\xi \in d\Phi_M(TM_m)^0$ if and only if $\xi_M(m) = 0$, and similarly for N. Hence, writing

$$\xi = \xi_1 + \xi_2$$

with $\xi_1 \in d\Phi_M(TM_m)^0$ and $\xi_2 \in d\Phi_N(TN_n)^0$ we see that $\xi_M(m) = \xi_{2M}(m)$, where $\xi_{2N}(n) = 0$. We have thus proved:

Theorem 26.3. *Let* G *have two Hamiltonian actions with moment maps* $\Phi_\mathrm{M}\colon \mathrm{M} \to \mathrm{g}^*$ *and* $\Phi_\mathrm{N}\colon \mathrm{N} \to \mathrm{g}^*$ *that intersect cleanly. Then* $\Phi_\mathrm{M}^{-1}(\Phi_\mathrm{N}(\mathrm{N}))$ *is a coisotropic submanifold of* M. *If* $\mathrm{m} \in \mathrm{M}$ *and* $\mathrm{n} \in \mathrm{N}$ *are such that*

$\Phi_M(m) = \Phi_N(n)$ *then the leaf of the null foliation through* m *is the orbit of* m *under* G_n^0, *the connected component of the isotropy group of* n.

If we take $N = \mathcal{O}$ and Φ the inclusion map of $\mathcal{O}$ in g^* we get Theorem 26.2 as a special case of Theorem 26.3. Actually, the crucial property of Φ_N that is really necessary in Theorem 26.3 is the fact that $\Phi_N(N)$ is an invariant submanifold of g^* under the coadjoint action of G. We can formulate a generalization of Theorem 26.3 that makes use only of the invariance property: Let $\mathscr{C}$ be an invariant submanifold of g^* (so that $\mathscr{C}$ is a set-theoretic union of orbits). At each point l of g^*, we have an identification of $(Tg^*)_l$ with g^* and hence of $(T^*g^*)_l$ with g. We can thus identify the normal space of $\mathscr{C}$ at l (i.e., the space $N\mathscr{C}_l$ of covectors vanishing on $T\mathscr{C}_l$) as a subspace of g. Explicitly,

$$\xi \in N\mathscr{C}_l \qquad \text{iff} \qquad \langle v, \xi \rangle = 0 \quad \forall v \in T\mathscr{C}_l. \tag{26.13}$$

Now since $\mathscr{C}$ is invariant, $\eta_{g*}(l) \in T\mathscr{C}_l$ for any $\eta \in g$. Then (26.7) and (26.13) imply that if $\xi \in N\mathscr{C}_l$, then

$$0 = \langle \eta_{g*}(l), \xi \rangle = \langle l, [\xi, \eta] \rangle = -\langle \xi_{g*}(l), \eta \rangle.$$

Since this must hold for all $\eta \in g$, we see that $\xi_{g*}(l) = 0$ (i.e., ξ is in the isotropy algebra of l). We have thus proved that

$$N\mathscr{C}_l \subset g_l.$$

Now the isotropy group G_l preserves l and $\mathscr{C}$ under the coadjoint action, and hence $T\mathscr{C}_l$ is an invariant subspace of g^* under the coadjoint action, and hence $N\mathscr{C}_l$ is an invariant subspace of g under the adjoint representation of G_l acting on g. In particular,

$$N\mathscr{C}_l \text{ is an ideal in } g_l \tag{26.14}$$

and hence a subalgebra of g. We will let $G_{\mathscr{C},l}$ denote the connected subgroup generated by $N\mathscr{C}_l$. We now assert

Theorem 26.4. *Let* $\mathscr{C}$ *be an invariant submanifold of* g*. *Let there be given a Hamiltonian action of* G *on a symplectic manifold* M *whose moment map* Φ *intersects* $\mathscr{C}$ *cleanly. Then* $\Phi^{-1}(\mathscr{C})$ *is a coisotropic submanifold of* M. *The leaf of the null foliation through a point* $m \in \Phi^{-1}(\mathscr{C})$ *is the orbit of* m *under the group* $G_{\mathscr{C},\Phi(m)}$.

Proof. Let $W = d\Phi_m^{-1}(T\mathscr{C}_{\Phi(m)})$. Then, as before, W contains $g_M(m) + g_M(m)^\perp$ and hence is coisotropic, with $W^\perp \subset g_M(m) \cap g_M(m)^\perp$. We must determine which ξ are such that $\xi_M(m) \in W^\perp$. By (26.2), this is the same as requiring that $\xi \in N\mathscr{C}_{\Phi(m)}$. Q.E.D.

As a corollary we can consider the special case $\mathscr{C} = \Phi(M)$, where the null foliation is trivial:

Corollary. *If* $\Phi(\mathrm{M}) = \mathscr{C}$ *and* $l = \Phi(\mathrm{m})$ *then* $\mathrm{G}_{\mathscr{C},l}$ *is contained in the connected component of the isotropy group* G_{m}. *If, in addition,* $d\Phi_{\mathrm{m}}(\mathrm{TM}_{\mathrm{m}}) = \mathrm{T}\mathscr{C}_l$ *then these groups are equal by (26.4).*

We shall let G_m^0 denote the connected component of this isotropy subgroup, G_m, so that $G_m^0 = G_{\mathscr{C},l}$ under both the hypotheses of the corollary. It follows from (26.13) that

$$G_m^0 \text{ is a normal subgroup of } G_l.$$

Suppose that the hypotheses of Theorem 26.3 are also satisfied, and let L_m denote the leaf of the null foliation of $\Phi^{-1}(\mathcal{O})$ through m. We know from Theorem 26.3 that $L_m = G_l^0 \cdot m$. If m' is some other point on L_m, so that $m' = am$ for some $a \in G_l^0$, then $G_{m'} = aG_m a^{-1}$ and, since G_m^0 is normal in G_l, we see that

$$G_{m'}^0 = G_m^0 \qquad \forall m' \in L_m.$$

It also follows immediately from the preceding discussion that G_l^0/G_m^0 acts locally freely on L_m. We will prove

Theorem 26.5. *If dim* $\mathcal{O}$ *is maximal among all* G *orbits in* $\mathscr{C}$ *then* $\mathrm{G}_l^0/\mathrm{G}_{\mathrm{m}}^0$ *is Abelian.*

Proof. From the definition of g_l it follows that if $\xi \in g_l$ then

$$\langle l, [\xi, \zeta] \rangle = 0 \qquad \forall \zeta \in g,$$

where $\langle\,,\rangle$ denotes the pairing between g^* and g. If dim $\mathcal{O}$ is maximal, this means that dim g_β is locally constant near $\beta = l$, and hence the g_β fit together to form a smooth vector bundle over $\mathscr{C}$ near l. Thus, given any smooth curve l_t in $\mathscr{C}$ with $l_0 = l$ and any ξ and $\eta \in g_l$, we can find smooth curves ξ_t and η_t, where

$$\xi_t \in g_{l_t}, \qquad \eta_t \in g_{l_t}$$

and

$$\xi_0 = \xi, \qquad \eta_0 = \eta.$$

Then

$$\langle l_t, [\xi_t, \eta_t] \rangle \equiv 0.$$

Let an overdot denote derivative at $t = 0$. Then differentiating the preceding equation gives

$$0 = \langle \dot{l}[\xi, \eta] \rangle + \langle l, [\dot{\xi}, \eta] \rangle + \langle l, [\xi, \dot{\eta}] \rangle.$$

The last two terms vanish since ξ and η are in g_l and $\dot{l}$ can be any tangent vector in $T\mathscr{C}_l$. Thus

$$[g_l, g_l] \subset T\mathscr{C}_l^0.$$

By the corollary to Theorem 26.4 this implies that

$$[g_l, g_l] \subset g_m,$$

proving the theorem.

Of course, we can forget about M and its moment map in the formulation of Theorem 26.5 and state it as

Theorem 26.5′. *If $\mathscr{C}$ is an invariant submanifold of* g^* *then at each point ℓ of $\mathscr{C}$ where dim $\mathcal{O}_\ell$ is maximal,* $[\mathrm{g}_\ell, \mathrm{g}_\ell] \subset \mathrm{N}\mathscr{C}_\ell$; *that is, the quotient algebra* $\mathrm{g}_\ell/\mathrm{N}\mathscr{C}_\ell$ *is Abelian.*

In the case that $\mathscr{C} = g^*$ this result is a theorem of Duflo and Vergne (1969). If the algebra g possesses a nondegenerate invariant bilinear form, we can identify the coadjoint action on g^* with the adjoint action on g. In this case, the theorem, with $\mathscr{C} = g^* = g$, asserts that the centralizer of a generic element is Abelian – a basic result in the theory of semisimple Lie algebras.

Let us now return to the moment map. One way of producing the situation in Theorem 26.4 is as follows: Suppose we have a Hamiltonian group action of G on M and another group H that acts on M, not necessarily preserving the symplectic structure, but commuting with the action of G. Suppose that we are also given a representation γ of H on g^* satisfying

$$\Phi(cm) = \gamma(c)\Phi(m), \qquad c \in H, m \in M. \tag{26.15}$$

This implies that $\gamma(c)$ commutes with the coadjoint action of G, at least on $im\ \Phi$. The principal case that we have in mind is where $H = \mathbb{R}^+$, so M is a "homogeneous" symplectic manifold, with every positive real number t acting on M, so as to satisfy $t^*\omega = t\omega$ (the $t\omega$ on the right means ordinary numerical multiplication of the form ω by the positive number t). Then we can also take $\gamma(t)$ to be multiplication by t on g^*, and (26.15) holds. If $\mathcal{O}$ is a G orbit, we can take $\mathscr{C} = \gamma(H) \cdot \mathcal{O}$ in Theorem 26.4. In the case that $H = \mathbb{R}^+$, the expression for $N\mathscr{C}_l$ given by (26.13) simplifies since in this case $T\mathscr{C}_l$ is spanned by the tangent space to the G orbit through l together with all multiples of l, so

$$N\mathscr{C}_l = g_l \cap l^0, \qquad \text{where } l^0 = \{\xi \in g \textit{ satisfying } \langle l, \xi \rangle = 0\}. \tag{26.16}$$

In each of Theorems 26.2–4 we have established that certain submanifolds of M are coisotropic and described their null foliations. If these foliations were fibrating, we could then form the reduced quotient manifolds, which would be new symplectic manifolds. We shall discuss sufficient conditions on Φ that guarantee that the foliations be fibrating in Section 27.

Here is another important variation on the theme of producing coisotropic submanifolds out of the moment map known as Marsden–Weinstein reduction. Let M and N be two symplectic manifolds with Hamiltonian G actions on them, whose moment maps are Φ_M and Φ_N. Then $M \times N$ is a Hamiltonian G space with moment map $\Phi_{M \times N}$ given by

$$\Phi_{M \times N}(m, n) = \Phi_M(m) + \Phi_N(n), \tag{26.17}$$

as can easily be checked. Also, if N is a symplectic manifold with form ω, let us write N^- for the same manifold, but with $-\omega_N$ taken to be the symplectic form. If we are given a Hamiltonian group action on N with moment map Φ, the same action gives a Hamiltonian group action on N^-, but with f_ξ replaced by $-f_\xi$ in (26.9), that is, with moment map $-\Phi_N$. Thus, if M and N are Hamiltonian G spaces, so is $M \times N^-$, and the corresponding moment map is given by

$$\Phi_{M \times N^-}(m, n) = \Phi_M(m) - \Phi_N(n). \tag{26.18}$$

In particular, we can apply this construction to the case where $N = \mathcal{O}$ is an orbit of G acting on g^*. For a coadjoint orbit $\mathcal{O}$ the moment map is just the injection of $\mathcal{O}$ as a submanifold of g^*, so (26.18) becomes

$$\Phi_{M \times \mathcal{O}^-}(m, l) = \Phi_M(m) - l. \tag{26.19}$$

We will now apply Theorem 26.2 to $M \times \mathcal{O}^-$ with the orbit in Theorem 26.2 taken to be the trivial orbit $\{0\}$. Then $\Phi_{M \times \mathcal{O}^-}$ intersects $\{0\}$ cleanly if and only if Φ_M intersects $\mathcal{O}$ cleanly and

$$\Phi_{M \times \mathcal{O}^-}^{-1}(0) = \{(m, l) \text{ with } \Phi_M(m) = l \text{ and } l \in \mathcal{O}\}.$$

Under the clean-intersection hypothesis, this is a coisotropic submanifold of $M \times \mathcal{O}^-$, and the null foliation through any point is the orbit of that point under the action of the connected component of the isotropy group of 0. The isotropy group of 0 is all of G, so if we assume that G is connected, the null foliation through any point (m, l) is just the orbit of (m, l) under the action of the full group G. The corresponding quotient space, $\Phi_{M \times \mathcal{O}^-}^{-1}(0)/G$ then carries a natural symplectic structure and this quotient space, together with its symplectic structure, is known as the Marsden–Weinstein reduced space of M at $\mathcal{O}$. Since G acts transitively on $\mathcal{O}$, we can (set-theoretically) identify

$$\Phi^{-1}_{M\times\mathcal{O}^-}(0)/G \qquad \text{with} \qquad \Phi_M^{-1}(l)/G_l \qquad \text{for any} \qquad l\in\mathcal{O}.$$

We shall denote the Marsden–Weinstein reduced space (together with its symplectic structure) by $M_{\mathcal{O}}$.

Theorem 26.2 associates a symplectic manifold to an orbit by taking the quotient of $\Phi_M^{-1}(\mathcal{O})$ by its null foliation, provided, of course, that this null foliation is fibrating. Let us temporarily, denote this quotient symplectic manifold by $Z_{\mathcal{O}}$. We can ask what is the relation between the quotient symplectic manifold $Z_{\mathcal{O}}$, provided by Theorem 26.2, and the Marsden–Weinstein reduced manifold, $M_{\mathcal{O}}$. We shall show that if the isotropy groups G_l are connected for some, and hence all, $l\in\mathcal{O}$, then

$$M_{\mathcal{O}}\times\mathcal{O}=Z_{\mathcal{O}}. \tag{26.20}$$

We now want to study in more detail the situation of Theorem 26.2. We assume that Φ_M intersects $\mathcal{O}$ cleanly, and that the null foliation of $\Phi_M^{-1}(\mathcal{O})$ is fibrating over the symplectic manifold $Z_{\mathcal{O}}$. Since $\Phi^{-1}(\mathcal{O})$ is invariant under the action of the group G, we know that each of the vector fields ξ_M is tangent to $\Phi_M^{-1}(\mathcal{O})$ and hence, by Theorem 25.3, the functions f_ξ are constant along the leaves of the null foliation, and therefore define functions, call them F_ξ on $Z_{\mathcal{O}}$. The group G preserves the null foliation on $\Phi_M^{-1}(\mathcal{O})$ and hence defines an action on $Z_{\mathcal{O}}$ preserving its symplectic structure. Furthermore, by the relation between the symplectic form on $Z_{\mathcal{O}}$ and the restriction of the symplectic form of M to $\Phi^{-1}(\mathcal{O})$, we see that the action of G on $Z_{\mathcal{O}}$ is Hamiltonian, with F_ξ being the function on $Z_{\mathcal{O}}$ corresponding to ξ in g. In other words, the moment map, $\Phi_{Z_{\mathcal{O}}}$ is constant on the fibers of $\Phi_M^{-1}(\mathcal{O})$ over $Z_{\mathcal{O}}$ and we have the commutative diagram

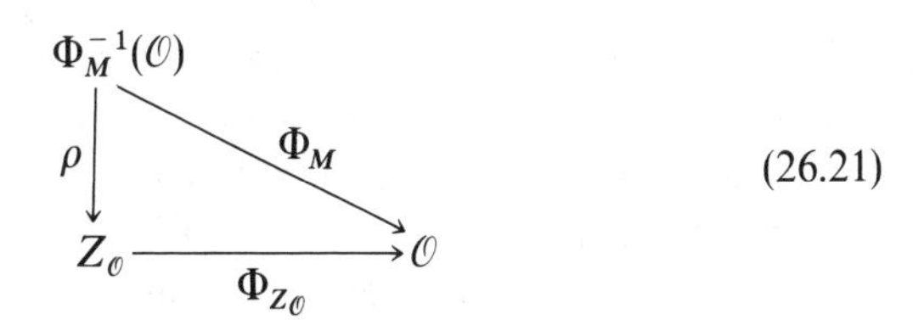

(26.21)

Now let z be a point of $Z_{\mathcal{O}}$, and let us denote its image $\Phi_{Z_{\mathcal{O}}}(z)\in\mathcal{O}$ by l. Let m be a point of $\Phi_M^{-1}(\mathcal{O})$ with $\rho(m)=z$, so that $\Phi_M(m)=l$. If $a\in G_l^0$, then $\rho(am)=\rho(m)=z$ by the description of the null foliation given by Theorem 26.2. Since $\rho(am)=a\rho(m)$, we conclude that

$$G_l^0\subset G_z.$$

On the other hand, since the map $\Phi_{Z_{\mathcal{O}}}$ is a G morphism, we know that

$$G_z \subset G_l.$$

Thus, if G_l is connected, we conclude that $G_z = G_l$ and hence that the map $\Phi_{Z_{\mathcal{O}}}$ gives a diffeomorphism of the G orbit, $G \cdot z$ through z and the orbit $\mathcal{O}$. On the other hand, let $F_l = \Phi_{Z_{\mathcal{O}}}^{-1}(l) = \Phi_M^{-1}(l)/G_l$, a manifold that we can identify with $M_{\mathcal{O}}$. We have the map

$$\gamma : \mathcal{O} \times F_l \to Z_{\mathcal{O}}, \qquad \gamma(al, z) = az,$$

which is well defined since $G_z = G_l$ for all $z \in F_l$. On the other hand, we have the map

$$m \to (m, \Phi(m)) \qquad \text{of} \qquad \Phi_M^{-1}(\mathcal{O}) \to \Phi_{M \times \mathcal{O}}^{-1}(0),$$

which induces a map of $Z_{\mathcal{O}}/G \to \Phi_{M \times \mathcal{O}}^{-1}(0)/G = M_{\mathcal{O}}$ and we can denote this induced map by π. So we have a map

$$(\Phi_{Z_{\mathcal{O}}} \times \pi) : Z_{\mathcal{O}} \to \mathcal{O} \times M_{\mathcal{O}}$$

whose inverse is γ, proving that γ is a diffeomorphism. We will let the reader verify, using (26.3) and (26.8), that γ is an isomorphism of symplectic structures. We thus have

Theorem 26.6. *Under the hypotheses of Theorem 26.2, assume that the null foliation is fibrating over a symplectic manifold,* $\mathrm{Z}_{\mathcal{O}}$*. Then* $\mathrm{Z}_{\mathcal{O}}$ *is a Hamiltonian* G *space with moment map* $\Phi_{\mathrm{Z}_{\mathcal{O}}}$ *given by (26.18). We can identify* $\mathrm{Z}_{\mathcal{O}}/\mathrm{G}$ *with the Marsden–Weinstein reduced space* $\mathrm{M}_{\mathcal{O}} = \Phi_{\mathrm{M} \times \mathcal{O}^-}^{-1}(0)/\mathrm{G} = \Phi_{\mathrm{M}}^{-1}(l)/\mathrm{G}_l$ *for any* $l \in \mathcal{O}$*. If* G_l *is connected for some, and hence all* $l \in \mathcal{O}$*, then we have a symplectic diffeomorphism of* $\mathrm{Z}_{\mathcal{O}}$ with $\mathcal{O} \times \mathrm{M}_{\mathcal{O}}$ *and this is a* G *morphism identification when we regard* $\mathrm{M}_{\mathcal{O}}$ *as a trivial* G *space.*

In the past few theorems we have been producing coisotropic manifolds out of the moment map. Under the hypothesis that the null foliation is fibrating, we then get the reduced manifold, so we have been producing new symplectic manifolds out of old ones using the moment map and reduction. We can also produce symplectic submanifolds of M directly out of the moment map by the following construction:

Theorem 26.7. *Let* α *be a point of* g^**, let* $\mathcal{O}$ *be the* G *orbit through* α*, and let* Z *be a submanifold of* g^* *that intersects* $\mathcal{O}$ *transversally at* α*. Thus*

$$T\mathcal{O}_\alpha \cap TZ_\alpha = 0 \qquad \text{and} \qquad T\mathcal{O}_\alpha + TZ_\alpha = Tg^*_\alpha = g^*.$$

Let m *be a point of* M *with* $\Phi(\mathrm{m}) = \alpha$. *Then there is a neighborhood,* U, *of* m *such that* $\Phi^{-1}(\mathrm{Z}) \cap \mathrm{U}$ *is a symplectic submanifold of* M.

Proof. By the equivariance of the moment map, we know that $\Phi(M) \supset \mathcal{O}$ and hence Φ is transversal to Z at m. Thus $\Phi^{-1}(Z)$ is a submanifold near m. To complete the proof of the theorem, it is enough to show that $T(\Phi^{-1}(Z))_m = d\Phi_m^{-1}(TZ_\alpha)$ is a symplectic subspace of TM_m. Now a subspace W of a symplectic vector space is a symplectic subspace if and only if $W \cap W^\perp = \{0\}$, so if we define $W = T(\Phi^{-1}(Z))_m = d\Phi_m^{-1}(TZ_\alpha)$ we must show that

$$W \cap W^\perp = \{0\}.$$

Let $V_\alpha = Im\, d\Phi_m \cap TZ_\alpha$ so that

$$W = d\Phi_m^{-1}(V_\alpha).$$

Since $W \supseteq \ker d\Phi_m$ it follows from (26.3) that $g_M(m) \supseteq W^\perp$. But $W \cap g_M(m) = \{\xi_M(m) \mid d\Phi_m(\xi_M(m)) \in V_\alpha\} = \{\xi_M(m) \mid d\Phi_m(\xi_M(m)) = 0\}$ since $V_\alpha \cap T\mathcal{O}_\alpha = \{0\}$. Thus, by (26.3),

$$W \cap g_M(m) = g_M(m) \cap g_M(m)^\perp.$$

Hence

$$\begin{aligned} W \cap W^\perp &= W \cap (W^\perp \cap g_M(m)) = (W \cap g_M(m)) \cap W^\perp \\ &= W^\perp \cap (g_M(m) \cap g_M(m)^\perp) \end{aligned}$$

and we must show that this last intersection is $\{0\}$. Now if $\xi \in g$ is such that $\xi_M(m) \in W^\perp$ it follows from (26.2) that

$$\xi \in V_\alpha^0.$$

On the other hand, if $\xi_M(m) \in g_M(m)^\perp$ then by (26.3)

$$\xi_{g^*}(\alpha) = d\Phi_m(\xi_M(m)) = 0 \text{ so } \xi \in g_\alpha.$$

But $g_\alpha^0 = T\mathcal{O}_\alpha$ so

$$\xi \in V_\alpha^0 \cap g_\alpha = (V_\alpha + T\mathcal{O}_\alpha)^0 = \{0\}$$

since $V_\alpha + T\mathcal{O}_\alpha = g^*$. Q.E.D.

We conclude this section by going back to the Hartree–Fock example of Section 23 (Proposition 23.3) and describe how to generalize the criterion (23.8). Let $G \times M \to M$ be a Hamiltonian action of G with moment map Φ. We will show that

Theorem 26.8. *The orbit of* G *through* m $\in$ M *is symplectic if and only if the stabilizer group of* m *is an open subgroup of the stabilizer group of* Φ(m). (If G is compact or semisimple the stabilizer group of $\Phi(m)$ is connected, so this implies that the two stabilizer groups are the same.)

Proof. Suppose that the orbit through m is symplectic. Then this orbit is a transitive Hamiltonian G space whose moment mapping is the restriction of Φ. Thus if $\mathcal{O}$ is the image of $G \cdot m$, $\Phi_{|G \cdot m}$ is a covering of $\mathcal{O}$, so the stabilizer group of m has to be open in the stabilizer group of $\Phi(m)$. Conversely, if the criterion is satisfied $\Phi|_{G \cdot x}$ is a covering of $\mathcal{O}$. If ι is the inclusion of Gm in M then the symplectic forms on M and $\mathcal{O}$ are related by

$$\Phi^* \omega_{\mathcal{O}} = \iota^* \omega$$

so $\iota^* \omega$ is nondegenerate.

27. Group actions and foliations

In the preceding section we stated many theorems under auxiliary hypotheses of a differential topological nature. Thus we assumed that the moment map was transversal to an orbit or a submanifold in g^*, we assumed that certain foliations were fibrating, etc. In order to be sure that the theorems are not vacuous, we want to provide conditions that guarantee, under reasonably general circumstances that these hypotheses are in fact fulfilled. For this we need to present some standard facts about group actions and foliations. For the most part, our treatment follows the exposition of Palais (1957a,b; 1960a,b). We do not intend to be thorough or state the results in their natural generality, referring the reader to the book by Bredon (1972) for a systematic exposition. The results of this section will be rather technical, and the reader may prefer to come back to this section as the results are used, or omit it altogether.

We begin with an analysis of the situation of Theorem 26.2; here our exposition follows Kazhdan, Kostant, and Sternberg (1978). So M is a Hamiltonian G space for some Lie group G with moment map $\Phi: M \to g^*$, we let $\mathcal{O}$ be a G orbit in g^* and set

$$\mathscr{V} = \Phi^{-1}(\mathcal{O})$$

with $x \in \mathscr{V}$. As in the discussion preceding Theorem 26.2, we set

$$W_x = d\Phi_x^{-1}(T\mathcal{O}_{\Phi(x)}).$$

We would like to know whether there is some neighborhood U of x such that $\mathscr{V} \cap U$ is a submanifold with $W_x = T(\mathscr{V} \cap U)_x$. (This would yield the local version of the hypotheses of Theorem 26.2. If we could guarantee these facts at all $x \in \mathscr{V}$ we would have the hypotheses of the theorem.) One way of being sure that $\mathscr{V}$ is a submanifold near x would be to find vector fields $\zeta_1, \ldots, \zeta_k$, defined near x, that are linearly independent at all points near x, whose values at x span W_x and such that the flow generated by the ζ_i carry $\mathscr{V}$ into itself. Indeed, suppose that we could find such vector fields. Applying exp $t\zeta_1$ to the point x we get a curve lying in $\mathscr{V}$. Applying exp $t\zeta_2$ to this curve we get a surface lying in $\mathscr{V}$, etc., until we get a k-dimensional manifold $\mathscr{V}' \subset \mathscr{V}$ passing through x with $T\mathscr{V}'_x = W_x$. Now dim $W_x = d + \dim \ker d\Phi_x$, where $d = \dim \mathcal{O}$, and dim ker $d\Phi$ is upper semicontinuous. Hence

$$\dim W_x \geqslant \dim W_y \qquad \forall y \in \mathscr{V} \text{ near } x$$

and, in particular, for all $y \in \mathscr{V}'$. By construction, $T\mathscr{V}'_y \subset W_y$ and dim $T\mathscr{V}'_y = \dim W_x$. So

$$W_y = T\mathscr{V}'_y$$

at all $y \in \mathscr{V}'$ (near x). Replacing $\mathscr{V}'$ by a smaller piece of it passing through x, if necessary, we find a k-dimensional manifold $\mathscr{V}'$ passing through x with

$$\Phi(\mathscr{V}') \subset \mathcal{O} \qquad \text{and} \qquad d\Phi_y^{-1}(T\mathcal{O}_{\Phi(y)}) = T\mathscr{V}'_y \qquad \forall y \in \mathscr{V}'.$$

But this clearly implies that we can find a tubular neighborhood of $\mathscr{V}'$ (and thus a neighborhood of x) with the property that no point not on $\mathscr{V}'$ is mapped into $\mathcal{O}$(i.e., $\mathscr{V}' \cap U = \mathscr{V} \cap U$ for a suitable neighborhood U of x). So the problem boils down to finding the vector fields $\zeta_1, \ldots, \zeta_k$. To this end we will try to apply (26.11). We can choose elements $\xi_1, \ldots, \xi_d$ in g so that the corresponding vector fields $\xi_{1g^*}, \ldots, \xi_{dg^*}$, when evaluated at $\Phi(x)$, form a basis for $T\mathcal{O}_{\Phi(x)}$. Then, if we set $\zeta_i = \xi_{iM}$ for $i = 1, \ldots, d$, the vector fields $\zeta_1, \ldots, \zeta_d$ are linearly independent at each point of M near x, and, when evaluated at x, form a basis of a subspace of TM_x, which, together with $g_M(x)^\perp$, spans all of W_x. So we must find vector fields that preserve $\mathscr{V}$ and whose value at x span $g_M(x)^\perp$. Now if f is a (locally defined) invariant function, so that $D_{\xi_M} f = 0$ for all $\xi \in g$, then ξ_f is orthogonal to $g_M(m)$ at all points. So the question becomes: Can we find (locally defined) functions $f_1, \ldots, f_r$ that are invariant under G, and such that their differentials at x,

$df_{1x}, \ldots, df_{rx}$ span the annihilator space of $g_M(x)$? If we can find such functions, we then set

$$\zeta_{d+1} = \xi_{f_1}, \ldots, \zeta_{d+r} = \xi_{f_r}$$

and then $\zeta_1, \ldots, \zeta_k$ with $k = d + r$ are our desired vector fields. Now we can certainly find such functions if the G orbit through x has the same dimension as the nearby G orbits, so that we can find a (local) cross section for the orbits. For then we take the f's to be coordinates on a cross-section of the orbits, so that the level surfaces of the f's are precisely the G orbits (locally). Thus, if $\dim G \cdot x = \dim G \cdot z$ for all z near x, then $\mathscr{V}$ is a submanifold near x. We claim that this is the case for an open dense set of x. Indeed, for any $y \in M$ let us set

$$d(y) = \dim G \cdot y$$

and

$$M_0 = \{x \mid d(y) \text{ is constant in some neighborhood of } x\}.$$

Our claim is that M_0 is an open dense subset of M. It is clearly open. To see that it is dense, define, for any $x \in M$ and any neighborhood W of x,

$$d(x, W) = \max_{y \in W} d(y)$$

and set

$$e(x) = \inf_{W} d(x, W).$$

We can choose some neighborhood W of x such that $d(x, W) = e(x)$. If now U is any neighborhood of x contained in W, then $\max_{y \in U} d(y) = e(x)$. Hence any $y \in U$ with $d(y) = e(x)$ lies in M_0, proving that x is in the closure of M_0 (i.e., that M_0 is dense). We have thus proved

Proposition 27.1. *There is an open dense subset of* M *with the property that for any* x *in this subset we can find a neighborhood* U *of* x *such that* $\mathscr{V} \cap \mathrm{U}$ *is a submanifold, where* $\mathscr{V} = \Phi^{-1}(\mathscr{O})$ *and* $\mathscr{O}$ *is the orbit in* g^* *containing* $\Phi(\mathrm{x})$. *In particular, if we take* M^* *to be the set of points whose orbits have maximal dimension then,* $\mathscr{V} \cap \mathrm{M}^*$ *is a closed submanifold of* M^*.

We can use some of the preceding discussion to provide an improvement of Theorem 26.2, and, in particular, give a description of the Marsden–Weinstein reduced space, $M_{\mathscr{O}}$ under certain circumstances. Suppose that we know that $\mathscr{V}$ is a closed submanifold of M with $W_x = T\mathscr{V}_x$ at all $x \in \mathscr{V}$. Let $l = \Phi(x)$. Then

$$\dim W_x = \dim \mathcal{O} + \dim \ker d\Phi_x$$
$$= \dim \mathcal{O} + \dim g_M(x)^\perp,$$

while
$$\dim \mathcal{O} = \dim g - \dim g_l$$

and
$$\dim g_M(x)^\perp = \dim M - (\dim g - \dim g_x)$$

so
$$\dim W_x = \dim M + \dim g_x - \dim g_l.$$

Now $\dim g_l$ is constant for $l \in \mathcal{O}$, and so the assumption that $W_x = T\mathscr{V}_x$, and hence has constant dimension, implies that $\dim g_x$ is constant, and hence that the orbits $G \cdot x$ have constant dimension for $x \in \mathscr{V}$. This means that we can find functions locally defined on $\mathscr{V}$, invariant under G and whose level surfaces are the G orbits of $\mathscr{V}$. We can extend these functions to be defined near $\mathscr{V}$, and then the corresponding vector fields span $g_M(x)^\perp$. For example, if the G action on $\mathscr{V}$ possessed a global cross section, then we could find functions $f_1, \ldots, f_r$, whose restriction to $\mathscr{V}$ is invariant under G, so that the corresponding vector fields span $g_M(x)^\perp$ at all points of $\mathscr{V}$. Finding a global cross section is very rare. But, in practice, it is sometimes not so hard to find a family of functions invariant under G or satisfying the weaker requirement of being invariant under G when restricted to $\mathscr{V}$. The set of functions on M invariant under G is a (usually infinite-dimensional) Lie algebra under Poisson bracket. It is sometimes possible to find a finite-dimensional algebra g' of such functions whose vector fields span $g_M(x)^\perp$ for all $x \in \mathscr{V}$. Suppose that the corresponding vector fields are globally integrable, and let G' denote the simply connected group corresponding to g'. Then $G \times G'$ acts locally transitively on $\mathscr{V}$ and hence transitively on the connected components, and preserves the restriction of the symplectic form of M to $\mathscr{V}$. It thus preserves the null foliation and, therefore, acts locally transitively on the quotient manifold S if the null foliation is fibrating. Then each component of S is a transitive symplectic $G \times G'$ space and hence covers an orbit of $G \times G'$ acting on $g^* + g'^*$. Such orbits are clearly products of G orbits with G' orbits, and the G orbit in question is clearly $\mathcal{O}$, so S looks locally like $\mathcal{O} \times \mathcal{O}'$, where $\mathcal{O}'$ is a G' orbit in g^*. We have thus identified the Marsden–Weinstein reduced space $M_{\mathcal{O}}$ as an orbit $\mathcal{O}'$ in g'^* for this case. We can summarize the previous discussion in the following theorem:

Theorem 27.1. *Suppose that* $\mathscr{V} = \Phi^{-1}(\mathcal{O})$ *is a submanifold and that* $\mathrm{W}_\mathrm{x} = \mathrm{T}\mathscr{V}_\mathrm{x}$ *at all* $\mathrm{x} \in \mathscr{V}$. *Then all the* G *orbits in* $\mathscr{V}$ *have the same dimension* *(locally). If* f *is a function that is* G*-invariant when restricted to* $\mathscr{V}$*, then the corresponding vector field* ξ_f *takes values in* $g_M(x)^\perp$ *at each* x *in* $\mathscr{V}$*. Suppose that* $g'_{\mathscr{V}}$ *is a finite-dimensional Lie algebra consisting of such vector fields, so that every element of* $g'_{\mathscr{V}}$ *is globally integrable, and such that* $g'_{\mathscr{V}}$ *spans* $g_M(x)^\perp$ *at each point. Let* $G'_{\mathscr{V}}$ *denote the corresponding simply connected Lie group. Then* $G \times G'_{\mathscr{V}}$ *acts transitively on the connected components of* $\mathscr{V}$*. Suppose that the null foliation of* $\mathscr{V}$ *is a fibration of* $\mathscr{V}$ *over the symplectic manifold* S*. Then* $G \times G'_{\mathscr{V}}$ *acts transitively on each connected component of* S *and each connected component of* S *is a covering space of* $\mathcal{O} \times \mathcal{O}'_{\mathscr{V}}$*, where* $\mathcal{O}'_{\mathscr{V}}$ *is an orbit of* $G'_{\mathscr{V}}$ *acting on* $g'^*_{\mathscr{V}}$*.*

We now return to purely group-theoretical considerations. In the preceding discussion, we made heavy use of the the notion of cross section for a group action. Roughly speaking, a local cross section through a point x will exist only if the G orbit though x looks the same as all nearby orbits, and the existence of a global cross section is a very stringent condition on a group action. There is a weaker notion than cross section that has a much broader range of existence, and that is the notion of a slice: Let X be a manifold upon which G acts differentiably and let x be a point of X and let G_x denote the isotropy group of x. A submanifold (or more generally a subset) S containing x is called a *slice* through x if

(i) S is closed in $G(S)$;
(ii) $G(S)$ is an open neighborhood of the orbit $G \cdot x$;
(iii) $G_x(S) = S$; and
(iv) $aS \cap S \neq \varnothing$ implies that $a \in G_x$.

Here are two examples of slices: Suppose that x is a fix point of G so that $G_x = G$. Then any G-invariant neighborhood of x (for instance, all of X) is a slice. A second, more interesting example is the following: Suppose that X is a vector bundle over a manifold M upon which G acts transitively, so $M = G/H$. We regard M as the zero section of the vector bundle X, and suppose that x is a point of this zero section. We then claim that the fiber through x is a slice. Indeed, condition (i) is clearly fulfilled. Condition (ii) holds because $G(S) = X$ in this case. Conditions (iii) and (iv) are obvious since elements a in G carry fibers into fibers and, hence, a will carry a point of the fiber over x into the fiber over ax. Similarly, if E were a vector bundle over M possessing a G-invariant positive definite scalar product, we could let X be a disk bundle of the form $X = \{v, \|v\| < r\}$. Then if x were a point of the zero section, as above, the fiber of X over x would again be a slice.

Now suppose that Y is a manifold upon which G acts, and suppose that Y carries a Riemannian metric invariant under G. Let x be a point of Y and set $M = G \cdot x$, the orbit through x. Let E be the normal bundle to M in Y. Then G acts on E as vector bundle automorphisms. We also have the exponential map Exp: $E \to M$, which is a G morphism, and is locally a diffeomorphism near each point of M. Suppose we could find some positive number r such that Exp is a diffeomorphism on the disk bundle of radius r. For instance, we could find such an r if M were compact. Then, since we can find a slice for the G action on the disk bundle, its image under Exp is a slice through x for the G action on Y. If G is a compact Lie group acting on Y we can always find an invariant Riemann metric by simply choosing any Riemann metric and then averaging over G. Also, every orbit $G \cdot x$ is compact. We have thus proved:

Proposition 27.2. *If a compact group* G *acts as diffeomorphisms of some manifold* Y *then there exists a slice* S *through every point* x *of* Y. *Furthermore we can choose coordinates on* S *so that* S *is an open invariant disk in a vector space upon which* G_x *acts linearly.*

(This is a special case, due originally to Koszul, of a theorem of Mostow, which asserts that slices exist through all points of any continuous action of a compact group.)

Suppose now that G, and hence G_x, is a compact Lie group and let G_x^0 denote the connected component of G_x. If G_x^0 acts trivially on S, this means that $G_x^0 \subset G_y$ for each $y \in S$, while $G_y \subset G_x$ by property (iv) of a slice. Thus, $\dim G_y = \dim G_x$ and hence $\dim G \cdot y = \dim G \cdot x$ and all nearby orbits to $G \cdot x$ have the same dimensions. If all nearby orbits do not have the same dimension as $G \cdot x$, then G_x^0 acts nontrivially; let us examine the possibilities in this case. Let V denote the vector space with the representation of G_x on it in which S lies, as provided by Proposition 27.2. Let us write

$$V = V_0 + V_1,$$

where V_0 denotes the subspace of vectors that are fixed by G_x^0, and V_1 is an invariant complement under G_x^0 upon which G_x^0 acts nontrivially. The points of S whose orbits have the same dimension as $G \cdot x$ are then precisely $S \cap V_0$, so the set of points near x whose orbits have the same dimension as $G \cdot x$ is a submanifold. Furthermore, we claim that $\dim V_1 \geqslant 2$. Indeed, the representation of G_x^0 on V_1 is nontrivial, and hence some one-parameter subgroup of G_x^0 acts nontrivially, so some ξ in the Lie algebra of G_x^0 acts nontrivially on V_1. But we can put an invariant scalar product on V_1, so ξ must be represented by a matrix with purely imaginary eigenvalues. Since V is real, these eigenvalues must occur in pairs if they are not zero, hence dim $V_1 \geqslant 2$. We have thus proved

Proposition 27.3. *Let* $r = \min \dim G_x$, *where* X *is a connected space upon which the compact group* G *acts differentiably. Then the set of points where* $\dim G_x > r$ *is a union of submanifolds of codimension at least 2, and hence its complement is connected.*

Thus the set of points whose orbits have smaller than maximal dimension has codimension at least 2. This is an improvement over the discussion preceding Proposition 27.1. Actually, a more refined analysis along the lines we have described will show that the set of points where dim $G_x > r + i$ has codimension greater than i, a result of Montgomery, Samuelson, and Zippin (1956).

For any point y in X, we have $G_{ay} = aG_y a^{-1}$, so the isotropy group at any point near x is conjugate to the isotropy group of a point of S. We can ask how many such conjugacy classes of subgroups of G can arise as isotropy groups. The answer is provided by the following special case of a theorem of Mostow (1957a) who proved the result for arbitrary actions of a compact group.

Proposition 27.4. *Let the compact Lie group* G *act differentiably on the manifold* X. *Then each point* x *of* X *has a neighborbood* U, *so that, up to conjugacy, only a finite number of subgroups of* G *arise as isotropy groups of points of* U. *In particular, if* X *is compact, only a finite number of conjugacy classes of subgroups of* G *can arise as conjugacy classes of isotropy groups of points of* X.

Proof. This is done by a triple induction: on the dimension of X, the dimension of G, and on the number of components of G. The theorem is trivially true if dim $X = 0$ or if $G = \{e\}$, so we have a starting point. By induction, we can assume that the proposition is true for any proper subgroup H of G, since such a subgroup must be either of lower dimension, or if of the same dimension have fewer components than G. We may also assume that the dimension of X is positive and that the theorem is true for the given G and all G spaces of dimension less than dim X. Let x be some point of X. There are two alternatives: x is or is not a fixed point of G. If x is not a fixed point of G, then G_x is a proper subgroup of G and we can find a slice S through x which is a differentiable G_x space. Since G_x is a proper subgroup of G, there is some neighborhood W of x in S for which only a finite number of subgroups of G_x appear, up to conjugacy as isotropy groups. But then $G(W)$ is a neighborhood of x in X all of whose points have isotropy groups conjugate to the isotropy groups of points in W. Now suppose that x is a fixed point of G. Using the exponential map from the tangent space to x into X, we can identify the G action near x with the action of G in a linear representation near the origin. The action preserves some metric. Let S be the unit sphere relative to this metric. If v is any nonzero vector in the representation space we clearly have $G_v = G_u$, where $u = v/\|v\|$, the unit vector in the direction of v. Thus the set of isotropy groups of nonzero vectors is the same as the set of isotropy groups for the action of G on the unit sphere S. But S has dimension less than X and is compact, so, by induction, gives rise to only a finite number of conjugacy classes of isotropy groups. Q.E.D.

If H is a compact Lie group acting differentiably on X, then the set of H-fixed points is a submanifold of X. Indeed, using an H-invariant Riemann metric, the set of H-fixed points near an H-fixed point x is just the image under the exponential map of the H-fixed vectors in the tangent space TX_x, and these form a linear subspace of TX_x. In particular, this applies when H is a closed subgroup of a compact Lie group G acting differentiably on X. On the other hand, it follows from the arguments leading to Proposition 27.3 that the set of points x whose isotropy group G_x strictly contains H will be a submanifold of the manifold of fixed points of H. We have thus proved

Proposition 27.5. *Let* G *be a compact Lie group acting differentiably on a manifold* X *and let* H *be a closed subgroup of* G. *Let* X_H *denote the set of points whose isotropy group is* H *that is,*

$$X_H = \{x \in X \text{ with } G_x = H\}.$$

Then X_H *is a submanifold of* X *and the tangent space* $T(X_H)_x$ *to* X_H *at any point* $x \in X_H$ *consists of the space of* H*-fixed vectors in* TX_x.

In the case that X is a symplectic manifold and G acts as symplectic diffeomorphism, we claim that X_H is a symplectic submanifold of X. Indeed, we need to show that at each point of X_H, the space of H-fixed vectors is a symplectic subspace of the symplectic vector space TX_x. So we need to prove the following.

Lemma 27.1. *Let* V *be a symplectic vector space and* H *a compact subgroup of* $Sp(V)$. *Then the space of* H*-fixed vectors is a symplectic subspace.*

Proof. Let $(,)$ denote the symplectic form on V and let $b(,)$ denote an H-invariant positive definite scalar product on V. There exists a unique linear map, $A: V \to V$ such that

$$b(u, v) = (u, Av), \qquad \forall\, u, v \text{ in } V.$$

and A clearly commutes with all elements of H. Let W denote the space of H-fixed vectors in V. We must show that $(,)$ is nondegenerate when restricted to W. Suppose that $u \in W$ satisfies $(u, v) = 0$ for all $v \in W$. Since $AW \subset W$, this implies that $b(u, v) = (u, Av) = 0$ for all $v \in W$, which implies that $v = 0$ since b is positive definite. Q.E.D.

Suppose that the action of G is Hamiltonian with moment map Φ, and $x \in X_H$. By (26.4), $d\Phi(TX_x) = h^0$, and so Φ maps each connected component of X_H into an affine subspace of the form $l + h^0$.

Suppose that H is a normal subgroup of G, so that X_H is a G-invariant symplectic manifold and the restriction of Φ to X_H is the moment map for the G action on X_H. Applying (26.4) to this restricted moment map shows that $d\Phi_x(TX_H)_x = h^0$, so the map Φ, when restricted to X_H, is a submersion.

In particular, if we take $H = G$, then X_G is the manifold of fixed points, a union of symplectic submanifolds of X, and Φ maps each connected component of X_G onto a single point of g^*. We have thus proved

Theorem 27.2. *Let the compact Lie group* G *act as a group of symplectic automorphisms of the symplectic manifold* X *and let* H *be a closed subgroup of* G. *Then* X_H *is a symplectic submanifold. If the* G *action is Hamiltonian with moment map* Φ *then* Φ *maps each component of* X_H *into an affine subspace of* g^* *of the form* $l + h^0$. *If* H *is a normal subgroup, then* X_H *is* G*-invariant, and the restriction of the moment map to each component is a submersion onto an open subset of the affine space. If we take* H = G *then* Φ *maps each component of* X_G *to a point. If* X *is compact, there will be finitely many components. The corresponding points in* g^* *are called the vertices of* Φ.

Suppose that X is compact and that $G = T$ is a torus. Then conjugation in T is trivial, and a finite number of subgroups of T can arise as isotropy groups of points of X. For each such subgroup H, X_H will be a submanifold with a finite number of components. Let us list all the various components of the various X_H as $X_1, \ldots, X_N$ and the corresponding subgroups as $T_1, \ldots, T_N$ so that

$$X = \bigcup_{i=1}^{N} X_i \qquad \text{(disjoint union)}$$

$$X_i \qquad \text{is a component of } X_{T_i}.$$

The T_i's are labeled with possible repetition. Let $v_1, \ldots, v_k$ be the vertices of Φ. We have

Theorem 27.3. *Each* X_i *is a* T-*invariant symplectic submanifold of* X. *For each* i *there is a vector* $l_i \in t^*$ *such that* $\Phi(X_i)$ *is an open subset of the affine subspace* $l_i + t_i^0$, *and* Φ_{X_i} *is a submersion onto its image. Each* $\Phi(X_i)$ *is the union of a finite number of open convex sets.*

Proof. The first two assertions follow immediately from Theorem 27.2. Let T_i have codimension m in T. So $\Phi(X_i)$ is a bounded open subset of the m-dimensional affine plane $l_i + t_i^0$. The boundary of $\Phi(X_i)$ lies in $\Phi(X)$ and a point of this boundary is the image of a point of X where $d\Phi$ has lower rank. Thus the boundary of $\Phi(X_i)$ will be a union of sets of the form $\Phi(X_j)$, where T_j strictly contains T_i. This shows, first of all, that the collection of vertices is nonempty. The theorem then follows by induction on the codimension of T_i.

We also state an immediate consequence of Proposition 27.3 for Hamiltonian group actions, using (26.4)

Theorem 27.4. *Let there be given a Hamiltonian group action of the compact Lie group* G *on the symplectic manifold* X *with moment map* Φ, *and suppose that min corank* $d\Phi = r$. *Then the set of points x, where corank* $d\Phi > r$ *is a union of submanifolds of codimension* >2. *Thus the set where corank* $d\Phi = r$ *is connected.*

We now turn to a description of some of the issues involved in deciding whether or not a foliation is fibrating, referring the reader to Reeb (1952) or Palais (1957), for more details. We begin with two examples of one-dimensional foliations in two dimensions that are not fibrations, and which exhibit the key problems. The first example is the straight line flow on the two-dimensional torus: Let $T = \mathbb{R}^2/\mathbb{Z}^2$ be the two-dimensional torus and let the group $\mathbb{R}$ act on T by letting $t \in \mathbb{R}$ send the point $[x, y]$ into $[x + at, y + bt]$. Here $[x, y]$ denotes the point on the torus corresponding to the point (x, y) in the plane (i.e., $[x, y] = (x, y) \bmod \mathbb{Z}^2$). Here a and b are real numbers, and it is clear that the action is well defined for any choice of a and b. If a/b is irrational, it is clear that every orbit is dense on the torus. Let B be the space of leaves of this foliation, and ρ: $T \to B$ the projection map sending each point of T into the leaf on which it lies. Suppose we want to put a topology on B that makes ρ continuous. If U is an open subset of B relative to this topology, then $\rho^{-1}(U)$ would have to be a saturated open set of T, that is, an open set that is a union of leaves. But every leaf intersects

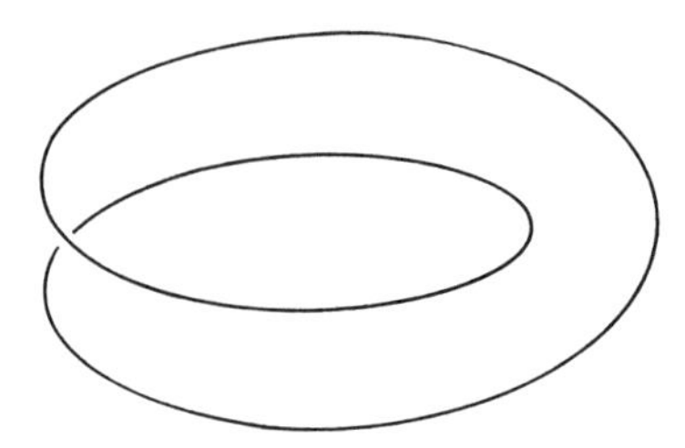

Figure 27.1.

any nonempty open set, so if $U \neq \varnothing, \rho^{-1}(U) = T$ and so $U = B$. Thus the only topology that we can put on B is the trivial topology in which the only open sets are B and $\varnothing$.

It is fairly clear that this sort of pathology cannot occur if all the leaves are compact. So let us examine what can go wrong when there are compact leaves, or, more important, where each leaf is the orbit of a compact Lie group. For this purpose we look at the horizontal foliation of the Möbius band: Define transformation $A: \mathbb{R}^2 \to \mathbb{R}^2$ by

$$A(x, y) = (x + 1, -y)$$

and use this to define a $\mathbb{Z}$ action on $\mathbb{R}^2$ by letting $n \in \mathbb{Z}$ act as A^n. Then define

$$M = \mathbb{R}^2/\mathbb{Z}.$$

Clearly M is fibered over the circle S^1 by the projection map π given by

$$\pi([(x, y)]) = x \bmod \mathbb{Z}$$

and $\pi^{-1}(\theta)$ is a line, for $\theta \in S^1$, but the lines get identified with a twist as we go around the circle, so M is a Möbius band (Figure 27.1). Now the action of $\mathbb{R}$ on $\mathbb{R}^2$ given by

$$f_t(x, y) = (x + t, y)$$

clearly commutes with A and hence defines an action F of $\mathbb{R}$ on M:

$$F_t\pi(x, y) = \pi(x + t, y).$$

Note that $F_2 = id$, so F defines an action of $\mathbb{R}/2\mathbb{Z} = G$ on M. Note also that $F_1\pi(x, y) = \pi(x, -y) \neq \pi(x, y)$ unless $y = 0$. So the orbits of the circle G acting on M wrap around twice; they are two-fold coverings of the center circle with $y = 0$ (except for the center circle itself, which is a single orbit). So the space of leaves looks like a half-line; it can be parametrized by all $y \geqslant 0$. This is a nice topological space, but is not a manifold near $y = 0$.

Working from this Möbius band example, we can formulate two closely related theorems – one for compact group actions and one for general foliations.

Suppose that M is a manifold with a foliation. To put a manifold structure on the space B of leaves, it is enough to show that through any point on any leaf L we can pass a submanifold S, transverse to the leaf such that S meets any leaf at most once. Then we can use the local coordinates on the various S's as local coordinates on B, giving B a manifold structure. (Of course, B need not be Hausdorff relative to the induced topology, but if M is Hausdorff and all the leaves are compact, then a straightforward topological argument shows that the quotient topology is Hausdorff as well.) So our problem is one of examining what happens near each leaf.

Suppose that G is a compact Lie group acting on the manifold M and that all orbits of G have the same dimension, so the orbits form leaves of a foliation on M. Let S be a slice through a point x, as provided by Proposition 27.2. Since all orbits have the same dimension, S is transversal to all nearby orbits. The isotropy group G_x leaves S invariant, so, in particular, does its connected component, G_x^0. The action of G_x^0 on S must be trivial, otherwise we could find a tangent vector to a G_x^0 orbit in S which would be both tangent to S and to a G orbit, contradicting the constancy of the orbit dimensions. Thus G_x^0 acts trivially. Let $\Gamma_x = G_x/G_x^0$. Then the action of G_x on S gives an action of Γ_x on S which, by Proposition 27.2, we know to be equivalent to a linear representation equivalent in fact, to the representation of Γ_x on the normal space to the orbit at x.

Proposition 27.6. *A necessary and sufficient condition for the* G *orbit foliation to be fibrating is that the representation of* Γ_x *on the normal space* N_x *to the orbit* $G \cdot x$ *be trivial for all* x.

(In the Möbius band example, if x is a point of the center circle, then $\Gamma_x = \mathbb{Z}_2$, and its representation is nontrivial.)

Proof. If the foliation is fibrating, then we can identify N_x with the cotangent space to B at $\rho(x)$ and so G_x and hence Γ_x acts trivially on N_x. Conversely, if the representation of Γ_x on N_x is trivial, then so is the action of G_x on S. We then have a map

$$\psi : S \times G/G_x \to X,$$

$$\psi(x', aG_x) = ax',$$

which is well defined and one-to-one since $G_x \subset G_{x'}$ and by property (iv) of a slice, it is continuous and a local diffeomorphism (and hence a global diffeomorphism) onto some neighborhood of $G \cdot x$. Q.E.D.

Thus, for example, in deciding whether or not the Marsden–Weinstein reduced space $M_{\mathcal{O}}$, is or is not a manifold, we are looking at foliations coming from G actions, and so must examine representations of finite groups in order to check. We shall see that in many cases the reduced space is not a manifold, but still, as the example of the Möbius band shows, the reduced space is not too bad. In fact the preceding argument shows in general that:

Proposition 27.7. *If* G *is a compact Lie group acting on the manifold* X *and if* S *is a slice at* x *then the map* $G \times S \to X$ *given by the action of* G *induces a homeomorphism of* S/G_x *with* $G(S)/G$.

We can also use the example of the Möbius band to introduce the notion of holonomy, which exists for any foliation, irrespective of any group action: Suppose we start with any point x' in a fiber of a point x in the center circle. Then there is a unique curve (in this case the leaf of the one-dimensional foliation) that passes through x' and projects onto the center circle. Going all the way around the circle brings us back to the same fiber over x but to another point x''. This transformation of x' into x'' measures the failure of the foliation to be fibrating. Now we mimic this construction in general. Of course, there may not, in general, be any transverse fibration, but we can put one on and see how much of the construction depends on the choice. So let M be a manifold with a foliation, F of codimension q, and let L be a leaf of this foliation. Let us put some Riemann metric on M; so we get an exponential map from the normal bundle $N(L)$ to the leaf into M, which will be a diffeomorphism on some neighborhood of the zero section, which we identify with L. Let x be a point of L and let γ be some piecewise-smooth path starting at x. Let $D_\epsilon(x)$ be the disk of radius ϵ in the normal space to L at x. If $\epsilon > 0$ is sufficiently small, there will be a unique curve γ' starting at any point x' in $D_\epsilon(x)$ that projects onto γ and whose image under the exponential map lies on a single leaf of the foliation. This gives a transformation of $D_\epsilon(x)$ into N_y, the normal space at y (where y is the end point of γ) by sending x' into the endpoint of γ'. It is easy to see that this transformation is smooth and depends only on the homotopy class of γ. In particular, if γ is a closed path, so that $y = x$, we get a transformation of $D_\epsilon(x)$ into N_x and thus we get a homomorphism of $\pi_1(L)$ into the group of germs of diffeomorphisms of N_x into itself at the origin. Let $G(q)$ denote the group of germs of diffeomorphism of $\mathbb{R}^q$ into itself at the origin. Choosing some coordinates on N_x we get a homomorphism of $\pi_1(L)$ into $G(q)$. This homomorphism depends on a number of choices, the choice of Riemann metric, of coordinates on N_x, etc. But it is easy to see that the homomorphisms coming from different choices differ by a conjugation in $G(q)$. This homomorphism, $h:\pi_1(L) \to G(q)$, determined up to a conjugacy, is called the holonomy homomorphism and its image in $G(q)$ is called the holonomy group of the foliation f at the leaf L. (There is a homomorphism of $G(q)$ onto $Gl(q)$ sending each germ into its Jacobian matrix at the origin. The composite of this homomorphism with h is called the linear holonomy homomorphism and its image the linear holonomy group of F at L.) The following is a special case of the Reeb Stability Theorem.

Proposition 27.8. *Let* F *be a smooth foliation and* L *be a compact leaf of* L *with trivial holonomy group. Then there exists a neighborhood* U *of* L, *consisting of a union of leaves, a disk* D *in* $\mathbb{R}^q$ *(where* q *is the codimension of the foliation), and a diffeomorphism* f *of* $L \times D$ *onto* U *with* $f_{L\times\{0\}} = id$. *In particular, if all the leaves of* F *are compact and simply connected (or, more generally, have trivial holonomy) then* F *is a fibration.*

Proof. Since L is compact, we can find some $\epsilon > 0$ such that the exponential map is a diffeomorphism on the portion of the normal bundle consisting of vectors of length less than ϵ. Also, the fundamental group $\pi_1(L)$ is finitely generated, so we can find a sufficiently small neighborhood of 0 in $\mathbb{R}^q$ so that the maps corresponding to generators of $\pi_1(L)$ are actually the identity (and not merely their germs). Thus, for a sufficiently small ϵ, the map going from x' to y' described above is well defined, and is independent of the path because of the vanishing holonomy. It provides the desired diffeomorphism, proving the proposition.

Given a foliation F, we can define the corresponding "relation," $\Sigma \subset X \times X$ by

$$\Sigma = \{(x, y) \text{ such that } x \text{ and } y \text{ lie on the same leaf of } F\}.$$

Proposition 27.9. *The relation* Σ *is an immersed submanifold of* $X \times X$. *It is an embedded submanifold if and only if all the leaves are closed and their holonomy groups are trivial.*

Proof. Let x and y be two points on the same leaf L of the foliation, and let γ be a curve in L joining x to y. As above, let $D(x)$ be a small disk transverse to L at x and $D(y)$ transverse to L at y so that the map φ_γ: $D(x) \to D(y)$

is defined. Let $y_1, \ldots, y_q$ be coordinate functions on $D(y)$ and define $x_i = \varphi_\gamma^* y_i$ $(i = 1, \ldots, q)$ to be functions on $D(x)$. Extend the functions $y_1, \ldots, y_q$ so as to be defined on a whole neighborhood V of y by making them constant on the leaves in V, and similarly extend the x's to be defined in some neighborhood U of x by making them constant along the leaves. Let π_1 and π_2 be the maps of $X \times X \to X$ given by projection onto the first and second components. Then $U \times V$ is a neighborhood of (x, y) in $X \times X$ and the equations

$$\pi_1^* x_i - \pi_2^* y_i = 0$$

give an immersed component, corresponding to φ_γ, of Σ in $U \times V$, and this procedure clearly gives all immersed components and proves that Σ is an immersed submanifold. If the holonomy is trivial, there is only one φ for a given (x, y) independent of γ, so Σ is embedded. Conversely, the functions $x_1, \ldots, x_q$ and $y_1, \ldots, y_q$ clearly determine φ_γ, so if Σ is embedded, the holonomy is trivial. Q.E.D.

28. Collective motion

In the next few sections we want to present several physical applications of the moment map; particularly to the construction and analysis of physical models, to the formulation of physical laws, and to the solution of certain equations of motion. In later sections we shall present some mathematical applications of the moment map; principally to the harmonic analysis of group representations. In this section we attempt to give a mathematical formulation to the notion of a "model" of a given mechanical system. An example of what we have in mind is to try to explain what we mean when we say that a system of point particles moves, under a given Hamiltonian, "as if it were a rigid body." More generally, what do we mean when we say that one physical system moves "as if it were" some other physical system? Thus, for example, in nuclear physics one has the "liquid drop model" of the nucleus in which it is assumed that an appropriate "model" for the mechanical system consists of N nucleons (considered as point particles) is a "liquid drop" – that the nucleus behaves "as if it were a liquid drop." Here, of course, we must explain what exactly we mean by the mechanical system consisting of a "liquid drop," but we also must explain, in more generality, what we mean by one system behaving as if it were another. To get a grip on this problem, let us examine the intuitive and familiar case of the rigid body. Suppose we had a system of N point particles. We can imagine that they are (almost) rigidly attached to one another, if there is some potential energy

$V = \sum V_{ij}$, where V_{ij} is a function of the distance between the ith and jth particle and takes on a very sharp minimum at certain specified distances r_{ij}. So we can imagine that the Hamiltonian of such a system has the form

$$V + K + K' + H,$$

where V is the potential energy described above, K is an internal kinetic energy involving the motion of the points relative to one another, K' is some kinetic energy of the overall center of mass, and H is a Hamiltonian (involving both kinetic and potential terms) but is a function solely of the total angular momentum and inertia tensor relative to the center of mass. (We shall discuss these concepts more fully below.) The contribution of K' is simply to give an overall linear motion to the center of mass, and so by introducing coordinates relative to the center of mass we can ignore it. (This is an elementary example of reduction relative to the action of the translation group $\mathbb{R}^3$ acting simultaneously on all particles. Here Marsden–Weinstein reduction and orbital reduction coincide, since $\mathbb{R}^3$ is commutative and so its orbits are points.) Therefore, we are looking at a Hamiltonian of the form

$$V + K + H.$$

Then the actual motion of the system would be described, approximately, as a "rigid-body motion" coming from H, together with a superimposed rapid oscillation coming from $V + K$. To a good approximation, we might expect to be able to ignore the rapid oscillation, which averages itself out over the rigid-body motion. The situation is even more convincing in quantum mechanics. There, one might expect that for relatively low energies one can assume that the internal state of the system, and so the observed spectrum, should look like that of a rigid rotor. The fact that one sees unmistakable rotational levels in complicated nuclear spectra supports this description of the system. Thus the rigid-body description comes from the property that H is a function of the total angular momentum and inertia tensor, or quadrupole moment: Let $q_1, \ldots, q_N$ be the position vectors (relative to the center of mass) and let $p_1, \ldots, p_N$ be the corresponding momenta. We can then define the functions L and Q on the phase space $\mathbb{R}^{6N}$ by

$$L = \sum p_i \wedge q_i \in \mathbb{R}^3$$

and

$$Q = \sum q_i \otimes q_i \in S^2(\mathbb{R}^3).$$

The assumption about H is that it is a function of L and Q (i.e., that $H = H(L, Q)$). Any such function can be said to define a "collective rigid-body Hamiltonian." Now the important point that we wish to make is that the maps $L: \mathbb{R}^{6N} \to \mathbb{R}^3 \sim o(3)^*$ and $Q: \mathbb{R}^{6N} \to S^2(\mathbb{R}^3)$ can be regarded as components of a moment map

$$\Phi = (L, Q),$$
$$\Phi: \mathbb{R}^{6N} \to g^*,$$

where G is the semidirect product of $SO(3)$ with $S^2(\mathbb{R}^3)$. Here we regard $S^2(\mathbb{R}^3)$ as the set of 3×3 symmetric matrices, and let any $B \in Gl(3, \mathbb{R})$ act on the symmetric matrix S by sending it into $(B^t)^{-1} S B^{-1}$. We will define a Hamiltonian action of G on $\mathbb{R}^{6N}$ and show that its moment map has the desired form. In fact, we shall define a homomorphism, $j: G \to Sp(6)$. The group $Sp(6)$ has a Hamiltonian action on $\mathbb{R}^6$, and hence on the direct product of N copies of $\mathbb{R}^6$ with itself, which we may identify with $\mathbb{R}^{6N}$. The homomorphism j will then induce a Hamiltonian action of G on $\mathbb{R}^{6N}$. We will then be able to compute its moment map by using the following principles:

(a) Let $j: G \to G_1$ be a homomorphism of Lie groups and let $i: g \to g_1$ be the induced homomorphism of the corresponding Lie algebras. Let $i^*: g_1^* \to g^*$ be the adjoint of i. Suppose that we are given a Hamiltonian action of G_1 on a symplectic manifold M with moment map $\Phi_1: M \to g_1^*$. Then the homomorphism j induces a Hamiltonian action of G on M and its moment map is given by

$$\Phi = i^* \circ \Phi.$$

This fact follows immediately from the definitions. We leave its verification to the reader.

(b) If we are given Hamiltonian actions of the group G on M and N with moment maps Φ_M and Φ_N then the action of G on $M \times N$ is Hamiltonian with moment map

$$\Phi_{M \times N}(m, n) = \Phi_M(m) + \Phi_N(n).$$

We have already mentioned this fact in Section 26.

We have computed the moment map for the action of the symplectic group on its symplectic vector space in Section 14, see (14.15). We found that the map

$$\Phi_1: V \to \mathrm{sp}(V)^*$$

is given by

$$\langle \Phi_1(v), \eta \rangle = \tfrac{1}{2}(\eta v, v) \qquad \eta \in \mathrm{sp}(V), v \in V.$$

Combining this with (a) and (b) will allow us to compute the moment map for the action of G on $\mathbb{R}^{6N}$ as

$$\langle \Phi(q_1, p_1, \ldots, q_N, p_N), \xi \rangle = \frac{1}{2} \sum_{k=1}^{N} \left((i(\xi)) \begin{pmatrix} q_k \\ p_k \end{pmatrix}, \begin{pmatrix} q_k \\ p_k \end{pmatrix} \right). \tag{28.1}$$

So everything hinges on the definition of the homomorphism $j: G \to Sp(6)$ and the associated homomorphism, $i: g \to sp(6)$.

We will define the homomorphism j in slightly greater generality, considering the case where G is the semidirect product of a subgroup H of $Gl(n)$ with the space of symmetric 2-tensors on $\mathbb{R}^n$ and where j is a homomorphism of G into $Sp(2n)$. The multiplication in G is given by

$$(A, S)(A', S') = (AA', S + (A^{-1})^t S A^{-1}).$$

We define

$$j((A, S)) = \begin{pmatrix} A & 0 \\ SA & (A^{-1})^t \end{pmatrix}.$$

The conditions for a matrix to be symplectic, given on p. 26, show that the image of j lies in $Sp(2n)$. Matrix multiplication shows that j is a homomorphism. We can write the typical element of g as (ξ, T), where T is a symmetric matrix and $\xi \in g$. Then

$$i((\xi, T)) = \begin{pmatrix} \xi & 0 \\ T & -\xi^t \end{pmatrix}.$$

Combined with our preceding formula for Φ this shows that

$$\langle \Phi(q_1, \ldots, p_N), (\xi, T) \rangle = \operatorname{tr} \xi J + \tfrac{1}{2} \operatorname{tr} TQ,$$

where

$$J(q_1, \ldots, p_N) = -\sum_k q_k \otimes p_k$$

and

$$Q(q_1, \ldots, p_N) = \sum q_k \otimes q_k.$$

Of course the J, as written, lies in the dual space of $gl(n)$. To get an element of h^*, we must apply the projection $\pi: gl(n)^* \to h^*$ to J. In the case that

$H = O(n)$, we may identify h and h^* with $\bigwedge^2(\mathbb{R}^n)$ and $L = \pi J$ is just the total angular momentum

$$L = \sum_i p_i \wedge q_i.$$

This completes the proof of our assertion that the total angular momentum L and the quadrupole moment Q can be regarded as components of the moment map. The crucial property of the collective rigid-body Hamiltonian $\mathscr{H}$ is that is can be written in the form

$$\mathscr{H} = F \circ \Phi, \tag{28.2}$$

where F is some smooth function defined in an open set containing the image of Φ in g^*. We can generalize from the rigid-body example: Let G be any Lie group with a Hamiltonian action on M whose associated moment map is Φ. A Hamiltonian $\mathscr{H}$ on M will be called *collective* if it is of the form (28.2).

For example, in the liquid drop model of the nucleus, one wants to imagine that the system of point particles behaves as an incompressible fluid, but that only the "quadratic approximation" to the shape of the liquid drop is what matters; in more mathematical language, what this means is that only the quadrupole moments matter. Thus, a possible configuration of a liquid drop is specified by a positive definite symmetric tensor $Q \in S^2(\mathbb{R}^3)$ and the group $Sl(3, \mathbb{R})$ acts on $S^2(\mathbb{R}^3)$ as above. The fact that we use $Sl(3, \mathbb{R})$ and not $Gl(3, \mathbb{R})$ is the expression of the fact that we are dealing with a "liquid" drop, and not a "gaseous" drop; in other words, that "volume" is preserved. In fact, $\det(AQA^t) = \det Q$ if $\det A = 1$, and so $\det Q$, thought of as the volume of the liquid drop is an invariant. If Q is positive definite, we can, by an appropriate A, bring Q to the form cI, where $c = (\det Q)^{1/3} > 0$. The isotropy group in $Sl(3, \mathbb{R})$ of cI is $SO(3)$, and so the corresponding orbit of Q in $S^2(\mathbb{R})$ is five-dimensional. Such an orbit N is the configuration space of the liquid drop of given volume; notice that cI is the unique point in N that is left fixed by $SO(3)$ – it corresponds to a spherical globule of liquid. In fact, the liquid drop model of the nucleus is one in which we take $H = Sl(3, \mathbb{R})$, instead of $SO(3)$ as in the rigid-body model, but where the Hamiltonian has the same form (28.2).

We now explain in general, what ingredients go into the solution of a "collective" Hamiltonian (those of the form (28.2)). For this purpose, we remind the reader that a smooth function F defined on an open subset U of a vector space (in this case the vector space is g^*) defines a map L_F of U into the dual space (in our case g) by the formula

$$\langle m, L_F(l)\rangle = dF_l(m)$$
$$= \frac{d}{dt}F(l + tm). \tag{28.3}$$

The map L_F is sometimes known as the Legendre transformation (see Sternberg 1983) associated with the function F. Given the function F and the moment map Φ, associated with the G action on M, we now have two ways of constructing a vector field on M: We can form the collective Hamiltonian $\mathscr{H} = F \circ \Phi$ and then, using the symplectic structure on M, the corresponding vector field ξ_H. Or, we can proceed as follows: At each $m \in M$ we apply the map Φ to get a point $\Phi(m) \in g^*$, and then the map L_F to get an element $L_F(\Phi(m))$ in g. Now each ζ in g gives rise to a vector field, ζ_M on M, which we can evaluate at the point m, to get $\zeta_M(m)$. In particular, we can do this for $\zeta = L_F(\Phi(m))$ to obtain the tangent vector $[L_F(\Phi(m))]_M(m)$ at m. Doing this at all m gives a vector field on M. We claim that these two ways of getting vector fields coincide; that is, that

$$\xi_H(m) = [L_F(\Phi(m))]_M(m), \qquad \forall m \in M. \tag{28.4}$$

To prove (28.4) it suffices to show that both sides have the same symplectic scalar product, relative to the symplectic bilinear form $(\,,)_m$, with any tangent vector $v \in TM_m$. But

$(\xi_H(m), v)_m = v(F \circ \Phi)$	by the definition of ξ_H and of H
$= d\Phi_m(v)F$	by the chain rule
$= \langle d\Phi_m(v), L_F(\Phi(m))\rangle$	by the definition of L_F
$= ([L_F(\Phi(m))]_M(m), v)_m$	by (26.2)

Q.E.D.

It follows from (28.4) and the equivariance of Φ that $d\Phi_m(\xi_H(m)) = L_F(\Phi(m))_{g^*}(\Phi(m))$, the value at $\Phi(m)$ of the vector field given by the coadjoint action associated to element $L_F(\Phi(m))$ in g. But this is just the value at $\Phi(m)$ of the vector field associated to the function F when restricted to the orbit $\mathcal{O}$ through $\Phi(m)$. (This can either be seen directly or by applying (28.4) to the orbit $\mathcal{O}$ thought of as a symplectic G space whose moment map is the injection of $\mathcal{O}$ into g^*.) Thus, if $m(t)$ denotes the trajectory of the Hamiltonian system ξ_H with $m(0) = m$, we see that $\Phi(m(t))$ lies entirely on the orbit $\mathcal{O}$ through $\Phi(m)$ and is a solution curve of the Hamiltonian system corresponding to $F_{\mathcal{O}}$. If $\gamma(t)$ denotes this curve in $\mathcal{O}$, then we can form the curve $\xi(t) = L_F(\gamma(t))$ and (28.4) says that $m'(t) = [\xi(t)]_M(m(t))$. So we can find the solution curve $m(t)$ by applying the following three steps:

(1) Find the orbit $\mathcal{O}$ through $\Phi(m)$;

(2) Find the solution curve to the Hamiltonian system on $\mathcal{O}$ corresponding to $F_{\mathcal{O}}$ passing through $\Phi(m)$ at $t = 0$. Call this curve $\gamma(t)$.

(3) Compute the curve $\xi(t) = L_F(\gamma(t))$. This is a curve in g. Solve the differential equations (i.e., find the curve in G satisfying),

$$a'(t) = \xi(t)a(t), \qquad a(0) = e.$$

Then $a(t)m$ is the desired solution curve.

Step 1 is purely kinematic; it depends solely on the Hamiltonian group action and has nothing to do with F.

Step 2 involves solving a Hamiltonian system with (usually) many fewer degrees of freedom than M. Thus, for example, in the liquid drop model, $\mathcal{O}$ is at most 12-dimensional, while $M = R^{6N}$ has dimension $6N$. This type of Hamiltonian equation has become popular in recent years in the study of mechanical systems associated with nonlinear partial differential equations. In case g has a nondegenerate invariant bilinear form (which is definitely *not* the case for the semidirect product groups above), one can identify g^* with g so that $\mathcal{O}$ is an orbit in g and the differential equations become $d\gamma/dt = [\gamma(t), F(\gamma(t))]$, which are known as Lax equations (see Lax 1968). If $g = sl(n)$ (or $u(n)$), the fact that $\gamma(t)$ lies on a fixed orbit means that the eigenvalues of the $\gamma(t)$ remain constant, so one speaks of an "isospectral deformation." However, the natural setting is on an orbit in g^*.

Step 3 can pose some interesting problems even if the solution of step 2 is trivial. For instance, suppose that F is an invariant. Then on each $\mathcal{O}$, $\gamma(t)$ is a constant, but the map L_F need not be trivial. Thus, $\xi(t)$ will be a constant element of g and $a(t)$ will be a one-parameter group. Thus the motion corresponding to $F \circ \Phi$ when F is an invariant is given by the action of a one-parameter group, the one-parameter group depending on m. (For the case of a spherical top, this is the spinning motion.) We might think of the solutions for noninvariant F as "generalized precessions or nutations."

Notice that step 3 simplifies if we are only interested in partial information about the trajectory at $m(t)$. For example, suppose that $M = T^*N$ and $m = (n, p)$, where $n \in N$; that is, m is a point in phase space whose corresponding point in configuration space is n. We might only be interested in the time evolution of n, and this may involve less than the full curve $a(t)$. In the case of the groups for the rigid body or liquid drop models it is only the $SO(3)$ or $Sl(3, \mathbb{R})$ component that acts on configuration space. The curve $\xi(t) \in g$ can be written as $\xi(t) = (\eta(t), \zeta(t))$, where $\eta \in O(3)$ or

$Sl(3, \mathbb{R})$ and $\zeta \in S^2(\mathbb{R}^3)$. Let $b(t)$ be the $SO(3)$, or $Sl(3, \mathbb{R})$ component of $a(t)$. We can find $b(t)$ by solving $b'(t) = b(t)\eta(t)$.

In Section 30 we shall indicate some methods that can be used in deciding which function F to choose on g^* to give the collective Hamiltonian. But first we need to develop some facts about the geometry of cotangent bundles and the moment map for semidirect products.

We can use the Legendre transformation (28.3) to make the space of smooth functions (or polynomials) defined on (an open subset of) g^* a Poisson algebra in the sense of Section 14. We simply define

$$\{f_1, f_2\}(l) = \langle l, [L_{f_1}(l), L_{f_2}(l)] \rangle \tag{28.5}$$

for any l in the common domain of definition of f_1 and f_2. The right-hand side of (28.5) makes sense since $L_f(l)$ is an element of g, and so we can compute the Lie bracket of the two Lie algebra elements $L_{f_1}(l)$ and $L_{f_2}(l)$, which we can then evaluate on l giving (28.5). In view of the discussion leading to the proof of (28.4) we know that

$$\{f_1, f_2\}(l) = \{f_{1|\mathcal{O}}, f_{2|\mathcal{O}}\}_{\mathcal{O}}, \tag{28.6}$$

where $\mathcal{O}$ is the G orbit through l and $\{\,,\}_{\mathcal{O}}$ is the Poisson bracket on $\mathcal{O}$ coming from its symplectic structure. This proves that (28.5) does indeed define a Poisson bracket, that is, satisfies the three conditions

$$\{f_1, f_2\} = -\{f_2, f_1\} \qquad \text{antisymmetry,}$$

$$\{f_1, f_2 f_3\} = \{f_1, f_2\} f_3 + f_2 \{f_1, f_3\} \qquad \text{derivation property,}$$

$$\{f_1, \{f_2, f_3\}\} = \{\{f_1, f_2\}, f_3\} + \{f_2, \{f_1, f_3\}\} \qquad \text{Jacobi's identity.}$$

Here is another direct proof that (28.5) satisfies these conditions: First of all, observe that if $\xi \in g$ is thought of as a linear function on g^*, then (28.3) says that the corresponding map L_ξ is the constant map that assigns ξ to every $l \in g^*$, (i.e., $L_\xi(l) \equiv \xi$). Next observe that

$$L_{f_1 f_2} = f_1 L_{f_2} + f_2 L_{f_1}$$

by Leibnitz's rule, so that $\{f_1, f_2 f_3\} = \{f_1, f_2\} f_3 + \{f_1, f_3\} f_2$. The bracket is clearly antisymmetric because the Lie bracket in g is. To prove the Jacobi identity, observe that it is true for f_1, f_2, and f_3 all linear (i.e., elements of g), since $\{\,,\}$ coincides with the Lie bracket $[\,,]$ for such elements and g is a Lie algebra. It follows from the derivation property that if it is true for f_3 and f'_3, then it is true for their product $f''_3 = f'_3 f_3$. Hence it is true for any polynomial f_3. Interchanging the roles of f_1 and f_3, and using antisymmetry, we conclude by the same argument that it is true whenever f_1, f_2, and

f_3 are polynomials. Hence, by approximation, it is true for all smooth functions.

The advantage to the definition (28.5) is that it is purely Lie algebra based, and does not involve the group-theoretical notion of orbit. Hence, it is of use in certain infinite-dimensional situations where the group-theoretical constructions are not available.

If M is symplectic with a Hamiltonian action of G on it, with moment map $\Phi: M \to g^*$, we can consider the corresponding map Φ^* from functions on g^* to functions on M,

$$\Phi^* f = f \circ \Phi,$$

which assigns to each function f the corresponding "collective Hamiltonian" on M. We leave it to the reader to verify that Φ^* is a homomorphism of Poisson structures; that is, that

$$\{\Phi^* f_1, \Phi^* f_2\}_M = \Phi^*\{f_1, f_2\}. \tag{28.7}$$

In Section 30 we shall give a prescription for choosing a collective Hamiltonian and, more particularly, a collective "kinetic energy" in various important cases. Here we discuss a method for verifying that a given Hamiltonian is indeed collective for a suitable group (which the reader should compare with the statement and proof of Theorem 27.1). Suppose that we are given Hamiltonian actions of the groups G and G' on the same symplectic manifold M that centralize each other. Thus, for each ξ' in the Lie algebra g' of G', the function $f_{\xi'}$, is invariant under the action of G, and hence the moment map $\Phi': M \to g'^*$ is invariant under the action of G. Thus, any function that is collective for the G' action (i.e., of the form $F \circ \Phi'$) will be G-invariant. Now it may turn out that the group G' is sufficiently large that all G-invariant functions are G'-collective.

Here is an important example: Take $M = T^*G$, with G acting on M by the induced action from left multiplication on G, and take G' to be another copy of G but acting by right multiplication. Under the left-invariant identification of $T^*G = M$ with $G \times g^*$ the actions and their corresponding moment maps are given by

$$l_a(c, \alpha) = (ac, \alpha), \qquad \Phi(c, \alpha) = c\alpha \tag{28.8}$$

and

$$r_b(c, \alpha) = (cb^{-1}, b\alpha), \qquad \Phi'(c, \alpha) = -\alpha. \tag{28.9}$$

[The reader can verify these for himself, or look ahead to Section 29. In fact the formulas for the moment maps follow from equivariance (eq. 26.6) and their evaluation at points of the form (e, α).] From (28.8) it follows that any left-invariant Hamiltonian must be a function of α alone (of the form $H(\alpha)$), and according to (28.9) such a Hamiltonian is collective (of the form $F \circ \Phi'$) with $F(\alpha) = H(-\alpha)$. Thus

any left-invariant Hamiltonian for T^*G is collective for the right action. (28.10)

Let us show how this remark, together with the integration procedure described earlier, gives a prescription for solving the equations of motion of a free rigid body. For the rigid body, the configuration space is taken to be $SO(3)$, the group of rotations of the body about its center of mass. We may identify $o(3)$ with $\mathbb{R}^3$ in the usual fashion so that Lie bracket gets identified with the "vector" or "cross" product $\times$, of $\mathbb{R}^3$ (where $u \times v = *(u \wedge v)$), so that (28.5) becomes

$$\{f_1, f_2\}(l) = l \cdot (\operatorname{grad} f_1(l) \times \operatorname{grad} f_2(l)) \tag{28.11}$$

where $\cdot$ denotes the scalar product in $\mathbb{R}^3$.

As we shall see in Section 30, the left-invariant Hamiltonian in this case is given by

$$H(c, \alpha) = H(\alpha) = \tfrac{1}{2} \sum_1^3 \frac{\alpha_i^2}{I_i}, \tag{28.12}$$

where the I_j are the "moments of inertia," and we have chosen a basis in which the inertia tensor is diagonal. [We are here using the notation of the standard texts. In the notation of Section 30, we should write α_{23} instead of α_1, etc. (this is our identification of $o(3)$ with $\mathbb{R}^3$) and $I_1 = \lambda_2 + \lambda_3$, etc.] The $SO(3)$ orbits in this case are spheres, and the problem of integrating the Hamiltonian system on each orbit (step 2 above) becomes easy since the flow lines must be level curves of H, so that the flow lines are obtained by intersecting the ellipsoids $H =$ constant with the spheres. The radius of the sphere is called the total angular momentum. For distinct moments of inertia $I_1 > I_2 > I_3$ the flow on the sphere of radius m has saddle points at $(0, \pm m, 0)$ and centers at $(\pm m, 0, 0)$ and $(0, 0, \pm m)$, corresponding to the critical points of H, restricted to the sphere. The saddles are connected by four "heteroclinic" orbits as indicated in Figure 28.1.

These curves tell us how the instantaneous "axis of rotation" is changing: We must apply step 3, which gives us (in this case) a linear map from g^* to g (sending $(\alpha_1, \alpha_2, \alpha_3)$ into $(\alpha_1/I_1, \alpha_2/I_2, \alpha_3/I_3)$), which, when applied to any orbit on the sphere, gives the instantaneous rotation. The actual motion of the rigid body is then obtained by applying step 3 above.

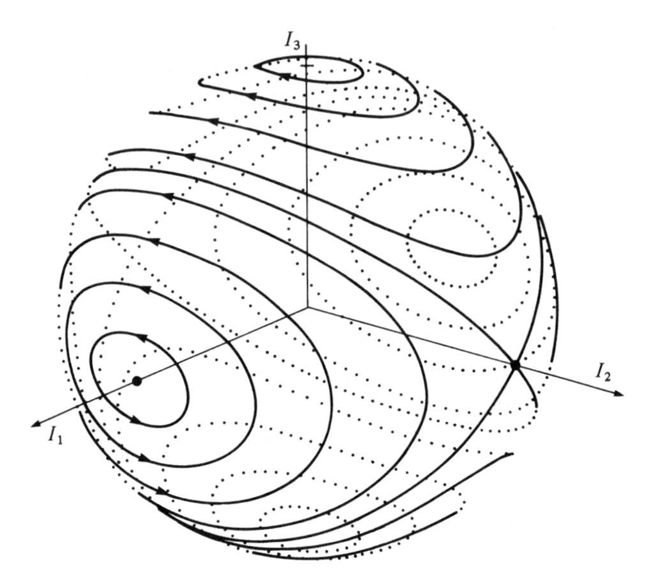

Figure 28.1.

29. Cotangent bundles and the moment map for semidirect products

Let N be a differentiable manifold and let T^*N denote its cotangent bundle. We recall (see Loomis and Sternberg, 1968, and Sternberg, 1983), that there is a canonically defined linear differential form θ on T^*N given by

$$\langle \theta_p, v \rangle = \langle p, d\pi_p v \rangle \qquad \text{for} \qquad v \in T(T^*N)_p,\ p \in T^*N,$$

where π denotes the projection of T^*N onto N. If $q^1, \ldots, q^n$ is a system of local coordinates on N, then this gives a basis of the tangent and cotangent spaces in the domain U of the coordinates and hence a system of coordinates $q^1, \ldots, q^n, p_1, \ldots, p_n$ on $\pi^{-1}(U)$. In terms of these coordinates

$$\theta = p_1 dq^1 + \cdots + p_n dq^n.$$

(All of this is explained in any standard text on differential geometry; e.g., those cited above.) Then $-d\theta$ is a symplectic form on T^*N, giving T^*N a canonical symplectic structure. If we think of N as the "configuration space" of some mechanical system, then T^*N with its symplectic structure is the associated "phase space." Any diffeomorphism (one-to-one differentiable map with differentiable inverse) φ of N induces a transformation of T^*N into T^*N as follows: A point (q, p) of T^*N is determined by a point $q \in N$ and a linear function p on the tangent space TN_q to N at q. The differential of φ at q, which we denote by $d\varphi_q$, maps TN_q into $TN_{\varphi(q)}$. Its transpose $d\varphi_q^*$ will map $T^*N_{\varphi(q)}$ to T^*N_q hence $d\varphi_q^{*-1}$ maps T^*N_q to $T^*N_{\varphi(q)}$. The induced transformation $\hat{\varphi}$ on T^*N is given by

$$\hat{\varphi}(q,p) = (\varphi(q), d\varphi_q^{*-1}p). \tag{29.1}$$

It follows directly from the definitions that the induced transformation $\hat{\varphi}$ preserves the form θ, and hence preserves the symplectic form $\omega = -d\theta$. Let ξ be a vector field on N. It generates a flow on N that induces a flow on T^*N. The infinitesimal generator of this flow will be a vector field $\hat{\xi}$ on T^*N. It is clear that

$$d\pi_p(\hat{\xi}(p)) = \xi(\pi(p)) \qquad \text{for any} \qquad p \text{ in } T^*N$$

so that the function $\langle \theta, \hat{\xi} \rangle$ is given by

$$\langle \theta, \hat{\xi} \rangle(p) = \langle p, \xi(\pi(p)) \rangle,$$

as follows from the definition of the form θ.

The flow generated by $\hat{\xi}$ preserves θ and hence $D_{\hat{\xi}}\theta = 0$. Thus

$$i(\hat{\xi})\, d\theta + di(\hat{\xi})\theta = 0$$

or, since $\omega = -d\theta$

$$i(\hat{\xi})\omega = df_\xi,$$

where

$$f_\xi(p) = \langle \theta, \hat{\xi} \rangle(p) = \langle p, \xi(\pi(p)) \rangle. \tag{29.2}$$

Thus the vector fields of the form $\hat{\xi}$ are Hamiltonian. Furthermore, if ξ is any vector field on N and φ is any diffeomorphism of N, the vector field $\varphi^{*-1}\xi$ is given by

$$(\varphi^{*-1}\xi)(q) = d\varphi_{\varphi^{-1}(q)}\xi(\varphi^{-1}(q)),$$

with

$$\begin{aligned} f_\xi(\hat{\varphi}^{-1}(p)) &= \langle \hat{\varphi}^{-1}(p), \xi(\pi(\hat{\varphi}^{-1}(p))) \rangle \\ &= \langle d\varphi^*_{\varphi^{-1}(\pi(p))}p, \xi(\varphi^{-1}(\pi(p))) \rangle \\ &= \langle p, d\varphi_{\varphi^{-1}(\pi(p))}\xi(\varphi^{-1}(\pi(p))) \rangle. \end{aligned}$$

So

$$(\hat{\varphi}^{-1})^* f_\xi = f_{(\varphi^{-1})^*\xi}. \tag{29.3}$$

Now $\xi \rightsquigarrow (\varphi^{-1})^*\xi$ is just the action of a diffeomorphism φ on the vector field ξ while $f \rightsquigarrow (\hat{\varphi}^{-1})^* f$ is the induced action on functions on T^*N. We can regard the algebra of all smooth vector fields on N as the Lie algebra of the group of all smooth diffeomorphisms of M. (This identification is of course in a formal sense, as we have not developed a theory of infinite-dimensional groups and algebras.) We have assigned to each diffeomorphism of N a symplectic transformation of T^*N and it is easy to check that this defines an action of the group $\mathrm{Diff}(N)$ on T^*N. Each vector field on N gives rise to a Hamiltonian vector field on T^*N whose associated function is given by (29.2). Equation (29.3) tells us that the map $\xi \rightsquigarrow f_\xi$ is a G morphism for $G = \mathrm{Diff}(N)$ (and hence by differentiation, or by direct verification) is a Lie algebra homomorphism from the Lie algebra $D(N)$ of all vector fields on N into the algebra of all functions on T^*N under Poisson bracket.

We have thus shown that the group $\mathrm{Diff}(N)$ of all diffeomorphisms of N, whose "Lie algebra" can, in a formal sense, be thought of as the algebra $D(N)$ of all smooth vector fields on N has a Hamiltonian action on T^*N. The "moment map" for this action is the rule that assigns to each point p in T^*N the linear function which, when evaluated on the vector field $\xi \in D(N)$, gives the value

$$\langle p, \xi(\pi(p))\rangle.$$

The functions f_{ξ_N} given by (29.2) are all (homogeneous) linear in p, with coefficients that are arbitrary functions of q. It is clear that the Poisson bracket of any two such functions is again of the same type. On the other hand, if f is a function of the q's alone (i.e., f is a function on N), then $\{f_\xi, f\} = -\xi_N f$ is again a function on N. Thus we can consider the Lie algebra $D(N) \times F(N)$ (semidirect product) with bracket relation

$$[(\xi^1, f_1), (\xi^2, f_2)] = ([\xi^1, \xi^2], \xi^2 f_1 - \xi^1 f_2)$$

and the map of $D(N) \times F(N)$ into $F(T^*N)$ sending (ξ_N, f) into $f_{\xi_N} + f$ is a homomorphism of Lie algebras. (Here by abuse of notation, we have used the same symbol f to denote a function on N and the same function thought of as a function on T^*N.) In what follows, it will be useful for us to describe the picture for the group whose "Lie algebra" is $D(N) \times F(N)$. This is the semidirect group product

$$\mathrm{Diff}\, N \times F(N),$$

where

$$(\phi_1, f_1)\cdot(\phi_2, f_2) = (\phi_1 \circ \phi_2, f_1 + \rho(\phi_1) f_2).$$

We wish to show that this group acts on T^*N. For this we observe that any function f on N determines a transformation t_f on T^*N defined by

$$t_f(q, p) = (q, p - df_q).$$

It is easy to see that this is a canonical transformation. If sf is a one-parameter family of functions, the corresponding infinitesimal generator is $-(\partial f/\partial q_1)(\partial/\partial p_1) - \cdots - (\partial f/\partial q_n)(\partial/\partial p_n)$, which corresponds to the function f on T^*N. If $\phi \in \mathrm{Diff}(N)$, then

$$\begin{aligned}
\hat{\phi} t_f \hat{\phi}^{-1}(q, p) &= \hat{\phi} t_f(\phi^{-1}(q), d\phi^*_{\phi^{-1}q} p) \\
&= \hat{\phi}(\phi^{-1}(q), d\phi^*_{\phi^{-1}q} p - df_{\phi^{-1}q}) \\
&= (q, p - d\phi^{*-1}_{\phi^{-1}(q)}\, df_{\phi^{-1}(q)}) \\
&= (q, p - d(f \circ \phi^{-1})),
\end{aligned}$$

the last equation following from the chain rule. Thus

$$\hat{\phi} t_f \hat{\phi}^{-1} = t_{\rho(\phi) f}, \tag{29.4}$$

and this implies that we have a Hamiltonian action of the semidirect product $\mathrm{Diff}(N) \times F(N)$ on T^*N.

The "moment map" for this Hamiltonian action is given by

$$\langle \Phi(p), (\xi, f) \rangle = \langle p, \xi(\pi(p)) \rangle + f(\pi(p)).$$

We shall apply this result in the following form. Suppose we have a representation of the Lie group H on a vector space V. Suppose we have a map $v \rightsquigarrow f_v$ of V into the space of functions, $F(N)$ on some manifold N. Suppose that we have an action $H \times N \to N$ of H on N, and suppose that

$$f_v(a^{-1} n) = f_{av}(n) \tag{29.5}$$

for all $a \in H$, $v \in V$, and $n \in N$.

It is easy to check that this implies that we have a homomorphism from the semidirect product of H and V into the semidirect product of $\mathrm{Diff}(N)$ and $F(N)$ and so we get a Hamiltonian action of the semidirect product $H \times V$ on T^*N, where $a \in H$ acts by $\hat{a}$ and $v \in V$ acts by t_{f_v}. The Lie algebra of $H \times V$ is $h \times V$, where h is the Lie algebra of H. For $\xi \in g$, let ξ_N denote the corresponding vector field on N. Then

$$f_\xi = f_{\xi_N}. \tag{29.6}$$

For $v \in V$ thought of as a Lie algebra element, the corresponding function is just f_v (thought of as a function on T^*N).

We shall let G denote the semidirect product of H and V. Let p be some fixed element of V^*. We let ap denote $a^{*-1}p$, that is

$$\langle ap, v \rangle = \langle p, a^{-1} v \rangle \qquad \text{for any} \qquad v \in V.$$

Let us take $N = H$ in the preceding. For $p \in V^*$ define the functions f_v^p on H by

$$f_v^p(b) = \langle bp, v \rangle, \qquad b \in H. \tag{29.7}$$

Then

$$f_v^p(a^{-1}b) = \langle a^{-1}bp, v \rangle = \langle bp, av \rangle = f_{av}(b).$$

Condition (29.5) is satisfied, and hence we get a moment map, $\Phi^p\colon T^*H \to g^* = h^* + V^*$. To describe this map we shall identify T^*H with $H \times h^*$ via left multiplication. That is, left multiplication by any $c \in H$ gives an identification of $TH_e = h$ with TH_c and hence of T^*H_c with h^*. If $\xi \in h$, left multiplication by $\exp t\xi$ is a one-parameter group of transformations on H whose infinitesimal generator we denote by ξ_H. Since

$$\begin{aligned}(\exp t\xi)c &= c(c^{-1} \exp t\xi c)\\ &= c \exp t \operatorname{Ad}_{c^{-1}}\xi,\end{aligned}$$

we see that if we use left multiplication to identify TH_c with h, then

$$\xi_H(c) = \operatorname{Ad}_{c^{-1}}\xi. \tag{29.8}$$

By definition, $\langle \Phi(c, \alpha), (\xi, 0) \rangle = \langle \alpha, \xi_H(c) \rangle = \langle c\alpha, \xi \rangle$ for $\xi \in h$. From this we see that the h^* component of $\Phi(c)$ is just $c\alpha$. On the other hand, $f_v^p(c) = \langle cp, v \rangle$ shows that the V^* component of $\Phi(c, \alpha)$ is cp. Thus

$$\Phi(c, \alpha) = (c\alpha, cp). \tag{29.9}$$

More generally, let N be some manifold upon which H acts transitively. Let $H_p = \{a \in H \mid ap = p\}$ denote the isotropy group of $p \in V^*$ and let $H_n = \{a \in H \mid an = n\}$ denote the isotropy group of $n \in N$. Suppose that $H_p \supset H_n$. (If $N = H$, then $H_n = \{e\}$. At the other extreme, we could take N to be the orbit through p of H in V^* and $n = p$, in which case $H_n = H_p$.) Then every point of N can be written as bn for some $b \in H$ and we define the function f_v^p on N by

$$f_v^p(bn) = \langle bp, v \rangle. \tag{29.10}$$

This is well-defined since $bn = b'n$ if and only if $b' = ba$ with $an = n$, from which $ap = p$, so $bp = b'p$. Then as before, it is clear that (29.5) is satisfied and so we get a moment map

$$\Phi^p : T^*N \to g^* = h^* \oplus V^*.$$

To describe the moment map we note that we have a map of h onto TN_n sending $\xi \in h$ into $\xi_N(n)$. The transpose of this map gives an injection

$\tau: T^*N_n \to h^*$. We can use the action of $b \in H$ to identify TN_n with TN_{bn} and hence to identify T^*N_{bn} with T^*N_n. Then arguing much as before, we see that with this identification (which depends on b)

$$\Phi^p(bn, \alpha) = (b\tau\alpha, bp). \tag{29.11}$$

In particular, if $N = H \cdot p$ is the orbit of p in V^*, then Φ_N^p maps T^*N into g^*.

Let us now compare this result with the description of the g^* orbits of a semidirect product given in Section 19. We know from there that all the g^* orbits are fibered over H orbits in V^*, and if we fix a point $p \in N$, where $N = H \cdot p$ is an H orbit in V^*, then the G orbits in g^* sitting over N are parametrized by the H_p orbits in h_p^*, where H_p is the isotropy group of p. In particular, there will be exactly one orbit corresponding to the trivial orbit $\{0\}$ and this is the G orbit through the point $(0, p) \in g^*$. Taking $\alpha = 0$ in (29.11) shows that the image of T^*N contains this orbit. On the other hand, the points of the G orbit through $(0, p)$, which sit over p, consist of all points of the form $(p \odot v, p)$ and, as we saw in Section 19,

$$p \odot V = h_p^0$$

and so $\dim p \odot V = \dim h - \dim h_p = \dim H/H_p = \dim N$. Thus we have proved

If we take $N = H \cdot p$ in (29.11) then $\Phi^p(T^*N)$ is a single G orbit in g^*, and is, in fact, the unique G orbit passing through the point $(0, p)$.

(29.12)

For example, in the liquid drop model, we take $H = Sl(3, \mathbb{R})$, and $V = S^2(\mathbb{R}^3)$. Suppose we take $p = Q$ to be a positive definite symmetric tensor. By moving along its orbit in V^* we may assume that we have chosen $Q = cI$. Then $H_Q = SO(3)$, and the orbit given in (29.12) is the cotangent bundle of the five-dimensional orbit through Q; that is, it is the phase space of the liquid drop of volume c^3. Notice, however, that the "typical" orbit over $(0, Q)$ will be twelve-dimensional, corresponding to a nontrivial, two-dimensional orbit of $SO(3)$ acting on $o(3)^*$. It corresponds to a liquid drop with an intrinsic "spin" or, as Buck, Biedenharn, and Cusson (1979) call it, a vortex motion. Let us return to the case where $N = H$ and show how to apply Theorem 27.1 to describe the Marsden–Weinstein reduced spaces of G acting on T^*H. We first remark that we can allow the group H to act on itself by right multiplication and that all right multiplications commute with all left multiplications. Recall that r_b denotes right multiplication by b^{-1}. By abuse of notation, we also denote by r_b the induced action on T^*H.

It is easy to check that in terms of the *left* identification of T^*H with $H \times h^*$ we have

$$r_b(c, \alpha) = (cb^{-1}, b\alpha) \tag{29.13}$$

and that the moment map $\Psi: T^*H \to h^*$ for this action is given by

$$\Psi(c, \alpha) = -\alpha. \tag{29.13)'$$

Now suppose that we are given a representation of H on V and a $p \in V^*$ so as to get the associated action of $H \circledS V$ on T^*H. Let H_p denote the isotropy group of p. We claim that right multiplication by H_p commutes with the action of $H \circledS V$. Indeed, right multiplication by H_p commutes with left multiplication by H, so it suffices to prove that the functions f_v^p are invariant under right multiplication by elements of H_p. But

$$f_v^p(cb^{-1}) = \langle cb^{-1}p, v\rangle = \langle cp, v\rangle \qquad \text{if } bp = p.$$

Let $\pi: h^* \to h_p^*$ be the projection dual to the injection of h_p into h as a subalgebra. Then the moment map, $\Phi': T^*H \to h_p^*$ is given by

$$\Phi' = \pi \circ \Psi;$$

that is,

$$\Phi'(c, \alpha) = -\pi\alpha. \tag{29.14}$$

Writing $G = H \circledS V$ and G' for H_p acting by right multiplication on T^*H, we claim that G acts transitively on $\Phi'^{-1}(\beta)$ for any $\beta \in h_p^*$. Indeed,

$$\Phi'^{-1}(\beta) = \{(c, \alpha) \qquad \text{such that} \qquad \pi\alpha = -\beta\}.$$

Now left multiplication by H acts transitively on the c component and does not move α. So it suffices to prove that V acts transitively on the set of all (e, α) with $\pi\alpha = -\beta$. Over the identity e, V acts by translation by $df_v^p(e)$. Now for any $\xi \in h$

$$df_v^p(e)(\xi) = \langle \xi p, v\rangle,$$

so, in the notation of Section 19,

$$df_v^p(e) = p \odot v.$$

But we have seen that

$$p \odot V = h_p^0 = \ker \pi, \tag{29.15}$$

proving our assertion. We are thus in the situation of Theorem 27.1, where

$G \times G'$ acts transitively on $\Phi'^{-1}(\mathcal{O}')$ for any orbit $\mathcal{O}'$ in h_p^*. We thus have

Let $\alpha \in h^*$ satisfy $\pi\alpha = -\beta$ and let $\mathcal{O}'$ be the H_p orbit $-H_p \cdot \beta$ in h_p^*. Then there is a symplectic diffeomorphism between (the connected components of) the Marsden–Weinstein reduced manifold $M_{\mathcal{O}}$, and the G orbit through (α, p) in g^*. (29.16)

For the case that $V = g$, this result was observed by Ratiu (1980a) (cf. also Holmes and Marsden, 1982).

We claim that (29.16) generalizes (29.12). Indeed, taking $\beta = 0$ in (29.16) gives $\Phi'^{-1}(0)/H_p$ as the reduced manifold, and it is easy to check that this is just $T^*(M)_{\mathcal{O}}$, where $M = H \cdot p$.

As we shall have many occasions to use the construction in (29.16) we describe it once again, in a more general and geometrical setting:

Let $P \to M$ be a principal bundle with structure group K. Thus K acts to the right on P and $P/K = M$. The right action of K on P induces an action on T^*P, which is Hamiltonian. If $\Phi' : T^*P \to k^*$ is the moment map for this action, and $\mathcal{O}'$ is a K orbit in k^*, then we can consider the corresponding Marsden–Weinstein reduced space,

$$M_{\mathcal{O}'} = \Phi_{T^*P \times \mathcal{O}'}^{-1}(0)/K,$$

where

$$\Phi_{T^*P \times \mathcal{O}'}(z, \beta) = \Phi'(z) - \beta.$$

This reduced manifold will play an important role for us later on in our description of particle mechanics in the presence of a Yang–Mills field.

In the case at hand, $M = H \cdot p$, $K = H_p$, and $P = H$ regarded as a principal bundle over M with H_p acting as right multiplication. The content of (29.16) is that G acts transitively on $\Phi'^{-1}(\beta)$ for each β in h_p^* and thus Theorem 27.1 applies to give the description of $M_{\mathcal{O}'}$. Thus (29.16) describes the most general coadjoint orbit of a semidirect product G as a reduced space.

We can turn the argument in (29.16) around to get an example of a principal bundle P where a subgroup of Aut P already acts transitively on Marsden–Weinstein reduced spaces (without having to introduce functions). Indeed, using the same notations as in (29.16), let us now take

$$K = H_p \circledS V$$

and

$$P = G = H \circledS V,$$

so that K is a subgroup of G and $G/K = H/H_p = H \cdot p$ as above. So we have taken the same base manifold, but a larger fiber. Since H_p leaves p fixed (i.e., $\{p\}$ is an H_p orbit in V^*), we know from the general theory of coadjoint orbits for semidirect products that $\mathcal{O}'' = \mathcal{O}' \times \{p\}$ is a K orbit if $\mathcal{O}'$ is a coadjoint orbit of H_p. Using the left-invariant identification, we can write a typical element of T^*G as

$$(c, u; \alpha, l) \qquad \text{where } c \in H, u \in V, \alpha \in h^*, l \in V^*.$$

Right multiplication by K induces a Hamiltonian action of K on T^*G with moment map Φ' given by

$$\Phi'(c, u; \alpha, l) = -(\pi\alpha, l).$$

In this situation, the group K acts transitively on the (α, p) component of $\Phi'^{-1}(\mathcal{O}'')$, while G (by left multiplication) acts transitively on the (c, u) component; so G acts transitively on $M_{\mathcal{O}''}$.

As another variation on the same theme, let us now take $N = H$ in (29.11) but this time with the right action of H on itself. We get an action of $G = H \circledS V$ on T^*H and this time the moment map is given by

$$\Phi^p(a, \alpha) = (-\alpha, a^{-1}p). \tag{29.17}$$

Let T be some quadratic function of α, say,

$$T(\alpha) = \tfrac{1}{2} \operatorname{tr} \alpha Q^{-1} \alpha$$

if the α are given as matrices, but where now Q, some nonsingular matrix, is given as some "external" information. For any $x \in V$ define the function F on g^* by

$$F(\alpha, p) = T(\alpha) - (p, x). \tag{29.18}$$

Then

$$\mathcal{H} = F \circ \Phi^p$$

is a Hamiltonian system on T^*H consisting of a left-invariant kinetic energy plus a "representation function," (see Kuperschmidt and Vinogradov, 1977). As we shall see, this is precisely the form of the equations of motion of a rigid body about a fixed point in the presence of a uniform (gravitational) field. But first we make some general remarks:

Let l_b denote left multiplication by b,

$$l_b a = ba.$$

Let L_b denote the induced action on $T^*H = H \times h^*$ so

$$L_b(a, \alpha) = (ba, \alpha).$$

Then

$$\Phi^p \circ L_b(a, \alpha) = (-\alpha, a^{-1}b^{-1}p)$$

and so

$$\Phi^p \circ L_b = \Phi^p \qquad \text{if} \qquad bp = p,$$

hence $\mathscr{H}$ is invariant under L_b if $bp = p$. On the other hand,

$$\Phi^p_0 R_b = b \cdot \Phi^p.$$

So, if both T and x are invariant under $b \in H$, then the Hamiltonian $\mathscr{H}$ is invariant under right multiplication by b. For the case of the rigid body, we take $H = SO(3)$ and $V = \mathbb{R}^3$ so that $G = E(3)$. The vector p will represent a constant force field in space and $x \in \mathbb{R}^3$, the position of the center of mass. From the point of view of "body coordinates," x is fixed and p is varying and the collective Hamiltonian corresponding to the function F is precisely the Hamiltonian of the rigid body with one point fixed in a uniform gravitational field, considered as a Hamiltonian system on $T^*(SO(3))$, cf. Holmes and Marsden (1982). As we know, the functions $\|p\|^2$ and $|\alpha \wedge p|$ are invariants of $e(3)^*$ and the generic orbit is four-dimensional. These functions must pull back to be invariants of motion of the flow on $T^*(SO(3))$ generated by H. The function $\|p\|^2$ measures the intensity of the gravitational field, while $|\alpha \wedge p|$ pulls back to give the angular momentum about the p axis; we call this function M_p. Suppose that the inertia tensor has some axis of symmetry and that the center of mass lies on this axis. Then if b is a rotation about this axis, b preserves both T and x, and hence right multiplication by b preserves H. The infinitesimal generator of this group then gives rise to a third integral. This is the case of Lagrange's top.

It is instructive to look at the flow generated by F on g^*. Let e_1, e_2, e_3 be an orthonormal basis of $o(3)$ in terms of which T is diagonal, and f_1, f_2, f_3 the corresponding basis of $\mathbb{R}^3$ so that if we consider $e_1, \ldots, f_3$ as functions on g^*

$$F = \tfrac{1}{2}\sum(e_j^2/A_j) - \sum x_j f_j.$$

Clearly

$$\{F, e_1\} = (-e_2e_3/A_2) + (e_3e_2/A_3) + x_x f_3 - x_3 f_2$$

$$\vdots$$

$$\{F, f_1\} = (e_2 f_3/A_2) - (e_3 f_2/A_3)$$

$$\vdots$$

If we let $\omega_i = e_i/A_i$ and think of ω_i and f_i as coordinates on g^*, the corresponding differential equations are

$$A_1(d\omega_1/dt) = (A_2 - A_3)\omega_2\omega_3 + x_2 f_3 - x_3 f_2$$
$$\vdots$$
$$(df_1/dt) = \omega_2 f_3 - \omega_3 f_2$$
$$\vdots$$

These are the Euler equations in Poisson form.

More generally, let $g = h + V$ be the Lie algebra of any semidirect product and let F be a function on g^* of the form (29.18). The corresponding flow on g^* will be called a *Euler–Poisson flow* and the corresponding differential equations will be called the *Euler–Poisson differential equations.* Let us describe these both. We begin with a discussion of the flow on g^* associated to an arbitrary function F, where g is an arbitrary Lie algebra: The function F determines the map $L_F: g^* \to g$. Thus $L_F(l)$ is an element of g for each l in g^*. The element $L_F(l)$ of g acts on g^* via the coadjoint representation and, in particular, acts on the element l, sending it into $L_F(l) \cdot l$. The vector field

$$l \to L_F(l) \cdot l$$

is the infinitesimal generator of the flow corresponding to F. For the case of a semidirect product, with F of the form (29.18), the map L_F is given by

$$L_F(\alpha, p) = (Q^{-1}\alpha, -x).$$

Writing

$$\hat{\alpha} = Q^{-1}\alpha$$

and using the form of the coadjoint action for a semidirect product, we see that the vector field corresponding to F is given by

$$(\alpha, p) \to (\hat{\alpha} \cdot \alpha + p \odot x, \hat{\alpha} p),$$

where $\hat{\alpha} \cdot \alpha$ denotes the action of $\hat{\alpha} \in h$ on $\alpha \in h^*$ and $\hat{\alpha} p$ denotes the action of $\hat{\alpha}$ on $p \in V^*$. Thus the Euler–Poisson equations are

$$\frac{d\alpha}{dt} = \hat{\alpha} \cdot \alpha + p \odot x,$$

$$\frac{dp}{dt} = \hat{\alpha} p. \tag{29.19}$$

If the quadratic function T in (29.18) is invariant under H, these equations simplify somewhat, since then

$$\hat{\alpha}\cdot\alpha = 0. \tag{29.20}$$

An immediate check shows that equations (29.19) do indeed specialize to the Euler equations in Poisson form written above for the case $V = \mathbb{R}^3$ and $h = o(3)$.[†]

We will conclude this section by describing in terms of the formalism above the three classical examples of "integrable tops." We will continue to use $e_1, \ldots, f_3$ for coordinates on g^* and continue to denote by (x_1, x_2, x_3) the coordinates of the center of mass of the rigid body in question. As we pointed out above, the pull-backs of the functions p^2 and $\alpha \wedge p$ are invariants of motion for every collective Hamiltonian on $T^*SO(3)$. In the $e_1, \ldots, f_3$ coordinates, these functions are given by

$$f_1^2 + f_2^2 + f_3^2$$

and

$$e_1 f_1 + e_2 f_2 + e_3 f_3.$$

The *Lagrange top* is, by definition, a top whose axis of symmetry is in alignment with its center of mass; so its motion is described by a collective Hamiltonian of the form $F \circ \Phi^P$, where

$$F = A(e_1^2 + e_2^2) + Be_3^2 + 2f_3.$$

(Here we have chosen $x = (0, 0, -2)$ as center of mass and $2/A$ and $2/B$ as moments of inertia.) We remarked above that the Lagrange top has a third integral of motion associated with rotation about its axis of symmetry. This is just $G \circ \Phi^P$, where G is the coordinate function e_3.

The two other known examples of integrable tops, the *Kowalevski top* and the *Tchapligine top*, also possess an axis of symmetry; but in these cases the center of mass of the top, instead of being in alignment with the axis of symmetry of the top, is in fact *in the plane perpendicular* to the axis of symmetry; that is, for both these tops

$$F = A(e_1^2 + e_2^2) + Be_3^2 + 2f_1.$$

To get a third integral of motion, we look for a complex-valued quadratic function

$$G = (e_1 + ie_2)^2 + \lambda(f_1 + if_2), \qquad \lambda \in \mathbb{R},$$

[†] The rest of this section can be omitted on first reading.

on g^* such that

$$\{G, F\} = i\beta G e_3, \qquad \beta \in \mathbb{R}. \tag{*}$$

Indeed, if such a function exists then, by (*) $\{|G|^2, F\} = 0$; so $|G|^2 \circ \Phi^P$ furnishes a third integral of motion. Setting $y = e_1 + ie_2$ and $z = f_1 + if_2$ we have

$$G = y^2 + \lambda z.$$

A straightforward computation shows that $(1/2i)\{G, F\}$ is:

$$[2(B - A)y^2 + \lambda Bz]e_3 - (2 + \lambda A)f_3 y.$$

Choosing $\lambda = -2/A$, this reduces to

$$\frac{1}{2i}\{G, F\} = [2(B - A)y^2 + \lambda Bz]e_3.$$

For the *Kowalevski top*, the moments of inertia are related by

$$B = 2A,$$

so this equation becomes

$$\frac{1}{2i}\{G, F\} = BGe_3;$$

that is, (*) holds with $\beta = 2B$.

The third classical example of an integrable top, the Tchapligine top, does not, strictly speaking, possess a third integral of motion. It admits a function that is constant along those trajectories of motion for which *the angular momentum about the axis of symmetry is zero*. To find such a function, we must look for a function G on g^* for which $\{G, F\}$ is a multiple of $e_1 f_1 + e_2 f_2 + e_3 f_3$. The pull-back of G to $T^*SO(3)$ will then have the desired property. We will look for a G of the form

$$G = e_3 y\bar{y} + \lambda(y + \bar{y})f_3.$$

A short computation shows that $\{G, F\}$ is equal to the imaginary part of the expression

$$-2y[-\bar{y}z + 2e_3 f_3 + \lambda A(\bar{z}y - \bar{y}z - 2e_3 f_3) + 2\lambda B e_3 f_3]. \tag{**}$$

For the Tchapligine top, the moments of inertia are related by

$$B = 4A$$

Setting $\lambda = -1/2A$, (**) becomes

$$y(\bar{y}z + \bar{z}y + 2e_3 f_3)$$

or

$$2y(e_1 f_1 + e_2 f_2 + e_3 f_3);$$

so $\{G, F\}$ is a multiple of angular momentum about the axis of symmetry, as desired.

30. More Euler–Poisson equations

A number of other interesting mechanical systems arise from the Euler–Poisson equations, either via the pull-back using the moment map (so as to give a collective Hamiltonian as in the case of a rigid body) or (what is really a special case of this construction) via restriction to an orbit in g^*. Let us list a few:

The spherical pendulum (see Duistermaat, 1980)

A point of unit mass swings on a rigid massless pendulum in the presence of a uniform (vertical) gravitational field. Here the configuration space is a sphere; say, the unit sphere if we choose our units of length so that the pendulum has length 1. The phase space is T^*S, which we may identify with an orbit of the Euclidean group $E(3)$ acting on the dual of its Lie algebra. If p denotes a point on the unit sphere and α an element of T^*S_p then the Hamiltonian for the spherical pendulum is

$$H(\alpha, p) = \tfrac{1}{2}\|\alpha\|^2 + p_3,$$

where the first term is the kinetic energy and the second term is the potential energy. This is just the restriction to the given coadjoint orbit of the function

$$F(\alpha, p) = \tfrac{1}{2}\|\alpha\|^2 + \langle p, e_3 \rangle,$$

which is of the form (29.18). As F is invariant under rotations about the e_3 axis the functions F and the angular momentum function about the e_3 axis

$$A(\alpha, p) = \alpha_1 p_2 - \alpha_2 p_1$$

Poisson commute. Restricted to T^*S, they are functionally independent at almost all points; that is, the spherical pendulum is "completely integrable." (We shall study the notion of complete integrability in Chapter IV.) Indeed,

$dH = 0$ implies $dp_3 = 0$, which can occur only at points of the cotangent bundle sitting over the north and south poles $p = (0, 0, \pm 1)$, and at such points $dH = 0$ only if $\alpha = 0$. Thus these are the only "equilibrium points" where $dH = 0$ – the unstable equilibrium at the north pole $p = (0, 0, 1)$, $\alpha = 0$, and the stable equilibrium at the south pole $p = (0, 0, -1)$, $\alpha = 0$. At all other points, the functional dependence of A and H means that

$$dA = \lambda\, dH,$$

or what amounts to the same thing,

$$\xi_A = \lambda \xi_H,$$

where ξ_A and ξ_H are the Hamiltonian vector fields corresponding to A and H in the symplectic structure of T^*S. Over the poles, at points where $\alpha \neq 0$, the vector field ξ_H has a nonzero projection onto the sphere, corresponding to the nonzero velocity of the pendulum, while ξ_A projects onto the zero vector, since rotation about the e_3 axis fixes the north and south poles. Thus no such dependence can take place. At all other points, the projection of ξ_A is a nonzero tangent to a horizontal circle and hence $\xi_A \neq 0$, so the $\lambda \neq 0$ as well. The above equation then says that ξ_H is tangent to the trajectory of ξ_A, which is just an orbit of the group of rotations about the e_3 axis acting on T^*S. If the above equation holds at one point, the invariance of H under these rotations implies that it holds along the entire trajectory, with λ constant along this trajectory. Thus the corresponding trajectories of the pendulum must be horizontal circles. These motions were first discovered by Huygens (see Duistermaat, 1980); the physical explanation being that the centrifugal force, together with the constraining force of the pendulum, exactly balances the force of gravity. To find these circular motions explicitly, we may use polar coordinates since we are staying away from the poles. Thus we introduce the variables θ and φ on the punctured sphere, where θ (mod 2π) gives the longitude and $0 < \varphi < \pi$ gives the latitude (i.e., $p_3 = \cos\varphi$). We denote the corresponding momenta by p_θ and p_φ so that the functions H and A are given in terms of these coordinates by

$$A = p_\theta$$

and

$$H = \tfrac{1}{2} p_\varphi^2 + \frac{1}{2 \sin^2 \varphi}\, p_\theta^2 + \cos\varphi.$$

Thus dH can be a multiple of dp_θ only if the coefficients of dp_φ and $d\varphi$ in dH

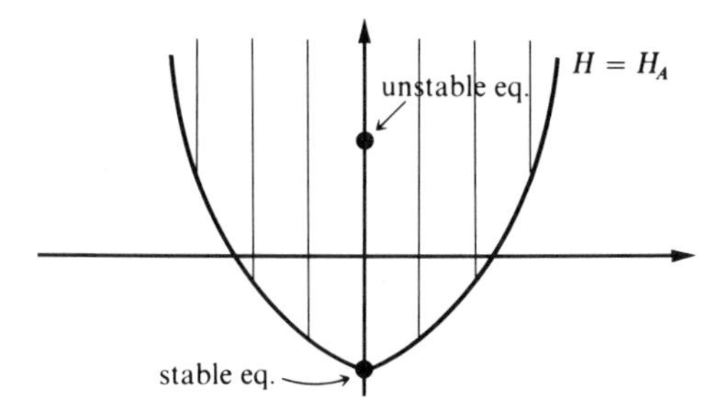

Figure 30.1.

vanish. Thus

$$p_\varphi = 0 \qquad \text{and} \qquad A^2(\sin\varphi)^{-3}\cos\varphi + \sin\varphi = 0. \qquad (*)$$

For the second equation to hold we must have $\cos\varphi < 0$. Then the above equations give

$$dH = \frac{A}{\sin^2\varphi}\, dA = (-\cos\varphi)^{1/2}\, dA.$$

Thus the motion is indeed a horizontal circle, traversed uniformly with period $(-\cos\varphi)^{1/2}$. For fixed A, equations (*) are exactly the equations for the critical points of H. There is thus a unique critical point and it is clearly a minimum; indeed for each fixed A, it is clear that H tends to ∞ as $p_\varphi \to \pm\infty$ or as $\cos\varphi \to \pm 1$. Let H_A denote the minimum value of H for fixed A.

Let $\rho: T^*S \to \mathbb{R}^2$ be the map

$$\rho = (A, H).$$

From the above discussion we see that the image of ρ consists of all points with $-\infty < A < +\infty$ and $H \geqslant H_A$. The critical set for the map ρ (i.e., the set where $d\rho$ fails to be surjective) is precisely the set where A and H are functionally dependent. The image of the critical set is called the set of critical values. It consists of the set $H = H_A$, the boundary of the image, together with the image of the two equilibria; these two image points being $(0, \pm 1)$. Now the point $(0, -1)$ is in fact a boundary value for ρ, while the point $(0, +1)$ lies in the interior of the image of ρ. Thus the image of ρ looks like Figure 30.1 with one interior critical value of $(0, 1)$.

For each regular value (A, H), its inverse image $\rho^{-1}(A, H)$ is a torus consisting of the product of the two circles, $p_\theta = A$ and θ arbitrary mod 2π and the smooth circle

$$\frac{1}{2}p_\varphi^2 + \frac{1}{2\sin^2\varphi}A^2 + \cos\varphi = H$$

in the $p_\varphi = \varphi$ plane. Thus the inverse image of the regular set is a bundle of tori. We shall see in Chapter IV that this is a general phenomenon, and also indicate the physical significance of the topology of the set of regular values.

The C. Neumann system of harmonic motion constrained to a sphere (Ratiu 1981b)

We first discuss the general idea of *constrained motion* (Moser 1980a). Let M be a symplectic manifold with form ω and let N be a symplectic submanifold. This means that at every point n of N we have

$$TM_n = TN_n \oplus TN_n^\perp.$$

If H is any function on M and ξ_H is the corresponding Hamiltonian vector field, then along N we can use the above decomposition to write

$$\xi_H = \xi_H^N + \xi_H^{N\perp}.$$

If v is any vector tangent to N then

$$\omega(\xi_H^N, v) = \omega(\xi_H, v) = \langle dH, v\rangle = \langle dH_{|N}, v\rangle$$

so

ξ_H^N is the Hamiltonian vector field on N corresponding to the restricted function $H_{|N}$ relative to the restricted symplectic form $\omega_{|N}$. (30.1)

Suppose that N is given as the zero set of $2r$ functions, so

$$N = \{F_1 = 0, \ldots, F_{2r} = 0\}.$$

It is clear that N is symplectic if and only if the matrix $(\{F_i, F_j\})$ is nondegenerate at all points of N. Since the $F_i = 0$ along N, it follows from (30.1) that $\xi_{F_i}^N = 0$ and so the ξ_{F_i} form a basis of $TN^\perp$; that is, the vector fields ξ_{F_i} are sections of the vector bundle $TN^\perp$, which give a basis of $TN_n^\perp$ at every point n. Thus

$$\xi_H^{N\perp} = \sum \lambda_i \xi_{F_i},$$

where the functions λ_i are determined along N. If we extend the functions λ_i so as to be defined near N then we can write

$$\xi_H^N = \xi_{H^*} \quad \text{along } N, \quad \text{where } H^* = H - \sum \lambda_i F_i, \tag{30.2}$$

The term $\sum \lambda_i F_i$ can be thought of as giving the "forces" which constrain the motion to the submanifold N. Note that along N

$$\xi_{H^*} F_k = \{H^*, F_k\} = 0$$

and $F_i = 0$ so

$$\{H, F_k\} = \sum \lambda_j \{F_j, F_k\}, \tag{30.3}$$

which we can solve for the λ_j in view of the nonsingularity at each point of N of the matrix $(\{F_j, F_k\})$.

An important special case of this construction is the following: Suppose that the equations

$$F_1 = 0, \ldots, F_r = 0$$

define a coisotropic submanifold C of M. Thus

$$\{F_i, F_j\} = 0, \qquad i, j = 1, \ldots, r. \tag{30.4}$$

Then C has a null foliation spanned by the vector fields ξ_{F_i}. Suppose that this null foliation is fibrating over some manifold Y. Thus $\rho: C \to Y$ and we know that Y is symplectic. If H is a function on M satisfying

$$\{H, F_i\} = 0$$

Then H is constant along the leaves of the null foliation and defines a Hamiltonian $\bar{H}$ on Y. This is the method of reduction that we have studied previously. Suppose that we are given a cross section $s: Y \to C$ of Y whose image is thus a symplectic submanifold N of C. Then the reduced Hamiltonian $\bar{H}$ on Y can be identified with the restricted Hamiltonian $H_{|N}$ on N under the symplectic diffeomorphism s. Suppose that N is (locally) given by

$$G_1 = 0, \ldots, G_r = 0.$$

The condition that these equations define a transverse section to the null foliation of C is that

$$(\{F_i, G_j\}) \qquad \text{be nonsingular along} \qquad F = 0, G = 0. \tag{30.5}$$

Thus the vector field ξ_H^N, corresponds to a Hamiltonian of the form

$$H^* = H - \left(\sum \lambda_i F_i + \sum \mu_j G_j\right).$$

Taking the Poisson bracket of this equation with F_j and using the fact that $\{H, F_j\} = 0$, $\{F_i, F_j\} = 0$ and (30.5) we see that

$$\mu_j = 0, \qquad j = 1, \ldots, r.$$

Thus, in this case, the constraining Hamiltonian has the form

$$H^* = H - \sum \lambda_i F_i, \qquad \text{where} \qquad \{H, G_j\} = \sum \lambda_i \{F_i, G_j\}. \tag{30.6}$$

For example, suppose that $M = T^*X$ for some manifold X, that W is a submanifold of X, and that $C = T^*X_{|W}$. Then C is coisotropic and its null foliation can be identified with the normal bundle N^*W consisting of all covectors which vanish on TW. The corresponding quotient symplectic manifold can be identified with T^*W. The choice of a Riemann metric on X allows the identification of T^*X with TX, and we can consider TW as a subbundle of $TX_{|W} = T^*X_{|W}$. Thus TW is then our cross section symplectic submanifold N.

For instance, if $X = \mathbb{R}^n$ and W is the unit sphere, then $M = \mathbb{R}^{2n}$ and C is given by $F = 0$, where

$$F = \|x\|^2 - 1,$$

while $N = TX$ is given by

$$F = 0, \qquad G = 0,$$

with

$$G(x, y) = \langle x, y \rangle.$$

Let

$$\begin{aligned} H &= \tfrac{1}{2}\langle Ax, x \rangle + \tfrac{1}{2}\|x \wedge y\|^2 \\ &= \tfrac{1}{2}\langle Ax, x \rangle + \tfrac{1}{2}(\|x\|^2\|y\|^2 - \langle x, y \rangle^2). \end{aligned} \tag{30.7}$$

Here A is a positive definite selfadjoint matrix, which we may assume to be diagonal. The restriction of H to N is the same as the restriction of the harmonic-oscillator Hamiltonian $\tfrac{1}{2}\langle Ax, x \rangle + \tfrac{1}{2}\|y\|^2$, but it is an immediate computation that $\{H, F\} = 0$, so the equation of "harmonic motion restricted to the sphere" is the same as that of

$$H^* = H - \lambda F.$$

Now we can write

$$H = \tfrac{1}{2}\|\alpha\|^2 + \tfrac{1}{2}\langle A, p \rangle, \tag{30.8}$$

where

$$p = x \otimes x \qquad \text{and} \qquad \alpha = x \wedge y. \tag{30.9}$$

Thus H has the form (29.18) when the values (30.9) are substituted. Here $g = o(n) + S^2(\mathbb{R}^n)$. We must explain the meaning of (30.9): The set of all $x \otimes x$ with $\|x\| = 1$ is an $O(n)$ orbit acting on the space of symmetric tensors $S^2(\mathbb{R}^n)$, which we may, of course, identify with the unit sphere in $\mathbb{R}^n$. From the general theory of coadjoint orbits for semidirect products we know that its cotangent bundle can be identified with the coadjoint orbit consisting of all

$$(\alpha, p) \qquad \text{with} \qquad p = x \otimes x, \alpha = p \odot v, \qquad v \text{ in } S^2(\mathbb{R}^n).$$

Now the set of all $p \odot v$ is a subspace of $o(n)$ invariant under the action of the isotropy group in $O(n)$ of p. In the present case, this isotropy group is exactly the isotropy group of x, the orthogonal group in n-1 dimensions. Now under $O(n-1)$ we have the decompositions into invariant subspaces

$$S^2(\mathbb{R}^n) = \mathbb{R} + \mathbb{R}^{n-1} + S^2(\mathbb{R}^{n-1}),$$

where $\mathbb{R}$ is identified as the set of all multiples of $x \otimes x$ and $\mathbb{R}^{n-1}$ is identified as the set of all $x \otimes y + y \otimes x$, where $y \perp x$. Similarly, we have the identification

$$o(n) = \mathbb{R}^{n-1} + o(n-1),$$

where the $\mathbb{R}^{n-1}$ is identified with the space of all $x \wedge y$ and the $o(n-1)$ is the orthogonal algebra of the space $x^{\perp}$. Since $o(n-1)$ is irreducible and not equivalent to any subrepresentation of $S^2(\mathbb{R}^n)$, we conclude that the image space $p \odot V$ consists of $\mathbb{R}^{n-1}$; that is, every α in $p \odot V$ can be written as $\alpha = x \wedge y$ with $\langle x, y\rangle = 0$. Thus we conclude that the C. Neumann problem is a Euler–Poisson system on a (minimal-dimensional) coadjoint orbit in $\big(o(n) + S^2(\mathbb{R}^n)\big)^*$. This result is due to Ratiu (1981b).

Geodesic flow on an ellipsoid

Let X be a Riemannian manifold. The geodesic flow on $T^*X \sim TX$ is given by the Hamiltonian $\frac{1}{2}\|y\|^2$, where y denotes a (co)tangent vector and $\|y\|$ its length in the given metric. If W is a submanifold, then this restricts to give the geodesic flow on TW. Let us examine explicitly what the equations (with constraints) look like for the case where W is a hypersurface in X given by $F = 0$ for some function F. We write

$$F_1(x, y) = F(x) \qquad \text{and} \qquad F_2(x, y) = \langle dF_x, y\rangle.$$

Then

$$\{F_1, F_2\} = \|dF_x\|^2,$$

which is strictly positive since we assume that $dF \neq 0$ anywhere on W. Then H^* is given by (30.2)

$$H^* = H - \lambda F_1 - \lambda_2 F_2,$$

with λ_1 and λ_2 determined by (30.3). Now

$$\{H, F_1\} = \langle dF_x, y\rangle = F_2 = 0 \qquad \text{on} \qquad N = TW$$

and

$$\{H, F_2\} = \langle d^2 F_x y, y\rangle,$$

so

$$\lambda_1 = -\langle d^2 F_x y, y\rangle \| dF_x \|^{-2}$$

and

$$\lambda_2 = 0.$$

Thus the constrained equations are

$$\frac{dx}{dt} = H_y^* = y,$$

$$\frac{dy}{dt} = -H_x^* = \lambda_1 dF_x.$$

Now suppose that W is an ellipsoid in $\mathbb{R}^n$ so that

$$F = \tfrac{1}{2}(1 + Q(x)),$$

where

$$Q(x) = -\langle A^{-1}x, x\rangle$$

and A is a positive definite matrix. Then

$$dF_x = -A^{-1}x$$

and the preceding differential equations become

$$\frac{dx}{dt} = y, \qquad \frac{dy}{dt} = -Q(y)\|A^{-1}x\|^{-2}A^{-1}x. \tag{30.10}$$

We will now (following Moser, 1980b) prove the following: Consider the same Lie algebra, $o(n) + S^2(\mathbb{R}^n)$ as for the Neumann problem and the Poisson-type function

$$P(\alpha, p) = \tfrac{1}{2}T(\alpha) - \langle A^{-1}, p\rangle, \tag{30.11}$$

where T is the quadratic form on $o(n) = \wedge^2(\mathbb{R}^n)$ induced from the scalar product $\langle A^{-1}x, y\rangle$ on $\mathbb{R}^n$. We will consider the restriction of P to the same orbit (T^*S^{n-1}) as before, and show that the trajectories of P along the submanifold $P = 0$ give rise to the same curves as the solutions of (30.10) along the unit tangent bundle $\|y\| = 1$. (The trajectories will be the same, but as we shall see, we will have to reparametrize them.) As a first step we shall use the following coordinates on the minimal g^* orbit,

$$p = y \otimes y, \qquad \|y\| = 1$$

and

$$\alpha = x \wedge y \qquad \text{where} \qquad \langle A^{-1}x, y\rangle = 0.$$

Notice that here y plays the opposite role than it did in the Neumann problem. Also, since A is positive definite, there is a unique $x = z + ty$ that satisfies $\langle A^{-1}x, y\rangle = 0$. In other words, the set of all pairs

$$(y, x) \quad \|y\| = 1, \qquad \langle A^{-1}x, y\rangle = 0$$

is a section of the null foliation of $T^*\mathbb{R}^n_{|S^{n-1}} \to T^*S^{n-1}$. We can thus identify our minimal coadjoint orbit as the symplectic submanifold N of $T^*\mathbb{R}^n$ defined by the constraints

$$f_1 = \tfrac{1}{2}(\|y\|^2 - 1) = 0 \qquad \text{and} \qquad f_2 = Q(x, y) = 0, \tag{30.12}$$

where we have written $Q(x, y) = -\langle A^{-1}x, y\rangle$.

Note that $\{f_1, f_2\} = Q(y) < 0$, giving another verification that N is symplectic. The induced symplectic structure on N is equivalent, via the projection, to the symplectic structure on T^*S^{n-1} and hence to that of the coadjoint orbit.

The function P, when restricted to the orbit, is equal to

$$K(x, y) = \tfrac{1}{2}(Q(x)Q(y) - Q(x, y)^2) + \tfrac{1}{2}Q(y).$$

Of course, K is defined on all of $\mathbb{R}^{2n}$, so we can consider the flow induced on our minimal orbit by P as being the constrained flow from $\mathbb{R}^{2n}$ coming from the Hamiltonian K and the constraints (30.12). The corresponding constrained Hamiltonian is

$$K^* = K - l_1 f_1 - l_2 f_2,$$

where the l's are determined by (30.3) as

$$l_1 = \|A^{-1}y\|^2(1 + Q(x))Q(y)^{-1} - \|A^{-1}x\|^2,$$
$$l_2 = 0.$$

Thus the constrained equations of motion are

$$\frac{dx}{ds} = K^*_y = -(1 + Q(x))A^{-1}y - l_1 y,$$

$$\frac{dy}{ds} = -K^*_x = -Q(y)A^{-1}x.$$

Now along the hypersurface $K = 0$, (30.12) implies that $1 + Q(x) = 0$ (so x lies on the ellipsoid) and the preceding equations simplify to

$$\frac{dx}{ds} = -l_1 y, \qquad \frac{dy}{ds} = -Q(y)A^{-1}x, \qquad l_1 = -\|A^{-1}x\|^2. \quad (30.13)$$

If we compare (30.13) with (30.10), we see that the right-hand sides differ by a factor of $-l_1$. Thus the trajectories are the same, and we can reduce (30.13) to (30.10) by making the time change $ds/dt = -l_1$ (which amounts to multiplying the Hamiltonian by a factor $-l_1$).

Thus, the problem of solving the equations for the geodesic flow on the ellipsoid is essentially equivalent to the problem of solving the Euler–Poisson equations with a function P given by (30.11).

31. The choice of a collective Hamiltonian

The collective motion description on a symplectic manifold M involves two assertions: a kinematical one, giving the Hamiltonian action of a Lie group G on M and the attendant moment map, and then a dynamical one, involving the choice of the function F on g^*. In this section we discuss various strategies for choosing F.

One procedure involves an auxiliary symplectic manifold X together with a family of Hamiltonian actions of G on X, giving a family of moment maps, $\Phi^\sigma: X \to g^*$. It might be the case that each such action also carries with it a natural Hamiltonian $\mathscr{H}^\sigma$ on X and that each such $\mathscr{H}^\sigma$ is constant on the inverse image $(\Phi^\sigma)^{-1}(l)$, for $l \in g^*$; more precisely, that there exists a function defined on some open set of g^* and such that

$$\mathscr{H}^\sigma = F \circ \Phi^\sigma$$

for all σ. Then we would take F as our desired function. We shall use this procedure for the construction of the collective kinetic energy for the rigid-body, compressible liquid drop, and incompressible liquid drop models of the nucleus. Thus G is the semidirect product of H with V, where $V = S^2(\mathbb{R}^3)$ and $H = SO(3)$, $Gl(3, \mathbb{R})$, or $Sl(3, \mathbb{R})$. (In the entire discussion we could replace 3 by n with little or no change.) Via the embedding of G in $Sp(6N)$, described above, we get a moment map from the phase space of N particles to g^*. In particular, each configuration $(\vec{q}_1, \ldots, \vec{q}_N)$, $\vec{q}_i \in \mathbb{R}^3$, of the N particles determines a $Q \in V^*$, where $Q = \sum \vec{q}_i \otimes \vec{q}_i$. Each such Q determines an action of G on $X = T^*H$, as described in Section 29. We shall refer to Q as the "initial configuration" and denote by $\Phi^Q: T^*H \to g^*$, the moment map of this action. Each such Q also determines a kinetic energy Hamiltonian $\mathscr{H}^Q$ on T^*H, as will be described below. For suitable choice of a family of initial configurations, we shall find a function F on g^* such that

$$\mathscr{H}^Q = F \cdot \Phi^Q.$$

Let $A(t)$ be some curve of matrices in H. Let $\vec{q} \in \mathbb{R}^3$ be a particle with unit mass. The curve $A(t)\vec{q}$ describes the evolution of the particle under the action of the one-parameter family $A(t)$. The associated kinetic energy is

$$\tfrac{1}{2}\| A'(t)q \|^2 = \tfrac{1}{2} \operatorname{tr} A'(t)\vec{q} \otimes \vec{q}^{\,\mathrm{t}} A'^{\mathrm{t}}(t). \tag{31.1}$$

Write

$$A(t+\tau) = A(t)A(t)^{-1}A(t+\tau) = A(t)B_t(\tau).$$

Then

$$A'(t) = A(t)\xi, \qquad \text{where} \qquad \xi = \left.\frac{dB_t(\tau)}{d\tau}\right|_{\tau=0}$$

This is the left-invariant identification we have been using to relate TH_e with $TH_{A(t)}$. The expression for the kinetic energy becomes $\frac{1}{2} \operatorname{tr} A\xi\vec{q} \otimes \vec{q}^{\,\mathrm{t}}\xi^{\mathrm{t}}A^{\mathrm{t}}$, where $A = A(t)$. Summing over all the particles (of equal unit mass) gives $\frac{1}{2}\operatorname{tr} A\xi Q\xi^{\mathrm{t}}A^{\mathrm{t}}$, where $Q = \sum \vec{q} \otimes \vec{q}^{\,\mathrm{t}}$ is the inertia tensor. This is the Lagrangian form of kinetic energy. We must pass from the Lagrangian to the Hamiltonian, to obtain a function on $T^*H \sim H \times h^*$. This is done via the Legendre transformation. Since the Lagrangian is quadratic in ξ, the procedure is particularly simple. We must find the linear map $\xi \rightsquigarrow \alpha$ of $h \to h^*$ such that

$$\langle \alpha(\xi), \eta \rangle = dL_\xi \eta,$$

where L is the function of ξ given by $L(\xi) = \frac{1}{2} \operatorname{tr} A\xi Q\xi^{\mathrm{t}}A^{\mathrm{t}}$. We must then invert this linear map so as to express ξ as a function of α and then

$$\mathscr{H}^Q(A, \alpha) = L\big(\xi(\alpha)\big) = \tfrac{1}{2}\langle \alpha, \xi(\alpha) \rangle,$$

where L, of course, depends on A and Q.

In the case of all three of our groups, H, the bilinear form of B on h defined by $B(\xi, \eta) = \operatorname{tr} \xi\eta$ is nondegenerate and invariant under the adjoint representation. We shall use it to identify h with h^*. We can thus think of α as a matrix, and the coadjoint representation is just given by conjugation, $A \in H$ sends α into $A\alpha A^{-1}$. Thus $\langle \alpha, \eta \rangle = \operatorname{tr} \alpha\eta$, and the Legendre transformation is determined by $\alpha = \alpha(\xi)$, where

$$\operatorname{tr} \alpha\eta = \tfrac{1}{2} \operatorname{tr} A\xi Q\eta^{\mathrm{t}}A^{\mathrm{t}} + \tfrac{1}{2} \operatorname{tr} A\eta Q\xi^{\mathrm{t}}A^{\mathrm{t}} \qquad \forall \eta \in h. \tag{31.2}$$

From here on, the three groups require different treatments.

For the case $H = SO(3)$ we have $A^t = A^{-1}$ and so L is independent of A – the Lagrangian is invariant under left multiplication. Equation (31.2) becomes

$$\operatorname{tr} \alpha\eta = \tfrac{1}{2} \operatorname{tr} \xi Q \eta^t + \tfrac{1}{2} \operatorname{tr} \eta Q \xi^t \qquad \forall \eta = -\eta^t,$$

together with the additional requirement that

$$\alpha = -\alpha^t,$$

so that $\alpha \in h \sim h^*$. It is clear that the solution of this equation is given by

$$\alpha = -\tfrac{1}{2}(Q\xi + \xi Q). \tag{31.3}$$

If Q is positive definite (which will happen if the q do not all lie in a subspace), we can invert this equation and solve for ξ as a function of α. Indeed, it is clear that for any $C \in SO(3)$, if

$$Q' = CQC^{-1} \qquad \text{and} \qquad \xi' = C\xi C^{-1},$$

then

$$\alpha' = C\alpha C^{-1} = -\tfrac{1}{2}(Q'\xi' + \xi' Q').$$

If we choose C so that Q' is diagonal with positive eigenvalues $\lambda_1, \lambda_2, \lambda_3$, this last equation becomes

$$\alpha'_{ij} = -\tfrac{1}{2}(\lambda_i + \lambda_j)\xi'_{ij},$$

which can clearly be solved for ξ', and hence we can solve for ξ in terms of α. Thus the Hamiltonian becomes

$$\mathscr{H}^Q(A, \alpha) = \tfrac{1}{2} \operatorname{tr} \xi_Q(\alpha) \cdot \alpha, \tag{31.4}$$

where $\xi_Q(\alpha)$ is obtained as a function of α by solving (31.3). For example, if Q were diagonal, $\xi_{ij} = -2(\lambda_i + \lambda_j)\alpha_{ij}$ and (31.4) becomes

$$\sum \frac{\alpha_{ij}^2}{\lambda_i + \lambda_j},$$

which, up to factors of 2, gives the standard expression for the kinetic energy in terms of the angular momentum and the "moments of inertia" $\lambda_i + \lambda_j$. (The reason for the factors of 2 is that we have used $\operatorname{tr} \xi \cdot \eta$ instead of the more conventional $-\frac{1}{2} \operatorname{tr} \xi \cdot \eta$ to identify $o(3)$ with $o(3)^*$.) Since we will be considering various Q's, we will need the general expression (31.4). Let us identify $S^2(\mathbb{R}^3)$ with the space of symmetric 3×3 matrices, so that $A \in SO(3)$ acts on Q by sending it to

$$AQA^t = AQA^{-1}.$$

By Section 29, we know that each Q determines a moment map

$$\Phi^Q: T^*SO(3) \to o(3)^* \oplus S^2(\mathbb{R}^3) = g^*$$

given by

$$\Phi^Q(A, \alpha) = (A\alpha A^{-1}, AQA^{-1}). \tag{31.5}$$

If $\Phi^Q(A, \alpha) = \Phi^Q(B, \alpha')$, then $\alpha' = C\alpha C^{-1}$ and $Q = CQC^{-1}$, where $C = B^{-1}A$. Then, by the preceding discussion, we know that $\xi' = C\xi C^{-1}$ and hence that

$$\mathscr{H}^Q(A, \alpha) = \mathscr{H}^Q(B, \alpha').$$

In other words, $\mathscr{H}^Q$ is constant on inverse images under Φ^Q of the same point in $o(3)^* \oplus S^2(\mathbb{R}^3) = g^*$. In fact, we can define the function F on g^* by

$$F(\alpha, Q) = \tfrac{1}{2} \operatorname{tr} \xi(Q, \alpha) \cdot \alpha, \tag{31.6}$$

where $\xi(Q, \alpha)$ is obtained by solving (31.3) for ξ in terms of α and Q. The function F is defined on the subset of g^* where Q is positive definite. Then

$$\begin{aligned}\Phi^Q(A, \alpha) = F(A\alpha A^{-1}, AQA^{-1}) &= \tfrac{1}{2} \operatorname{tr} \xi(AQA^{-1}, A\alpha A^{-1}) \cdot A\alpha A^{-1} \\ &= \tfrac{1}{2} \operatorname{tr} A\xi(Q, \alpha)A^{-1}A\alpha A^{-1} \\ &= \tfrac{1}{2} \operatorname{tr} \xi(Q, \alpha) = \mathscr{H}^Q(A, \alpha).\end{aligned}$$

In other words

$$\mathscr{H}^Q = F \circ \Phi^Q. \tag{31.7}$$

Now let M be any symplectic manifold upon which G, the semidirect product of $SO(3)$ with $S^2(\mathbb{R}^3)$ acts, with moment map $\Phi: M \to g^*$. For instance, we may take $M = \mathbb{R}^{6N}$ as described above. Then (on the subset of M mapping onto the domain of definition of F) we can define $\mathscr{H} = F \circ \Phi$, and this defines the "collective rigid-body kinetic energy." Of course, we might still want to add some potential energy term; for example, we might want to consider the motion of a rigid body in the presence of a gravitational field. As we have seen in the last section, this particular problem (for uniform gravitational fields) can also be formulated and solved for the rigid body by use of the moment map and the theorems of the preceding section. Let us now consider some of our other examples of collective Hamiltonians, where, as we shall see, the functions that we get on T^*H are not necessarily left-invariant.

Let us consider the compressible liquid drop, where $H = Gl(3, \mathbb{R})$. An $A \in H$ acts on $Q \in S^2(\mathbb{R}^3)$ by sending it into AQA^t. The only invariants under this action are the rank and signature of Q. Thus all positive definite

Q are equivalent to one another under this action. The group G in this case is $(9 + 6 =)$ 15-dimensional and the analysis provided in Section 19 shows that the generic orbit in g^* is 14-dimensional, the next highest orbit being 12-dimensional. Equation (31.2) now becomes

$$\begin{aligned}\operatorname{tr} \alpha\eta &= \tfrac{1}{2} \operatorname{tr} A^{t}A\xi Q\eta^{t} + \tfrac{1}{2} \operatorname{tr} \eta Q\xi^{t}A^{t}A \\ &= \operatorname{tr} \eta Q\xi^{t}A^{t}A \qquad \forall\eta\end{aligned}$$

with no further conditions on α. This clearly has the solution

$$\alpha = Q\xi^{t}A^{t}A$$

or

$$\xi = A^{-1}(A^{t})^{-1}\alpha^{t}Q^{-1}$$

and so

$$\mathcal{H}^{Q}(A, \alpha) = \tfrac{1}{2} \operatorname{tr} (A^{t})^{-1}\alpha^{t}Q^{-1}\alpha A^{-1}. \tag{31.8}$$

Notice that this Hamiltonian on the 18-dimensional manifold $T^*Gl(3, \mathbb{R})$ is not invariant under left multiplication. The moment map Φ^{Q}: $T^*Gl(3, \mathbb{R}) \to g^*$ is given by

$$\Phi^{Q}(A, \alpha) = (A\alpha A^{-1}, AQA^{t}),$$

and one can check that, in general, $\mathcal{H}^{Q}$ will not be constant on the inverse image of the points in g^*. However, this will be the case if we choose our initial configuration of particles to be spherical, that is, if we choose $Q = cI$ where I is the identity matrix. As the choice of c amounts to a choice of unit of volume, we might as well take $c = 1$ so that

$$\Phi^{I}(A, \alpha) = (A\alpha A^{-1}, AA^{t}) \tag{31.9}$$

and

$$\mathcal{H}^{I}(A, \alpha) = \tfrac{1}{2} \operatorname{tr} (A^{t})^{-1}\alpha^{t}\alpha A^{-1}. \tag{31.10}$$

Then, defining F on g^* by

$$F(\beta, Q) = \tfrac{1}{2} \operatorname{tr} \beta^{t}Q^{-1}\beta, \tag{31.11}$$

we see that

$$\mathcal{H}^{I}(A, \alpha) = F \circ \Phi^{I}. \tag{31.12}$$

We might want to add to $\mathcal{H}^{I}$ some collective potential energy terms which would measure the energy involved in compression or distortion. That is, we might want to choose

$$F(\beta, Q) = \tfrac{1}{2}\operatorname{tr}\beta^{t}Q^{-1}\beta + \lambda(\det Q - 1)^2 + \mu\operatorname{tr}(Q - cI)^2, \qquad c = (\det Q)^{1/3}, \tag{31.13}$$

where λ and μ are physical parameters. Then, if M is any symplectic manifold with a Hamiltonian G action and moment map Φ, the collective Hamiltonian would be $\mathscr{H} = F \circ \Phi$.

Now let us turn to the case where $H = Sl(3, \mathbb{R})$. Equation (31.2) now becomes

$$\operatorname{tr} \alpha\eta = \operatorname{tr} \eta Q \xi^{t} A^{t} A \qquad \forall\eta \qquad \text{with} \qquad \operatorname{tr} \eta = 0$$

and

$$\operatorname{tr} \alpha = 0.$$

Thus

$$\alpha = Q\xi^{t}A^{t}A - \tfrac{1}{3}(\operatorname{tr} Q\xi^{t}A^{t}A)I$$

and so

$$\xi = A^{-1}(A^{t})^{-1}\alpha^{t}Q^{-1} - \left(\frac{\operatorname{tr} Q^{-1}\alpha A^{-1}(A^{t})^{-1}}{\operatorname{tr} Q^{-1}A^{-1}(A^{t})^{-1}}\right)A^{-1}(A^{t})^{-1}Q^{-1}$$

yielding

$$\mathscr{H}^{Q}(A, \alpha) = \frac{1}{2}\operatorname{tr} \xi\alpha = \frac{1}{2}\operatorname{tr} (A^{t})^{-1}\alpha^{t}Q^{-1}\alpha A^{-1} - \frac{1}{2}\frac{(\operatorname{tr} Q^{-1}\alpha A^{-1}(A^{t})^{-1})^2}{\operatorname{tr} Q^{-1}A^{-1}(A^{t})^{-1}}. \tag{31.14}$$

Once again, this Hamiltonian is neither left-invariant, nor, for general Q, constant on the pre-images of points under Φ^{Q}. But again, things work out if we choose the initial configuration to be spherical, so that $Q = cI$. (We can no longer choose $c = 1$ since the elements of $Sl(3, \mathbb{R})$ preserve $\det Q$.) Define the function F on g^* by

$$F(\beta, S) = \frac{1}{2}\operatorname{tr} \beta^{t}S^{-1}\beta - \frac{1}{2}\frac{(\operatorname{tr} \beta S^{-1})^2}{\operatorname{tr} S^{-1}} \tag{31.15}$$

and, taking $Q = cI$ in (31.14) and $\Phi^{cI}(A, \alpha) = (A\alpha A^{-1}, cAA^{t})$ we get

$$\mathscr{H}^{cI} = F \circ \Phi^{cI}. \tag{31.16}$$

One could also add a "collective potential term"

$$U(\beta, S) = U(S) = \mu\operatorname{tr} (S - cI)^2, \qquad c = \det S,$$

where μ is some physical parameter associated to resistance to deformation. Then $F + U$ would be a candidate for a collective Hamiltonian when pulled back to any symplectic manifold upon which G acts with moment map Φ: $M \to g^*$. Notice that $F + U$ is invariant under the action of the subgroup $SO(3)$ acting on g^*. On 10-dimensional orbits (with positive definite S) there is only one fixed point under $SO(3)$, namely, the point $(0, cI)$ corresponding, under our identification of these orbits as cotangent bundles of the liquid drop configurations, to the zero covector at the spherical configuration. Any Hamiltonian invariant under $SO(3)$ can have a strict minimum only at this point. In particular, the restriction of $F + U$ to such an orbit has an absolute minimum at this point if $\mu > 0$. The most general tangent vector to the 10-dimensional orbit at this point can be written as (γ, β), where $\gamma \in sl(3, \mathbb{R})/o(3)$ can be thought of as a symmetric matrix. Thus the quadratic approximation to the $F + U$ Hamiltonian will be of the form

$$\tfrac{1}{2}a \operatorname{tr} \gamma^2 + b \operatorname{tr} \beta^2. \tag{31.17}$$

with suitable a and b. This is the Bohr–Mottleson Hamiltonian.

Another model, whose physical significance is unclear but which has interesting mathematical properties is given by the function

$$F(\alpha, Q) = \tfrac{1}{2} \operatorname{tr} \alpha Q \alpha^{t} Q^{-1}. \tag{31.18}$$

This function has the property of being invariant under $Sl(3, \mathbb{R})$. Indeed

$$\begin{aligned} F(A\alpha A^{-1}, AQA^{t}) &= \tfrac{1}{2} \operatorname{tr} A\alpha A^{-1} AQA^{t}(A^{t})^{-1}\alpha^{t} A^{t}(A^{t})^{-1} Q^{-1} A^{-1} \\ &= \tfrac{1}{2} \operatorname{tr} A\alpha Q\alpha^{t} Q^{-1} A^{-1} \\ &= \tfrac{1}{2} \operatorname{tr} \alpha Q \alpha^{t} Q^{-1} \\ &= F(\alpha, Q). \end{aligned}$$

Furthermore, we claim that the Hamiltonian system corresponding to this P on each of the 10-dimensional orbits is completely integrable. That is, we claim that there exist five independent Poisson commuting functions on the phase space T^*N of the liquid drop, one of the functions being the restriction of the F given by (31.18) to T^*N, where we have embedded T^*N as a 10-dimensional orbit in the dual of the Lie algebra of $Sl(3) \circledS S^2(\mathbb{R}^3)$, as described in Section 29. The following proof of complete integrability was suggested to us in a conversation with Kostant.

Let $C_2 \in U$ be the second-order Casimir operator of $Sl(3, \mathbb{R})$. Here $U = U(sl(3, \mathbb{R}))$ denotes the universal enveloping algebra of $sl(3, \mathbb{R})$. Every element of U acts as a partial differential operator on the space of C^∞ functions on N, where N is any manifold on which $Sl(3, \mathbb{R})$ acts. In particular, C_2, being in the center of U, gives rise to an operator $\hat{C}_2$, which commutes with the action. In particular, the symbol of $\hat{C}_2$ (i.e., the expression obtained by replacing $\partial/\partial q_j$ by p_j in $\hat{C}_2$ and only retaining the terms of highest, i.e., second, order), which we shall denote by $\tau(\hat{C}_2)$ is invariant under the action of $Sl(3, \mathbb{R})$. We claim that, up to a constant multiple, $\sigma(\hat{C}_2)$ coincides with the restriction F to T^*N, where F is given by (31.8). Indeed, both functions are invariant under $Sl(3, \mathbb{R})$ so it suffices to compare them at T^*N_{cI}. Now C_2 gives rise to the function on g^* sending α into $\frac{1}{2}\operatorname{tr}\alpha^2$. On the other hand, $\Phi(T^*N_{cI})$ consists of all (α, CI) satisfying $\alpha = \alpha^t$. For these α, (31.18) gives $\frac{1}{2}\operatorname{tr}\alpha^2$.

To prove complete integrability, it certainly suffices to find five independent elements in U that mutually commute and one of which is C_2. This can be done by the method of Gel'fand and Cetlin (1950): Consider the string of subgroups $Gl(1, \mathbb{R}) \subset Gl(2, \mathbb{R}) \subset Sl(3, \mathbb{R})$. Let C_2 and C_2' be the two independent Casimirs in $Sl(3, \mathbb{R})$. They commute with all of U and hence with the two Casimir operators in $Gl(2)$, which, in turn, commute with the Casimir operator of $Gl(1)$. The symbols of these five operators then give five commuting functions on the phase space of the liquid drop. We shall explain this construction in more generality in Chapter IV.

32. Convexity properties of toral group actions

Consider the group $Sp(2)$ of all linear symplectic transformations of the plane. We know that its Lie algebra can be identified with the space of all homogeneous quadratic polynomials. We also know from Chapter I that the quadratic polynomial $\frac{1}{2}(p^2 + q^2)$ is the infinitesimal generator of the group of rotations in the q-p plane. If we let T_1 denote this one-dimensional torus (the circle group) and t_1 its Lie algebra, then t_1 has a preferred generator, ξ such that $\exp 2\pi\xi = id$ and the image of ξ is the vector field corresponding to the function $\frac{1}{2}(p^2 + q^2)$. Thus we may identify t_1 with $\mathbb{R}^1$ and hence t_1^* with $\mathbb{R}^1$ as well. The moment map is thus given by

$$\Phi(q, p) = \tfrac{1}{2}(p^2 + q^2)$$

under these identifications; that is,

$$\langle \Phi(q, p), a\xi \rangle = \tfrac{1}{2}a(p^2 + q^2).$$

We can consider the product T^n of n copies of T_1 and the corresponding product action on the symplectic vector space $\mathbb{R}^{2n}$. Identifying t_n^* with $\mathbb{R}^n$, as above, we see that the moment map

$$\Phi\colon \mathbb{R}^{2n} \to \mathbb{R}^n$$

is given by

$$\Phi(q_1,\dots,q_n;p_1,\dots,p_n) = \left(\tfrac{1}{2}(p_1^2 + q_1^2),\dots,\tfrac{1}{2}(p_n^2 + q_n^2)\right). \tag{32.1}$$

Notice that the image of the moment map is the convex cone consisting of all points $(x_1,\dots,x_n)$ with $x_i \geqslant 0$ (the "positive 2^n-tant").

Now let T be any r-dimensional torus and suppose that we are given a linear symplectic T action on some symplectic vector space V. Since T is compact, we can put a T-invariant positive definite scalar product on V, which, together with the symplectic form, determines a complex structure on V and a T-invariant Hermitian form whose imaginary part is the given symplectic structure. We can then find an orthonormal basis consisting of simultaneous eigenvectors (weight vectors) corresponding to the simultaneous eigenvalues (weights) $\alpha_1,\dots,\alpha_n$ of the elements of t. Here the α_i are linear functions on t. In other words, we can find a symplectic linear isomorphism of V with $\mathbb{R}^{2n}$ (where $\dim V = 2n$) such that the linear symplectic action of t on V is given by the homomorphism

$$\rho: t \to t_n \sim \mathbb{R}^n,$$
$$\rho(\zeta) = (\alpha_1(\zeta),\dots,\alpha_n(\zeta)).$$

The moment map is then the composition of the moment map in (32.1) with the adjoint $\rho^*: t_n^* \to t^*$, and so is given by

$$\Phi(q_1,\dots,q_n;p_1,\dots,p_n) = \tfrac{1}{2}(p_1^2 + q_1^2)\alpha_1 + \cdots + \tfrac{1}{2}(p_n^2 + q_n^2)\alpha_n. \tag{32.2}$$

We thus have proved

Proposition 32.1. *The image of Φ is the convex region*

$$S(\alpha_1,\dots,\alpha_n) = \left\{\sum_{i=1}^{n} s_i\alpha_i, s_1,\dots,s_n \geqslant 0\right\}$$

in t*, *the α_i's being the weights of the representation of* T *on* V.

So far we have been discussing linear torus actions. But we shall now show that for any compact Lie group, we can use the Darboux–Weinstein theorem (Theorem 22.1) to show that near a fixed point, the moment map for a compact Lie group has the same image as the moment map for a linear action. Indeed, let X be a symplectic manifold with symplectic form Ω, G a compact Lie group and $G \times X \to X$ a Hamiltonian G action. Let $\Phi: X \to g^*$ be the associated moment mapping and let x be a fixed point of G. We will use the Darboux–Weinstein theorem to show that Φ is right-equivalent to the moment mapping associated with the linear action of G on the tangent space $V = T_x X$. Let X be equipped with a G-invariant Riemannian metric and let exp: $V \to X$ be the exponential map defined by this metric. This map intertwines the action of G on X and the linear action of G on V, and it maps a G-invariant neighborhood U_0 of zero in V diffeomorphically onto a neighborhood U of x in X. Let Ω_1 be the linear symplectic form on V and let $\Omega_0 = (\exp)^*\Omega$. For U_0 sufficiently small there exists, by Theorem 22.1 a G-equivariant mapping Ψ: $(U_0, 0) \to (V, 0)$ such that $\Psi^*\Omega_1 = \Omega_0$. Let $\Phi_0 = \Phi \circ \exp$ and let Φ_1 be the moment mapping associated with the linear action of G on V and the linear symplectic form, Ω_1. From the defining property (26.9, 26.10) of the moment map we conclude

Theorem 32.1. *The mappings* Φ_0: $U_0 \to g^*$ *and* $\Phi_1 \circ \Psi$: $U_0 \to g^*$ *differ by a translation.*

Corollary. *Let* $U_1 = \Psi(U_0)$. *Then the image of* Φ: $U_0 \to g^*$ *is, up to a translation, identical with the image of* Φ_1: $U_1 \to g^*$.

We can now return to the case where $G = T$ is a torus, and combine Proposition 32.1 with Theorem 32.1 to obtain

Theorem 32.2 (Local Convexity Theorem). *Let* X *be a symplectic manifold,* T *an* r*-dimensional torus,* $T \times X \to X$ *a Hamiltonian action of* T *on* X, *and* $\Phi: X \to t^*$ *the associated moment mapping. Let* x *be a fixed point of* T *and let* $p = \Phi(x)$. *Then there exists a neighborhood* U *of* x *in* X *and a neighborhood* U′ *of* p *in* t^* *such that*

$$\Phi(U) = U' \cap (p + S(\alpha_1, \ldots, \alpha_n)),$$

where $\alpha_1, \ldots, \alpha_n$ *are the weights associated with the linear isotropy representation of* T *on the tangent space of* X *at* x.

We will need a relative form of this theorem later on. As above, let X be a symplectic manifold, T an r-dimensional torus, $T \times X \to X$ a Hamiltonian action of T on X and x a point of X. Let T_1 be the stabilizer group of x. The inclusion of t_1 in t dualizes to give a linear mapping the other way

$$\pi: t^* \to t_1^*.$$

Let $\alpha_1, \ldots, \alpha_n$ be the weights of the representation of T_1 on the tangent space of X at x. Let $S(\alpha_1, \ldots, \alpha_n)$ be the convex region in t_1^* and let

$$S'(\alpha_1, \ldots, \alpha_n) = \pi^{-1} S(\alpha_1, \ldots, \alpha_n) \tag{32.3}$$

in t^*.

Theorem 32.3. *There exists a neighborhood* U *of* x *in* X *and a neighborhood* U′ *of* p = Φ(x) *in* t* *such that*

$$\Phi(U) = U' \cap (p + S'(\alpha_1, \ldots, \alpha_n)).$$

Proof. Let V be the tangent space to X at x. We will make a few simplifying assumptions. We will assume: (i) X is an open subset of V; (ii) x is the origin in V; (iii) the action of T_1 is the linear isotropy action; and (iv) the symplectic form on X is the restriction to X of the linear symplectic form on V. These assumptions are warranted by Theorem 32.1. Let X_1 be the subset of X consisting of all points in X that have T_1 as stabilizer group. It is clear that X_1 is the intersection of X with a linear subspace W of V. By Theorem 27.2 $\Phi: X \to t^*$ maps X_1 submersively onto an open subset of $p + t_1^\perp$. In particular,

$$d\Phi_0 : W \to t_1^\perp \to 0. \tag{32.4}$$

Let $\Phi_1 = \pi \circ \Phi =$ the moment map associated with the linear action of T_1 on V. The action of T_1 on W is trivial; so Φ_1 is constant on the affine subspaces

$$W + a, \qquad a \in V, \tag{32.5}$$

of V. Therefore Φ maps the affine subspace (32.5) into the affine subspace

$$\pi^{-1}(\Phi_1(a)) = t_1^\perp + q, \qquad q = \Phi(a). \tag{32.6}$$

By (32.4) this mapping is a submersion near $a = 0$; so the image of (32.5) contains an open neighborhood of q in (32.6). This being true for all points a sufficiently close to zero, the image of $\Phi: X \to t^*$ contains an open neighborhood of $\Phi(x)$ in $\pi^{-1}(\Phi_1(X))$. Q.E.D.

Definition. *Let* X *be a manifold and* f: X → ℝ *a smooth function. A real number* a *is a local maximum of* f *if there exists a point* $x_0 \in X$ *and a neighborhood* U *of* x *such that* $f(x_0) = a$ *and* $f(x) \le a$ *for* $x \in U$.

Let X be a connected compact symplectic manifold, G a compact Lie group, $G \times X \to X$ a Hamiltonian action of G on X, and $\Phi: X \to g^*$ the associated moment mapping. For each $\xi \in g$, let ϕ^ξ be the ξth component of Φ: that is,

$$\varphi^\xi = \langle \Phi, \xi \rangle.$$

The following lemma is crucial:

Lemma 32.1. *The function* φ^ξ *has a unique local maximum.*

Before proving this lemma we will make an application of it. Let $G = T =$ an r-torus.

Theorem 32.4. *The image of the moment mapping* $\Phi\colon X \to t^*$ *is a convex polytope (Atiyah, 1982; Guillemin and Sternberg, 1982a).*

Proof. Let p be a point of the boundary of $\Phi(X)$ and let $x \in X$ be a pre-image point of p. Let T_1 be the stabilizer group of x and let

$$\alpha_i \in t_i^* \qquad i = 1,\ldots,n,$$

be the weights of the isotropy representation of T_1 on the tangent space to X at x. By Theorem 32.2 there exists a neighborhood U of x and a neighborhood, U' of p such that $\Phi(U) = U' \cap (p + S'(\alpha_1,\ldots,\alpha_n))$. For each $\xi \in t$, let $e_\xi\colon t^* \to \mathbb{R}$ be the linear functional, $f \in t^* \to f(\xi)$. Let S_i be a boundary component of $S'(\alpha_1,\ldots,\alpha_n)$. We can choose $\xi \in t$ such that $e_\xi = 0$ on S_i and e_ξ is negative on the interior of $S'(\alpha_1,\ldots,\alpha_n)$. Then if $e_\xi(p) = a$,

$$\phi^\xi(x) = (e_\xi \circ \Phi)(x) \leqslant a$$

for $x \in U$; so a is a local maximum of ϕ^ξ. By Lemma 32.1 $\phi^\xi(X) \leqslant a$. Applying this argument to all the faces S_i of $S'(\alpha_1,\ldots,\alpha_n)$ we conclude that $\Phi(X) \subseteq p + S'(\alpha_1,\ldots,\alpha_n)$. We already know (Theorem 27.3) that $\Phi(X)$ is a finite union of convex sets, and we now know that $\Phi(X)$ looks like a convex polytope near each of its boundary points, hence $\Phi(X)$ is a convex polytope. Q.E.D.

To prove Lemma 32.1 we will need to review some elementary Morse theory. Let X be a connected compact n-dimensional manifold and f: $X \to \mathbb{R}$ a smooth functions. Let C_f be the critical set of f.

Definition. f *is* clean *if each connected component of* C_f *is a submanifold of* X *and if in addition at each critical point* $x \in C_f$, *the Hessian,* d^2f_x *of* f *is nondegenerate in directions normal to* C_f *at* x. *If* C_f *consists of isolated points then* f *is called a Morse function.*

Let C be a connected component of C_f. If f is clean, the index of d^2f_x is constant along C and its value is called the *index of the critical set* C.

Suppose X is equipped with a Riemannian metric. Then f defines a gradient vector field on X. Let $\psi_t\colon X \to X$, $-\infty < t < \infty$ be the flow generated by this gradient vector field; and, for each component C_i of C_f let

$$W_i = \{x \in X, \psi_t(x) \to C_i \text{ as } t \to +\infty\}.$$

One of the fundamental theorems of Morse theory is the following. (See, for instance, Bott 1957.)

Theorem 32.5. *If* f *is clean then each* W_i *is a cell bundle over* C_i *with fiber dimension equal to the index of* C_i *and*

$$X = \bigcup W_i \qquad \text{(disjoint union)}$$

is a decomposition of X *into cell bundles.*

Corollary 32.1. *Let* $f: X \to \mathbb{R}$ *be clean. Suppose that for all the critical manifolds* C_i, *dimension* C_i *and index* C_i *are even. Then* f *has a unique local maximum.*

Proof. Let $C_1, \ldots, C_k$ be the critical manifolds of index 0 and $C_{k+1}, \ldots, C_N$ the remaining critical manifolds. Then $f = \text{constant} = a_i$ on C_i, and $a_1, \ldots, a_k$ are the local maxima of f. By the theorem quoted above

$$X = W_1 \cup \cdots \cup W_k \cup W_{k+1} \cup \cdots \cup W_N$$

with W_i open for $i \leqslant k$ and W_i of codim $\geqslant 2$ for $i > k$. But if W_i is of codimension $\geqslant 2$ it cannot disconnect X, so

$$X - \bigcup_{i>k} W_i$$

is connected. Hence $k = 1$. Q.E.D.

If f is Morse in the corollary above, then W_1 is actually a cell. Let x_0 be a point of W_1 and γ a closed curve passing through x_0. By the transversality theorem one can perturb γ so that it intersects the W_i's, $i > 1$, transversally. But since codim $W_i \geqslant 2$ for $i > 1$, this means this intersection is empty. Thus γ lies in W_1 and can be contracted to x_0. This proves a result that we will use later on in this section.

Corollary 32.2. *If* f *is Morse and the indices of all its critical points are even then* $\pi_1(X, x_0) = 0$.

Returning to Lemma 32.1, it is enough to prove:

Theorem 32.6. *The function* ϕ^ξ *is clean and the indices and dimensions of its critical manifolds are all even.*

Proof. Let $\xi^\#$ be the vector field on X associated with ξ and let $\rho_t: X \to X$, $-\infty < t < \infty$, be the flow generated by $\xi^\#$. If x is a critical point of ϕ^ξ, then $\xi^\#(x) = 0$ by (26.9) and hence $\rho_t(x) = x$ for all t. Let V be the tangent space to X at x and let $L: V \to V$ be the infinitesimal generator of the linearized flow

$$(d\rho_t)_x : V \to V. \tag{32.7}$$

It is easy to check that the Hessian of ϕ^ξ at x is the quadratic form

$$v \in V \to \Omega_x(Lv, v). \tag{32.8}$$

Let X be equipped with a G-invariant Riemannian metric and let

$$\exp : V \to X \tag{32.9}$$

be the exponential map. This map intertwines the flow ρ_t on X and the linearized flow (32.7) on V. Therefore, if W is the subspace of V on which $L = 0$, (32.9) maps a neighborhood of the origin in W diffeomorphically onto a neighborhood of the zero set of $\xi^\#$ on X. By (26.9), the zero set of $\xi^\#$ is identical with the critical set of ϕ^ξ; so we have proved that the connected components of the critical set are manifolds. Moreover, by (32.8) $d^2\phi_x^\xi$ is nondegenerate on V/W; so the critical manifold is clean at x. Finally it is clear from (32.2) that the index of $d^2\phi_x^\xi$ is even. Q.E.D.

We will illustrate Theorem 32.4 with a simple example. Let $X = \mathbb{C}P^N$ and let T be the $(N + 1)$-dimensional torus group

$$\{(e^{i\theta_1}, \ldots, e^{i\theta_{N+1}}), \theta_1, \ldots, \theta_{N+1} \in \mathbb{R}\}.$$

T acts on $\mathbb{C}P^N$ as a group of collineations, and it is easy to see that it preserves the Kaehler form associated with the Hermitian inner product $\langle z, w\rangle = z_1\bar{w}_1 + \cdots + z_{N+1}\bar{w}_{N+1}$ on $\mathbb{C}^{N+1}$. The one-parameter subgroup $T_r = \{(0, \ldots, e^{i\theta_r}, \ldots, 0)\}$ is globally Hamiltonian, and in terms of homogeneous coordinates its generating function is $|z_r|^2/|z|^2$; so the action of T itself is globally Hamiltonian and its associated moment mapping, $\Phi : X \to \mathbb{R}^{N+1}$ is

$$(z_1, \ldots, z_{N+1}) \to (|z_1|^2, \ldots, |z_{N+1}|^2)/|z|^2.$$

Its image is the simplex

$$\{(s_1, \ldots, s_{N+1}), s_1 + \cdots + s_{N+1} = 1, s_1, \ldots, s_{N+1} \geqslant 0\},$$

and the vertices of this simplex are the images of the fixed points $x_i = (0, \ldots, z_i, \ldots, 0)$ of T.

We now describe some partial results for the case of a compact connected Lie group G, not necessarily a torus. Let G be a connected, compact Lie group and T a Cartan subgroup of G. Let G act on g^* by its coadjoint action and let t^* be the subspace of g^* consisting of vectors in g^* that are stabilized by T. Let t^*_{reg} be the set of vectors in g^* whose stabilizer group is precisely T. It is easy to see that

$$t^*_{\text{reg}} = t^* - \text{a finite number of hyperplanes.}$$

We will denote by t_0^* one of the connected components of t_{reg}^* (for instance, the interior of the positive Weyl chamber) and by t_+^* its closure in t^*.

Now let X be a compact connected manifold upon which G acts in a Hamiltonian fashion, and let $\Phi: X \to g^*$ be the associated moment mapping. Let H be a closed connected subgroup of T and let

$$X_H = \{x \in X, \text{the stabilizer group of } x \text{ in } G \text{ is } H\}.$$

We will prove the following theorem

Theorem 32.7. *$\Phi(\mathrm{X_H}) \cap \mathrm{t}_+^*$ is a finite union of r-dimensional convex polytopes where $\mathrm{r} = \dim \mathrm{T/H}$.*

The first step in the proof will be to describe $\Phi(X_H)$ more explicitly. Let h be the Lie algebra of H and let $h^\perp$ be the annihilator of h in g^*. Let k^* be the subspace of g^* consisting of all vectors $v \in g^*$ that are stabilized by H.

Theorem 32.8. *Let $\mathrm{X_1}, \ldots, \mathrm{X_m}$ be the connected components of $\mathrm{X_H}$. Then there exist points $\mathrm{p_1}, \ldots, \mathrm{p_m}$ in k^* such that Φ maps $\mathrm{X_i}$ submersively onto an open subset of $k^* \cap (\mathrm{p_i} + \mathrm{h}^\perp)$.*

Proof. Let $\alpha_1, \ldots, \alpha_n$ be the positive roots of the Lie algebra g. Then g^* can be decomposed into a direct sum of T-invariant subspaces

$$g^* = t^* + \sum_{i=1}^{n} g_i^* \tag{32.10}$$

each g_i^* being two-dimensional and the representation of t on g_i^* being the representation

$$\xi \in t \to \alpha_i(\xi)\begin{pmatrix} 0 & 1 \\ -1 & 0 \end{pmatrix}.$$

Let $\phi_i: X \to g_i^*$ be the composition of the moment mapping and the projection of g^* onto g_i^*. We can think of ϕ_i as a complex-valued function that transforms under T according to the rule

$$D_{\xi^\#}\phi_i = \sqrt{-1}\,\alpha_i(\xi)\phi_i. \tag{32.11}$$

If $\xi \in h$, then $\xi^\# = 0$ on X_H; so by (32.11) $\phi_i \equiv 0$ on X_H if $\alpha_i(\xi) \neq 0$. This shows that the image of X_H lies in k^*. To prove the second half of the theorem, let K be the centralizer of H in G. We can canonically identify k^* with the dual of the Lie algebra of K. Since K preserves X_H we can view X_H as a Hamiltonian K-space. By Theorem 27.2, for each connected component X_i of X_H, there exists a point p_i in k^* such that Φ_K maps X_H submersively onto an open subset of $k^* \cap (p_i + h^\perp)$. Q.E.D.

Corollary. *If the centralizer of* H *is* T *there exist* $p_1, \ldots, p_m \in t^*$ *such that* Φ *maps* X_i *onto an open, convex polyhedral subregion of* $t^* \cap (p_i + h^\perp)$.

To prove Theorem 32.7 we observe first of all that if the centralizer of H in G is T, Theorem 32.7 is a special case of the corollary to Theorem 32.8. If the centralizer of H is bigger than T but is not equal to G, we can assume the theorem is true by induction. (Replace G by K in Theorem 32.7.) Therefore we are reduced to proving the theorem when G centralizes H. Furthermore, by Theorem 32.8 we can replace G by G/H in Theorem 32.7 or, in other words, can assume H is trivial. For the theorem to be nonvacuous, there has to exist some point $p \in X$ such that the stabilizer group of p is trivial. If such a point exists, then by Theorem 27.1 the set of points for which this is not true is of codimension $\geqslant 2$ in X. Therefore to prove Theorem 32.7 it is enough to prove

Theorem 32.9. *Let* X *be a compact connected Hamiltonian* G*-space with the property that there exists some point* $p \in X$ *such that the stabilizer of* p *in* G *is trivial. Then* $\Phi(X) \cap t_+^*$ *is a finite union of convex polytopes.*

An element, $\xi \in g^*$ is *regular* if its stabilizer is a Cartan subgroup of G. Every regular element of g^* is conjugate to a unique element of t_0^*. Let g_{reg}^* be the set of all regular elements of g^* and let $X_{\mathrm{reg}} = \Phi^{-1}(g_{\mathrm{reg}}^*)$.

Lemma 32.2. *The complement of* X_{reg} *in* X *is of codim* $\geqslant 2$. *In particular* X_{reg} *is dense.*

Proof. By Theorem 27.1, the complement of g_{reg}^* in g^* is of codimension $\geqslant$ 2. Let X_0 be the set of points $p \in X$ such that the stabilizer group of p in G is discrete. By Theorem 27.1 the complement of X_0 is of codimension $\geqslant 2$ and $\Phi: X_0 \to g^*$ is a submersion; so the complement of $X_0 \cap X_{\mathrm{reg}}$ is of codimension $\geqslant 2$. Q.E.D.

Let $Y = \Phi^{-1}(t_0^*)$. By definition, Y is contained in X_{reg}. We will show it is a kind of "symplectic cross section" to the action of G on X_{reg}.

Theorem 32.10. Y *has the following properties:*

(*i*) *It is a* T*-invariant symplectic submanifold of* X.
(*ii*) *Every* G *orbit in* X_{reg} *intersects* Y.
(*iii*) $\Phi(Y)$ *is a dense subset of* $\Phi(X) \cap t_+^*$.

Proof. It is obvious that Y is T-invariant and that every G orbit in X_{reg} intersects it. To see that it is a submanifold of X, observe that if $p \in t_0^*$, the coadjoint orbit of G through p intersects t_0^* transversally; so if $\Phi(x) = p$, Φ is transversal to t_0^* at x. Since Y is the pre-image of t_0^* it has to be a submanifold of X. To see that $\Phi(Y)$ is dense in $\Phi(X) \cap t_+^*$, let p be a point in $\Phi(X) \cap t_+^*$ and let $p = \Phi(x)$. Since X_{reg} is dense in X there exists a sequence of points x_i in X_{reg} such that x_i converges to x. The orbit through x_i intersects Y in some point y_i and since X is compact we can assume the y_i's converge to a point $z \in X$. Since the y_i's are conjugate to the x_i's, z is conjugate to x, and $q = \Phi(z)$ is conjugate to p. But p and q are both in t_+^* so they are Weyl-group conjugate. Replacing the y_i's by their images with respect to an appropriate Weyl-group element we can assume $\Phi(y_i)$ converges to p.

The fact that Y is symplectic follows from Theorem 26.8.

Let $\Phi_T: X \to t^*$ be the moment mapping associated with the action of T on X. It is clear that $\Phi = \Phi_T$ on Y. By Theorem 31.9, Y is a Hamiltonian T-space and $\Phi_T: Y \to t^*$ is its moment map. Moreover

$$\Phi(X) \cap t_+^* = \text{the closure of } \Phi_T(Y); \tag{32.12}$$

so if Y were compact, Theorem 32.9 would be an immediate consequence of Theorem 27.2. Since Y is not compact, we will get at the conclusions of Theorem 27.2 in a slightly roundabout way. If H is a closed subgroup of T, let $X'_H = \{x \in X, H \text{ is the stabilizer group of } x \text{ in } T\}$. By Theorem 27.1, X'_H is a symplectic submanifold of X. Since $X'_H \cap Y = \{x \in Y, H \text{ is the stabilizer group of } x \text{ in } T\}$, we can apply Theorem 27.2 to Y to conclude:

Theorem 32.11. $X'_H \cap Y$ *is a symplectic submanifold of* Y.

Let

$$T_1, \ldots, T_N \tag{32.13}$$

be a list of all subgroups of T that occur as stabilizer groups of points. (Since X is compact, there are finitely many such groups by Proposition 27.4.) Let $X_i = \{x \in X, T_i \text{ is the stabilizer group of } x \text{ in } T\}$. The X_i's are finitely connected by Proposition 27.5 so, allowing for duplications on the list (32.13), we can assume X_i is connected. The disjoint union

$$X = \bigcup_{i=1}^{N} X_i$$

is a stratification of X by connected symplectic submanifolds. Moreover, by Theorem 32.7, there exists an open set, O, in t^* and a point $a_i \in t^*$ such that $\Phi_T: X \to t^*$ maps X_i submersively onto $O_i \cap (a_i + t_i^\perp)$. Let $Y_i = X_i \cap Y$. By

Theorem 31.11 the disjoint union

$$Y = \bigcup_{i=1}^{N} Y_i \tag{32.14}$$

is a stratification of Y by symplectic submanifolds. Moreover, by applying Theorem 27.2 directly to Y_i, we get

Theorem 32.12. *There exists an open set* O'_i *in* t_0^* *such that* Φ_T *maps* Y_i *submersively onto* $\mathrm{O}'_i \cap (\mathrm{a}_i + \mathrm{t}_i^\perp)$.

Consider the set of all affine subspaces of t^* that are intersections of $a_i + t_i^\perp$'s and walls of t_+^*. Let $v_1, \ldots, v_M$ be the "vertices" or zero-dimensional subspaces in this set.

Theorem 32.13. *The closure of* $\Phi_T(\mathrm{Y}_i)$ *in* $\mathrm{a}_i + \mathrm{t}_i^\perp$ *is the union of a finite number of convex sets, each of which is the convex hull of a collection of* v_i*'s.*

The proof, an induction on dim $t_i^\perp$, will be left to the reader. In view of (32.10) and the fact that $\bigcup Y_i = Y$ we obtain

Theorem 32.14. $\Phi(\mathrm{X}) \cap \mathrm{t}_+^*$ *is the union of a finite number of convex sets each of which is the convex hull of a collection of* v_i*'s.*

We conjecture that this intersection is always a single convex polytope. As of this writing, we are unable to prove this conjecture[†] in general, but can prove it for the case of a Kaehlerian Hamiltonian action. This proof appears in Guillemin and Sternberg (1982a).

As a last application to the techniques of this section we will prove the following facts about the coadjoint action of a Lie group G on g^* when G is compact and connected.

Theorem 32.15. *The orbits of* G *in* g* *are simply connected.*

Theorem 32.16. *For every element* α, *of* g* *the stabilizer group of* α *in* G *is connected.*

Notice first of all that Theorem 32.15 implies Theorem 32.16. Indeed if $\mathcal{O}$ is the orbit through α and H is the stabilizer group, then the map

$$\pi: G \to \mathcal{O}, \qquad g \to \mathrm{Ad}(g)^* \alpha$$

is a fiber mapping with base $\mathcal{O}$, total space G, and fiber H. Since $\pi_1(\mathcal{O})$ and

[†] Added in page proof: This conjecture has just been proved by Frances Kirwan in a paper to appear in the *Inventiones Math.*

$\pi_0(G)$ are trivial, we conclude from the long exact sequence in homotopy

$$\cdots \to \pi_1(\mathcal{O}) \to \pi_0(H) \to \pi_0(G) \cdots$$

that $\pi_0(H)$ is trivial; that is, H is connected. To prove Theorem 32.15 we make use of the following standard transversality result (see, for instance, Guillemin and Pollack (1974)).

Proposition. *Let* M *be a manifold contained in a real vector space* V. *Then for almost all* $\xi \in V^*$ *the linear function* $l^\xi: V \to \mathbb{R}$ defined by $v \in V \to (v, \xi)$ *restricts to be a Morse function on* M.

Apply this to the manifold, $\mathcal{O} \subseteq g^*$. In this case the restriction of l^ξ to $\mathcal{O}$ is, as we saw in Section 26, the ξth component of the moment mapping. Therefore Theorem 32.15 follows from Theorem 32.5 and Corollary 32.2 of the theorem preceding it.

We showed in Section 26 that if G is a connected Lie group, every Hamiltonian G-space upon which G acts transitively is the covering of a coadjoint orbit. By Theorem 32.14 we can improve this result in the compact case.

Theorem 32.17. *If* G *is connected and compact, every Hamiltonian* G*-space upon which* G *acts transitively is a coadjoint orbit.*

33. The lemma of stationary phase

The applications we have discussed so far in this chapter have to do with classical mechanical systems. In the next two sections we will discuss some quasi-classical applications of the moment mapping. The term "quasi-classical" refers, in general, to asymptotic properties of quantum-mechanical systems for which Planck's constant can be treated as negligibly small compared with other parameters of the system. Typically, much can be deduced about the quasi-classical properties of quantum-mechanical systems from properties of the corresponding classical systems. For instance for simple systems like the hydrogen atom in the presence of a magnetic field, it was already observed in the 1920s that symmetries of the classical system explain the structure of the band spectra for the corresponding quantum system. The theorem we will prove in the next section will, in a sense, be a result of this sort.

A mathematical result that plays a fundamental role in almost all areas of quasi-classical analysis is the lemma of stationary phase. This will be the case here as well; however, we will need a refinement of the usual form of this lemma due to Duistermaat and Heckman (1982). Let us first recall the statement of the usual form of this lemma: Let X be a compact oriented n-dimensional manifold, $f: X \to \mathbb{R}$ a Morse function, and μ a volume form on X. The lemma of stationary phase describes the asymptotic behavior of the oscillatory integral, $\int e^{itf}\mu$ for t very large. Specifically it says (Guillemin and Sternberg 1977, p. 6) that

$$\left(\frac{t}{2\pi}\right)^{n/2}\int e^{itf}\mu = \sum c(p)\, e^{itf(p)} + R(t), \tag{33.1}$$

the sum taken over the critical points of f. The remainder term $R(t)$ is of order $O(t^{-1})$, and the constants $c(p)$ can be computed explicitly from the Hessian of f at p by the formula:

$$c(p) = \exp\left(i\frac{\pi}{4}\operatorname{sgn} d^2 f_p\right)[\det d^2 f_p(e_i, e_j)]^{-1/2}, \tag{33.2}$$

$e_1, e_2, \ldots, e_n$ being a basis of T_p for which $\mu(e_1, \ldots, e_n) = 1$. (It is clear that (33.2) does not depend on the choice of such a basis.)

In some rare instances, the remainder term $R(t)$ in (33.1) is not merely small for large t but is identically zero. For instance, we have already seen in Section 8 that this is the case when $X = \mathbb{R}^n$, $\mu =$ Lebesgue measure, and f is a quadratic function of $x_1, \ldots, x_n$; that is,

$$f(x) = f(0) + \sum \frac{\alpha_i}{2} x_i^2, \tag{33.3}$$

(In this case X is not compact, but the left-hand side of (33.1) is still well defined as an improper integral.) The theorem of Duistermaat and Heckman describes another example of an oscillatory integral that can be evaluated exactly by stationary phase. Let T be an r-dimensional torus, X a compact symplectic manifold of dimension $2n$ with symplectic form Ω, $T \times X \to X$ a Hamiltonian action of T on X, and $\Phi: X \to t^*$ the associated moment mapping. For simplicity we will assume for the rest of this section that the fixed-point set of T is finite. Let ξ be an element of t and let $\hat{\xi}$ be the vector field on X associated with it. By the definition of the moment mapping

$$i(\hat{\xi})\Omega = d\phi_\xi, \qquad \text{where} \qquad \phi_\xi = \langle \Phi, \xi \rangle. \tag{33.4}$$

We will say that ξ is *nondegenerate* if $\hat{\xi}$ is zero only at the fixed points of T.

By Theorem 32.5, if ξ is nondegenerate then ϕ_ξ is a Morse function. Let

$$\sigma = \frac{\Omega^n}{n!} \tag{33.5}$$

be the canonical symplectic volume form on X. We will prove

Theorem 33.1 (Theorem of stationary phase of Duistermaat and Heckman). *For ξ a nondegenerate element of* t, $\mathrm{f} = \phi_\xi$ *and* $\mu = \sigma$, $\mathrm{R(t)} \equiv 0$ *on the right-hand side of (33.1).*

The proof of this theorem, which we will give below, is due to Berline and Vergne (1983a). (There is a similar proof due to Raoul Bott (unpublished), in which the explicit constructions described below are replaced by general arguments in equivariant cohomology theory.)

To start with, let X be an arbitrary m-dimensional manifold and ξ a vector field on X. We will suppose that ξ is the Killing vector field for some Riemannian metric, ds^2 on X. (For instance, this is always the case if ξ is the vector field $\hat{\xi}$ in (33.4), since, as we pointed out in Section 32, there exists a T-invariant metric on X.) Let $\mathscr{A} = \mathscr{A}^0 + \cdots + \mathscr{A}^m$ be the DeRham complex on X, $\mathscr{A}^i$ being the space of exterior i-forms, and let $d: \mathscr{A}^i \to \mathscr{A}^{i+1}$ be exterior differentiation. Let δ_ξ be the operator $d + i(\xi)$. This operator does not, of course, carry $\mathscr{A}^i$ into $\mathscr{A}^{i+1}$, but it does map the space of even forms into the space of odd forms and vice versa. Morever, by Weil's identity,

$$\delta_\xi^2 = di(\xi) + i(\xi)d = D_\xi, \tag{33.6}$$

D_ξ being Lie differentiation by the vector field ξ (see Section 21). Let

$$\mathscr{A}_{\text{inv}} = \{\omega \in \mathscr{A}, D_\xi\omega = 0\}$$

represent the space of forms invariant with respect to the one-parameter group of diffeomorphisms generated by ξ. The map δ_ξ sends this space into itself, and by (33.6), $\delta_\xi^2 = 0$ on this space. We will prove

Lemma 33.1. *Let $\mu = \mu_\mathrm{m} + \mu_{\mathrm{m}-2} + \cdots$ be an element of $\mathscr{A}_{\text{inv}}$ of the same parity as* m, *which is δ_ξ-closed. Then μ_m is* d*-exact on the open set* $\mathrm{X}^0 = \{\mathrm{x} \in \mathrm{X}, \xi(\mathrm{x}) \neq 0\}$.

Proof. We will first of all show that there exists a 1-form, θ, on X^0 such that

$$D_\xi\theta = 0 \qquad \text{and} \qquad \theta(\xi) = 1. \tag{33.7}$$

In fact, using the Riemannian metric ds^2, we can define

$$\theta_x(v) = (\xi(x), v)_x/(\xi(x), \xi(x))_x$$

for $v \in TX^0$, $(,)_x$ being the inner product on TX^0 associated with ds^2. It is clear that θ has the properties (33.7). We also note that

$$i(\xi)\, d\theta = 0. \tag{33.8}$$

Indeed, $i(\xi)\, d\theta = D_\xi\theta - di(\xi)\theta = 0$, by (33.7). Now set

$$v = \theta \wedge (1 + d\theta)^{-1} \wedge \mu,$$

where $(1 + d\theta)^{-1}$ is defined by the Neumann series $1 - d\theta + d\theta \wedge d\theta \ldots$ We will show

$$i(\xi)\, dv = i(\xi)\mu. \tag{33.9}$$

Indeed,

$$\begin{aligned} dv &= d\theta \wedge (1 + d\theta)^{-1} \wedge \mu - \theta \wedge (1 + d\theta)^{-1} \wedge d\mu \\ &= d\theta \wedge (1 + d\theta)^{-1} \wedge \mu + \theta \wedge (1 + d\theta)^{-1} \wedge i(\xi)\mu, \end{aligned}$$

since $\delta_\xi\mu = d\mu + i(\xi)\mu = 0$. Applying $i(\xi)$ to the expression on the right we get, in view of (33.8) and (33.7),

$$\begin{aligned} i(\xi)\, dv &= d\theta \wedge (1 + d\theta)^{-1} \wedge i(\xi)\mu + (1 + d\theta)^{-1} \wedge i(\xi)\mu \\ &= i(\xi)\mu, \end{aligned}$$

proving (33.9). Now on X^0, $i(\xi): \mathscr{A}_m \to \mathscr{A}_{m-1}$ is injective; so (33.9) implies that $\mu_m = dv_{m-1}$, proving the lemma.

Coming back to the proof of Theorem 33.1, if we set $\beta = \phi_\xi - \Omega \in \mathscr{A}^0 + \mathscr{A}^2$, then by (33.4) β is δ_ξ-closed and so is $\mu = \exp it\beta$. But

$$\begin{aligned} \mu &= \exp it\phi_\xi \exp it(-\Omega) \\ &= e^{it\phi_\xi}\left(1 - it\Omega + \frac{(it)^2}{2!}\Omega \wedge \Omega + \cdots\right), \end{aligned}$$

and up to constant multiple $\mu_{2n} = \exp(it\phi_\xi)\sigma$. If C_ξ is the critical set of ϕ_ξ, then by the lemma there exists a $2n - 1$-form v_{2n-1} on $X - C_\xi$ such that $\exp(it\phi_\xi)\sigma = dv_{2n-1}$. Now assume that ξ is nondegenerate so that C is just the fixed point set of T, and let B_p be a small ball centered about the critical point p. By Stokes' theorem

$$\int_X \exp(it\phi_\xi)\sigma = \sum_p \left(\int_{B_p} \exp(it\phi_\xi)\sigma - \int_{\partial B_p} v_{2n-1} \right). \tag{33.10}$$

By the canonical form theorems proved in Section 32 we can find symplectic coordinates $x_1, \ldots, x_n$, $y_1, \ldots, y_n$ on B_p such that $\sigma = \sigma^0 = dx_1 \wedge \cdots \wedge dx_n \wedge dy_1 \wedge \cdots \wedge dy_n$ and

$$\phi_\xi = \phi_\xi^0 = \phi_\xi(p) + \sum_{j=1}^{n} \alpha_j(\xi, p)(x_j^2 + y_j^2)/2 \tag{33.11}$$

the α_j's being constants that depend on ξ and p. If we compute the terms on the right-hand side in (33.10) using these coordinates, and make a second application of Stokes' theorem (strictly speaking we should replace t by $(t + i\epsilon)$ here and let $\epsilon \to 0$ as in Section 8) we get

$$\int_{B_p} \exp(it\phi_\xi^0)\sigma^0 - \int_{\partial B_p} v_{2n-1}^0 = \int_{\mathbb{R}^{2n}} \exp(it\phi_\xi^0)\sigma^0.$$

However, we have already observed that for this integral the formula of stationary phase does not have a remainder term. Therefore the same is true for $\exp(it\phi_\xi)\sigma$. Q.E.D.

The right-hand side of (33.1) can be somewhat simplified in our case. If p is a fixed point of T, then $d(\exp \xi)_p$ is a linear mapping of T_p into itself. The eigenvalues of this mapping are all of modulus 1, that is, of the form λ_j, $\bar{\lambda}_j$, $j = 1, \ldots, n$, where

$$\lambda_j = \exp[i\alpha_j(\xi, p)]. \tag{33.12}$$

It is easy to see in fact that the $\alpha_j(\xi, p)$'s in (33.12) are the same as the $\alpha_j(\xi, p)$'s in (33.11). A simple computation shows that the constant $c(p)$ in (33.2) is just

$$(i)^n \left(\prod_{j=1}^{n} \alpha_j(\xi, p) \right)^{-1};$$

so we get from (33.1), with $R(t) = 0$ and $2n$ in place of n,

$$\frac{1}{(2\pi i)^n} \int_X e^{i\langle \Phi, \xi \rangle} \sigma = \sum_p \frac{e^{i\langle \Phi(p), \xi \rangle}}{\prod_{j=1}^{n} \alpha_j(\xi, p)}, \tag{33.13}$$

the sum taken over the fixed points of T. The left-hand side of this formula has the following interesting interpretation. Let $|\sigma|$ be the measure on X associated with the volume form σ; that is, for any measurable subset A of X set

$$|\sigma|(A) = \int_A \sigma.$$

Since measures are covariant objects, we get from $|\sigma|$ a measure $\Phi_*|\sigma|$ on t^*, the direct image of $|\sigma|$ with respect to the moment map $\Phi: X \to t^*$. The integral on the left-hand side of (33.13) is clearly just the Fourier transform of this measure:

$$\int_X e^{i\langle\Phi,\xi\rangle}\sigma = \widehat{\Phi_*|\sigma|}(\xi). \tag{33.14}$$

Since we know this Fourier transform explicitly by (33.13), we would expect to be able to determine the measure $\Phi_*|\sigma|$ itself fairly explicitly. This turns out to indeed be the case. We proved in the last section that the image of X in t^* is a convex polytope Δ. Let Δ_r be the set of regular values of Φ inside Δ and let

$$\Delta_r = \bigcup \Delta_i$$

where the Δ_i's are the connected components of Δ_r. It follows rather easily from the proof Theorem 32.7 that each Δ_i is itself a convex polytope, and, from the fact that Δ_i consists of regular values, it is immediately clear that $\Phi_*|\sigma|$ is a smooth function f_i times the Lebesgue measure on Δ_i. Duistermaat and Heckman have proven that *these f_i's are in fact polynomials of degree equal to $n - r$.* One special case of their result is no doubt well known to the reader: Let $X = S^2$ with its standard volume form and let $T = S^1$ acting as rotation about the z axis. The moment mapping in this case is just the projection map, $(x, y, z) \in S^2 \to z \in \mathbb{R}$ and the direct image of $|\sigma|$ is the Lebesgue measure on the interval $[-1, 1]$.

We conclude this section by mentioning an important theorem in harmonic analysis that is easily derivable from (33.13). Let G be a semisimple Lie group with T as its maximal torus and let $X = \mathcal{O} =$ a coadjoint orbit in g^*. Since X is a Hamiltonian G space, it is automatically a Hamiltonian T space, and the associated moment map $\mathcal{O} \to t^*$ turns out to be just the projection of g^* onto t^* restricted to $\mathcal{O}$. Thus the right-hand side of (33.13) gives a formula for

$$\int_{\mathcal{O}} e^{i\langle x,\xi\rangle}\,\sigma,$$

the Fourier transform of the delta function of the orbit $\mathcal{O}$. This formula was first proved by HarishChandra using techniques entirely different from those described here.

34. Geometric quantization

The theory of geometric quantization was developed in the early 1970s, independently by Kostant and Souriau, in order to bring into focus some vaguely preceived analogies between representation theory and quantum mechanics that had for a long time been part of the "folklore" of modern physics. There are several systematic accounts of this theory in the literature. (See, for instance, Simms and Woodhouse, 1976; Woodhouse, 1980; Sniatycki, 1980; and Guillemin and Sternberg 1977.) We attempt here only a brief and schematic introduction to the theory, emphasizing the role played by the moment mapping.

Let X be a symplectic manifold with symplectic form Ω. Suppose the cohomology class $[\Omega]$ in $H^2(X, \mathbb{R})$ is *integral*, that is, is the image of a cohomology class in $H^2(X, Z)$. Then there exists on X a line bundle L whose Chern class is $[\Omega]$. Moreover, one can equip L with a connection ∇ such that the curvature form of ∇ is Ω. Now suppose G is a Lie group and $G \times X \to X$ a Hamiltonian action of G on X. Let $\Phi: X \to g^*$ be the associated moment mapping. From this data one gets an infinitesimal representation of G on the space of sections of L in the following way. To each ξ in the Lie algebra of G there corresponds a vector field $\hat{\xi}$ on X. If s is a section of L we let

$$\xi s = (\nabla_{\hat{\xi}} + 2\pi i\langle\Phi, \xi\rangle)s. \tag{34.1}$$

It is relatively easy to check that (34.1) defines a representation of the Lie algebra g by linear operators on the space of smooth sections of L. We will say that our data are *prequantizable* if there is a global representation of G on this space extending this infinitesimal representation. (For example, this is automatically the case if X is compact and G is simply connected.) Supposing such a representation exists, lets look at its behavior at a fixed point p of G. Since $\hat{\xi}(p) = 0$, the first term in (34.1) vanishes and we get for $\exp \xi \in G$

$$(\exp \xi)s(p) = e^{2\pi i\langle\Phi(p), \xi\rangle}s(p).$$

In other words, the representation of G on the fiber L_p of L at p is the representation

$$\exp \xi \to \text{multiplication by } e^{2\pi i\langle\Phi(p), \xi\rangle}. \tag{34.2}$$

This formula will turn out to play an important role later on in this section.

To "quantize" the symplectic manifold X we need, in addition to the prequantum data described above, a polarization of X. By definition, a polarization is an integrable subbundle F of the complexified tangent bundle of X such that at each point $p \in X$, F_p is a complex Lagrangian subspace of the complex symplectic space $T_p \otimes \mathbb{C}$. Given a polarization of X, a local section $s: \mathscr{U} \to L$ is said to be *polarized* if $\nabla_\xi s = 0$ for all complex vector fields ξ on $\mathscr{U}$ for which $\overline{\xi(p)} \in F_p$ for all $p \in \mathscr{U}$. The polarized sections form a sheaf $\mathscr{L}$, and we can consider its cohomology groups:

$$H^i(X, \mathscr{L}), \qquad i = 0, 1, \dots. \tag{34.3}$$

These are the basic "quantum data" in the Kostant–Souriau theory. If the action of G on X preserves F, then it acts on these spaces; so we get a string of representations of G on abstract vector spaces. Among the goals of the theory are to show that, in all cases of interest,

(i) the construction above is, up to isomorphism, independent of the polarization;
(ii) all but one of the spaces (34.3) is zero;
(iii) the representation of G on the non-zero space in the sequence (34.3) is unitarizable.

Unfortunately the theory is far from achieving these goals. However, in the last few years there have been some auspicious developments. (See, for instance, Rawnsley, Schmid, and Wolf 1983.)

Two kinds of polarizations that are particularly important are *real* polarizations: $F = \bar{F}$, and *complex* polarizations: $F \cap \bar{F} = \{0\}$. Here we will concentrate exclusively on complex polarizations. We first note that if X is equipped with a complex polarization then, by the Newlander–Nirenberg theorem, it is a complex manifold. Indeed, define a complex-valued function f on X to be *holomorphic* if $\bar{\xi} f = 0$ for all sections ξ of F. By the Newlander–Nirenberg theorem, one can show that locally about any point p of X enough such functions exist to form a complex coordinate system at p. Similarly one can show that there exists for every point p a neighborhood $\mathscr{U} \ni p$ and a polarized section $s_{\mathscr{U}}: \mathscr{U} \to L$. Moreover, if $s_{\mathscr{U}}$ and $s_{\mathscr{V}}$ are polarized sections, then $s_{\mathscr{U}} = f_{\mathscr{U}\mathscr{V}} s_{\mathscr{V}}$, where the transition function $f_{\mathscr{U}\mathscr{V}}$ is holomorphic on $\mathscr{U} \cap \mathscr{V}$; so L has the structure of a holomorphic line bundle. In this case the cohomology groups (34.3) are the usual Deaubault cohomology groups. In particular, if X is compact they are all finite-dimensional.

A polarization F is called *Kaehlerian* if the Hermitian form

$$H_p(v, w) = -i\Omega(v, \bar{w}), \qquad v, w \in F_p$$

is positive definite at all points $p \in X$. We note that if F is Kaehlerian, it is automatically complex.

We will next discuss the constructions above from the quasi-classical point of view. Traditionally physicists write the symplectic form Ω with an $\hbar^{-1}$ in front. In quasi-classical analysis, one tries to study the effect on one's data of letting $\hbar$ tend to zero; so here we will want to see what happens when we apply the procedure of geometric quantization to the sequence of symplectic manifolds

$$(X, \hbar^{-1}\Omega), \qquad \hbar = 1, \tfrac{1}{2}, \tfrac{1}{3}, \ldots.$$

(Note that $\hbar$ has to be the reciprocal of an integer for $\hbar^{-1}\Omega$ to be integral.) We first observe that if we replace Ω by $k\Omega$, the moment map Φ gets replaced by $k\Phi$, the line bundle L by $\bigotimes^k L$, and the connection ∇ by the connection induced by ∇ on $\bigotimes^k L$.

Suppose now that X is compact. If the polarization F is Kaehlerian then by the Kodaira vanishing theorem all the cohomology groups

$$H^i(X, \bigotimes^k \mathscr{L}) = 0, \qquad i > 0, \tag{34.4}$$

vanish for k sufficiently large, $\bigotimes^k \mathscr{L}$ being the sheaf of polarized sections of $\bigotimes^k L$. Thus only the group

$$H^0(X, \bigotimes^k \mathscr{L}) = \Gamma\left(\bigotimes^k \mathscr{L}\right), \tag{34.5}$$

that is, the space of global holomorphic sections of $\bigotimes^k L$, figures asymptotically in the quantum picture. Our goal is to study the asymptotic behavior of the group G on the space (34.5). For simplicity we will assume from now on that $G = T =$ an r-dimensional torus. Then every irreducible representation of T is one-dimensional and is of the form

$$\exp \xi \in T \to e^{i\mu(\xi)}, \qquad \xi \in t, \tag{34.6}$$

where μ is an element of t^*. Not all μ's define a representation of T. Those that do form a lattice Γ in t^*, the so-called weight lattice. Let ρ_k be the representation of T on the space (34.5). This representation is completely determined by the multiplicities $m(\mu, \rho_k)$ with which the representations (34.6) occur in it. A convenient way of "collating" these multiplicities is by means of the measure

$$\nu_k = \frac{1}{k^r} \sum_{\mu \in \Gamma} m(\mu, \rho_k)\delta\left(\frac{\mu}{2\pi k}\right) \tag{34.7}$$

on t^*. The factors of k occurring in (34.7) are scaling factors inserted to ensure that ν_k is bounded and compactly supported uniformly in k. If we ignore these factors ν_k is simply a sum of delta functions concentrated on the lattice Γ and assigning to each $\mu \in \Gamma$ the weight $m(\mu, \rho_k)$.

In the previous section we encountered another measure on t^*, the measure $\nu = \Phi_*|\sigma|$, where $|\sigma|$ is the canonical symplectic measure on X. The rest of this section will be devoted to proving the following quasi-classical result:

$$\nu_k \to \nu \qquad \text{weakly as} \qquad k \to \infty. \tag{34.8}$$

The proof we will give of (34.8) is based on the Duistermaat–Heckman formula and the Atiyah–Bott fixed-point formula. The first we have already discussed in the previous section. We will briefly discuss the second: Let X be a compact manifold and f a smooth map of X into itself. f is called a *Lefshetz map* if it has a finite number of fixed points and if at each fixed point p, $(I - df_p): T_p \to T_p$ is bijective. Now let X be a compact complex manifold, f a holomorphic mapping of X into itself, and L a holomorphic line bundle. By a *lifting* of f we will mean a rule assigning to each $p \in X$ a linear map

$$f_p^{\#}: L_{f(p)} \to L_p$$

and depending holomorphically on p. Given a lifting of f, we get an induced map $f^{\#}$ on holomorphic sections of L, namely,

$$f^{\#} s(p) = f_p^{\#} s(f(p)),$$

and this in turn induces a map

$$\hat{f}_i: H^i(X, \mathscr{L}) \to H^i(X, \mathscr{L})$$

on the Deaubault cohomology groups. The Atiyah–Bott fixed-point formula says that if f is Lefshetz,

$$\sum(-1)^i \operatorname{tr} \hat{f}_i = \sum_p \frac{\operatorname{tr} f_p^{\#}}{\det_{\mathbb{C}}(1 - df_p)}, \tag{34.9}$$

the sum taken over the fixed points of f. (By $\det_{\mathbb{C}}(1 - df_p)$ we mean the determinant of $1 - df_p$ regarded as complex linear mapping of the space of holomorphic tangent vectors at p into itself.) For the proof of this theorem we refer to Atiyah and Bott (1967).

We now return to the proof of (34.8). If ξ is a nondegenerate element of t and is sufficiently small, $f = \exp \hat{\xi}$ is Lefshetz and its fixed points are just the fixed points of T. Applying (34.9) to f with L replaced by $\bigotimes^k L$ we get for the left-hand side of (34.9), $\sum m(\mu, \rho_k) e^{i\mu(\xi)}$, in view of (34.6). Notice that because of the Kodaira vanishing theorem only the $\hat{f}_0$ contribute to the alternating sum on the left. For the right-hand side of (34.9) we get, by (33.12) and (34.2),

$$\sum_p \frac{\exp(2\pi i k \langle \Phi(p), \xi \rangle)}{\prod_{j=1}^{n} (1 - \exp[i\alpha_j(\xi, p)])};$$

so we have established, for ξ nondegenerate and sufficiently small, the identity

$$\sum m(\mu, \rho_k) e^{i\mu(\xi)} = \sum_p \frac{\exp(2\pi i k \langle \Phi(p), \xi \rangle)}{\prod_{j=1}^{n} (1 - \exp[i\alpha_j(\xi, p)])}. \tag{34.10}$$

Let $\eta = 2\pi k \xi$ in this identity. Then

$$\sum m(\mu, \rho_k) \exp[i(\mu/2\pi k)(\eta)] = \sum \frac{\exp(i\langle \Phi(p), \eta \rangle)}{\prod_{j=1}^{n} (1 - \exp\{i[\alpha_j(\eta, p)/2\pi k]\})}.$$

The right-hand side of this identity is equal to

$$k^n (2\pi i)^n \sum \frac{e^{i\langle \Phi_1(p), \eta \rangle}}{\prod_{j=1}^{n} \alpha_j(\eta, p)} + O(k^{n-1});$$

so by (33.13):

$$\frac{1}{k^n} \sum m(\mu, \rho_k) \exp[i(\mu/2\pi k)(\eta)] = \int e^{i\langle \Phi, \eta \rangle} \sigma + O\left(\frac{1}{k}\right). \tag{34.11}$$

Let e_η be the function $x \in t^* \to e^{i\langle x, \eta \rangle}$ on t^*. Then the left-hand side of (34.11) is just the measure v_k evaluated on e_η and, by (33.14), the integral on the right is just the measure v evaluated on e_η. Thus we have proved:

$$v_k(e_\eta) \to v(e_\eta) \tag{34.12}$$

for every nondegenerate element $\eta \in t$. Since every continuous function on t^* can be uniformly approximated on compact sets by finite linear combinations of the e_η's, this proves (34.8).

Remark. For the special case where T is the Cartan subalgebra of a compact Lie group G and $X = O$ is a coadjoint orbit of G, (34.8) was first proved by Guillemin, Kashiwara, and Kawai (1979) using Fourier integral operator techniques. In his thesis, Heckman (1980) gave a purely group-theoretical proof of the same result using the formula of HarishChandra, mentioned at the end of Section 33, and the Weyl character formula. The proof given here is essentially Heckman's proof, but with the HarishChandra formula replaced by the Duistermaat–Heckman formula and the Weyl character formula replaced by the Atiyah–Bott fixed-point formula.

It is natural to ask whether the quasi-classical result we have just described can be improved upon. In other words, suppose that ρ is the representation of T on the space of holomorphic sections of L. Can one get from the symplectic data associated with X and the action of G on X an *exact* formula for the multiplicities $m(\mu, \rho)$? It turns out that there are some results in this direction. Namely, suppose $\mu \in t^*$ is a regular value of the moment mapping. Then one can form the reduced space:

$$X_\mu = \Phi^{-1}(\mu)/T.$$

From the prequantum data on X, one gets an induced set of prequantum data on X_μ, and from the Kaehlerian polarization F, a Kaehlerian polarization of X_μ. In particular, one obtains cohomology groups

$$H^i(X_\mu, \mathscr{L}_\mu), \qquad i = 0, 1, \ldots,$$

analogous to the cohomology groups (34.3). One can prove that $H^0(X_\mu, \mathscr{L}_\mu)$ is isomorphic to the subspace of $H^0(X, \mathscr{L})$ consisting of those vectors that transform according to representation $e^{i\mu}$. In particular,

$$\dim H^0(X_\mu, \mathscr{L}_\mu) = m(\mu, \rho). \tag{34.13}$$

If the Kaehler form on X_μ is sufficiently positive, the left-hand side of (34.13) can be computed by the Hirzebruch–Riemann–Roch formula, and this, in turn, gives a recipe for $m(\mu, \rho)$, which is entirely expressible in terms of symplectic data. (Incidentally, this result admits a formulation that is true for non-Abelian compact groups as well.) For details we refer to Guillemin and Sternberg (1982c).

III

Motion in a Yang–Mills field and the principle of general covariance

This chapter is devoted to the study of the formulation and solution of the equations of motion of particles and, to some extent, continuous media, in the presence of a Yang–Mills field. In the first two sections, we study the symplectic structure on T^*P and its various reduced spaces, where $P \to X$ is a principal bundle. A connection on P allows us to pull any Hamiltonian function on T^*X back to T^*P. This procedure is the natural generalization of the method of *minimal coupling* in classical electrodynamics. We then discuss the solution of these equations in the presence of additional symmetry. Then we justify these equations using the principle of general covariance, already discussed briefly at the end of Chapter I. This justification will treat gravitational forces differently from all the others. We therefore also present an alternative derivation that is more geometrical in character and treats all forces on an equal footing.

35. The equations of motion of a classical particle in a Yang–Mills field

In Section 29 we considered the following construction: Let $\pi : P \to X$ be a principal bundle with structure group K (not necessarily compact). By definition, this means that we are given an action of K on P that is usually denoted by right multiplication. To fix the notation, we shall let R_a denote the action of $a^{-1} \in K$ on P, which we also write as

$$R_a(p) = pa^{-1}, \qquad p \in P \text{ and } a \in K.$$

The action of K on P induces a corresponding action on T^*P that is Hamiltonian with moment map Φ: $T^*P \to k^*$. (Here we are changing notation slightly and writing Φ instead of Φ' and will write $\mathcal{O}$ instead of $\mathcal{O}'$.)

The fact that P is a principal bundle means that around every point of X we can find a coordinate neighborhood U such that $\pi^{-1}(U)$ is isomorphic as a K space to $U \times K$, with K acting only on the second factor by right multiplication. (See Greub, Halperin, and van Stone 1972, or Sternberg, 1983, for the basic definitions and facts concerning principal bundles and connections.) This isomorphism depends upon (and is equivalent to) the choice of a section of P over U. In terms of such a local trivialization, we may identify $T^*(\pi^{-1}U)$ with $T^*U \times T^*K$, which we may further identify with $T^*U \times K \times k^*$ using the left identification of T^*K with $K \times k^*$. In terms of this identification, the moment map Φ is given by

$$\Phi(z; c, \alpha) = -\alpha, \qquad z \in T^*U, c \in K, \alpha \in k^*. \tag{35.1}$$

From (35.1) we see that Φ is a submersion. For each orbit $\mathcal{O}$, we can form the Marsden–Weinstein reduced space, $(T^*P)_{\mathcal{O}} = \Phi_{T^*P \times \mathcal{O}^-}^{-1}(0)/K$. In terms of the local coordinates used above, a point (z, c, α, β) belongs to $\Phi_{T^*P \times \mathcal{O}^-}^{-1}(\mathcal{O})$ if and only if $\beta \in \mathcal{O}$ and $\alpha = -\beta$. A local cross section for the K action is obtained by requiring $c = e$, which then leaves $-\alpha$ free to range over $\mathcal{O}$, so

$$\begin{aligned} \dim\,(T^*P)_{\mathcal{O}} &= \dim T^*X + \dim \mathcal{O} \\ &= 2 \dim X + \dim \mathcal{O}. \end{aligned} \tag{35.2}$$

We shall see that $(T^*P)_{\mathcal{O}}$ is the natural "phase space" for a particle with "internal symmetry group K" and "charge type $\mathcal{O}$." For example, if $K = U(1)$, its coadjoint action is trivial, an orbit $\mathcal{O}$ consists of a point e of $u(1)^*$. If we identify $U(1)$ with the structure group of electromagnetic theory (the group of local electromagnetic gauges), then the elements of $u(1)^*$ can be identified with electric charge. Thus, a choice of an orbit in this case amounts to a choice of electric charge, and the choice of a coadjoint orbit in general is the generalization, to arbitrary structure group, of the notion of "charge."

In this section, we shall present several (slightly different) descriptions of the reduced space $(T^*P)_{\mathcal{O}}$. These will involve the geometry of connections on P (or, as they are known in the physics literature, Yang–Mills fields), and various other basic constructions such as pull-backs and associated bundles. As we do not assume a detailed knowledge on the part of the reader of these matters and in order to establish our notation (in an area of varying conventions), we shall present all of these matters *ab initio*. This will involve a considerable detour, so let us give a brief description (for the reader who is familiar with these matters) of where we are heading. The details and definitions will follow, and the uninitiated reader is advised to skip ahead to the next paragraph. Let $P \to X$ be a principal bundle with structure group K. The projection of $T^*X \to X$ defines a pull-back bundle, which we shall denote by $\pi^\# : P^\# \to T^*X$. Since the orbit $\mathcal{O}$ has a (left) K action, we can form the associated bundle, $\mathcal{O}(P^\#)$, which is just the quotient of $P^\# \times \mathcal{O}$ by the (diagonal) action of K. We let $\rho : P^\# \times \mathcal{O} \to \mathcal{O}(P^\#)$ denote the projection of each point onto its K orbit. A connection θ (that is a k-valued form on P) induces a connection $\theta^\#$ on $P^\#$ that we can consider as a k-valued linear differential form on $P^\# \times \mathcal{O}$ (depending only on the $P^\#$ factor). Let $\Phi_\mathcal{O} : \mathcal{O} \to k^*$ be the injection of $\mathcal{O}$ as an orbit in k^* and $\omega_\mathcal{O}$ its symplectic form. We can consider both $\Phi_\mathcal{O}$ and $\omega_\mathcal{O}$ as being defined on $P^\# \times \mathcal{O}$ (depending only on the $\mathcal{O}$ factor). Finally, we let α_X denote the canonical 1-form on T^*X. On $P^\# \times \mathcal{O}$, we can define the closed 2-form ν (see (35.19) below) by

$$\nu = d(\langle \Phi_\mathcal{O}, \theta^\# \rangle + (\pi^\#)^* \alpha_X) + \omega_\mathcal{O}.$$

(In this equation, $\theta^\#$ is a 1-form with values in k, and $\Phi_\mathcal{O}$ is a function with values in k^* so that $\langle \Phi_\mathcal{O}, \theta^\# \rangle$ is an ordinary $\mathbb{R}$-valued 1-form.) We shall show that the leaves of the null foliation of ν are precisely the K orbits and so there is a symplectic form Ω on the associated bundle $\mathcal{O}(P^\#)$ such that

$$\nu = \rho^* \Omega$$

(cf. Proposition 35.4 below). We shall also show that the connection θ determines a symplectic diffeomorphism of $(T^*P)^-_\mathcal{O}$ with $\mathcal{O}(P^\#)$ (with its symplectic form Ω). We shall also express Ω in terms of the curvature on $\mathcal{O}$. From the fact that $\mathcal{O}(P^\#)$ is a bundle over T^*X, we can pull back any function $\mathcal{H}$ on T^*X to $\mathcal{O}(P^\#)$. The corresponding equations of motion on $\mathcal{O}(P^\#)$ are known as the equations of minimal coupling. We shall give some examples of these equations in the next section The solution curves to these equations project down to give a family of curves on X. In Section 38 we shall get these curves from the principle of general covariance, provided that we take $\mathcal{H}$ to be the Hamiltonian determined by a choice of Riemann (or Lorentzian) metric on X. This will give a generalization of the method of Einstein, Infeld, and Hoffman described at the end of Chapter I. We now proceed to the details.

We shall let Aut P denote the group of automorphisms of the principal bundle P. Thus Aut P consists of all diffeomorphisms ϕ of P with itself that satisfy

$$\phi(pa) = \phi(p)a, \qquad \forall a \in K. \tag{35.3}$$

Such a ϕ satisfies $\pi\phi(p) = \pi\phi(pa)$ and hence determines a diffeomorphism $\bar{\phi}$ of X defined by

$$\bar{\phi}(\pi p) = \pi\phi(p). \tag{35.4}$$

We thus have a homomorphism Aut $P \to$ Diff X. The kernel of this diffeomorphism is called the "gauge group" of P and is denoted by Gau P. Thus ϕ is an element of Gau P if ϕ is a diffeomorphism of $P \to P$ that satisfies (35.3) and, in addition,

$$\pi\phi(p) = \pi p. \tag{35.5}$$

To understand the significance of Gau P, let us first examine the case where X is a point. Here, of course, Gau P = Aut P since nothing can happen in X. The principal bundle P is now a "principal homogeneous K spaces", that is, K acts transitively and freely on P. So fixing a point p in P, every other point is of the form pa for some a in K, and $pa = pb$ if and only if $a = b$. Thus, having fixed a point p in P we get a K morphism, F_p: $P \to K$ determined by $F_p(q) = a$ if $q = pa$. This mapping depends on the choice of p, but once having chosen p, we have $F_p(qb) = F_p(q)b$, so we can identify P with K and the action of K on P with the action of K on itself given by right multiplication. For any group, the only transformations of K into itself that commute with right multiplication are the left multiplications. Thus, in this case, $\phi \in$ Gau P if and only if there is some element $c_p \in K$ such that

$$F_p(\phi(q)) = c_p F_p(q).$$

The element c_p depends upon the choice of p because F_p does. Let us examine this p dependence. Suppose that we replace p by $p' = pb$. Then,

$$q = pa \Rightarrow q = p'b^{-1}a$$

so

$$F_{pb}(q) = b^{-1}F_p(q)$$

and hence

$$F_{pb}(\phi(q)) = b^{-1}F_p(\phi(q)) = b^{-1}c_p F_p(q) = b^{-1}c_p b F_{pb}(q)$$

so

$$c_{pb} = b^{-1}c_p b.$$

Let A denote the conjugation action of K on itself so that $A_b(c) = bcb^{-1}$ for all b and c in K. Let $\mathscr{C}: P \to K$ denote the function defined by $\mathscr{C}(p) = c_p$. Then the preceding equation can be written as

$$\mathscr{C} \circ R_b = A_b \circ \mathscr{C}. \tag{35.6}$$

This last equation makes sense for any principal bundle. Since A_b is an automorphism of K, the collections of functions $\mathscr{C}$ satisfying (35.6) is closed under pointwise multiplication, that is, under the operation $(\mathscr{C}_1\mathscr{C}_2)(p) = \mathscr{C}_1(p)\mathscr{C}_2(p)$ (the right-hand side being pointwise multiplication in K).

Thus the set of functions $\mathscr{C}$ from P to K satisfying (35.6) forms a group. We claim that this group is, in fact, isomorphic to the gauge group Gau P. Indeed, for each $\mathscr{C}$ satisfying (35.6), define the map $\Phi_{\mathscr{C}}:P \to P$ by

$$\phi_{\mathscr{C}}(p) = p\mathscr{C}(p). \tag{35.7}$$

Then

$$\phi_{\mathscr{C}}(pb^{-1}) = (pb^{-1})\mathscr{C}(pb^{-1}) = pb^{-1}b(p)b^{-1} = \phi_{\mathscr{C}}(p)b^{-1}$$

so (35.3) holds and clearly so does (35.5). Thus, if $\mathscr{C}$ is a smooth map satisfying (35.7), $\phi_{\mathscr{C}}$ is an element of Gau P. It is easy to check that

$$\phi_{\mathscr{C}_1\mathscr{C}_2} = \phi_{\mathscr{C}_1} \circ \phi_{\mathscr{C}_2}.$$

(The reader should check that the order of multiplication works out!) Finally, if $\phi:P \to P$ is a diffeomorphism satisfying (35.5), then we can use (35.7) to define $\mathscr{C}$ and then (35.3) implies (35.6). We have thus proved

Proposition 35.1. *The correspondence* (35.7) *gives an isomorphism between the group of smooth functions* $\mathscr{C}$:P $\to K$ *satisfying* (35.6) *and Gau* P.

Proposition 35.1 has a nice reformulation in terms of the notion of associated bundle, which we now recall, and which we will use often in what follows. Suppose that we are given a smooth (left) action of the structure group K on some smooth manifold F. We will denote the action of $b \in K$ on F by B_b, so that

$$B_b f = bf \qquad \text{for } b \in K \text{ and } f \in F.$$

We can then form the space $P \times F$, which is fibered over X by the P factor, and upon which K acts diagonally: $b(p, f) = (pb^{-1}, bf)$. The quotient space of the $P \times F$ under the action of K is called the bundle associated to P by F and will be denoted by $F(P)$ or by $F_B(P)$ when the action B needs to be made explicit. We will let ρ denote the projection of $P \times F$ onto $F(P)$ that assigns to each point (p, f) in $P \times F$ its K orbit. (If F and B need to be made explicit, we shall denote this map by ρ_F or ρ_B as necessary.) Over a neighborhood $\pi^{-1}(U) \times F$, where we have chosen a trivialization $\pi^{-1}(U) \sim U \times K$, the set of points of the form (u, e, f), $u \in U$, $f \in F$ gives a smooth cross section for the K action. This gives a local trivialization to $F(P)$ over U, that is, an identification of $\pi_F^{-1}(U)$ with $U \times F$, where π_F denotes the projection of $F(P)$ onto X. This shows that $\pi_F: F(P) \to X$ is a smooth bundle (and we shall frequently denote π_F by π when there is no possibility of confusion). A section of $F(P)$ is a map $s:X \to F(P)$ satisfying $\pi_F \circ s = id$. There is an identification between such sections and functions $h: P \to F$ that satisfy

$$h(pb^{-1}) = bh(p) \qquad \forall p \in P \text{ and } b \in K. \tag{35.8}$$

Indeed, given any such h define $s(\pi p) = \rho(p, h(p))$. Condition (35.8) guarantees that this is well defined so defines the section s. Conversely, starting with s, this same equation uniquely defines h which satisfies (35.8). Also, it is immediate that s is smooth if and only if h is. If F has some algebraic structure that is preserved by the action B, then the bundle $F(P)$ carries this additional structure. For example, if F is a vector space and B is a linear representation of K on F, then $F(P)$ is a vector bundle. Suppose that we take $F = K$ and B to be the conjugation action $B = A$. Then $K_A(P)$ is a bundle of groups and the sections of $K_F(P)$ have a multiplication that makes the space of all sections of $K_A(P)$ into a group. We can therefore reformulate Proposition 35.1 as

Proposition 35.2. *The group Gau* (P) *is isomorphic to the group of all smooth sections of* $\mathrm{K_A(P)}$.

We now want to show that a choice of a connection for the principal bundle P allows us to consider the reduced phase space $(T^*P)_\mathcal{O}$ as being fibered over T^*X, a result due to Weinstein (1978). For this purpose we will want to use various formulations of the notion of connection. So, in order to establish our notation and have the different definitions at our disposal, we review several of the definitions here. The simplest definition is that a connection of a principal bundle P with structure group K is a choice of horizontal subspace at each point of P, the choice being invariant under the action of K. A choice of horizontal subspace at $p \in P$ is the same as projection $\mathbf{V}_p: TP_p \to TP_p$, where the image of $\mathbf{V}_p$ is the tangent space to the fiber of P at p. The condition of K-invariance means that

$$dR_b \circ \mathbf{V}_p = \mathbf{V}_{pb^{-1}} \circ dR_b. \tag{35.9}$$

The action of K on P defines for each $p \in P$ a map†

$$U_p: K \to P$$

$$U_p(c) = pc^{-1} \qquad \forall c \in K.$$

For any $b \in K$ we have

$$R_b \circ U_p(c) = pc^{-1}b^{-1} = U_p(bc).$$

† The map U_p is nothing other than the map we denoted by ψ_p in Section 23 for a general group action. It will be convenient to have this special notation for principal bundles.

Writing l_b for the operation of left multiplication of b on K, we can write this last equation as

$$R_b \circ U_p = U_p \circ l_b.$$

Let $u_p: k \to TP_p$ be the differential of U_p at e, so

$$u_p(\xi) = \frac{d}{dt} U_p(\exp t\xi).$$

Now

$$R_b \circ U_p(\exp t\xi) = p(\exp t\xi)b^{-1} = pb^{-1}b \exp t\xi b^{-1} = U_{R_b(p)}(b \exp t\xi b^{-1})$$

and differentiating at $t = 0$ gives

$$dR_b(u_p(\xi)) = u_{R_b(p)}(\mathrm{Ad}_b \xi). \tag{35.10}$$

Since u_p maps k bijectively onto the tangent space to the fiber, we can form

$$\theta_p = u_p^{-1} \circ \mathbf{V}_p. \tag{35.11}$$

So $\theta_p: TP_p \to k$ is a k-valued linear differential form on P known as the *connection form.* By (35.9) and (35.10), it satisfies

$$\theta_{R_b p} \circ dR_b = \mathrm{Ad}_b \theta_p. \tag{35.12}$$

Thus a connection is a k-valued form θ on P that satisfies (35.12) and

$$\theta_p \circ u_p = \mathrm{id} \qquad \forall\, p \in P. \tag{35.13}$$

Given θ_p we recover the horizontal subspace at p as $\ker \theta_p$ and the projection V_p from (35.11) as $\mathbf{V}_p = u_p \circ \theta_p$.

There is still another way of looking at connections that we will need to use. Let aut (P) denote the Lie algebra consisting of all smooth vector fields ξ on P that are K-invariant, that is, that satisfy

$$R_b^* \xi = \xi \qquad \forall\, b \in K. \tag{35.14}$$

The reason for our notation is that if ξ is such a vector field, then the flow it generates consists of transformations belonging to Aut P; in other words, (35.14) is the "infinitesimal version" of (35.3), so we can think of aut P as being the "Lie algebra" of Aut P.

At a fixed point x of X we can consider the set of sections of TP defined over $\pi^{-1}(x)$ and satisfying (35.14) there. These form a vector space that we shall denote by E_x, and the set of all such vector spaces fit together to form a smooth vector bundle that we shall denote by E. Then we can regard an element ξ of aut P as a smooth section of the vector bundle E. Given an element η in E_x, we can form $d\pi_p(\eta(p))$ for any $p \in \pi^{-1}(x)$, and $d\pi_p(\eta(p))$ is independent of the choice of p. We thus get a linear map $\sigma_x : E_x \to TX_x$ and these fit together smoothly to give a vector bundle morphism

$$\sigma : E \to TX.$$

The kernel of this map σ is the vector bundle V of vertical invariant vector fields, and we clearly have the exact sequence

$$0 \to V \to E \xrightarrow{\sigma} TX \to 0 \tag{35.15}$$

of vector bundles over X. A smooth section of V is then a vertical K-invariant vector field on P and so can be regarded as an infinitesimal generator of Gau P. Thus, taking sections of (35.15) gives rise to the exact sequence of Lie algebras

$$0 \to \text{gau } P \to \text{aut } P \xrightarrow{\nu} D(X), \tag{35.16}$$

where $D(X)$ denotes the algebra of smooth vector fields on X.

Now suppose we are given a connection. The projection onto the vertical $\mathbf{V}_p$, being K-invariant, defines a splitting

$$\mu : E \to V$$

of the sequence (35.15); and conversely, any such splitting clearly defines a connection. Thus, we can regard a connection on P as being a splitting of the sequence (35.15).

Suppose we are given a connection θ on P, and let p be a point of P. The tangent space TP_p splits into a direct sum

$$TP_p = \mathbf{V}_p(TP_p) \oplus \ker \mathbf{V}_p$$

of a vertical and horizontal component, and we may identify the horizontal component $\ker \mathbf{V}_p$ with the tangent space TX_x to the base at $x = \pi p$ so we can write

$$TP_p = V_p(TP_p) \oplus TX_x.$$

This, in turn, gives an identification of the dual space

$$TP_p^* = (V_p P)^* \oplus T^*X_x,$$

and hence, in particular, a projection

$$\kappa_p : T^*P_p \to T^*X_x$$

onto the second component. Let $z \in TP_p$ and $n \in T^*P_p$ and let

$$z = v + w \qquad \text{and} \qquad n = l + m$$

be their decompositions relative to the direct sum decompositions given above. Then $\langle n, z\rangle = \langle l, v\rangle + \langle m, w\rangle$. If we write $v = u_p\xi$ for $\xi \in k$, then

$$\begin{aligned}\langle l, u_p\xi\rangle &= \langle z, u_p\xi\rangle \\ &= \langle \Phi(p, n), \xi\rangle\end{aligned}$$

by the definition of the moment map $\Phi: T^*P \to k^*$ in the cotangent bundle situation for any K action on a manifold. (See Section 28.) Also,

$$\langle m, w\rangle = \langle \kappa_p(n), d\pi_p(z)\rangle.$$

Let $\eta \in T(T^*P)_{(p,n)}$ and z be its image in TP_p. Recall that the fundamental 1-form α_p of T^*P is given by

$$(\alpha_P)_{(p,n)}(\eta) = \langle n, z\rangle = \langle \Phi(p, n), \xi\rangle + \langle \kappa_p(n), d\pi_p(z)\rangle$$

if $v = u_p\xi$.

Let α_X denote the fundamental 1-form on T^*X. Let $\kappa: T^*P \to T^*X$ be defined by

$$\kappa(p, n) = (\pi(p), \kappa_p(n)). \tag{35.17}$$

Then

$$\begin{aligned}(\alpha_X)_{\kappa(p,n)}(d\kappa_{(p,n)}(\eta)) &= \langle \kappa(p, n), d\pi_p(z)\rangle \\ &= \langle \kappa_p(n), d\pi_p(z)\rangle.\end{aligned}$$

Thus we can express the fundamental 1-form of T^*P as

$$\alpha_P = \langle \Phi, \theta\rangle + \kappa^*\alpha_X. \tag{35.18}$$

We will use (35.18) to give an alternative description for the reduced phase space $(T^*P)_\mathcal{O}$ introduced at the beginning of this section. We first recall the notion of the pull-back of a principal bundle: Let $P \to X$ be a principal bundle, and let $f: Y \to X$ be a smooth map. We then define the bundle $f^\# P$ over Y as the fiber product over X of Y and P:

$$f^\# P = \{(y, p) \text{ with } f(y) = \pi p\}.$$

It is easy to check that $f^\# P$ is a smooth manifold and, defining $\pi^\#: f^\# P \to Y$ to be projection onto the first factor makes $f^\# P$ into a principal K bundle over Y. Projection onto the second factor gives a K map, $\tilde{f}: f^\# P \to P$ making the diagram

$$\begin{array}{ccc} f^\# P & \xrightarrow{\tilde{f}} & P \\ {\scriptstyle \pi^\#}\downarrow & & \downarrow{\scriptstyle \pi} \\ Y & \xrightarrow{f} & X \end{array}$$

commute. The map $\tilde{f}$ carries fibers into fibers, and is, in fact, a K morphism.

If θ is a connection form on P, then $\tilde{f}^*\theta$ is a connection form on $f^{\#}P$. All of this is perfectly general. We will want to apply it to the case where $Y = T^*X$ and the map f is just the projection of T^*X onto X. Writing $P^{\#}$ instead of $f^{\#}P$ in this case, we have the diagram

$$\begin{array}{ccc} P^{\#} & \longrightarrow & P \\ \downarrow & & \downarrow \\ T^*X & \longrightarrow & X \end{array}$$

where a point of $P^{\#}$ is a triple of the form (x, y, p) with $y \in T^*X_x$ and $\pi p = x$. We can use the map κ to define a map $\sigma: T^*P \to P^{\#}$ by

$$\sigma(p, n) = \big(\pi p, \kappa_p(n), p\big)$$

giving the commutative diagram

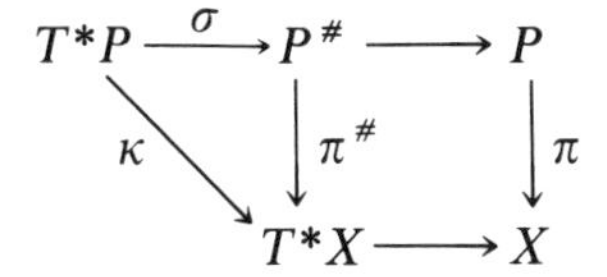

(the maps σ and κ depend on the connection). Now let Q be a Hamiltonian K space with moment map Φ_Q. (For example, we might take Q to be an orbit of K in k^* with Φ_Q the identification of Q as an orbit in k^*.) We can form the Hamiltonian K space $T^*P^- \times Q$ whose symplectic form is $d\alpha_P + \omega_Q$, where ω_Q is the symplectic form of Q, and whose moment map is $\bar{\Phi} = \Phi_{T^*P^- \times Q} = \Phi_Q - \Phi$. In particular, the inverse image of 0 under the moment map, $\bar{\Phi}^{-1}(0)$ consists of all (p, n, q) with $\Phi(p, n) = \Phi_Q(q)$. Now $\sigma \times \text{id}$ maps $T^*P \times Q$ onto $P^{\#} \times Q$. We claim that the restriction of $\sigma \times \text{id}$ to $\bar{\Phi}^{-1}(0)$ defines a diffeomorphism of $\bar{\Phi}^{-1}(0)$ with $P^{\#} \times Q$. Indeed, for fixed (p, q), the map $\sigma \times \text{id}: T^*P_p \times \{q\} \to T^*X_x \times \{(p, q)\}$ sends the covector $n \in T^*P_p$ to $\kappa_p(n)$. Thus the horizontal component of n is determined by its image. But the vertical component of n is determined by $\Phi(p, n) = \Phi_Q(q)$. Thus the restricted map is one-to-one and its inverse is clearly smooth and is thus a diffeomorphism. Let χ denote this diffeomorphism. If we now substitute $\Phi = \Phi_Q$ into (35.17) and use (35.18), we obtain

Proposition 35.3. *Let χ denote the restriction of $\sigma \times id$:* $\mathrm{T^*P \times Q} \to \mathrm{P^{\#} \times Q}$ *to* $\Phi^{-1}_{\mathrm{T^*P^- \times Q}}(0)$. *Then χ is a* K*-equivariant diffeomorphism. Furthermore, if we define the closed 2-form v on* $\mathrm{P^{\#} \times Q}$ by

$$v = \mathrm{d}(\langle \Phi_{\mathrm{Q}}, \theta^{\#} \rangle + \pi^{\#*}\alpha_{\chi}) + \omega_{\mathrm{Q}}, \tag{35.19}$$

then $\chi^*\nu$ is the restriction of the symplectic form of $\mathrm{T^*P^-} \times \mathrm{Q}$ *to the coisotropic submanifold* $\Phi^{-1}_{\mathrm{T^*P^-} \times \mathrm{Q}}(0)$.

Now by our general principles of reduction, we know that the symplectic manifold associated to the coisotropic $\Phi^{-1}_{\mathrm{T^*P^-} \times \mathrm{Q}}(0)$ is just its quotient under the action of the isotropy group at 0 in k^*, that is, under all of K. But, for $P^\# \times Q$, the quotient space for the K action is exactly the associated bundle $Q(P^\#)$. Thus we have proved

Proposition 35.4. *The leaves of the null foliation of the form ν on* $\mathrm{P^\#} \times \mathrm{Q}$ *are precisely the orbits of the* K *action. In particular, there exists a unique symplectic form Ω on the associated bundle* $\mathrm{Q(P^\#)}$ *such that*

$$\nu = \rho^*\Omega, \tag{35.20}$$

where $\rho:\mathrm{P^\#} \times \mathrm{Q} \to \mathrm{Q(P^\#)}$ *is the natural projection. The diffeomorphism χ induces a symplectic diffeomorphism $\bar{\chi}$ of* $(\mathrm{T^*P_Q})^-$ *with* $\mathrm{Q(P^*)}$.

(Originally, the form ν was defined directly by (35.19) and Proposition 35.4 was proved in Sternberg (1977). The fact that it could be obtained by the reduction procedure was observed by Weinstein (1978)).

It will be useful and instructive for us to have a direct proof of (35.20) in a slightly more general setting: Let $P^\# \to X^\#$ be a principal K bundle (so we are not assuming for the moment that $X^\#$ is a cotangent bundle). Let Q be a Hamiltonian K space with moment map Φ_Q and let $\theta^\#$ be a connection on $P^\#$. We can then define the form

$$\nu = d\langle \Phi_Q, \theta^\# \rangle + \omega_Q \tag{35.21}$$

on $P^\# \times Q$. It is clearly closed and invariant under the action of K. We claim that the K orbits are isotropic submanifolds for ν. Indeed, if η is a vector field on $P^\# \times Q$ that is the image of an element of k, the invariance of the 1-form, $\langle \Phi_Q, \theta^\# \rangle$ implies that

$$0 = D_\eta \langle \Phi_Q, \theta^\# \rangle = i(\eta)\, d\langle \Phi_Q, \theta^\# \rangle + d\langle \Phi_Q, i(\eta)\theta^\# \rangle.$$

But, by one of the defining properties of a connection, (35.13),

$$i(\eta)\theta^\# = \eta$$

so

$$i(\eta)\langle \Phi_Q, \theta^\# \rangle = d\langle \Phi_Q, \eta \rangle,$$

while

$$i(\eta)\omega_Q = d\langle \Phi_Q, \eta \rangle,$$

proving that

$$i(\eta)v = 0. \tag{35.22}$$

We have thus proved

Proposition 35.5. *Let* $P^{\#} \to X^{\#}$ *be a principal* K *bundle with connection,* $\theta^{\#}$, *and let* Q *be a symplectic manifold with a Hamiltonian* K *action and associated moment map* Φ_Q, *and symplectic form* ω_Q. *If we define the form* v *on* $P^{\#} \times Q$ *by (35.21), then* v *is* K*-invariant and the* K *orbits are isotropic for* v *and so there is a closed 2-form* ω *defined on the associated bundle* $Q(P^{\#})$ *such that*

$$v = \rho^*\omega. \tag{35.23}$$

In general, there is no reason to expect ω to be of maximal rank (although it is so in certain important cases to be studied later on). In (35.19), where $X^{\#} = T^*X$ is a cotangent bundle, the addition of the term $d\pi^{\#} * \alpha_X$ is what guaranteed that Ω in (35.20) is symplectic. Why this is so will become clear once we give an alternative description of the forms ω and Ω in terms of the curvature of the connection $\theta^{\#}$. For this purpose, we review the basic concepts in the next section.

36. Curvature

Let $P \to X$ be a principal bundle with structure group K and connection form θ. The corresponding projection onto the vertical at each $p \in P$ is denoted by $\mathbf{V}_p$. Let

$$\mathbf{h}_p = \text{id} - \mathbf{V}_p$$

so that $\mathbf{h}_p$ is the projection onto the "horizontal subspace" $\ker \theta_p$. We can define the operator h on the space of differential forms on P by

$$(h\sigma)_p(v_1, \ldots, v_k) = \sigma_p(\mathbf{h}_p v_1, \ldots, \mathbf{h}_p v_k),$$

where σ is a differential form of degree k and $v_1, \ldots, v_k$ are tangent vectors at p. Thus h takes forms of degree k into forms of degree k. The form σ can either be a scalar-valued form, or take values in a vector space. For example, $h\theta = 0$ since $\mathbf{h}_p v$ lies in the kernel of θ_p. More generally, the form $h\sigma$ is "horizontal" in the sense that $(h\sigma)_p(v_1, \ldots, v_k) = 0$ if any one of the vectors v_i are tangent to the fiber of the projection of P onto X, because any such "vertical" tangent vector is in the kernel of h_p. It is clear that the operator h

preserves all algebraic operations; for example,

$$h(\sigma_1 \wedge \sigma_2) = h\sigma_1 \wedge h\sigma_2,$$

and that it is a projection, that is,

$$h^2 = h.$$

It is also K-invariant; that is,

$$R_b^* h\sigma = hR_b^*\sigma$$

for any k form σ and any $b \in K$.

The projection h allows us to define the important operator δ called the "covariant exterior derivative" by

$$\delta\sigma = h\, d\sigma.$$

So δ maps k-forms to $(k+1)$-forms. Because of the properties of h and of exterior differentiation d it satisfies

$$\delta(\sigma_1 \wedge \sigma_2) = \delta\sigma_1 \wedge \sigma_2 + (-1)^{\deg \sigma_1}\sigma_1 \wedge \delta\sigma_2$$

$$R_b^*\delta = \delta R_b^* \qquad \forall\, b \in K,$$

and

$$i(\xi) \circ \delta = 0 \qquad \text{for any vertical vector field } \xi. \tag{36.1}$$

(In particular, in (36.1) we can take ξ to be a vector field associated to an element of the Lie algebra k by the action of K on P.)

The curvature form $F(\theta)$ of the connection θ is the g-valued 2-form on P defined by

$$F(\theta) = \delta\theta. \tag{36.2}$$

It follows from (36.1) that $F(\theta)$ is horizontal:

$$i(\xi)F(\theta) = 0 \qquad \text{for any vertical vector field } \xi. \tag{36.3}$$

It follows from (35.12) that

$$R_b^* F(\theta) = \mathrm{Ad}_b F(\theta). \tag{36.4}$$

Let ζ_1 and ζ_2 be two horizontal vector fields. Applying the Lie derivative D_{ζ_1}, to the equation $\langle\theta,\zeta_2\rangle \equiv 0$ we get

$$0 = \langle D_{\zeta_1}\theta, \zeta_2\rangle + \langle\theta, D_{\zeta_1}\zeta_2\rangle = \langle i(\zeta_1)\, d\theta, \zeta_2\rangle - \langle v, [\zeta_1, \zeta_2]\rangle,$$

since

$$D_{\zeta_1}\theta = i(\zeta_1)\, d\theta + di(\zeta_1)\theta,$$

$$i(\zeta_1)\theta = 0,$$

and

$$D_{\zeta_1}\zeta_2 = -[\zeta_1, \zeta_2].$$

Thus, for horizontal vector fields ζ_1 and ζ_2, we have

$$\langle F(\theta), \zeta_1 \wedge \zeta_2 \rangle = \langle \theta, [\zeta_1, \zeta_2] \rangle, \qquad \zeta_1, \zeta_2 \text{ horizontal.} \tag{36.5}$$

The connection θ determines a splitting of the sequence (35.15) and thus a map λ that assigns a horizontal vector field $\lambda\eta$ to each vector field η on X. In other words, θ determines a linear map $\lambda: D(X) \to \text{aut } P$, which is a linear splitting of the Lie algebra sequence (35.16). We shall see that the curvature is a measure of the failure of λ to be a Lie algebra homomorphism. Indeed, let $\mu: \text{aut } P \to D(X)$ denote the last map in (35.16) so that μ is a Lie algebra homomorphism and $\mu \circ \lambda = \text{id}$. Thus

$$\mu(\lambda[\xi_1, \xi_2]) = [\xi_1, \xi_2]$$

and

$$\mu([\lambda\xi_1, \lambda\xi_2]) = [\mu\lambda\xi_1, \mu\lambda\xi_2] = [\xi_1, \xi_2]$$

for any ξ_1 and ξ_2 in $D(X)$. This implies that

$$\lambda([\xi_1, \xi_2]) - [\lambda\xi_1, \lambda\xi_2]$$

lies in the kernel of μ (i.e., is a vertical vector field). Now λ when applied to any vector field on X gives a horizontal vector field by definition. So applying θ to the preceding expression gives, by (36.5),

$$\langle \theta, \lambda[\xi_1, \xi_2] - [\lambda\xi_1, \lambda\xi_2] \rangle = \langle F(\theta), \lambda\xi_1 \wedge \lambda\xi_2 \rangle. \tag{36.6}$$

In words, (36.6) says that the vertical vector field $\lambda[\xi_1, \xi_2] - [\lambda\xi_1, \lambda\xi_2]$, which measures the failure of λ to be a homomorphism, is, in fact, the vertical vector field corresponding to the element ξ in the Lie algebra k, where $\xi = \langle F(\theta), \xi_1 \wedge \xi_2 \rangle$.

The expression $\lambda[\xi_1, \xi_2] - [\lambda\xi_1, \lambda\xi_2]$ lies in the kernel of μ (i.e., in gau P). We have identified gau P with the space of sections of the associated bundle $k(P)$ to the adjoint representation of K on k. Thus the curvature maps any two vector fields on X into a section of $k(P)$. In fact, we can see directly that $F(\theta)$ can be thought of as a 2-form on X with values in the associated bundle $k(P)$. Indeed, at each point $x \in X$, choose some point $p \in \pi^{-1}(x)$. Any tangent vector at x corresponds to a unique horizontal tangent vector to P at p. So, if v_1 and v_2 are tangent vectors at x and w_1, w_2 the corresponding horizontal tangent vectors at p, we can consider the element $(p, \langle F(\theta), w_1 \wedge w_2 \rangle)$ of $P \times k$. The transformation law (36.4) for $F(\theta)$ shows that the image of this element under the map $\rho: P \times k \to k(P)$

is independent of the choice of p. We shall write $F^*(\theta)$ for this $k(P)$-valued 2-form on X.

We shall now use the curvature form $F^*(\theta)$ to give an alternative expression to the form ω occurring in (35.23) and the form $\chi^*\sigma$ occurring in Proposition 35.1. For this we first observe that a choice of connection on P induces a connection, that is, a complement to the vertical at each point of $Q(P)$ for any associated bundle. Indeed, if $\rho: P \times Q \to Q(P)$ denotes the projection onto K orbits, then $d\rho_p(h_p(TP)_p \times \{0\}) \subset \rho(p,q)$ is independent of the choice of the representative (p,q) on account of the invariance of the horizontal subspaces $h_p(TP_p)$ under the action of K. We thus get, at each point z of $Q(P)$, a well-defined projection, V_z, of $T(Q(P))_z$ onto the subspace of vertical vectors. In particular, if Q is a symplectic manifold with a symplectic structure preserved by K, the symplectic form ω_Q induces an antisymmetric bilinear form on the vertical tangent vectors assigning to two vertical vectors u_1 and u_2 the value $\omega_{Qq}(u_1, u_2)$, where $u_i \in TQ_q$ and $d\rho_{(p,q)}(0, u_i) = u_i$, $\rho(p,q) = z$. The invariance of ω_Q under K guarantees that this is independent of the choice of (p,q) with $\rho(p,q) = z$. The projection V_z allows us to extend this to an antisymmetric form, $\tilde{\omega}_{Qz}$, defined on all of $T(Q(P))_z$, i.e.,

$$\omega_{Qz}(w_1, w_2) = \omega_{Qq}(u_1, u_2), \tag{36.7}$$

where $u_i \in TQ_q$ and $d\rho_{(p,q)}(0, u_i) = V_z(w_i)$, $\rho(p,q) = z$. We thus get a smooth 2-form $\tilde{\omega}_Q$, defined on $Q(P)$. (Or, to conform to the notations of Section 35, on $Q(P^\#)$ if our principal bundle is $P^\#$.) The map $\Phi: Q \to k^*$ is a K morphism and hence induces a map, $\tilde{\Phi}: Q(P^\#) \to k^*(P^\#)$ of the associated bundles. Finally, the form $F(\theta^\#)$ is a $k(P^\#)$-valued 2-form on $X^\#$. If $\pi: Q(P^\#) \to X^\#$ denotes the projection onto the base, we can construct the pull-back via π of $F(\theta^\#)$, in particular, $\pi^* F(\theta^\#)$. This is a $k(P^\#)$-valued 2-form on $Q(P^\#)$. To avoid a messy accumulation of symbols, we shall simply denote it by F. So

$$F = \pi^* F(\theta^\#). \tag{36.8}$$

We can pair the $k(P^\#)$-valued 2-form F with the $k^*(P^\#)$-valued function $\tilde{\Phi}$ to obtain the scalar-valued 2-form $\langle \tilde{\Phi}, F \rangle$ on $Q(P^\#)$. We claim that the closed 2-form ω of Proposition 35.5 is given by the formula

$$\omega = \langle \tilde{\Phi}, F \rangle + \tilde{\omega}_Q. \tag{36.9}$$

To prove (36.9), we must show that (35.23) holds if we define ω to be the right-hand side of (36.9) and define v by (35.21). For this it suffices to verify equality at every point (p,q) of $P^\# \times Q$. Now both $\rho^*\omega$ and v have the tangent vectors to the K orbits on $P^{\#} \times Q$ as null directions. It therefore suffices to verify (36.9) for three types of pairs of tangent vectors at (p, q):

$$(u_1, 0), \qquad (u_2, 0), \tag{a}$$

$$(u, 0), \qquad (0, v), \tag{b}$$

$$(0, v_1), \qquad (0, v_2), \tag{c}$$

where the u's are horizontal tangent vectors in TP_p and the v's are elements of TQ_q.

Expanding the differential $d\langle \Phi_Q, \theta^{\#} \rangle$ occurring in (35.21) as

$$d\langle \Phi_Q, \theta^{\#} \rangle = \langle d\Phi_Q \wedge \theta^{\#} \rangle + \langle \Phi_Q, d\theta^{\#} \rangle,$$

we see that the "mixed" term $\langle d\Phi_Q \wedge \theta^{\#} \rangle$ vanishes on all three types of pairs, since $\theta^{\#}$ vanishes on horizontal vectors. We must therefore verify that

$$\langle \Phi_Q, d\theta^{\#} \rangle + \omega_Q = \rho^*(\langle \tilde{\Phi}, F \rangle + \tilde{\omega}_Q)$$

when we evaluate on the three types. For type (a), the ω_Q term makes no contribution to either side and the first terms give the same answer in view of the definition (36.2) of the curvature and the form F. For type (b), both sides give zero since no "mixed" terms occur on either side. For type (c), equality follows from the definition of $\tilde{\omega}_Q$ and the fact that images of the vectors under $d\rho$ are vertical.

We can now summarize the results of the preceding discussion together with the content of Propositions 35.3 and 35.4:

Proposition 36.1. *Let* $\mathrm{P} \to \mathrm{M}$ *be a principal bundle with structure group* K, *and let* $\mathrm{P}^{\#} \to \mathrm{T^*M}$ *be the pull-back of* P *to the cotangent bundle* $\mathrm{T^*M}$ *of* M. *Let* θ *be a connection on* P *and* $\theta^{\#}$ *the induced connection on* $\mathrm{P}^{\#}$. *Let* Q *be a Hamiltonian* K *space and* $\mathrm{Q(P^{\#})}$ *the associated bundle. The moment map,* $\Phi: \mathrm{Q} \to \mathrm{k}^*$, *induces a function,* $\tilde{\Phi}: \mathrm{Q(P^{\#})} \to \mathrm{k^*(P^{\#})}$. *The symplectic form* ω_{Q} *of* Q, *together with the connection* θ, *induces a 2-form,* $\tilde{\omega}_{\mathrm{Q}}$, *on* $\mathrm{Q(P^{\#})}$ *and the curvature of* $\theta^{\#}$ *gives rise to a* $\mathrm{k(P^{\#})}$*-valued 2-form* $\mathrm{F}^{\#}$ *on* $\mathrm{Q(P^{\#})}$. *Then*

$$\Omega = \mathrm{d}\pi^{\#*}\alpha_{\mathrm{M}} + \langle \tilde{\Phi}, \mathrm{F}^{\#} \rangle + \tilde{\omega}_Q$$

is a symplectic form on $\mathrm{Q(P^{\#})}$, where $\pi^{\#}$ *denotes the projection onto* $\mathrm{T^*M}$ *and* α_{M} *is the fundamental 1-form on* $\mathrm{T^*M}$. *This form* Ω *coincides with the symplectic form of Proposition 36.2. In particular, there is a symplectic diffeomorphism of* $(\mathrm{T^*P_Q})^-$ *with* $(\mathrm{Q(P^{\#})}, \Omega)$.

As we have seen, a connection on P induces a projection of T^*P_Q onto T^*M. Any Hamiltonian $\mathscr{H}$ on T^*M then pulls back to T^*P_Q and hence determines a Hamiltonian vector field ξ_H on T^*P_Q. This is the principle of minimal coupling. Suppose that $K = U(1)$ and Q is taken to be a point in k^*, which is a coadjoint orbit in the case of an Abelian group. The term $\tilde{\omega}_Q$ then does not occur. The reader should check that for $M = \mathbb{R}^3$, we get the modification of the symplectic structure associated to a magnetic field, as in Subsection B of Section 20. If we take $M = \mathbb{R}^4$, $\mathscr{H}(q, p) = \frac{1}{2}\|p\|^2$, then we get the Lorentz equations as in Subsection D of Section 20. Thus the equations of the principle of minimal coupling generalize the Lorentz equations to an arbitrary Yang–Mills field.

Let us see what these equations look like in general. In view of the fact that the Ω in Proposition 36.1 is the negative of the symplectic form on T^*P_Q, the vector field ξ_H is determined by the equation $i(\xi_H)\Omega = -d\mathscr{H}$.

Let us write

$$\xi_H = \lambda\left(a\frac{\partial}{\partial q}\right) + \lambda\left(b\frac{\partial}{\partial p}\right) + w$$

in terms of local coordinates on T^*M, where $\lambda(v)$ is the horizontal vector corresponding to the vector v in T^*M, and where w is vertical. Then

$$i(w)\Omega = i(w)\tilde{\omega}_Q,$$

since the first two terms in the expression for Ω vanish when evaluated on a vertical vector. Since $d\mathscr{H}$ has no vertical component, we conclude that $w = 0$. The coefficient of dp in $i(\xi_{\mathscr{H}})\Omega$ is $-a$, so we see that $a = \partial\mathscr{H}/\partial p$. The dq term in $i(\xi_{\mathscr{H}})\Omega$ is

$$bdq + \left[\tilde{\Phi}, i\left(a\frac{\partial}{\partial q}\right)F\right],$$

which must equal $-(\partial\mathscr{H}/\partial q)dq$.

Now $\langle\tilde{\Phi}, i(a\,\partial/\partial q)F\rangle$ is a covector that we can consider as a point of T^*M. Thus we obtain the equations

$$\frac{dq}{dt} = \frac{\partial\mathscr{H}}{\partial p}, \qquad \frac{dp}{dt} + \left[\tilde{\Phi}, i\left(\frac{dq}{dt}\right)F\right] = -\frac{\partial\mathscr{H}}{\partial q}. \tag{36.10}$$

These equations, together with the equation $w = 0$, are the Hamiltonian equations corresponding to an $\mathscr{H}$ that comes from T^*M. Notice that $w = 0$ says that the solution curves are all horizontal (i.e., *obtained by parallel transport from their projections onto* T^*M).

In the case that $Q = \mathcal{O}$ is an orbit in k^*, we can give a more intrinsic interpretation to these equations: The connection θ induces a projection of

T^*P onto T^*M. The Hamiltonian $\mathscr{H}$ then pulls back to a function on T^*P, which is K-invariant. The corresponding flow leaves invariant $\Phi_P^{-1}(l)$ for any l in k^*, where Φ_P is the moment map for the K action on T^*P. We know that this determines a Hamiltonian flow on the reduced space $\Phi_P^{-1}(l)/K_l$, which is just $T^*P_{\mathcal{O}}$, where $\mathcal{O}$ is the orbit through l, by the results of Section 29. A simple check of all the identifications involved shows that the equations obtained are exactly those of minimal coupling described above.

There is another way of writing equations (36.10) in terms of local coordinates that proves useful. Suppose we are given two connections θ and θ_0 on P. Then

$$\theta - \theta_0 = \tilde{A}$$

is a 1-form on P that vanishes on vertical vectors and that satisfies

$$R_b^* \tilde{A} = \mathrm{Ad}_b \tilde{A}.$$

Let $A^\#$ denote the pull-back of A to T^*P. Then $\langle \Phi, \tilde{A} \rangle$ is an invariant horizontal 1-form on T^*P and can be thought of as a function from T^*P to T^*M. If κ and κ_0 are the maps from $T^*P \to T^*M$ corresponding to θ and θ_0, then

$$\kappa = \kappa_0 + \langle \Phi, \tilde{A}^\# \rangle.$$

Let us apply this to the case where P is (locally) given as $P_{|U} = U \times K$ and we take θ_0 to be the flat connection corresponding to the direct product splitting, so $\kappa_0 : T^*P = T^*U \times T^*K \to T^*U$ is just projection onto the first factor. The form $\tilde{A}$ will be given at a point $(x, c) \in U \times K$ as

$$\tilde{A}_{(x,c)} = \mathrm{Ad}_{c^{-1}} \tilde{A}_{(x,e)}.$$

We let A denote the k-valued form

$$A = s^* \tilde{A} = s^* \theta,$$

where $s(x) = (x, e)$. So we can write

$$A_{(x,c)} = \mathrm{Ad}\, c^{-1} A_x.$$

Finally, in terms of the left-invariant identification of T^*K with $K \times k^*$, the moment map Φ is given by $\Phi(c, \beta) = -\beta$. So

$$\begin{aligned} \kappa(x, \xi; c, \beta) &= (x, \xi - \langle \beta, \mathrm{Ad}_{c^{-1}} A_x \rangle) \\ &= (x, \xi - \langle c\beta, A_x \rangle), \end{aligned}$$

where $x \in M$ and $\xi \in T^*M_x$, $c \in K$, and $\beta \in k^*$. So

$$\kappa^* \mathscr{H} = \mathscr{H}(x, \xi - \langle c\beta, A_x \rangle).$$

If $q_1, \ldots, q_n$ and $\xi^1, \ldots, \xi^n$ are canonical coordinates on T^*U, the corresponding Hamiltonian equations become

$$\frac{dq^i}{dt} = \frac{\partial \mathscr{H}}{\partial \xi_i}$$

$$\frac{d\xi_i}{dt} = \frac{\partial \mathscr{H}}{\partial q^i} + \frac{\partial \mathscr{H}}{\partial \xi_i}\frac{\partial \langle c\beta, A_x\rangle}{\partial q^i}.$$

In order to carry out an integration by parts calculation in Section 40, we shall want more explicit form of these equations for the case where $\mathscr{H}(x, \xi) = \frac{1}{2}\|\xi\|^2$. In local coordinates

$$\mathscr{H}(q, \xi) = \tfrac{1}{2}\sum g^{ij}(q)\xi_i\xi_j.$$

Let us set $\gamma = c\beta$. (This amounts to choosing the representative $(e, c\beta)$ equivalent to (c, β) in the associated bundle.) Let us also write

$$p_i = \xi_i - \langle \gamma, A_i\rangle,$$

where

$$A = \sum A_i dq^i.$$

So

$$\frac{dq^i}{dt} = \sum g^{ij} p_j \tag{36.11}$$

and

$$\frac{dp_i}{dt} = \frac{d\xi_i}{dt} - \frac{d}{dt}\langle \gamma, A_i\rangle$$

$$= -\tfrac{1}{2}\sum \frac{\partial g^{\mu\nu}}{\partial q_i} p_\mu p_\nu + \sum g^{\mu\nu} p_\mu \left\langle \gamma, \frac{\partial A_\nu}{\partial q_i}\right\rangle - \frac{d}{dt}\langle \gamma, A_i\rangle$$

$$= \tfrac{1}{2}\sum \frac{\partial q^{\mu\nu}}{\partial q_i} p_\mu p_\nu + \left\langle \gamma, \sum \frac{dq_\nu}{dt}\frac{\partial A_\nu}{\partial q_i} - \frac{dA_i}{dt}\right\rangle - \left\langle \frac{d\gamma}{dt}, A_i\right\rangle. \tag{36.12}$$

We now make some remarks about integrating these equations of motion in the presence of additional symmetry. Suppose that we are given a Lie group G of automorphisms of the principal bundle P. Thus $G = \mathrm{Aut}(P)$. Then G has a Hamiltonian action on T^*P that commutes with the K action. If Q is some Hamiltonian K space we can consider Q as a trivial G space and thus get a Hamiltonian action of G on $T^*P \times Q^-$ which commutes with the K action and hence descends to give a Hamiltonian G action on T^*P_Q. Now suppose that S is some Hamiltonian G space and that the moment map for G actions on T^*P_Q and on S intersect cleanly in g^*. We can then perform Marsden–Weinstein reduction with respect to these G actions getting $(T^*P_Q)_S$.

The group G acts on the base space X and hence on T^*X. Suppose that the Hamiltonian $\mathscr{H}$ is G-invariant. Suppose also that the connection on P is G-invariant. Then the induced Hamiltonian on T^*P is both G- and K-invariant. If we take Q to be a coadjoint orbit of K and S to be a coadjoint orbit of G, then we get an induced flow on the space $(T^*P_Q)_S$. The original flow on T^*P can then be reconstructed purely group-theoretically; see, for example, Araham and Marsden (1978), p. 305. It turns out (Shnider and Sternberg, 1983) that under special hypotheses about the G action, one can give a different geometric construction of this reduced space, related to "dimensional reduction" a construction that has recently been discovered in conjunction with the Weinberg–Salam model, and other "unified models" in elementary particle physics. Let us give a brief introduction to this construction.

In the quantum theory of fields, the starting point is always a Lagrangian, which is a (local) functional on the space of classical fields. For example, in a "pure Yang–Mills" theory, the "classical fields" are just the Yang–Mills fields, that is, the connections on a principal bundle. The functional that is chosen is constructed as follows: Each connection Θ on P has a curvature F that can be thought of as a 2-form on M with values in the associated bundle $k(P)$. Suppose that we are given a (pseudo) Riemann metric on M. We then have the Hodge $*$ operator, and $*F$ is an $(n-2)$-form with values in $k(P)$, where $n = \dim M$. Let us also suppose that we are given an invariant scalar product on k. Using this scalar product, we can form $F \wedge *F$, which is a scalar-valued n-form, which we will denote simply by F^2. Then $\int_M F^2$ is the Yang–Mills functional of Θ – the Lagrangian is given by $\mathscr{L}(\Theta) = F^2$. In actual theories one deals not with pure Yang–Mills theories, but with so-called Yang–Mills–Higgs fields, where the data consist of a *Higgs field*, which is just a section of an associated vector bundle, in addition to the Yang–Mills field Θ. The Yang–Mills–Higgs Lagrangian is an expression of the form $\mathscr{L}(\Theta, s) = aF^2 + bB(d_\Theta s) + cC(s)$, where $d_\Theta s$ denotes the covariant differential of the section s, where B is a quadratic function and C a quartic function. Here a, b, and c are parameters to be adjusted by the empirical data. Whereas the pure Yang–Mills functional has an obvious intrinsic geometrical significance, the meaning of the Yang–Mills–Higgs is not so transparent. The method of dimensional reduction, as developed by Manton (1979), Chapline and Manton (1981), and Harnad, Shnider and Tafel (1980) is, roughly speaking, the following: One starts with a pure Yang–Mills theory associated with a principal bundle with a "large"

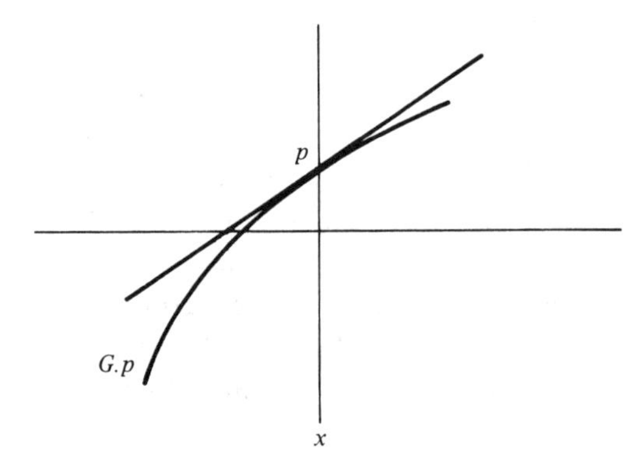

Figure 36.1.

structure group and over a higher-dimensional base manifold, but looks only at Yang–Mills fields that are invariant under an automorphism group $G \subset \text{Aut } P$ of this bundle. Under suitable hypotheses, the set of invariant Yang–Mills fields coincides with a set of Yang–Mills–Higgs data associated with a smaller principal bundle Q, over the quotient manifold M/G, and the Yang–Mills functional on P becomes the Yang–Mills–Higgs functional on the reduced data. To see why this is so, let us look at what happens when the quotient space M/G is a point (i.e., when G acts transitively on the base). In this case, we claim that the set of all G-invariant connections on P can be identified with the set of all elements in a finite-dimensional affine space, and the Yang–Mills functional (for a G-invariant Riemann metric on M) becomes a quartic polynomial on this affine space. The first of these assertions is, in fact, a theorem of Wang (1958) characterizing invariant connections. Let us sketch the formulation and proof of this theorem. The details will be given later on in the more general context. So we assume that $P \to M$ is a principal bundle with structure group K and that $G \subset \text{Aut } P$ acts transitively on M. Pick some point p in P, and let $x = p$ be its image in M. Let $B = G_x$ be the isotropy group of x. Let g denote the Lie algebra of G, b the Lie algebra of B, and k the Lie algebra of K. The action of G determines a map of g into the tangent space to the G orbit through p (i.e., a linear map of g into TP_p, see Figure 36.1), and this maps b into vertical tangent vectors. Composing the map of b into vertical vectors with the map u_p of vertical vectors into k gives a map λ_p of $b \to k$. A G-invariant connection determines, and is completely determined by, a complement to the vertical through p. Indeed, by the property of K-invariance, such a complement to the vertical is determined on the whole fiber through p, and by G-invariance, it is then determined on all of p. Any such complement to the vertical determines, and is determined by, a linear map, $s:g \to k$ (which agrees with λ when restricted to b and which satisfies some equivariance conditions, which we shall write down later). Thus the set of all G-invariant connections is parametrized by the set of all such s, which form an affine space. This is the idea behind the theorem of Wang. It is also easy to check that the curvature of the connection associated with s is given by the quadratic expression $q(s) \in \mathrm{Hom}(g \wedge g, k)$, where

$$q(s)(\xi \wedge \eta) = s([\xi, \eta]) - [s(\xi), s(\eta)].$$

Now q is a quadratic function of s. The Yang–Mills functional is quadratic in the curvature, and hence quartic in s. (The integration in the Yang–Mills expression just pulls out as a constant, since everything is assumed to be G-invariant.) The problem of minimizing the Yang–Mills functional among all G-invariant connections thus reduces to the problem of finding the minima of the quartic polynomial in s. This is the rough idea. Now let us discuss the general case, where, for the moment all we assume is that G is a Lie group of automorphisms of a principal bundle $P \to M$.

Let G_x denote the isotropy group of a point $x \in M$. Each $a \in G_x$ maps the fiber $\pi^{-1}(x)$ into itself. In particular, each $p \in \pi^{-1}(x)$ determines a map

$$\Lambda_p : G_x \to K$$

by

$$ap = p(\Lambda_p(a)) \tag{36.13}$$

and Λ_p is clearly a homomorphism:

$$\begin{aligned}(a_1 a_2)p &= a_1(a_2 p) = a_1(p\Lambda_p(a_2)) \\ &= (a_1 p)\Lambda_p(a_2) \qquad \text{since} \qquad a_1 \in \mathrm{Aut}\, P \\ &= p\Lambda_p(a_1)\Lambda_p(a_2).\end{aligned}$$

Also

$$\begin{aligned}a(pb^{-1}) &= p\Lambda_p(a)b^{-1} \\ &= pb^{-1}b\Lambda_p(a)b^{-1}\end{aligned}$$

or

$$\Lambda_{pb^{-1}}(a) = b\Lambda_p(a)b^{-1}$$

or

$$\Lambda_{R_b(p)} = A_b \Lambda_p, \tag{36.14}$$

where A_b denotes conjugation by b. In particular, *the set of all p' with $\Lambda_{p'} = \Lambda_p$ consist of all p' of the form $p'b$, where $b\Lambda_p(a)b^{-1} = \Lambda_p(a)$ for all $a \in G_x$, that is, of all b in the centralizer of $\Lambda_p(G_x)$.*

If G were compact, then the theorem of Mostow (Proposition 27.4) would say that there are, locally, only a finite number of orbit types on M. Suppose we make the drastic assumption that there is only a single orbit type – that there is a global cross section for the action of G on M. More precisely, let us assume that

$$M = N \times (G/B),$$

where G does not act at all on N and where B is some closed subgroup of G. Thus on each G orbit of M there is a unique point corresponding to the coset of the identity in G/B, and the set of G orbits in M is thus parametrized by points of N. The isotropy group at our distinguished points are all equal to B.

If the groups B and K are compact, we would know that the set of conjugacy classes of homomorphisms of B into K would be discrete. Then, if M and K were connected, there would be only one conjugacy class of homomorphisms Λ_p of B into K. Let us adopt this also as an additional hypothesis. So we assume that there is a single conjugacy class of homomorphisms of B into K with all Λ_p, for $p \in \pi^{-1}(N \times B)$, belonging to this conjugacy class. But this means that we can fix one homomorphism $\Lambda: B \to K$, and consider the set of all p for which $\Lambda_p = \Lambda$. Any two such p and p lying over the same point $M \times B$ differ from one another by right multiplication by some $b \in \mathrm{Cent}\,(\Lambda(B))$. Thus,

Under the above hypotheses, we get a principal bundle, $Q \to N$, whose *structure group is* $\mathrm{Cent}\,(\Lambda(B))$.

The passage from $P \to M$ with structure group K to the smaller bundle $Q \to N$ with structure group $\mathrm{Cent}\,(\Lambda(B))$ is one-half of the geometrical part of the process of dimensional reduction. It now turns out that a G-invariant connection on P gives rise to and is determined by a connection on this reduced bundle, and also a section of an associated bundle on N, and the "pure" Yang–Mills functional on invariant connections becomes the Yang–Mills–Higgs functional on the reduced bundle. See, for example Harnad, Shnider, and Tafel (1980).

Now H is a subgroup of G and is mapped, via Λ into K. Thus both E and F become Hamiltonian H spaces with moment maps $\Phi_H^E: E \to h^*$ and $\Phi_H^F: F \to h^*$. Assuming that the maps intersect cleanly in h^*, we can form the reduced space

$$E_F = (\Phi_H^E - \Phi_H^F)^{-1}(0)/H.$$

This is a symplectic manifold. Since C is a subgroup of K, C has a Hamiltonian action on E with moment map $\Phi_C^E: E \to c^*$. Considering F as a trivial C space then gives a Hamiltonian action of C on E_F, so we can form the symplectic manifold

$$T^*Q_{(E_F)}.$$

We claim *there is a canonical symplectic diffeomorphism between* $(T^*P_E)_F$ *and* $T^*Q_{(E_F)}$.

To see this diffeomorphism, we first observe that the space $(T^*P_E)_F$ can be described as follows: Let $\Phi_G: T^*P \to g^*$ be the moment map for the induced action of G on T^*P and consider the subset of $T^*P \times E \times F$ consisting of all (z, e, f) satisfying

$$\Phi_K(z) = \Phi_K^E(e) \qquad \text{and} \qquad \Phi_G(z) = \Phi_G^F(f). \tag{36.15}$$

Then $(T^*P_E)_F$ is the quotient space of this submanifold under the action of $G \times K$. By the action of an appropriate element of $G \times K$ we can arrange that z lies over a point of Q; that is, that $z \in T^*P_{|Q}$. (Notice that $H \times C$ are still free to act on such z.) We let L denote the subset of $T^*P_{|Q} \times E \times F$ satisfying (36.15). Let L' denote the subset of $T^*Q \times E \times F$ consisting of points satisfying

$$\Phi_H^E(e) = \Phi_H^F(f) \tag{36.16}$$

and

$$\Phi_C(y) = \Phi_C^E(e), \qquad y \in T^*Q, \qquad e \in E. \tag{36.17}$$

For each $p \in Q$ we have the restriction map $T^*P_p \to T^*Q_p$ and this, together with the identity on the E and F components, induces a smooth map $r: L \to L'$ since (36.16) is a consequence of both equations in (36.15) and (36.17) is a consequence of the first equation in (36.15). We claim that r is, in fact, a diffeomorphism. Indeed, the full tangent space TP_p is spanned by TQ_p together with the images of elements of g and k via their respective actions. But equations (36.15) determine the value of $z \in T^*P_p$ on precisely these two types of tangent vectors, showing that r is an injective immersion. Conversely, given $y \in T^*Q_p$, and so defined as a linear function on the subspace TQ_p and given e and f, we can try to extend y to be defined on all of TP_p by (36.15). The consistency conditions for doing so are exactly (36.16) and (36.17) completing the proof that r is a diffeomorphism.

If Y is a smooth submanifold of a manifold W we have the injection map $i: T^*W_{|Y} \to T^*W$ and the restriction map $\rho: T^*W_{|Y} \to T^*Y$. If α_W and α_Y denote the canonical 1-forms of T^*W and T^*Y it follows from their definition that

$$i^*\alpha_W = \rho^*\alpha_Y.$$

From this observation it follows that the diffeomorphism r carries the presymplectic form on L' back to the presymplectic form on L. Passing to the quotient of the action of $C \times H$ shows that r induces a symplectic diffeomorphism between $(T^*P_E)_F$ and $T^*Q_{(E_F)}$. In Shnider and Sternberg (1983) it is shown how to use this dimensional reduction identification to solve G-invariant Hamiltonian systems.

37. The energy-momentum tensor and the current

In Section 36 we wrote down equations of motion of point particles in the presence of an external Yang–Mills field. In the present section we set up some formalism that allows us to discuss the behavior of continuous media. In particular, we shall write down the analogs of the equations expressing the conservation of charge and the Einstein identity involving the covariant divergence of the energy-momentum tensor when the electromagnetic field is replaced by an arbitrary Yang–Mills field. The justification for these equations will be given in the next section, and the reader might prefer to look at that section first, before delving into the formalism developed here. The material in this was, for the most part, first presented in Sternberg (1978).

We need to introduce some differential complexes associated to the bundle P and the connection Θ. We will be considering differential forms with values in the associated vector bundles $k(P)$ and $k^*(P)$.

Let E_1, E_2, and E_3 be vector bundles over the same manifold M and suppose that we are given a bilinear map $\beta: E_1 \times E_2 \to E_3$. Then the tensor product of β with exterior multiplication induces a map $E_1 \otimes \wedge^k T^*M \times E_2 \otimes \wedge^l T^*M \to E_3 \otimes \wedge^{k+l} T^*M$. Taking sections, we see that β induces an "exterior multiplication" of E_1-valued forms with E_2-valued forms to obtain E_3-valued forms. If B_1, B_2, and B_3 are vector spaces on which we are given linear representations of K, then each of them gives rise to associated bundles $E_1 = B_1(P)$, etc. A K-equivariant map from $B_1 \times B_2 \to B_3$ gives a $\beta: E_1 \times E_2 \to E_3$ and hence a corresponding exterior multiplication. We wish to use this construction for the following choices:

(i) $B_1 = B_2 = B_3 = k$ and the map of $k \times k \to k$ is given by Lie bracket;

(ii) $B_1 = k$ and $B_2 = B_3 = k^*$ and $k \times k^* \to k^*$ is given by the coadjoint representation of k acting on k^*;

(iii) $B_1 = k$, $B_2 = k^*$, $B_3 = \mathbb{R}$ and the map is evaluation.

Let $\bigwedge^r(M)$ denote the space of smooth scalar-valued exterior differential forms on M of degree r

$\mathscr{A}^r(M)$ the space of smooth $k(P)$-valued r-forms, i.e., of smooth sections of

$$g(P) \otimes \bigwedge^r T^*M$$

and

$\mathscr{A}^{*r}(M)$ the space of smooth $k^*(P)$-valued r-forms. Since k acts on itself via the adjoint representation and on k^* by the coadjoint representation, we get pairings

$$\mathscr{A}^r \times \mathscr{A}^l \to \mathscr{A}^{r+l} \text{ denoted by } \omega^r \otimes \omega^l \to [\omega^r, \omega^l],$$
$$\mathscr{A}^r \times \mathscr{A}^{*l} \to \mathscr{A}^{*(r+l)} \text{ denoted by } \omega^r \otimes \sigma^l \to \omega^r \# \sigma^l,$$

and since we have the evaluation map $k^* \times k \to \mathbb{R}$

$$\mathscr{A}^r \times \mathscr{A}^{*l} \to \bigwedge^{(r+l)} \text{ denoted by } \omega^r \otimes \sigma^l \to \omega^r \wedge \sigma^l.$$

The direct sum $\mathscr{A} = \oplus \mathscr{A}^r$ is a Lie superalgebra† under the bracket and it is represented on the space $\mathscr{A}^* = \oplus \mathscr{A}^{*r}$. Further, one has the identity

$$[\omega^p, \omega^q] \wedge \sigma^r = \omega^p \wedge (\omega^q \# \sigma^r).$$

We have seen in Section 35 that the "Lie algebra" aut (P) of Aut (P) consists of all smooth vector fields ξ on P satisfying $R_a^* \xi = \xi$ and that these can be regarded as sections of a vector bundle, which we now denote by $E(P)$. Furthermore we have the exact sequence of vector bundles

$$0 \to k(P) \xrightarrow{i} E(P) \xrightarrow{\dot{\pi}} T(M) \to 0.$$

Taking smooth sections of these bundles gives the sequence

$$0 \to \mathscr{A}^0(M) \to \operatorname{aut}(P) \to \mathscr{V}(M),$$

where $\mathscr{V}(M)$ denotes the algebra of all smooth vector fields on M. This shows that we can regard $\mathscr{A}^0(M)$ as gau (P), the "Lie algebra" of the gauge group Gau (P).

Giving a connection Θ on P is equivalent to giving a splitting of the sequence $0 \to k(P) \to E(P) \to T(M) \to 0$. That is, giving Θ is the same as

† See Corwen, Nèeman, and Sternberg (1975) for the basic properties of Lie superalgebras (which are called graded Lie algebras there).

giving two vector bundle maps $l: E(P) \to k(P)$ and $\lambda: T(M) \to E(P)$ that satisfy

$$li = \text{id} \qquad \text{and} \qquad \lambda\dot{\pi} = \text{id},$$

so we have the diagram

$$0 \to k(P) \overset{i}{\to} E(P) \overset{\dot{\pi}}{\to} T(M) \to 0.$$

(with $l: E(P) \to k(P)$ and $\lambda: T(M) \to E(P)$ drawn as curved arrows back)

Two connections, Θ and Θ' give us two maps λ and λ' that satisfy $\dot{\pi}(\lambda - \lambda') = 0$, so $\lambda - \lambda' = i\tau$, where $\tau: T(M) \to R(P)$ so $\tau \in \mathscr{A}^{-1}(M)$. Conversely, given Θ and $\tau \in \mathscr{A}^1(M)$, we get a new connection Θ'. Thus the set of all connections is an affine space associated to the linear space $\mathscr{A}^1(M)$. In particular, we may identify the space of infinitesimal connections," the "tangent space to the space of connections at Θ," with $\mathscr{A}^1(M)$.

An element J of $\mathscr{A}^{n-1*}(M)$ is called a current, where $n = \dim M$. Let $\mathscr{A}^1_0(M)$ denote the subspace of $\mathscr{A}^1(M)$ consisting of forms with compact support. Then, if M is oriented (as we shall assume, for simplicity from now on) we can regard $J \in \mathscr{A}^{n-1*}(M)$ as defining a linear function on $\mathscr{A}^1_0(M)$ by integration:

The value of J on $\tau \in \mathscr{A}^1(M)$ is given by

$$\langle J, \tau \rangle = \int_M J \wedge \tau.$$

A connection Θ defines covariant differentials, $d_\Theta: \mathscr{A}^r \to \mathscr{A}^{r+1}$ and $d_\Theta: \mathscr{A}^{*r} \to \mathscr{A}^{*r+1}$, whose definition is similar to the δ defined in Section 36. It is defined as follows: Let E be any vector space with a given linear representation of K on E. We can form the associated bundle $E(P)$ and consider the space $\wedge^r(M, E(P))$ consisting of r-forms on M with values in the vector bundle $E(P)$. We can identify the space $\wedge^r(M, E(P))$ with the set of E-valued r-forms $\tilde{\Omega}$ on P with the properties

$$i(v)\tilde{\Omega} = 0 \qquad \text{for any vertical vector field}$$

and

$$R_a^*\tilde{\Omega} = a \cdot \tilde{\Omega} \qquad \forall a \in K.$$

For any Ω in $\wedge^r(M, E(P))$ with corresponding form $\tilde{\Omega}$ on P, the form $d_\Theta\Omega$ in $\wedge^{r+1}(M, E(P))$ is the $(r+1)$-form corresponding to $\delta\tilde{\Omega}$. Taking $E = k$ or k^* we get $d_\Theta: \mathscr{A}^r \to \mathscr{A}^{r+1}$ and $d_\Theta: \mathscr{A}^{*r} \to \mathscr{A}^{*r+1}$. It is easy to check that

$$d(\omega^r \wedge \sigma) = d_\Theta\omega^r \wedge \sigma + (-1)^r\omega^r \wedge d_\Theta\sigma$$

for $\omega^r \in \mathscr{A}^r$ and $\sigma \in \mathscr{A}^*$.

Now let us consider an $s \in \mathscr{A}^0$. We can think of s as a vertical vector field on P that satisfies

$$R_a^* s = s.$$

Equally well, we can think of s as giving a k-valued function, call it $\tilde{s}: P \to g$ satisfying

$$R_a^* \tilde{s} = (\mathrm{Ad_a})\tilde{s}.$$

The relation between s and $\tilde{s}$ is clearly given by

$$\tilde{s} = i(s)\Theta$$

(for any connection form Θ).

Let D_s denote Lie derivative with respect to the vector field s. Then

$$\begin{aligned} D_s\Theta &= d(i(s)\Theta) + i(s)\, d\Theta \\ &= d\tilde{s} + i(s)\, d\Theta. \end{aligned}$$

We claim that for any vertical vector field s we have

$$i(s)\, d\Theta = [\tilde{s}, \Theta]_k$$

(where the bracket on the right is the bracket in k; i.e., $[\tilde{s}, \Theta](\xi) = [\tilde{s}, \Theta(\xi)]$ for any vector field ξ). To prove this equality, it is enough to prove it at every point, and since both sides depend purely algebraically on s, it is enough to prove it in the special case where $s = \hat{\xi}$, the vector field on P coming from the right action of $\xi \in k$ on P. Now

$$i(\hat{\xi})\Theta = \xi, \qquad \text{a constant element of } k,$$

so

$$d(i(\hat{\xi})\Theta) = 0,$$

while

$$D_{\hat{\xi}}\Theta = [\xi, \Theta].$$

But $D_{\hat{\xi}}\Theta = di((\hat{\xi})\Theta) + i(\hat{\xi})\, d\Theta$ proving that $i(\hat{\xi})\, d\Theta = [\xi, \Theta]$. Thus

$$D_s\Theta = d\tilde{s} + [\tilde{s}, \Theta].$$

But for any element of $\mathscr{A}$ it is easy to check that $d_\Theta\omega = d\omega + [\omega, \Theta]$. Thus we have proved that

$$D_s\Theta = d_\Theta s.$$

Now suppose that s has a compact support. Then

$$\begin{aligned} J \wedge D_s\Theta &= J \wedge d_\Theta s \\ &= d(J \wedge s) - d_\Theta J \wedge s \end{aligned}$$

and so

$$\int J \wedge D_s\Theta = -\int d_\Theta J \wedge s.$$

In the next section we shall present a geometrical argument which suggests that a reasonable condition to impose on a current J is that it be orthogonal to all $\tau \in \mathscr{A}_0^1(M)$ of the form $\tau = D_s\Theta$ for s of compact support. From the above equation we see that this is equivalent to the condition

$$d_\Theta J = 0,$$

This equation is the generalization to the Yang–Mills case of the equation $dJ = 0$, which asserts the conservation of charge in electromagnetic theory.

It is a well-known theorem of Levi-Civita (see Sternberg, 1983) that a (pseudo) Riemman metric on M induces a (unique torsionless) connection on the tangent bundle $T(M)$ and hence on all tensor bundles. The corresponding exterior covariant differentiation operator will be denoted by $\hat{d}$. It maps any smooth section of $R(M)$ into a smooth section of $R(M) \otimes T^*(M)$, where $R(M)$ is any tensor bundle.

In the theory of general relativity, the matter-energy-momentum tensor is required to satisfy a condition known as Einstein's identity on the vanishing of its covariant divergence. If the local coordinates of the tensor T are T^{ij}, this condition is usually written as $T^{ik}_{\ |k} = 0$, where the $T^{ij}_{\ |k}$ denote the components of the covariant differential of T, and the summation convention is used. We can write this equation in coordinate-free notation as follows: The metric g allows us to identify $T^*(M) \otimes T^*(M)$ with $T^*(M) \otimes T(M)$. Let us denote the section of $T^*M \otimes TM$ corresponding to T by $\bar{\bar{T}}$. The volume form on M determined by g gives an identification of TM with $\wedge^{n-1} T^*M$, where $n = \dim M$. Thus a section of $T^*M \otimes TM$ corresponds to a section of $T^*M \otimes \wedge^{n-1} T^*M$; the section corresponding to $\bar{\bar{T}}$ will be denoted by $\bar{T}$. Thus the symmetric tensor field T corresponds to a section of $T^*M \otimes \wedge^{n-1} T^*M$. Now covariant differentiation induces a covariant exterior derivative $\hat{d}$ from sections of $T^*M \otimes^k T^*M$ to sections of $T^*M \otimes \wedge^{k+1} T^*M$. It is easy to check that the Einstein identity is equivalent to $\hat{d}\bar{T} = 0$ as a section of $T^*M \otimes \wedge^n T^*M$. In any event, this is the form in which we shall use this condition.

Let V be a vector field on M. We can form the Lie derivative $D_V g$ of the metric g with respect to the vector field V, so that $D_V g$ is a symmetric tensor field. In particular, we can take the scalar product (pointwise) of T with $D_V g$ to obtain a function $T \cdot D_V g$. On the other hand, if S is a section of $T^*M \otimes \wedge^n T^*M$ we can contract the T^*M component with the vector field V so as to obtain $S \cdot V$, which is a section of $\bigwedge^n T^*M$. We wish to prove the following fact: For any symmetric tensor field T and any vector field V of compact support we have

$$\int \tfrac{1}{2} T \cdot D_V g(\text{vol}) = -\int (\hat{d}\bar{T}) \cdot V,$$

where (vol) denotes the volume form associated to the metric.

Let V be any vector field on M and $\check{V}$ the linear differential form associated to V by the metric, thus, in local coordinates, if

$$V = \sum_i V^i \frac{\partial}{\partial x^i},$$

then

$$\check{V} = \sum_j V_j \, dx^j,$$

where

$$V_j = \sum_i g_{ij} V^i.$$

Let $\mathcal{S}$ denote the symmetrization operator on $T^* \otimes T^*$ so $\mathcal{S}(u \otimes v) = u \otimes v + v \otimes u$. We claim that the Lie derivative of g with respect to V, $D_V g$, is given by

$$D_V g = \mathcal{S}(\hat{d}\check{V}).$$

Indeed, in local coordinates we have

$$D_V g = \sum \left(\frac{\partial g_{ij}}{\partial x^k} V^k \, dx^i \, dx^j + g_{ij} \frac{\partial V^i}{\partial x^k} dx^k \, dx^j + g_{ij} \frac{\partial V^j}{\partial x^k} dx^i \, dx^k \right).$$

Let us choose normal coordinates so that at a given point we have

$$\frac{\partial g_{ij}}{\partial x^k} = 0$$

and

$$\begin{aligned} \hat{d}(V) &= \hat{d}\left(\sum g_{ij} V^i dx^j\right) \\ &= \sum g_{ij} \frac{\partial V^i}{\partial x^k} \, dx^k dx^j. \end{aligned}$$

Comparing the two preceding equations establishes our formula.

Let T be a symmetric tensor field, that is, a section of $S^2(T(M))$. We can write

$$\tfrac{1}{2} T \cdot D_V g = T \cdot \hat{d}\check{V}.$$

Let $\bar{\bar{T}}$ denote the section of $T(M) \otimes T^*(M)$ equivalent to T under the isomorphism of $T(M)$ with $T^*(M)$ determined by the metric. Then we can write the preceding equation as

$$\tfrac{1}{2}T \cdot D_V = \bar{\bar{T}} \cdot \hat{d}V$$

(since $\hat{d}V$ as section of $T(M) \otimes T^*M$ corresponds to $\hat{d}\check{V}$). Finally let $\bar{T}$ denote the section of $T^*(M) \otimes \wedge^{n-1} T^*(M)$ defined by $\bar{T} = \bar{\bar{T}}(\text{vol})$. Thus

$$\tfrac{1}{2}\bar{\bar{T}} \cdot D_V g(\text{vol}) = \bar{T} \wedge \hat{d}V.$$

Now $\hat{d}T$ is a section of $T^*M \otimes \wedge^n T^*M$, and hence $\hat{d}T \cdot V$ is an n-form, and we have

$$\hat{d}\bar{T} \cdot V + \bar{T} \wedge \hat{d}V = d(\bar{T} \cdot V).$$

Thus, if V has compact support we can write

$$\int (\tfrac{1}{2}\bar{T} \cdot D_V g)(\text{vol}) = -\int \hat{d}\bar{T} \cdot V.$$

In general relativity, in the absence of an electromagnetic field, the Einstein identity concerning the vanishing of the covariant divergence of the energy-momentum can be formulated, as we have seen, as $\hat{d}\bar{T} = 0$. We can think of an infinitesimal variation υ in the pseudo-Riemannian metric g as being a section of $S^2 T^*M$, and we can think of the symmetric tensor T as defining a linear function on the space of symmetric tensor fields of compact support by setting

$$\langle T, \upsilon \rangle = \int \tfrac{1}{2} T \cdot \upsilon(\text{vol}).$$

We see from the above discussion that the Einstein identity $\hat{d}\bar{T} = 0$ is equivalent to the assertion that

$$\langle T, D_V g \rangle = 0$$

for all vector fields V of compact support. In the presence of an electromagnetic field the Einstein identity is modified so as to read

$$\hat{d}\bar{T} + i(\hat{J})F(\text{vol}) = 0.$$

Here J is an $(n-1)$-form and $\hat{J}$ denotes the vector field corresponding to J by the relation $i(\hat{J})(\text{vol}) = J$. Since F is a 2-form, the interior product $i(\hat{J})F$ is a 1-form and thus $i(\hat{J})F(\text{vol})$ is a section of $T^*M \otimes \wedge^n T^*M$ as is $\hat{d}\bar{T}$. When we replace the electromagnetic field by an arbitrary Yang–Mills field, the same equation makes sense, where now F is a section of

$k(P) \otimes \wedge^2 T^*M$ and $J \in \mathscr{A}^{n-1*}(M)$ so that $\hat{J}$ is a section of $k(P) \otimes TM$. Thus the contraction or "interior product" $i(\hat{J})F$ makes sense, and we can write the preceding equation with this new interpretation.

We can also describe the equation $\hat{d}\bar{T} + i(J)F(\text{vol}) = 0$ as an orthogonality condition. Indeed, for any vector field V of compact support on M, let $\tilde{V}$ denote the horizontal lifting of V to P. Then $\tilde{V}$ is an element of aut (P) and thus $D_{\tilde{V}}\Theta = \tau_V \in \mathscr{A}^1(M)$ is an "infinitesimal condition." We claim that

$$J \wedge \tau_V = (-i(\hat{J})F \cdot V)(\text{vol}).$$

Indeed, $D_{\tilde{V}}\Theta = i(\tilde{V})\,d\Theta + di(\tilde{V})\Theta = i(\tilde{V})\,d\Theta$ since $i(\tilde{V})\Theta = 0$ because $\tilde{V}$ is horizontal. But for any vector fields V and W on M, $F(V, W)$ is that section of $k(P)$ corresponding to the k-valued function $d\Theta(\tilde{V}, \tilde{W})$. From this it is easy to see that the preceding formula holds.

Now let us consider the pair (T, J) as defining a linear function on the space of all pairs (υ, τ) where υ is a compactly supported section of S^2T^*M and $\tau \in \mathscr{A}^2(M)$ by

$$\langle (T, J), (\upsilon, \tau) \rangle = \langle T, \upsilon \rangle + \langle J, \upsilon \rangle = \int \tfrac{1}{2} T \cdot \upsilon(\text{vol}) + \int J \wedge \tau.$$

For any vector field V on M let us set

$$D_{\tilde{V}}(g, \Theta) = (D_V g, D_{\tilde{V}}\Theta).$$

Then the preceding computations show that for V of compact support we have

$$\langle (T, J), D_{\tilde{V}}(g, \Theta) \rangle = -\int \big(\hat{d}\bar{T} + i(\hat{J})F(\text{vol})\big) \cdot V.$$

Thus the condition that (T, J) be orthogonal to all $D_{\tilde{V}}(g, \Theta)$ is equivalent to the condition $d\bar{T} + \big(i(J)\big)F(\text{vol}) = 0$.

In the case of general relativity, it was pointed out by Einstein, Hoffman, and Infeld (1938) that the equations of motion of a particle can be derived as a limiting form of the equation $d\bar{T} = 0$ if we let T approach a generalized tensor field concentrated along a curve. In other words, we can get the equations for a point particle if the energy-momentum tensor becomes concentrated into a point mass (which in space-time will be represented by its worldline, which is a curve). The same holds true in the Yang–Mills case. But, following Souriau (1974), it is more instructive to show how the equations

$$d_\Theta J = 0$$

and

$$\hat{d}\bar{T} + i(J)F(\text{vol}) = 0$$

of the present section and the equations of motion described in Section 36 both derive from a common geometrical principle, which we shall explain in the next section.

38. The principle of general covariance

The goal of this section is to derive the equations of the past three sections from the principle of general covariance, as described at the end of Chapter I. For the convenience of the reader, we repeat the main points.

Let $\mathscr{G}$ be a group acting on a space $\mathfrak{X}$. In what follows, we shall pretend that $\mathscr{G}$ is a Lie group acting on a manifold $\mathfrak{X}$; but in the applications that we have in mind, the group $\mathscr{G}$ will be infinite-dimensional (for example, $\mathscr{G} = \text{Aut}(P)$) and so will $\mathfrak{X}$ (for example, $\mathfrak{X}$ might be the space of all pairs consisting of connections on P and metrics on M). By a *covariant theory* we shall mean a (smooth) function $\mathscr{F}$ defined on $\mathfrak{X}$, that is invariant under the action of $\mathscr{G}$, so

$$\mathscr{F}(\mathfrak{a}\mathscr{X}) = \mathscr{F}(\mathscr{X}) \qquad \mathfrak{a} \in \mathscr{G} \qquad \text{and} \qquad \mathscr{X} \in \mathfrak{X}.$$

Let $\mathscr{X}$ be a point of $\mathfrak{X}$ and let $\mathscr{B} = \mathscr{G} \cdot \mathscr{X}$ be the orbit through $\mathscr{X}$. Let μ be an element of $T^*\mathfrak{X}_{\mathscr{X}}$ so that μ is a linear function on the tangent space $T\mathfrak{X}_{\mathscr{X}}$ to $\mathfrak{X}$ at the point $\mathscr{X}$ (see Figure 38.1).

If $\mu = d\mathscr{F}_{\mathscr{X}}$ then it clear that μ vanishes when evaluated on all vectors tangent to $\mathscr{B}$; that is:

$$\mu \in T\mathscr{B}^0_{\mathscr{X}}. \tag{I}$$

This equation is, as we shall see, the prototype of generalized Einstein identities written at the end of the preceding section and the various equations of motion derived in the preceding paragraphs.

Now let us suppose that the tangent bundle $T\mathfrak{X}$ has a given global trivialization. Thus we have a definite identification of all the various tangent spaces $T\mathfrak{X}_{\mathscr{X}}$ with some fixed vector space Z. Then given $\mu \in Z^*$, we can try to find a $\mathscr{X}$ such that

$$d\mathscr{F}_{\mathscr{X}} = \mu. \tag{S}$$

This equation for $\mathscr{X}$ is, as we shall see, the prototype of the various "source equations" or "field equations." Naturally, (I) is a necessary condition for (S) so the "identities" are always a consequence of the field equations. Since we will be dealing with infinite-dimensional spaces, it will sometimes be

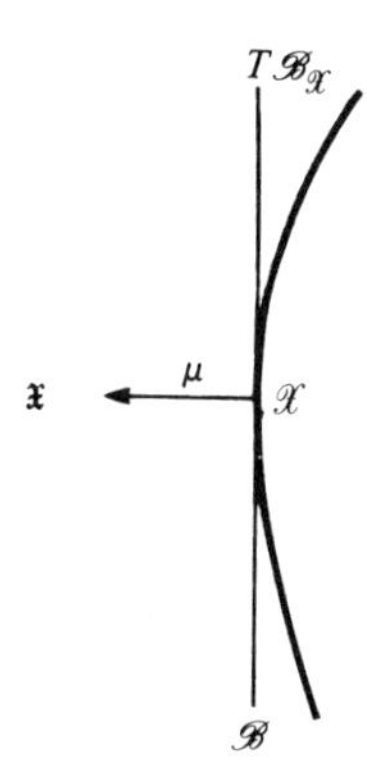

Figure 38.1.

necessary to consider situations where μ will not be defined on all of Z but only on some subspace Z_0, which contains all the $T\mathscr{B}_{\mathscr{X}}$ so that (I) still makes sense. Similarly, it may be that $\mathscr{F}$ is not defined on all of $\mathfrak{X}$, and yet (S) makes good sense where both sides are regarded as linear functions on Z_0. The extraction of information from (I) usually involves some integration by parts.

As the details of the integration by parts calculation can get somewhat involved, let us begin with an example that illustrates the method, and where the integration by parts is transparent. Let X be a symplectic manifold with symplectic form ω. Let $\mathscr{G}$ be the group of symplectic diffeomorphisms with compact support. Thus $\mathscr{G}$ consists of diffeomorphisms φ of X that satisfy $\varphi^*\omega = \omega$ and that equal the identity outside some compact subset. The "Lie algebra" of $\mathscr{G}$ consists of Hamiltonian vector fields of compact support. We shall assume, for simplicity, that the topology of X is such that every Hamiltonian vector field is of the form ξ_f, where $i(\xi_f)\omega = df$ where f is a smooth function on X. Then, for any function h on X, we have $D_{\xi_f}h = \{h, f\}$, where $\{,\}$ denotes Poisson bracket. We can identify the "Lie algebra" of $\mathscr{G}$ with the space of smooth functions of compact support under Poisson bracket. We will take $\mathfrak{X}$ to be the vector space consisting of all smooth functions on X. A point $\mathscr{X}$ of $\mathfrak{X}$ is just a smooth function H and since $\mathfrak{X}$ is a vector space, we may identify all the tangent spaces $T\mathfrak{X}_H$ with $\mathfrak{X}$ itself. Thus we take $Z = \mathfrak{X}$ and take Z_0 to consist of smooth functions of compact support. We can write the most general continuous linear function μ on Z_0 as

$$\langle \mu, f \rangle = \int uf\omega^n,$$

where $\dim X = 2n$ so that ω^n denotes Liouville measure on X and where u is a generalized function on X. The "tangent space to the orbit" $T\mathscr{B}_H$ consists of all $D_{\xi_f}H = \{f, H\}$, where f has compact support. Condition (I) becomes

$$\int u\{f, H\}\omega^n = 0$$

for all f of compact support. Since $D_{\xi_H}\omega^n = 0$, this last expression is the same as $-\int \{u, H\} f\omega^n$. As this is to vanish for all f of compact support, we see that (I) is equivalent to

$$\{u, H\} = 0,$$

that is, that u be invariant under the flow generated by H. For example, if u is a generalized function concentrated on a curve, this curve must be a trajectory of H.

Complications come in when we want to apply this argument with a "smaller" group or rather a different group replacing the group of symplectic diffeomorphisms. For example, suppose that $X = T^*M$ and we want to take $\mathscr{G}$ to be $\mathrm{Diff}_0(M)$. Each diffeomorphism of M induces a diffeomorphism of T^*M and hence acts on the space of functions on $X = T^*M$. But these induced symplectic diffeomorphisms will not have compact support vertically. We could rectify this by requiring μ to have compact support in the vertical direction. A more serious problem is that the Lie algebra of $\mathscr{G}$ is too small. The functions f on T^*M that correspond to vector fields on M are all linear in p. Thus, to say that $\{u, H\}$ is orthogonal to all such f is not much of a condition in general. On the other hand, if we are given that $\{u, H\}$ is of the form $F(q)[\partial\delta(p)/\partial p]$, then $\int\{u, H\}f(q)p\,dp\,dq = \int F(q)f(q)\,dq$. So if this is to vanish for all f of compact support, we conclude that $\{u, H\} = 0$. Now $\{u, H\}$ will have this desired form if u is of the form $u = T(q)[\partial^2\delta(p)/(\partial p)^2]$, where T is a symmetric tensor field on M and if H is a quadratic function of p. If H is the energy function associated to a non-degenerate pseudo-Riemannian metric g on M, then it is not hard to verify that the condition $\{u, H\} = 0$ is equivalent to the Einstein condition $\hat{d}_g T = 0$. We will, in fact, derive the condition $d_g T = 0$ directly from (I) by taking $\mathscr{G}$ to be $\mathrm{Diff}_0(M)$ and $\mathfrak{X}$ to be the space of all (pseudo-) Riemannian metrics on M.

Let us give another illustration of the principle of general covariance, this time a "quantum mechanical" version, but in a purely formal sense. Suppose we take $\mathscr{G}$ to be the group of all unitary operators on some Hilbert space, and $\mathfrak{X}$ to be the space of all self-adjoint operators on the same Hilbert space, with $\mathscr{G}$ acting on $\mathfrak{X}$ by conjugation. Since $\mathfrak{X}$ is a linear space, we can identify the tangent space to at any point with $\mathfrak{X}$ itself. The tangent space to the orbit through H is, at least formally, the set of all commutators $XH - HX$ as X ranges over the space of all skew-adjoint operators. In a formal sense, we can regard μ as being a trace class self-adjoint operator F, with $\mu(A) = \operatorname{tr} Z \cdot A$. Then condition (I) becomes

$$[H, F] = 0.$$

It would be interesting to speculate whether one can give a quantum version of the preceding discussion involving the passage from the full symplectic group on T^*M to Diff M. That is, take the Hilbert space to be $L^2(M)$ upon which Diff M acts by unitary transformations, and replace the full unitary group by the image of Diff M.

We now describe the data associated to a Yang–Mills theory. Let $\pi: P \to M$ be a principal bundle with structure group G. We take $\mathscr{G} = \operatorname{Aut}_0(P)$, the group of all automorphisms φ of P that equal the identity outside $\pi^{-1}C_\varphi$ where C_φ is a compact subset of M. We have a homomorphism $\operatorname{Aut}_0(P) \to \operatorname{Diff}_0(M)$ sending any $\varphi \in \mathscr{G} = \operatorname{Aut}_0(P)$ into the transformation $\bar{\varphi}$ that it induces on M. We let $\mathfrak{X}$ consist of all pairs $\mathscr{X} = (g, \Theta)$, where g is a pseudo-Riemannian metric on M (of suitable signature) and Θ is a connection on P. The group $\mathscr{G}$ acts on $\mathfrak{X}$ by

$$\varphi \cdot (g, \Theta) = (\bar{\varphi}^{-1*} g, \varphi^{-1*}\Theta).$$

The tangent space $T\mathfrak{X}_{\mathscr{X}}$ at any point $\mathscr{X}$ can be identified with $Z = \Gamma S^2 T^*M \oplus \mathscr{A}^1(M)$, as we have seen, and we shall take $Z_0 = \Gamma_0 S^2 T^*M \oplus \mathscr{A}^1_0(M)$. A linear function μ on this space will thus be a pair $\mu = (T, J)$, where T is a "generalized tensor field" and J is a "generalized element of $\mathscr{A}^{n-1*}(M)$." If T is a smooth tensor field and J is an actual element of $\mathscr{A}^{n-1*}(M)$ (i.e., J is a smooth section of $k^*(P) \otimes \wedge^{n-1} T^*M$), then we say that $\mu = \mu_{T,J}$ is smooth and we have

$$\langle \mu_{T,J}, (\upsilon, \tau) \rangle = \int_M \left(\tfrac{1}{2} T \cdot \upsilon (\mathrm{vol})_g + J \wedge \tau\right).$$

The "Lie algebra" of $\mathscr{G}$ is $\operatorname{aut}_0(P)$, the set of all vector fields ξ on P such that $R_a^* \xi = \xi$ and $\xi = 0$ outside $\pi^{-1}(C)$ for some compact subset C of M.

Any vector field ξ in aut(P) induces a vector field $\bar{\xi}$ on M (by projection). The "tangent space to the orbit" $T\mathscr{B}_{(g,\Theta)}$ consists of all (υ, τ) satisfying

$$\upsilon = D_{\bar{\xi}} g, \tau = D_\xi \Theta \qquad \text{for some } \xi \in \operatorname{aut}(P).$$

We can write every $\xi \in \operatorname{aut}_0(P)$ as a sum of its vertical and horizontal components,

$$\xi = \xi_{\text{vert.}} + \xi_{\text{horiz.}}$$

both belonging to $\text{aut}_0(P)$ with

$$\xi_{\text{horiz.}} = \tilde{\bar{\xi}},$$

the horizontal lift of the vector field $\bar{\xi}$ on M. As we have seen in the preceding section, if $\mu = \mu_{(T,J)}$ is smooth, then the condition that μ be orthogonal to all $D_{\xi_{\text{horiz.}}}(g, \Theta) = (D_{\bar{\xi}} g, D_{\tilde{\bar{\xi}}}\Theta)$ is the same as $\hat{d}\bar{T} + i(\tilde{J})F(\text{vol})_g = 0$. Now $D_{\xi_{\text{vert.}}}(T, J) = (0, D_{\xi_{\text{vert.}}}\Theta)$ and we have also seen in the last section that the condition that J be orthogonal to all $D_{\xi_{\text{vert.}}}\Theta$, which is the same as that μ be orthogonal to all $D_{\xi_{\text{vert.}}}(g, \Theta)$ is equivalent to the condition $d_\Theta J = 0$. Thus for smooth μ the condition (I) is equivalent to the Einstein–Yang–Mills identities

$$d_\Theta J = 0 \qquad \text{and} \qquad \hat{d}\bar{T} + i(\hat{J})F(\text{vol})_g = 0.$$

Now suppose that μ is given by a smooth section T of S^2T^*M and a smooth section $\tilde{J}$ of $k^*(P) \otimes T(M)$ concentrated along a curve γ so that

$$\langle \mu, (\upsilon, \tau) \rangle = \int_\gamma \left(\tfrac{1}{2} T \cdot \upsilon + \hat{J} \cdot \tau\right) ds,$$

where $\hat{J} \cdot \tau$ denotes the obvious contraction of a section of $k^*(P) \otimes T(M)$ with a section of $g(P) \otimes T^*(M)$. We shall now show that for such a μ, condition (I) implies the following: (a) that $J(s) = l(s) \otimes \gamma'(s)$, where γ' denotes the tangent vector to γ at $\gamma(s)$ and l is a section of $k^*(P)$; (b) that there is one fixed orbit $\mathcal{O}$, in k^* so that l is a section (over γ) of the subbundle $\mathcal{O}(P)$; (c) that $T(s) = p(s) \otimes p(s)$; and (d) that with suitable parametrization $q(s)$ of γ the curve $(q(s), p(s), l(s))$ is a solution curve of the equations of Section 36 with the Hamiltonian $H = \frac{1}{2}\|p\|^2$ determined by the metric g. Thus the law of "conservation of internal quantum number" and equations (36.11) and (36.12) are consequences of (I).

It will be convenient to work in a fixed trivialization $P_{|U} = U \times K$ over some neighborhood $U \subset M$. Let s denote the corresponding section, so that $S(x) = (x, e)$ in terms of the local product description. Any element φ of $\text{Diff}_0(U)$ induces an element of $\text{Aut}_0(P)$ sending (x, a) into $(\varphi(x), a)$, which we will denote by $\tilde{\varphi}$. It is clear that $\tilde{\varphi} \circ s = s \circ \varphi$ and, hence, if

$$A = s^*\Theta$$

then

$$\varphi^* A = s^* \tilde{\varphi}^* \Theta.$$

Let $\mathcal{F}(U,K)$ denote the space of functions from U to K, and $\mathcal{F}_0(U,K)$ those functions ψ that satisfy $\psi(x) \equiv e$ outside a compact subset of U. Each such function ψ determines an element $\hat{\psi} \in \mathrm{Aut}_0(P)$ given by

$$\hat{\psi}(x,a) = (x, \psi(x)a).$$

Then

$$s^*\hat{\psi}^*\Theta = s_\psi^*\Theta,$$

where s_ψ is the section of $U \times K$ given by $s_\psi(x) = (x, \psi(x))$. But

$$s_\psi^*\Theta = \mathrm{Ad}\,\psi(\Theta) + \psi^*(\theta_K),$$

where θ_K is the left-invariant form on K that identifies TK_a with $TK_e = k$. In short, we see that we have identified the semidirect product $\mathrm{Diff}_0(U) \times \mathcal{F}_0(U,K)$ with a subgroup of $\mathrm{Aut}_0(P)$, and the action of this group on the space of all connection forms A is determined by

$$(\phi, A) \rightsquigarrow \phi^*A \qquad \text{and} \qquad (\psi, A) \rightsquigarrow \psi^*(\theta_K) + \mathrm{Ad}\,\psi(A).$$

The "Lie algebra" of the group $\mathrm{Diff}_0\, U + \mathcal{F}_0(U,K)$ consists of all (V,r), where V is a compactly supported vector field on U and r is a compactly supported function from U to g. The infinitesimal version of the preceding equations gives the action of this Lie algebra on the space of (local) connection forms as

$$V \cdot A = D_V A \qquad \text{(Lie derivative)}$$

and

$$r \cdot A = dr + [r, A].$$

Within the space $\mathscr{e}$ we can consider the subset $\mathscr{e}_U$ consisting of those $(\bar{g}, \bar{\Theta})$, where $\bar{g} = g$ and $\bar{\Theta} = \Theta$ outside U. We may then identify the "big tangent space" $T(\mathscr{e}_U)_w$ with the set of all pairs (β, B) and the tangent space to the orbit with the set of all

$$\begin{aligned} \beta &= D_V g, \\ B &= D_V A + dr + [r, A]. \end{aligned}$$

We can (locally) write the covector μ as $\mu = (T, J)$, where J is a generalized section of $k^* \otimes T(M)$ and where T is a generalized section of $S^2T^*(M)$.

Let us suppose that the element ξ of $\mathrm{aut}_0(P)$ is supported over some neighborhood U on which we have given a section of $P_{|U}$. Then we can write $\xi = V + v$, where V is a vector field on M and v is a function from M to g. Then the "tangent vector to the orbit" corresponding to ξ is given as

$(D_V g, D_V A + dv + [v, A])$. Suppose that T and J are smooth measures along some curve Γ of M. Let us choose $V = 0$ and $v = \phi\bar{v}$, where ϕ vanishes on Γ. Then the condition (I) reduces to

$$\int \langle J \cdot \bar{v}, d\phi \rangle = 0,$$

where $J \cdot \bar{v}$ is a section of $T(M)$ along Γ since v is a k-valued function and J is a section of $k^* \times T(M)$. But for this to hold for all ϕ vanishing on Γ implies that $J \cdot \bar{v}$ is tangent to Γ. Since this is to hold for all $\bar{v}$, we see that we can write $J(s) = \gamma(s) \times \Gamma'(s)$, where γ is a k^*-valued function. Taking $V = 0$ but v now arbitrary, we get

$$\int (\gamma \times \Gamma') \cdot (dv + [v, A])\, ds = 0 \qquad \forall v.$$

We can write this last integral as

$$\int \gamma \cdot \left(\frac{dv}{ds} + [v, A(\Gamma'(s))] \right) ds = 0,$$

or, integrating by parts

$$\int (\gamma' + A(\Gamma'(s))\gamma) \cdot v\, ds = 0.$$

Since this is to hold for all v, we conclude that

$$\gamma' = -A(\Gamma'(s))\gamma,$$

that is, that γ is determined by parallel transport along Γ, and, in particular that $\gamma(s)$ lies on a fixed coadjoint orbit. We may now take $v = 0$ and look at the equations coming from V. Since

$$(D_V g)^{\mu\nu} = \sum \left(V^i \frac{\partial g^{\mu\nu}}{\partial q^i} - 2 \frac{\partial V^\mu}{\partial q^i} g^{i\nu} \right)$$

and

$$(D_V A)_\mu = \sum \left(\frac{\partial A_\mu}{\partial q^i} V^i + A_i \frac{\partial V_i}{\partial q^\mu} \right),$$

our condition becomes

$$\sum \int_\Gamma T_{\mu\nu} \left(V^i \frac{\partial g^{\mu\nu}}{\partial q^i} - 2 \frac{\partial V^\mu}{\partial q^i} g^{i\nu} \right) + J^\mu \cdot \left(\frac{\partial A_\mu}{\partial q^i} V^i + A_i \frac{\partial V_i}{\partial q^\mu} \right) = 0. \quad (38.1)$$

We first take $V = \phi\bar{V}$, where ϕ is a function vanishing along Γ. Since $J = \gamma \times \Gamma'$, the contribution of the second term disappears and we conclude that the vector b_μ with components $T_{\mu\nu}g^{\nu j}$ is tangent to Γ for each μ. So we can write

$$\sum T_{\mu\nu}g^{\nu j} = p_\mu \frac{dq^j}{ds},$$

or, by the symmetry of T,

$$T_{\mu\nu} = p_\mu p_\nu, \qquad \text{where} \qquad p_\mu = \sum g_{\mu j}\frac{dq^j}{ds}.$$

This proves (36.11). Substituting the known form of T and J into (38.1) gives

$$0 = \sum \int_\Gamma \frac{1}{2}p_\mu p_\nu\left(V^i\frac{\partial g^{\mu\nu}}{\partial q^i} - 2\frac{\partial V^\mu}{\partial q^i}g^{i\nu}\right) + \gamma\cdot\left(\frac{dq^\mu}{ds}\frac{\partial A_\mu}{\partial q^i}V^i + A_i\frac{dV^i}{ds}\right).$$

Integration by parts (and the fact that V has compact support) gives

$$0 = \sum \int \left\{\frac{1}{2}p_\mu p_\nu\frac{\partial g^{\mu\nu}}{\partial q^i} + \frac{dp_i}{ds} + \gamma\cdot\left(\frac{dq^\mu}{ds}\frac{\partial A_\mu}{\partial q^i} - \frac{dA_i}{ds}\right) - \frac{d\gamma}{ds}\cdot A_i\right\}V^i.$$

As this holds for all i, we get (36.12). We have thus proved our desired result.

In the preceding discussion we identified the (generalized) symmetric tensor field T and the (generalized) element J of $\mathscr{A}^{n-1*}(M)$ with the energy-momentum tensor and the current density, respectively. The only real justification we gave for these identifications was that the generalized Einstein identities hold and, in particular, if T and J are concentrated along a curve, we could identify this curve with the worldline of a point particle. From the point of view of physical interpretation, we would like to be able to associate conserved quantities to T and J. For the case of point particles, we showed in the preceding section that (I) implies conservation of total internal quantum number; that is, that we are dealing with the Hamiltonian equations of motion associated to a fixed orbit in k^*. For Hamiltonian systems we know how to associate conservation laws to symmetries of the system. We would like to have a similar procedure for continuous T and J. Since we have not derived the equations of our theory from a Lagrangian, we do not have Noether's theorem at our disposal to relate infinitesimal symmetries to conservation laws. We will instead, following Souriau (1974), use the identity (I) directly to relate infinitesimal symmetries to conserved quantities. (It would be interesting and useful to analyze the relation between Souriau's procedure and that of Noether.)

We will suppose that our space-time manifold M is so constructed as to admit some global time variable (i.e., some function $t: M \to \mathbb{R}$), so that it makes sense to talk of a "cosmic future," consisting of all points in M for which $t > b$ and a "cosmic past," consisting of all points where $t < a$. We will let $Z_{a,b}$ denote the set of (ν, τ) such that supp ν and supp τ are both contained in $t^{-1}[a, b]$, where $a < b$. Thus $Z_{a,b}$ consists of those "infinitesimal variations" of (g, Θ) that are temporally limited to the interval $a \leqslant t \leqslant b$. We have $Z_0 \subset Z_{a,b} \subset Z$. We shall assume that our linear function μ is "spatially compactly supported"; that is, we assume that supp $\mu \cap t^{-1}[a, b]$ is compact for any interval $[a, b]$ of time. Then μ, which was initially defined as a linear functional on Z_0, can be unambiguously extended so as to be a linear functional on $Z_{a,b}$ for any $[a, b]$.

Now let ξ be an element of aut(P) such that $D_\xi(g, \Theta) = 0$. Thus ξ is an infinitesimal symmetry of (g, Θ) (and, in particular, the vector field $\bar{\xi}$ on M is a Killing vector field for the metric g). Let $\lambda : M \to \mathbb{R}$ such that $\lambda = 0$ for $t < a$ and $\lambda = 1$ for $t > b$. Then we can form the vector field $\lambda\xi$, which also belongs to aut(P) and consider the variation $D_{\lambda\xi}(g, \Theta)$. Since $\lambda\xi = 0$ for $t < a$, we see that $D_{\lambda\xi}(g, \Theta) = 0$ for $t < a$; and since $\lambda\xi = \xi$ for $t > b$, we see that $D_{\lambda\xi}(g, \Theta) = 0$ for $t > b$. Thus $D_{\lambda\xi}(g, \Theta) \in Z_{a,b}$ and we can evaluate μ on $D_{\lambda\xi}(g, \Theta)$; that is we can form the quantity $\langle \mu, D_{\lambda\xi}(g, \Theta) \rangle$.

Note that since $\lambda\xi$ does not necessarily belong to aut$_0$(P), it does not follow from (I) that this expression vanishes. However, we claim that it *does* follow from (I) that $\langle \mu, D_{\lambda\xi}(g, \Theta) \rangle$ is independent of λ, a, and b. Indeed, by definition $\langle \mu, D_{\lambda\xi}(g, \Theta) \rangle = \langle \mu, D_{\varphi\lambda\xi}(g, \Theta) \rangle$ where φ is any smooth function that is identically 1 on the support of μ and that is spatially compact. If we made a second choice φ' and λ', then $\lambda\varphi - \lambda'\varphi'$ has compact support and hence $\lambda\varphi\xi - \lambda'\varphi'\xi$ belongs to aut$_0$(P). Hence, it follows from (I) that $\langle \mu, D_{\lambda\xi}(g, \Theta) \rangle = \langle \mu, D_{\lambda'\xi}(g, \Theta) \rangle$. Let us denote $\langle \mu, D_{\lambda\xi}(g, \Theta) \rangle$ by $\mu(\xi)$. The assertion that is independent of λ can be construed as a conservation law. In fact, we may choose λ to pass from 0 to 1 in an arbitrarily small neighborhood of some space like hypersurface S_0 (for example, we might choose S_0 as given by $t = t_0$). If we replace S_0 by some other space like surface S_1 (say $t = t_1$) we get the same value for $\mu(\xi)$. We thus get conservation under arbitrary time displacements.

For example, let ζ be an element of the center of the Lie algebra k and take $\xi = \zeta_P$, the infinitesimal generator of the one-parameter group consisting of right multiplication by the group generated by ζ. Then $\bar{\xi} = 0$ and $D_\xi \Theta = 0$ for any connection Θ. Thus ζ gives rise to a conserved quantity for any (g, Θ). In the case of electromagnetism, where $K = U(1)$ is Abelian, its Lie algebra is its own center. The conservation law corresponding to a generator of $u(1)$ is charge conservation. If P is a trivial bundle and Θ the trivial connection, every Killing vector field of g will give rise to conserved quantities. For the case of special relativity, the Lie algebra of Killing vector fields is the Poincaré algebra. In this case, the invariants associated by T to the elements of the Poincaré algebra by T are precisely the invariants given by the energy-momentum tensor in special relativity.

39. Isotropic and coisotropic embeddings

In the next section we shall present a generalization of the constructions of Section 35, which we shall use together with the results of this section, to give a normal form for the moment map in Section 41. In this section we shall prove some general results in symplectic geometry, due to Weinstein (1981) and Gotay (1982) and Marle (1982). These results rely, in part, on the constructions of Section 35. As the results of this section and the next two are primarily mathematical in character, the more physically inclined reader may prefer to go directly to Section 42.

Let M be a symplectic manifold and $i: X \to M$ an isotropic embedding of some other manifold X into M. This means that i is an embedding and that at each $x \in X$, $di_x(TX_x)$ is an isotropic subspace of $TM_{i(x)}$. Thus

$$di_x(TX_x) \subset (di_x(TX_x))^{\perp} \subset TM_{i(x)}.$$

We shall write this more simply as

$$TX \subset TX^{\perp} \subset TM_{|X}.$$

We can form the vector bundle

$$E = (TX)^{\perp}/TX,$$

which Weinstein calls the symplectic normal bundle [to (X, i)]. If $i_1: X \to M_1$ and $i_2: X \to M_2$ are two isotropic embeddings and if f is a symplectic diffeomorphism of some neighborhood of $i_1(X)$ into M_2 satisfying

$$i_2 = f \circ i_1, \tag{39.1}$$

then f defines a symplectic bundle isomorphism, $L_f: E_1 \to E_2$ (where E_1 and E_2 denote the corresponding symplectic normal bundles). The map sending f to L_f is functorial in the obvious sense. We claim that given i_1, i_2, and a linear symplectic isomorphism $L: E_1 \to E_2$, we can find a symplectic diffeomorphism f defined in some neighborhood of $i_1(X)$ such that $L = L_f$. Indeed, we can proceed as follows: Let i and M stand for either i_1 or i_2 and M_1 and M_2. Choose a Riemann metric on M. This will then determine a complement W to TX in $TX^{\perp}$. By Lemma 26.1 (with $U = TX^{\perp}$) we know that W is a symplectic subbundle of $TM_{|X}$. So $W^{\perp}$ is also a symplectic subbundle and $TX \subset W^{\perp}$. In fact, it is clear that TX is a Lagrangian subbundle of $W^{\perp}$. Choose a Lagrangian subbundle Z in $W^{\perp}$ complementary to TX. (This is always possible since the space of all Lagrangian subspaces of a symplectic vector space that are complementary to a given Lagrangian space is contractable – in fact, it is an affine space (cf. Proposition 2.3, on p. 118 of Guillemin and Sternberg (1977).) Of course we can identify Z with TX^*. We thus have

$$TM_{|X} = W \oplus (TX \oplus TX^*)$$

as symplectic vector bundles, where W is isomorphic to E. Thus L, together with the identity map on TX, induces a symplectic isomorphism

$$L': TM_{1_{|X}} \to TM_{2_{|X}}.$$

Now we can identify $U \oplus Z$ with the normal bundle to $i(X)$ in M; in fact, we can choose a Riemann metric such that $U \oplus Z$ is in fact the metric normal bundle to $i(X)$. Then, using the exponential map for this metric, we can identify a neighborhood of the zero section with a tubular neighborhood of $i(X)$. Thus we can find neighborhoods of $i_1(X)$ and $i_2(X)$ and a diffeomorphism g between them such that $dg = L'$ at all points of $i_1(X)$. Thus $g^*\omega_2$ is a symplectic form defined near $i_1(X)$ and $g^*\omega_2 = \omega_1$ at all points of $i_1(X)$. By the Darboux–Weinstein theorem (Theorem 22.1) we can find some diffeomorphism h defined on some neighborhood of $i_1(X)$ that restricts to the identity on $i_1(X)$ and such that $h^*g^*\omega_2 = \omega_1$. Then $f = g \circ h$ is our desired symplectic diffeomorphism. Notice that in the preceding constructions, if we are given an action of a compact Lie group K that preserves ω_1 and ω_2, then we get induced actions of K on E_1 and E_2; the Riemann metrics can be made K invariant and hence, by the equivariant form of the Darboux–Weinstein theorem, the map f can be chosen to be K-equivariant.

Suppose that we start with a symplectic vector bundle $E \to X$. Following Weinstein (1981) we shall show how to construct, canonically, an isotropic embedding of X whose symplectic normal bundle is E. Let $P \to X$ be the bundle of symplectic bases of E. Then P is a principal $Sp(2n)$ bundle where $2n$ is the fiber dimension of E. Thus, if $Q = \mathbb{R}^{2n}$ with its standard symplectic structure, then the associated bundle $Q(P)$ is just E. We have seen in Section 35 that the Marsden–Weinstein reduced space T^*P_Q, which carries a canonical symplectic structure, can be identified with $Q(P)$, which is just $E^\#$, the pull-back of E to T^*X; but this identification depends on a choice of connection. We shall see that X has an isotropic embedding in T^*P_Q that does not depend on a choice of connection: By definition, the Marsden–Weinstein reduced space is $\bar{\Phi}^{-1}(0)/Sp(2n)$, where $\bar{\Phi}: T^*P \times \mathbb{R}^{2n-} \to sp(2n)^*$

is the moment map for the product action of $Sp(2n)$. Contained in $\bar{\Phi}^{-1}(0)$ is the set of all points of the form $(z, 0)$, where $z \in T^*P$ satisfies $\Phi(z) = 0$, with Φ the moment map for the $Sp(2n)$ action on T^*P. By the formula for the moment map for cotangent actions (29.2), we see that $\Phi(z) = 0$ if and only if z vanishes on all vertical tangent vectors. Thus $z = d\pi_p^* l$ for some $l \in T^*X_{\pi(p)}$. So $\Phi^{-1}(0)/Sp(2n)$ is canonically identified with T^*X, and it is easy to check that with this identification, T^*X together with its standard symplectic structure is a symplectic submanifold of T^*P. But X is embedded in T^*X as the zero section, and this is clearly an isotropic embedding. A choice of connection shows that the full Marsden–Weinstein reduced space T^*P_Q can be identified with the pull-back, $E^\#$ of E to T^*X. From this we see that the symplectic normal bundle to X in T^*P_Q can be identified with our original symplectic vector bundle E. Again, if we are given a K action on E, the symplectic manifold T^*P_Q inherits a K action and the embedding is K-equivariant. We have thus proved:

Theorem 39.1 (The isotropic embedding theorem). *Any isotropic embedding,* $\mathrm{i}: \mathrm{X} \to \mathrm{M}$ *determines a symplectic normal bundle* $\mathrm{E} \to \mathrm{X}$ *where* $\mathrm{E} = \mathrm{TX}^\perp/\mathrm{TX}$. *If* f *is a symplectic diffeomorphism defined on some neighborhood of* $\mathrm{i}_1(\mathrm{X})$ *and satisfying (39.1), where* i_1 *and* i_2 *are isotropic embeddings of* X, *then* f *induces a symplectic isomorphism,* $\mathrm{L_f}: \mathrm{E}_1 \to \mathrm{E}_2$ *of the corresponding symplectic normal bundles. Conversely, given any symplectic isomorphism,* $\mathrm{L}: \mathrm{E}_1 \to \mathrm{E}_2$, *there exists an* f *with* $\mathrm{L} = \mathrm{L_f}$. *Given any symplectic vector bundle* $\mathrm{E} \to \mathrm{X}$ *there exists a standard isotropic embedding of* X *whose symplectic normal bundle is* E, *namely, the embedding of* X *as the zero section in* $\mathrm{T^*X}$, *where* $\mathrm{T^*X}$ *is regarded as a symplectic submanifold of* $\mathrm{T^*P_Q}$ *as described above. In the presence of a compact group* K *of automorphisms, all of the above assertions are true in the category of* K *morphisms.*

Theorem 39.2 (The coisotropic embedding theorem). *Let* τ *be a closed 2-form of constant rank on a differentiable manifold* Z. *Then there exists a symplectic manifold* X, *with symplectic form* ω *and an embedding* $\mathrm{i}: \mathrm{Z} \to \mathrm{X}$ *such that*
(i) $\mathrm{i(Z)}$ *is a coisotropic submanifold of* X

and

(ii) $\mathrm{i}^*\omega = \tau$.

If $(\mathrm{X}_1, \omega_1, \mathrm{i}_1)$ *and* $(\mathrm{X}_2, \omega_2, \mathrm{i}_2)$ *are two such coisotropic embeddings then there exist neighborhoods*

$$\mathrm{U}_1 \textit{ of } \mathrm{i}_1(\mathrm{Z}) \textit{ in } \mathrm{X}_1$$

and

$$U_2 \text{ of } i_2(Z) \text{ in } X_2$$

and symplectic diffeomorphism $f: U_1 \to U_2$ *such that*

$$i_2 = f \circ i.$$

If we are given an action of a compact Lie group K *on* Z, *which preserves* τ, *then we can choose* (X, ω, i) *such that* K *has a symplectic action on* X *and so that* i *is a* K *morphism. If* (X_1, ω_1, i_1) *and* (X_2, ω_2, i_2) *are two such* K-*equivariant coisotropic embeddings then the neighborhoods* U_1, U_2 *can be chosen to be* K-*invariant and the diffeomorphism* f *can be chosen to be a* K *morphism.*

Proof. The form τ determines a vector bundle $TZ^\perp \subset TZ$ (the tangent bundle to the null foliation of τ). Let

$$Y = (TZ^\perp)^*,$$

so Y is a vector bundle over Z; the fiber at each point $z \in Z$ is the dual space to $TZ_z^\perp$. We let $i: Z \to Y$ be the embedding of Z into Y as the zero section. If we have a K action on Z then it induces a K action on TZ, hence on $TZ^\perp$ (if the K action preserves τ) and hence on Y. The embedding i is clearly K-equivariant.

There is no canonical symplectic form on Y. But we can define one in a neighborhood of Z as follows: First choose a Riemann metric on Z. (In the case of a given K action choose it to be K-invariant.) Let $W_z \subset TZ_z$ be the orthocomplement to $TZ_z^\perp$. The restriction of τ to W_z is nondegenerate. Now at $i(z) \in Y$ we have the decomposition of the tangent space into a direct sum of the tangent to the fiber and the tangent to the zero section. As the fiber is a vector space, we may identify the tangent to the fiber with the fiber itself, that is with $(TZ_z^\perp)^*$. Thus we have the direct sum decomposition

$$TY_{i(z)} = (TZ_z^\perp)^* \oplus TZ_z^\perp \oplus W_z.$$

The first two summands are just a vector space and its dual, and so $(TZ_z^\perp)^* \oplus TZ_z^\perp$ carries a symplectic structure, and W_z has one as well. We give $TY_{i(z)}$ the direct sum symplectic structure. We thus have a nondegenerate symplectic form defined on the bundle $TY_{|i(z)}$ (which is K-invariant). Extend it so as to be defined (and K-invariant) on all of Y. We thus have a 2-form Ω defined on all of Y with the property that

$$i^*\Omega = \tau,$$

and is K-invariant. We now want to modify Ω so that we get a *closed* 2-form with the same property. Let

$$\mu = d\Omega.$$

So

$$d\mu = 0 \tag{39.2}$$

and, since $i^*\Omega = \tau$ and $d\tau = 0$,

$$i^*\mu = 0. \tag{39.3}$$

Let $\varphi_t\colon Y \to Y$ be multiplication by t so that φ_t is a retraction of Y onto Z, that is

$$\varphi_0\colon Y \to Z, \qquad i \circ \varphi_t = i, \qquad \varphi_0 \circ i = \mathrm{id},$$

and

$$\varphi_1 = \mathrm{id}.$$

Let ξ_t be the corresponding one-parameter family of vector fields. Notice that

$$\xi_t(i(z)) \equiv 0. \tag{39.4}$$

Recall from Section 21 that we defined an operator I on differential forms given by

$$Iv = \int_0^1 \varphi_t^*(i(\xi_t)v)\, dt, \tag{39.5}$$

and that

$$v - \varphi_0^* v = \int_0^1 \frac{d}{dt} \varphi_t^* v = dIv + I\, dv.$$

But $i \circ \varphi_0 = \varphi_0$ so

$$\varphi_0^* v = \varphi_0^* i^* v$$

so

$$v - \varphi_0^* i^* v = dIv + I\, dv.$$

Also, from (39.4) it follows that

$$(Iv)_{i(z)} = 0$$

at each point of Z. We can thus assert

Lemma 39.1. *Let φ_t be a retraction of* Y *onto* i(Z). *Let* ν *be a closed* k *form on* Y *such that* $i^*\nu = 0$. *Then there exists a* (k − 1)*-form* $I\nu$ *on* Y *such that*

$$\nu = d(I\nu)$$

and

$$(I\nu)_{i(z)} = 0 \tag{39.6}$$

at all points of Z.

Applying the lemma to $\mu = d\Omega$ we see that

$$\omega = \Omega - I\mu$$

is closed, and

$$\omega_{i(z)} = \Omega_{i(z)}$$

is nonsingular at all $z \in Z$. (Also ω is K-invariant if we are given a K action.) Hence, in some neighborhood of $i(Z)$ (which we may take to be K-invariant) the form ω is closed and nondegenerate. This neighborhood will be our X. This concludes the existence part of the theorem.

Now to the uniqueness. If we are given a coisotropic embedding $i_1 : Z \to X_1$ we can construct the geometric normal bundle $TX/di_1(TZ) = N_1$. This bundle can be identified with $N = (TZ^{\perp})^*$. Thus given two coisotropic embeddings, we get two vector bundles

$$\begin{array}{ccc} N_1 & & N_2 \\ \downarrow & \text{and} & \downarrow \\ Z & & Z \end{array}$$

and a vector bundle isomorphism

$$A : N_1 \to N_2.$$

(These vector bundles admit K actions and A is canonical and hence a K morphism in the presence of a K action.) By choosing (K-invariant) Riemann metrics on N_1 and N_2 we can use the exponential maps to identify neighborhoods of the zero section in N_1 and N_2 with tubular neighborhoods of $i_1(z)$ and $i_2(z)$ in X_1 and X_2. This allows us to regard A as a diffeomorphism between these neighborhoods. Let

$$\bar{\omega}_1 = A^*\omega_2.$$

Then $\bar{\omega}_1$ is a symplectic form and

$$(\bar{\omega}_1)_{i_1(z)} = (\omega_1)_{i_1(z)} \qquad \forall z \in Z.$$

By the K-invariant Darboux theorem, we can find some neighborhood U' of $i_1(z)$ and a diffeomorphism of U' into X with

$$g^*\bar{\omega}_1 = \omega_1.$$

Then $f = A \circ g$ (restricted to a small enough neighborhood) satisfies

$$f^*\omega_2 = \omega_2,$$

completing the proof of the theorem.

40. Symplectic induction

In this section we will generalize the constructions of Section 35, using the coisotropic embedding theorem. We shall look at the following special case of that theorem: Let M be a symplectic manifold and let

$$P \xrightarrow{\pi} M$$

be a principal K bundle over M. If ω_M denotes the symplectic form on M we shall take $Z = P$ and

$$\sigma = \pi^*\omega_M.$$

The null leaves of σ are just the fibers of π. Theorem 39.2 says that we can embed P as a coisotropic submanifold of a symplectic K space and that this embedding is locally unique (up to K-equivariant symplectic diffeomorphisms defined near $\iota(P)$). We shall now show that a choice of connection on P will give us an explicit construction of X that will show that the action of K on X is in fact Hamiltonian. The construction will also make X into a fiber bundle over M. This fibration of X over M is *not* canonical however; it will depend on the choice of connection.

Let θ be a connection on P. Thus θ is a k-valued linear differential form on P. Set

$$Y = P \times k^*.$$

By abuse of language, we can think of θ as being a k-valued linear differential form on Y, depending only on the first factor. Let $\pi_2 : P \times k^*$ denote projection onto the second factor, so that π_2 is a k^*-valued function on Y. Let $\langle\, , \rangle$ denote the pairing between k^* and k. Thus $\langle \pi_2, \theta \rangle$ is a scalar-valued linear differential form on Y. Define the 2-form ω on Y by

$$\omega = \pi^*\omega_M + d\langle \pi_2, \theta \rangle. \tag{40.1}$$

It is clear that ω is closed. We let K act on $Y = P \times k^*$ (diagonally) by the given action on P and the coadjoint action on k^*. The transformation properties of a connection guarantee that $\langle \pi_2, \theta \rangle$ and hence ω is K-invariant. The injection

$$\iota : P \to P \times k^*, \qquad \iota(p) = (p, 0) \tag{40.2}$$

is clearly K-equivariant and

$$\iota^*(\omega) = \pi^* \omega_M = \sigma. \tag{40.3}$$

We claim that ω is symplectic (i.e. nondegenerate) at all points of $\iota(P)$ and hence in some neighborhood of $\iota(P)$. Indeed, at a point $(p, 0)$, where $\pi_2 = 0$ we have

$$d\langle \pi_2, \theta \rangle = \langle d\pi_2 \wedge \theta \rangle.$$

Also, the connection allows us to split TP_p as $TP_p = H_p + V_p$, where H_p and V_p are the horizontal and vertical subspaces and to identify H_p with $TM_{\pi p}$ and V_p with k. Thus we may identify

$$TY_{(p,0)} \qquad \text{with} \qquad TM_{\pi p} + k + k^*.$$

Under this identification, the form $\omega_{(p,0)}$ is given by $(\omega_M)_p$ on $TM_{\pi p}$ and the standard antisymmetric form on $k + k^*$, and the spaces $TM_{\pi p}$ and $k + k^*$ are orthogonal under $\omega_{(p,0)}$. This proves that ω is symplectic at $\iota(P)$ and hence in some neighborhood of $\iota(P)$. In case K and M are compact, we can be a little more precise about the neighborhood we choose. Indeed, for fixed $m_0 \in M$, we can, since K is compact, choose some neighborhood $\mathscr{V}_0$ of 0 in k^* such that ω is symplectic at all points in $\pi^{-1}(m_0) \times \mathscr{V}_0$. Hence also at all points of $\pi^{-1}(m) \times \mathscr{V}_0$, where m is in some neighborhood of m_0. As M is compact we may cover M by finitely many such neighborhoods and let $\mathscr{V}$ be the intersection of the corresponding $\mathscr{V}_0$'s. Thus, if K and M are compact we may find some neighborhood $\mathscr{V}$ of 0 in k^* so that ω is symplectic on $P \times \mathscr{V}$.

The action of K on $P \times k^*$ is not only symplectic, it is Hamiltonian. Indeed, if $\xi \in k$ let ξ_P be the corresponding vector field on P, and ξ_Y the corresponding vector field on Y. Then

$$i(\xi_P)\pi^*\omega_M = 0$$

and

$$i(\xi_P)\theta = \xi$$

by the defining property of a connection. Hence by (40.1), the invariance of

$\langle \theta, \pi_2 \rangle$, and the preceding equation we have

$$\begin{aligned} i(\xi_Y)\omega &= i(\xi_Y)\, d\langle \theta, \pi_2 \rangle \\ &= D_{\xi_Y}\langle \pi_2, \theta \rangle - di(\xi_Y)\langle \pi_2, \theta \rangle \\ &= -d\langle \pi_2, \xi \rangle. \end{aligned}$$

Thus the action is Hamiltonian with moment map π_2. To summarize:

Proposition 40.1. *Let* $\mathrm{P} \overset{\pi}{\to} \mathrm{M}$ *be a principal* K *bundle over a symplectic manifold with form* ω_{M}. *Let* θ *be a connection on* P. *Then (40.1) defines a symplectic form on some neighborhood of* $i(\mathrm{P}) \subset \mathrm{P} \times \mathrm{k}^*$ *and (40.3) holds. The action of* K *on* $\mathrm{P} \times \mathrm{k}^*$ *is Hamiltonian with moment map* π_2. *If* K *and* M *are compact we can find some neighborhood* $\mathscr{V}$ *of* 0 *in* k^* *so that* ω *is symplectic on* $\mathrm{P} \times \mathscr{V}$.

To relate Proposition 40.1 to the constructions in Section 35, let F be a Hamiltonian K space with moment map $\Phi_F : F \to k^*$. Suppose that ω is symplectic on $P \times \mathscr{V}$, where $\mathscr{V}$ is a neighborhood of 0 in k^* as in Proposition 40.1 and that

$$\Phi_F(F) \subset \mathscr{V}.$$

We can then form the space

$$X \times F^-$$

with moment map

$$\Psi = \pi_2 - \Phi_F$$

and the Marsden–Weinstein reduced space

$$\Psi^{-1}(0)/K.$$

We can identify $\Psi^{-1}(0)$ with $P \times F$ since the k^* component in $P \times k^* \times F$ is determined by $\pi_2 = \Phi_F(f)$. It is then clear that (up to a sign) the form defined on $\Psi^{-1}(0)$ is the same as that given in Section 35, since $P \times F/K$ is the associated bundle $F(P)$. In particular, the form on $F(P)$ is symplectic. Notice that this depended on $\Phi_F(F)$ being "small" that is, in $\mathscr{V}$. The size of $\mathscr{V}$ depends, of course, on θ and ω_M. For compact K and M and fixed θ, $\mathscr{V}$ can be made large by replacing ω_M by some large multiple, $a\omega_M$, where $a \gg 0$ is some large real number. This was observed by Weinstein (1977, Remark (3), p. 242).

Now suppose that we are given a Lie subgroup $G \subset \text{Aut } P$ and suppose that the induced action of G on M is Hamiltonian with moment map

$$\Phi_M : M \to g^*.$$

Also suppose that the connection θ is G-invariant. Then by letting G act trivially on the k^* component we get an action of G on $P \times k^*$, which preserves ω. (If G and K are compact then we could choose our symplectic neighborhood X of $P \times \{0\}$ to be G-invariant.) The action of G on $P \times k^*$ is in fact Hamiltonian. Indeed if ζ_P and ζ_M are the vector fields on P and M corresponding to $\zeta \in g$ then

$$\begin{aligned} i(\zeta_P)\, d\pi^*\omega_M &= \pi^* i(\zeta_M)\omega_M \\ &= \pi^* d\langle \Phi_M, \zeta \rangle. \end{aligned}$$

Also, since $D_{\zeta_P}\theta = 0$ and ζ_P has no k^* component

$$i(\zeta_P)\, d\langle \pi_2, \theta \rangle = -d\langle \pi_2, i(\zeta_P)\theta \rangle.$$

Now at each $p \in P$, the k-valued form θ defines a linear map of $g \to k$ given by

$$\zeta \to (i(\zeta_P)\theta)_p.$$

The transpose of this map is a linear map

$$\Phi_{\theta,p} : k^* \to g^*,$$

and we can write the preceding equation as

$$i(\zeta_P) d\langle \pi_2, \theta \rangle = d\langle \Phi_\theta, \zeta \rangle,$$

where

$$\Phi_\theta : P \times k^* \to g^*,$$

is given by

$$\Phi_\theta(p, \beta) = \Phi_{\theta,p}(\beta).$$

$$\langle \zeta, \Phi_{\theta,p}(\beta) \rangle = \langle \beta, (i(\zeta_P)\theta)_p \rangle. \tag{40.4}$$

Thus we have proved:

Proposition 40.2. *If* G $\subset$ *Aut* P *preserves* θ *and induces a Hamiltonian action on* M *with moment map* Φ_{M}, *then the trivial action of* G *on* k* *gives a Hamiltonian action of* G *on* P $\times$ k* *with moment map*

$$\Phi_{\mathrm{G}} = \pi^*\Phi_{\mathrm{M}} + \Phi_\theta, \tag{40.5}$$

where Φ_θ *is given by* (40.4).

Let us now give two rather different applications of Proposition 40.2. In the first of these we shall take $M = \mathcal{O}$ to be an orbit of G in g^* and $K = G_\alpha$ to be the isotropy subgroup of some point α of $\mathcal{O}$. We shall take $P = G$ regarded as a principal K bundle over $\mathcal{O} = G/K$, with G acting as automorphisms via left multiplication. A G-invariant connection on P is given by the choice of an (Ad K)-invariant complement p to k in g. (This determines the horizontal space of the connection at the identity $e \in G = P$, and G-invariance then determines it everywhere. The (Ad K)-invariance is required for it to be a connection.) If K is compact we can always make this choice by putting an (Ad K)-invariant metric on g. (For the case that G is compact we shall study this situation in more detail and soon show that then there is a canonical choice of p.) The splitting

$$g = k + p$$

gives a splitting

$$g^* = k^* + p^*$$

and hence a linear map $l = l_p$ of $k^* \to g^*$. Then (40.5) becomes

$$\Phi_G(a, \beta) = a \cdot (\alpha + l(\beta)). \tag{40.6}$$

Indeed, this is exactly the content of (40.5) at the point $a = e$, and hence by the G-equivariance of the moment map it is true everywhere.

As a second illustration of the construction of Proposition 40.1, let us consider the following situation: let

$$\rho : Q \to N$$

be a principal K bundle, where N is any differentiable manifold, not necessarily symplectic. The group K has a natural Hamiltonian action on the cotangent bundle T^*Q. Inside T^*Q there is a K-invariant coisotropic submanifold P, consisting of those covectors which vanish when evaluated on any vertical tangent vector of Q. A covector in T^*Q belongs to P if and only if it is of the form $d\rho_q^*(\kappa)$ where $\kappa \in T^*N_n$ with $n = \rho(q)$. We thus have a natural projection

$$\pi : P \to T^*N,$$

where

$$\pi(d\rho_q^*(\kappa)) = \kappa.$$

It is easy to check that this makes P into a principal K bundle over T^*N and

the diagram

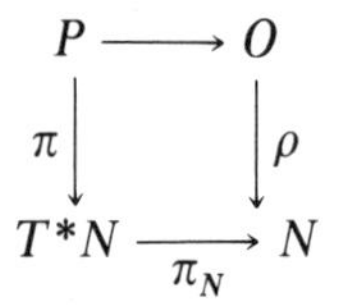

commutes, where $\pi_N: T^*N \to N$ is the standard projection of the cotangent bundle onto the base and the map from $P \to Q$ is the restriction of the standard projection $T^*Q \to Q$ to P. Thus we can regard P as being the pull-back of the bundle $Q \to N$ to T^*N, in other words, $P = \pi_N^\#(Q)$. We could start with this abstract definition of P as a pulled-back bundle, over $M = T^*N$ and use the embedding $\iota: P \to T^*Q = X$ to give the coisotropic embedding. If $N = G/K$ is a homogeneous G space, then G acts in Hamiltonian fashion on T^*N, and the construction of Proposition 40.2 applied to any Hamiltonian K space gives rise to a Hamiltonian G space. This is the symplectic analog of the induced representation [cf. Kazhdan, Kostant, and Sternberg (1978) and Guillemin and Sternberg (1981b)].

41. Symplectic slices and moment reconstruction

The purpose of this section is to describe a local normal form for Hamiltonian actions of a compact Lie group: Let G be a compact Lie group with a Hamiltonian action on a symplectic manifold M and with moment map Φ. Let p be a point of M. We shall show that Φ is completely determined in a G-invariant neighborhood of p by the following data:
(1) $\alpha = \Phi(p)$;
(2) the isotropy group G_p; and
(3) the linear representation of G_p on TM_p.
Let us first prove this result under the auxiliary hypothesis that α is left-fixed by all of G. Later on in this section we shall show how to reduce the general case to this one. Now $(\xi_M(p), \eta_M(p))_p = \langle \xi \cdot \alpha, \eta \rangle$ by (26.2) and $\xi \cdot \alpha = 0$ for all $\xi \in g$ by hypothesis. Thus $g_M(p)$ is an isotropic subspace of TM_p and hence

$$Z = G \cdot p$$

is an isotropic submanifold of M. By the isotropic embedding theorem (Theorem 39.1) the manifold M can be uniquely recovered (as a symplectic G space) up to symplectic diffeomorphism from the symplectic normal bundle $E \to Z$. The group G acts as automorphisms of this bundle and acts transitively on Z. Hence, as a homogeneous vector bundle, E is completely determined by the action of G_p on E_p. But $E_p = (TZ_p)^\perp/TZ_p$ and so the representation of G_p on E_p is completely determined by the action of G_p on T_pM. We can thus recover a G-invariant neighborhood of Z in M from the data listed above. In fact, if we make a small change in our construction of the canonical isotropic embedding, we can obtain an explicit form for the moment map: In our proof of the existence part of the isotropic embedding theorem, we took P to be the bundle of frames of the symplectic bundle E. Let us now choose a smaller P. Let us take $P = G$, considered as a G_p bundle over $Z = G/G_p$. For convenience in notation, let us set

$$H = G_p$$

and

$$i: h \to g$$

the corresponding injection of Lie algebras. So

$$i^*: g^* \to h^*$$

is the dual projection. Also, let

$$V = E_p = TZ_p^\perp/TZ_p.$$

Thus $E = V(G) = G \times V/H$ is the bundle associated to the principal H bundle G by the representation of H on V. The proof of Theorem 39.1 now shows that we have an isotropic embedding of Z in the Marsden–Weinstein reduced space T^*G_V, whose symplectic normal bundle is E. (By the uniqueness part of the isotropic embedding theorem all isotropic embeddings are equivalent.) Let us examine what this Marsden–Weinstein reduced space (and its G moment map) looks like in terms of the left-invariant identification of T^*G with $G \times g^*$. The moment map $\Phi_H: T^*G \to h^*$ for the right action of H on T^*G is given, in terms of this identification, by

$$\Phi_H(c, \beta) = -i^*\beta \qquad c \in g \quad \text{and} \quad \beta \in g^*.$$

The moment map $\Phi_V: V \to h^*$ is the quadratic map given by

$$\langle \Phi_V(v), \eta \rangle = \tfrac{1}{2}(\eta v, v)_V \qquad v \in V, \eta \in h,$$

where $(\,,)_V$ denotes the symplectic form on V. Thus $\bar{\Phi} = \Phi_V - \Phi_H: T^*G^- \times V \to h^*$ is given by

$$\bar{\Phi}(c, \beta, v) = i^*\beta + \Phi_V(v).$$

Thus $\bar{\Phi}^{-1}(0)$ consists of all (c, β, v) such that

$$i^*\beta = -\Phi_V(v). \tag{*}$$

The group H acts on $\bar{\Phi}^{-1}(0)$ by

$$r_b(c, \beta, v) = (cb^{-1}, b\cdot\beta, bv), \qquad b \in H.$$

The space T^*G_V is the quotient space of $\bar{\Phi}^{-1}(0)$ by this H action. Let $\mathscr{V} \subset g^* \times V$ be the quadratic variety consisting of all pairs (β, v) satisfying (*). We thus see that T^*G_V can be identified as the associated bundle $\mathscr{V}(G)$. The symplectic action of G on T^*G_V is that induced from left multiplication on G and the trivial action on V:

$$l_a(c, \beta, v) = (ac, \beta, v).$$

The moment map coming from this coadjoint action is

$$\Phi_G(c, \beta, v) = c\cdot\beta.$$

This is clearly constant on H orbits and hence descends to give a moment map $\Phi': T^*G_V \to g^*$. Now the symplectic structure and G action is determined by the isotropic embedding theorem, but the moment map (when it exists) is only determined up to a constant. The map Φ' has the property that it sends Z into 0. Our original moment map sent Z into α. Thus we have proved that a normal form for our moment map is given in terms of the above description of T^*G_V is given by

$$\Phi([(c, \beta, v)]) = \alpha + c\cdot\beta,$$

where (*) holds and where [ ] denotes the equivalence class, that is, the quotient modulo the H action.

So far, we have been discussing the case where α is fixed under all of G. We now turn to the general case. We will first need a construction that is interesting in its own right. Let G be a connected Lie group with Hamiltonian action on M and moment map Φ. Let $\mathcal{O}$ be the coadjoint orbit through $\alpha = \Phi(p)$. Let Y be a submanifold of g^* satisfying

$$g^* = T\mathcal{O}_\alpha \oplus TY_\alpha. \tag{41.1}$$

According to Theorem 26.7 there is a neighborhood U of p such that $\Phi^{-1}(Y) \cap U$ is a symplectic submanifold of U. If G happens to be a compact Lie group this theorem can be sharpened by making a particularly nice choice of Y. To see this we need first to recall some elementary facts about compact connected Lie groups. Fix a point $\alpha \in g^*$ and let K be the stabilizer group of α in G. It is clear that K is closed and, according to

Theorem 32.15, we know that K is connected. Let M be the center of K, and consider in g^* the subspaces

$$k^{\#} = \text{the elements of } g^* \text{ stabilized by } M$$

and

$$\mathcal{M}^{\#} = \text{the elements of } g^* \text{ stabilized by } K.$$

Clearly $\mathcal{M}^{\#} \subseteq k^{\#}$ and $\alpha \in \mathcal{M}^{\#}$. It is also clear that the coadjoint action of K on g^* preserves $k^{\#}$. We will prove the following standard facts about this situation.

Proposition 41.1.

(1) *The orbit $\mathcal{O}$ through α intersects $\mathrm{k}^{\#}$ transversally at α and at this intersection (41.1) is satisfied, with $\mathrm{Y} = \mathrm{k}^{\#}$.*

(2) *For points $\beta \in \mathrm{k}^{\#}$ near α the stabilizer of β in* G *is contained in* K.

(3) *The canonical projection, $\mathrm{g}^* \to \mathrm{k}^*$ maps $\mathrm{k}^{\#}$ bijectively onto k^*.*

(4) *The canonical projection, $\mathrm{g}^* \to \mathcal{M}^*$ maps $\mathcal{M}^{\#}$ bijectively onto $\mathcal{M}^*$.*

(5) *Every coadjoint orbit in g^* intersects $\mathrm{k}^{\#}$ in a finite number of* K *orbits.*

(6) *There is a neighborhood, $\mathrm{B}_\epsilon(\alpha)$ of α in $\mathrm{k}^{\#}$ such that every coadjoint orbit intersects $\mathrm{B}_\epsilon(\alpha)$ in exactly one* K *orbit.*

Before proving these assertions we will make a few remarks: Parts (1) and (2) of the theorem simply say that $k^{\#}$ is a slice for the action of G on g^* in the sense of Section 27. Part (3) says that k^* can be canonically imbedded in g^* and as such is a complementary space to

$$k^0 = \{f \in g^*, \langle f, \xi \rangle = 0 \text{ for all } \xi \in k\}.$$

Part (4) is a similar assertion about $\mathcal{M}^*$.

Proof. Fix a positive definite G-invariant bilinear form on g^*. Associated with this form one gets a bijective G-equivariant map $\rho: g^* \to g$. We claim

$$\rho(k^{\#}) = k \qquad \text{and} \qquad \rho(\mathcal{M}^{\#}) = \mathcal{M}. \tag{41.2}$$

To see this, let $\xi = \rho(\alpha)$. By definition

$$k = \{\eta \in g, [\xi, \eta] = 0\}. \tag{41.3}$$

Since $[\xi, \xi] = 0$, ξ is in k and, therefore, by (41.3) ξ is in the center, $\mathcal{M}$, of k. Since $\rho(\mathcal{M}^{\#})$ is the subset of g stabilized by K, $[\xi, \rho(\mathcal{M}^{\#})] = 0$, so $\rho(\mathcal{M}^{\#})$ is contained in k by (41.3) and hence is equal to the center of k (i.e., $\rho(\mathcal{M}^{\#}) = \mathcal{M}$). Similarly, since $\rho(k^{\#})$ is the subset of g stabilized by M, $[\xi, \rho(k^{\#})] = 0$; so $\rho(k^{\#})$ is contained in k. On the other hand, it is obvious that $\rho(k^{\#}) \supset k$; so $\rho(k^{\#}) = k$. This proves (41.2).

Parts (3) and (4) of the proposition follow immediately from (41.2). To prove part (1) let $p = (k^{\#})^0$ in g. Then

$$g = p \oplus k \qquad \text{(orthogonal decomposition).} \tag{41.4}$$

Moreover, $ad(\xi): g \to g$ preserves (41.4) mapping k into zero and p bijectively onto itself. This shows that $p = [\xi, g] =$ the tangent space to the orbit through ξ proving (1). To prove (2), let η be an element of k. Then $ad(\eta)$ preserves (41.4). Moreover, since $ad(\xi): p \to p$ is bijective, $ad(\eta): p \to p$ is bijective if η is sufficiently close to ξ. Thus the centralizer of η in g is contained in k. To prove (5) let t be a Cartan subalgebra of g containing ξ. Then by (41.3) t is contained in k; so it is also a Cartan subalgebra of k. Let η_1 and η_2 be elements of k on the same G-orbit. Then they have K-conjugates η_1' and η_2' in t. However, if two elements of t are conjugate by G they are Weyl-group conjugate. (See, for instance, Helgason, 1982.) Thus every G-orbit intersects k in at most n K-orbits, where n is the cardinality of the Weyl group.

Now choose a neighborhood $B_\epsilon(\alpha)$ of α so small that it contains no other G conjugate of α. As in the proof of (5) we may assume that α lies in t^* and that we are looking at a G orbit that intersects t in two points. These two points are conjugate by an element of G that normalizes t and fixes α. Thus the element in question lies in K and normalizes t. Hence the two points are K-conjugate. This proves (6) and completes the proof of the proposition.

Now let M be a symplectic manifold on which G acts in a Hamiltonian fashion and let $\Phi: M \to g^*$ be the associated moment map. By Theorem 26.7, we can find a K-invariant neighborhood U of $\Phi^{-1}(\alpha)$ such that

$$W = \Phi^{-1}(B_\epsilon(\alpha)) \cap U$$

is a symplectic submanifold of M. We note the following properties of W:

Proposition 41.2.

(a) W *is* K*-invariant.*

(b) *The action of* K *on* W *is Hamiltonian and the associated moment map is just the map* $\Phi: \mathrm{W} \to \mathrm{k}^{\#}$ *composed with the identification* $\mathrm{k}^{\#} \to \mathrm{k}^*$.

(c) *Every* G*-orbit in* X *that intersects* W, *intersects* W *in a single* K*-orbit.*

Proof. Parts (a) and (b) are obvious. To prove (c), let p_1 and p_2 be G-related points in W. Then $\Phi(p_1)$ and $\Phi(p_2)$ are G-related in $B_\epsilon(\alpha)$; so they are K-related. Hence without loss of generality we can assume $p_1 = gp_2$ and $\Phi(p_1) = \Phi(p_2) \in \beta_\epsilon(\alpha)$. Thus g is in the stabilizer of $\Phi(p_1) = \Phi(p_2)$; so by Part (2) of the theorem $g \in K$. Q.E.D.

Let M_0 be the union of all points p such that Gp intersects W. M_0 is an open G-invariant subset of M.

We will show shortly that the manifold M_0, can be reconstructed from W in a canonical way. Before we do this, however, let us show how the cross-section construction can be used to give an alternative version of an induction construction presented in Section 40. (The material below is essentially due to Weinstein (1982), §9.)

Let $\Phi_R : T^*G \to g^*$ be the moment map associated with the right action of G on T^*G and consider $Z = \Phi_R^{-1}(k^\#)$. If we identify T^*G with $G \times g^*$ as above then Φ_R is just the negative of the projection of $G \times g^*$ onto g^*; so $Z = G \times k^\#$. By the cross-section theorem, Z is a symplectic submanifold of T^*G at least in the $(G \times K)$ invariant neighborhood

$$Z_\alpha = G \times B_\epsilon(\alpha)$$

of $G \times \{\alpha\} = \Phi_R^{-1}(\alpha)$. Now in the preceding section we considered the orbit $\mathcal{O}$, through α and the principal K-bundle

$$\pi : G \to \mathcal{O}, \qquad g \to g\alpha.$$

The splitting $g = p \oplus k$ with $p = (k^\#)^0$ defines a connection on this bundle, and we showed how to use this connection to define a symplectic structure on a neighborhood

$$Z_0 = G \times B_\epsilon(0)$$

of $G \times \{0\}$ in $G \times k^*$. It is easy to see that the symplectic structure on Z_α given by the cross-section construction and the symplectic structure on Z_0 given by the induction construction are essentially the same. In fact, let $\tau : Z_0 \to Z_\alpha$ be the map $\tau(g, \beta) = (g, \alpha + \beta)$. This map is $(G \times K)$-equivariant since α is K-fixed; and it is not hard to show that it is a $(G \times K)$-equivariant symplectomorphism.

Coming back to M_0 and W we will now indicate how M_0 can be reconstructed from W in a canonical way. The moment mapping associated with the Hamiltonian action of K on Z_α is just the projection map

$$Z_\alpha = G \times B_\epsilon(\alpha) \to B_\epsilon(\alpha)$$

by Proposition 41.2, so we can form the product of Hamiltonian K spaces

$$Z_\alpha \times W^-$$

and reduce with respect to the zero orbit in k^*. This is exactly the symplectic induction construction discussed in Section 40. Let us denote the resulting space by M_1.

Theorem 41.1. M_0 *and* M_1 *are isomorphic as Hamiltonian* G *spaces.*

Proof. As an abstract set

$$M_1 = (G \times W)/K.$$

Map W into M_1 by the mapping

$$i: W \to M_1, \qquad w \to \text{equivalence class of } (e, w).$$

It is easy to check that i is an imbedding and is K-equivariant. There is a unique way of extending i to a G-equivariant map of M_0 onto M_1. Namely, if $x \in M_0$, pick elements $g \in G$ and $w \in W$ such that $x = gw$ and set $\sigma(x) = gi(w)$. Let us show that this unambiguously defines a map $\sigma: M_0 \to M_1$. Suppose we also have $x = g_1 w_1$. Then $g^{-1} g_1 w_1$ and w_1 are on the same G orbit in W, so they are on the same K orbit; that is, there exists $k_0 \in K$ such that $k_0 g^{-1} g$ stabilizes w_1. This implies that $k_0 g^{-1} g$ stabilizes the point $\Phi(w_1)$ in $B_\epsilon(\alpha)$, and, therefore, has to be in K. Thus there exists an element $k \in K$ such that

$$g_1 = gk.$$

In particular, $g_1 w_1 = gkw_1 = gw$, or

$$kw_1 = w.$$

Since i is k-equivariant, $i(w) = ki(w_1)$, so

$$gi(w) = gki(w_1) = g_1 i(w_1).$$

This proves that σ is well defined. Since σ is smooth on W and G-equivariant, it is smooth everywhere. It is easy to check that it is a symplectomorphism at all points $w \in W$ and so, by G-equivariance, at all points of M_0.

We have now seen how the Hamiltonian G space, M_0 can be reconstructed from the Hamiltonian K space, W. But, for $p \in W$ and $\alpha = \Phi(p)$, K leaves α fixed. We are thus back in the situation of the beginning of this section: W, as a Hamiltonian K space, can be reconstructed from α and the linear isotropy representation of K on TW_p. We have thus completed the proof of the main result of this section:

Theorem 41.2. *Let* G *be a compact Lie group. Let* M *be a Hamiltonian* G *space and* $p \in M$. *Then* M *is completely determined near* p *(as a Hamiltonian* G *space) by* $\Phi(p)$, *the isotropy group,* G_p, *and the representation of* G_p *on* TM_p.

42. An alternative approach to the equations of motion

In section 35 we wrote down a recipe for obtaining the equations of motion of a particle in the presence of a Yang–Mills field. This involved the choice of a Hamiltonian. In the justification of these equations by the principle of general covariance, we took this Hamiltonian to be the quadratic energy function associated to the Riemann metric. Thus, for example, in the case of general relativity, the Hamiltonian H on T^*M was determined by the Lorentzian metric of space-time M, while the projection of T^*P onto T^*M was determined by the connection (i.e., the Yang–Mills field). In this description, the "forces" coming from the Yang–Mills field (generalizing the electromagnetic forces) play a role completely different in the theory than did the "gravitational forces" arising from the Lorentzian metric. In the present section we develop an alternative approach to the equations of mechanics, based on some of the geometrical ideas of Elie Cartan in the theory of connections; the key notion being that one must not only consider an internal symmetry group K, but also larger "motion groups" G that contain it. The results of this section first appeared in Sternberg and Ungar (1978).

Let us give a brief description of Cartan's viewpoint. In the classical theory of connections, there were two approaches. One, associated primarily with Levi-Civita, stressed the role of the tangent bundle as a vector bundle, that is, as a collection of vector spaces, and a connection as providing a means of parallel transport along curves. This notion of parallel transport then gives rise to the concept of covariant derivative, which is so familiar in tensor analysis and general relativity. The principal bundle connected with this approach is the bundle of orthonormal frames, that is, the bundle of all orthonormal bases of all tangent spaces (the set of *vierbeine* in general relativity). The structure group of this bundle is the orthogonal (or Lorentz) group H, which is a subgroup of $Gl(d)$, where d is the dimension of the base manifold (4 in general relativity). In the classical theory of surfaces, there was another notion intimately related to parallel transport, and that is the development of a surface on a plane along a curve. Intuitively speaking, one "rolled" the surface on the plane along the curve, maintaining first-order contact. This gives an identification of the tangent space to the surface at each point of the curve with the plane. Then parallel translation in the Euclidean geometry of the plane gives the parallel transport in the sense of Levi-Civita along the curve (cf., e.g., Sternberg, 1964, Chapter V, especially Sections 4 and 6). From this point of view the notion of development is central, and the crucial property of the plane is that it is a homogeneous space in the sense that it admits a transitive group of isometries that includes the full rotation group as the isotropy group of a point. The plane is $E(2)/O(2)$, where $E(2)$ denotes the group of Euclidean motions of the plane. From this point of view, we might also want to study the development of a surface onto a sphere along a curve, where now the sphere is $O(3)/O(2)$, or the development onto hyperbolic space $Sl(2, \mathbb{R})/SO(2)$. This is the starting point for Cartan's approach. We must not only be given the structure group H, but also a larger group G containing H as a closed subgroup so that G/H is a homogeneous "model" space with which we would like to compare our manifold M. Thus, in general relativity, where H is the Lorentz group, we would take G to be the Poincaré group if we were comparing space time with Minkowski space and G to be the de Sitter group $O(1, 4)$ if we were comparing space-time with de Sitter space. The general construction envisaged by Cartan is the following: H is some closed subgroup of $Gl(d)$, where d is the dimension of M and is also a closed subgroup of G, where $Z = G/H$ has the same dimension as M. We are given a principal bundle P_G over M with structure group G together with a reduction of G to a principal H bundle P_H. We can form the bundle $Z(P_G)$ with fiber Z associated to P_G and the fact that we are given a reduction of P_G to P_H is the same as saying that we are given a section s of $Z(P_G)$ over M, which we can think of as a choice of origin in each fiber of $Z(P_G)$. We can consider the tangent bundle, $T(M)$, and the bundle $V(Z)_s$ of vertical tangent vectors to $Z(P_G)$ along s as d-dimensional vector bundles. They both admit H as structure group. Suppose that we are given an identification of these two bundles. In other words, we have not only identified each point of M with a "base point" of the fiber sitting over it, but have made this identification up to first order. Then a connection on P_G which has no null vectors when restricted to P_H is called a Cartan connection and can be used to formulate the intuitive notion of development described above. Although Cartan wrote his fundamental papers in the 1920s, the precise modern mathematical definition of a Cartan connection was first given by Ehresmann (1950). For further developments of the theory see the papers by Kobayashi and Shoshichi (1956, 1957). In what follows we will be making use of two principal bundles P_H and P_G with H a closed subgroup of G, with P_H a restriction of G and a connection on P_G. We shall drop, however, the requirement that G/H have the same dimension as M. For example, think of H as the Lorentz group and G as either the Poincaré or the conformal group. We want to consider the following situation: H is a closed subgroup of a Lie group G, and we are given a principal H bundle, $P_H \to M$. We can form the associated bundle,

$G(P_H)$ given by the left action of H on G. Thus $G(P_H)$ admits right multiplication by G, and it is easy to see that it becomes a principal G bundle over M. Another way of saying the same thing is that we start out with a principal G bundle P_G and a given reduction to a principal H bundle P_H. Let Q be a Hamiltonian G space and construct the associated bundle, $Q(P_G)$. We let ρ denote the projection of $P_G \times Q \to Q(P)$.

Let Θ be a connection on P_G. We are thus in the situation of Proposition 35.5, with some slight changes in notation, in that we are dropping the # and calling the structure group G instead of K. Let σ_Q now denote the form that was denoted by ω in Proposition 35.5, so that equations (35.21) and (35.22) now become

$$\rho^*\sigma_Q = d\langle \Phi, \Theta \rangle + \Omega. \tag{42.1}$$

Now let U be a manifold upon which H acts and let $f: U \to Q$ be a smooth map that is equivariant relative to the action of H. We can form the associated bundle $U(P_H)$. The map f induces a map $f: U(P_H) \to Q(P_H) = Q(P_G)$. We can then pull σ_Q back to $U(P_H)$; that is, form

$$\omega_U = f^*(\sigma_Q). \tag{42.2}$$

We can now describe the structure of the equations of mechanics. Suppose that the form ω_U has nullity 1. This means that at each point of $U(P_H)$ the space of tangent vectors that satisfy

$$i(\eta)\omega_U = 0 \tag{42.3}$$

is one-dimensional. This describes a system of ordinary differential equations on $U(P_H)$ whose solution curves form the null foliation of ω_U. The projection of each of these curves onto M will be a curve on M describing a trajectory of a "classical particle." More generally, if the rank of ω_U does not change, we obtain a (multidimensional) null foliation on $U(P_H)$; that is, through each point of $U(P_H)$ there will pass a submanifold whose tangent space at every point consists of solutions to the differential system (42.3). The case where the leaves of (42.3) project onto higher-dimension submanifolds of M corresponds to the situation where the classical particle cannot be localized in space. For example, this happens for the relativistic photon. Suppose that the set of all leaves of the null foliation forms a manifold $\mathscr{C}$, the so-called quotient manifold, and let $\Pi: U(P_H) \to \mathscr{C}$ denote the projection. Then ω_U determines a symplectic form, $\omega_{\mathscr{C}}$ on $\mathscr{C}$ such that

$$\Pi^*(\omega_{\mathscr{C}}) = \omega_U.$$

The space $\mathscr{C}$ is what Souriau calls "l'espace de mouvements" or "space of motion," while $U(P_H)$ is "l'espace d'evolution" or evolution space. The situation is summarized by the following diagram:

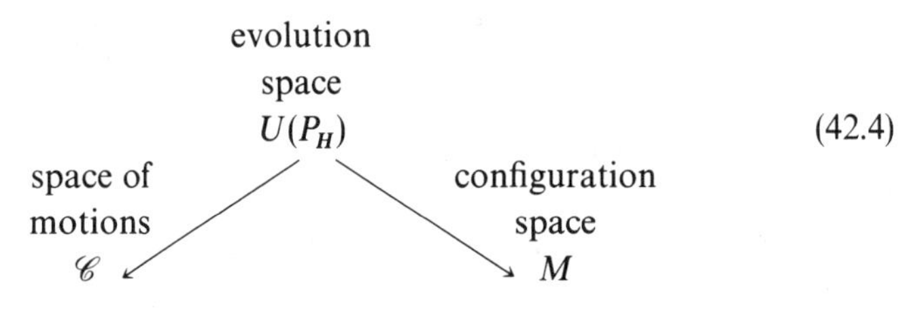

(42.4)

Frequently, it is convenient to choose a subbundle $\mathscr{E}$ of $U(P_H)$ such that the restriction of ω_U to $\mathscr{E}$ has a one-dimensional nullity. This gives trajectories on M and we have the diagram

$$\begin{array}{ccccc} & & U(P_H) & & \\ & \swarrow & \downarrow & \nwarrow & \\ C & \longleftarrow & & \longleftarrow & \mathscr{E} \\ & & \downarrow & \swarrow & \\ & & M & & \end{array} \tag{42.5}$$

However, the bundle $\mathscr{E}$ need not have the full symmetry group of $U(P_H)$. One must "break the symmetry" in order to "localize the particle." We shall illustrate this a little further on for the case of the photon.

Suppose that H is a closed subgroup of two different groups G_1 and G_2. This induces homomorphism of $h \to g_1$ and $h \to g_2$, together with a linear representation of H on g_1 and g_2. Suppose that we are given a linear map $\lambda: g_1 \to g_2$ that is equivariant for the action of H and such that the diagram

$$\begin{array}{ccc} & \nearrow & g_1 \\ h & & \downarrow \lambda \\ & \searrow & g_2 \end{array} \tag{42.6}$$

commutes. Let Q_1 and Q_2 be presymplectic manifolds with forms Ω_1 and Ω_2 with Hamiltonian actions of G_1 and G_2, respectively, and let f_1 and f_2 be H-equivariant maps of manifolds U_1 and U_2 upon which H acts into Q_1 and Q_2. We say that f_1 and f_2 are consistent with an H map $\phi: U_1 \to U_2$ if

$$f_1^*\Omega_1 = \phi^* f_2^* \Omega_2$$

and the diagram

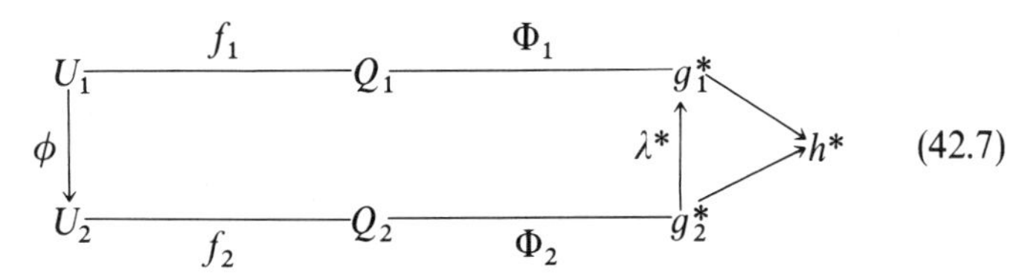

(42.7)

commutes. (Note that the right-hand triangle in this diagram automatically commutes, and that, if Q_1 and Q_2 are orbits of G_1 and G_2 acting on g_1^* and g_2^* so that the Φ_i determine the Ω_i, then this second condition implies the first.) Let P_H be a principal H bundle and let P_{G_1} and P_{G_2} be extensions of this bundle to G_1 and G_2. We say that the connections Θ_1 and Θ_2 on P_{G_1} and P_{G_2} are consistent if

$$\Theta_2 = \lambda\Theta_1 \qquad \text{when restricted to } P_H.$$

Notice that on $P_H \times U_1$ the above consistency conditions imply that $f_1^*\Omega_1 = \phi^* f_2^*\Omega_2$ and

$$(\Phi_1 \circ f_1)\cdot\Theta_1 = (\lambda^*\Phi_2 \circ f_2 \circ \varphi)\cdot\Theta_1 = (\Phi_2 \circ f_2 \circ f)\cdot\Theta_2.$$

From these two equations it clearly follows that the induced map ϕ: $U_1(P_H) \to U_2(P_H)$ satisfies

$$\phi^*\omega_{U_2} = \omega_{U_1}. \tag{42.8}$$

We now examine the above construction in the special case where the Lie algebra h has a vector space complement n in g that is invariant under the adjoint action of h. Thus we assume that

$$g = h + n \tag{42.9}$$

and

$$[h, n] \subset n. \tag{42.10}$$

Then we can write $\Theta = \Theta_h + \Theta_n$, where Θ_h denotes the h component of Θ and Θ_n the n component. Both of these forms are defined on P_G, but note that the restriction of Θ_h to P_H induces a connection on P_H. We shall let $\mathscr{H}$ denote projection onto the horizontal component of this connection. Since the tangent space to the fiber of P_H is spanned by vectors coming from h we see that Θ maps vertical tangent vectors into h and hence that Θ_n vanishes on vectors tangent to the fiber of P_H. From this we see that Θ_n, when restricted to P_H, determines a 1-form $\hat{\Theta}_n$ on M with values in the vector bundle $n(P_H)$. We shall let F^h denote the curvature of Θ_h so that F^h is a 2-form on M, with values in $h(P_H)$, that corresponds to the h-valued 2-form $d\Theta_h \circ \mathscr{H}$ on P_H. (Notice that F^h need not be equal to the h component of F because a horizontal vector for Θ_h in P_H only satisfies the condition $\Theta_h(\eta) = 0$ and hence need not be horizontal for Θ, which would require the additional condition $\Theta_n(\eta) = 0$.) The connection Θ_h induces a covariant (exterior) derivative on any vector bundle associated to P_H, which we shall denote by d_{Θ_h}. Thus, for example, we can form $d_{\Theta_h}\hat{\Theta}_n$, which will be a 2-form on M, with values in $n(P_H)$, that corresponds to the n-valued 2-form $d\Theta_n \circ \mathscr{H}$ on P_H.

We can also decompose the function Φ into components, $\Phi = \Phi_h + \Phi_n$ and thus $\tilde{\Phi} = \tilde{\Phi}_h + \tilde{\Phi}_n$. We claim that the following formula for ω_U holds:

$$\omega_U = \tilde{\Phi}_h F^h + d\tilde{\Phi}_n \wedge \hat{\Theta}_n + \tilde{\Phi}_n d_{\Theta_h} \hat{\Theta}_n + \tilde{\Omega}_h. \tag{42.11}$$

In this formula F^h, $\hat{\Theta}_n$, and $d_{\Theta_h}\hat{\Theta}_n$ are vector-bundle-valued forms on M, pulled back to $U(P_H)$. The form $\tilde{\Omega}_h$ is obtained from Ω via the projection, as described above, but this time relative to the connection Θ_h. To prove (42.11), let $\rho_H : P_H \times U \to U(P_H)$ denote the projection. We must show that ρ_H^*, when applied to the right-hand side of the equation (42.11) gives the restriction of $d(\Phi \cdot \Theta) + \Omega$ to $P_H \times U$. Since vectors tangent to the fiber of the map ρ_H give zero, it is sufficient to verify this equality for three types of pairs of vectors

η_1, η_2 (a)

η, ζ (b)

ζ_1, ζ_2 where the η's are horizontal and the ζ's are tangent to U. (c)

Notice that

$$d(\Phi \cdot \Theta) = d\Phi_h \wedge \Theta_h + d\Phi_n \wedge \Theta_n + \Phi_h \cdot d\Theta_h + \Phi_n \cdot d\Theta_n,$$

and the first term on the right vanishes on any of the above three combinations, so we may, in proving the formula, replace $d(\Phi \cdot \Theta) + \Omega$ by

$$\Phi_h \cdot d\Theta_h + d\Phi_n \wedge \Theta_n + \Phi_n \cdot d\Theta_n + \Omega.$$

In case (a) the second and last term of this expression vanish as do the second and last term in the formula for ω_U, and the other two terms of $\rho_H^* \omega_U$ give the same result as the corresponding terms above. In case (b) only the second term is nonzero for both expressions and they give the same answer. In case (c) it is only the last term that survives and it also yields the same answer. This completes the proof of formula (42.11) for ω_U. (In the special case where $n = \{0\}$, $g = h$, and $U = Q$ this gives a proof of the equality of the two definitions of σ in (42.1) and (42.2).)

Suppose we start with a principal bundle P_H and connection Θ_h. We can enlarge the bundle to $P_G = G(P_H)$, where H acts on G by left multiplication. Suppose that $\hat{\Theta}_n$ is a 1-form on M with values in $n(P_H)$. Then Θ_n corresponds to a 1-form Θ_n on P_H with values in n that vanish on vertical vectors and that satisfy $R_a^* \Theta_n = Ad_a \Theta_n$ for all $a \in H$. For any $p \in P_H$ we have the direct sum decomposition $(TP_G)_p = (TP_H)_p + \tilde{n}_p$, where $\tilde{n}_p$ consists of vertical tangent vectors corresponding to elements of n. Extend the definition of Θ_h to all of $(TP_G)_p$ by setting $\Theta_h(\tilde{\eta}) = 0$ for $\eta \in n$ and extend the definition of Θ_n by setting $\Theta_n(\tilde{\eta}) = \eta$. Then set $\Theta = \Theta_h + \Theta_n$. Here Θ is first defined as a map of $(TP_G)_H \to g$ but can clearly be uniquely extended, by right multiplication, so as to be defined as a connection form on P_G. We have thus shown: *Given P_H and Θ_h, the connections Θ on P_G that reduce to Θ_h are in one-to-one correspondence with the set of $n(P_H)$-valued 1-forms on M.*

In particular, given a connection Θ_h on P_H, an $n(P_H)$-valued 1-form $\hat{\Theta}_n$ on M and an H-invariant submanifold U of a presymplectic Hamiltonian G manifold Q, we get a presymplectic structure on $U(P_H)$. Notice that if F is any scalar-valued function on M, then $F\hat{\Theta}_n$ is another $n(P_H)$-valued 1-form. Of course nondegeneracy-type requirements will put some constraint on the class of admissible F's, but this still indicates a degree of flexibility in the choice of symplectic structure.

Two cases of interest are the following:

(a) G is the Poincaré group and H its Lorentz subgroup. Then h is the Lorentz algebra and n the four-dimensional algebra of translations. In this case P_H will be the bundle of frames and Θ_h the Levi-Civita connection. The form Θ_n identifies $n(P_H)$ with the tangent bundle $T(M)$ and the vanishing of the torsion gives $\hat{d}_{\Theta_h}\hat{\Theta}_n = 0$.

(b) G is the conformal group $SU(2,2)$ and $H = Sl(2,C)$. Then P_H is the double covering of the bundle of orthonormal frames, where we assume that M admits a spin structure. Then Θ_h will still be given by the Levi-Civita connection, but $\hat{\Theta}_n$ will be given by Dirac matrices, as was shown in Sternberg (1978a).

Case (a) is a special case of a semidirect product, so let us begin by formulating the theory in this setting. We assume that $G = H \times V$, where H is a Lie group acting on the vector space V. Then recall from section that $g^* = h^* + V^*$ and the action of an element $\begin{pmatrix} a & v \\ 0 & 1 \end{pmatrix}$ of G on an element (S, p) of g^* is given by

$$\mathrm{Ad}^{\#}_{\begin{pmatrix} a & v \\ 0 & 1 \end{pmatrix}}(S, p) = (\mathrm{Ad}^{\#h}_a S + a^{*-1}p \odot v, a^{*-1}p),$$

where $\mathrm{Ad}^{\#h}$ denotes the coadjoint action of H on h^*. The G orbits in g^* are fibered over the H orbits in V^*. Let $\mathcal{O}$ be a G orbit in g^* and $(S, p) \in \mathcal{O}$, and let U be the H orbit through (S, p). Then

$$\mathcal{O} = \{(\mathrm{Ad}^{\#h}_a S + a^{*-1}p \odot v, a^{*-1}p), a \in h, v \in V\}, \qquad (42.12)$$

while

$$U = \{(\mathrm{A}_a^{\#h} S, a^{*-1}p), a \in H\}. \tag{42.13}$$

If $P_H \to M$ is any principal H bundle over any manifold M and Θ_h is a connection on P_H and $\hat{\Theta}_n$, a $V(P_H)$-valued form on M, then the procedure described above gives rise to a closed 2-form ω_U on $U(P_H)$. If the quotient of $U(P_H)$ by the null foliation of ω_U is a manifold, then we have denoted this manifold by $\mathscr{C}$ and observed that it has a symplectic structure. Let us apply this whole procedure to the case where $M = V$, where P_H is the trivial bundle $P_H = V \times H$ with flat connection and $\hat{\Theta}_n(\xi) = \xi$ under the standard identification of $T(V)_x$ with V valid for any vector space. Then, by (42.11),

$$\omega_U = \Omega_{|U} + dp \wedge dv, \tag{42.14}$$

where Ω is the symplectic form on $\mathcal{O}$ and p is the function on $\mathcal{O}$ that assigns the value p to the point (S, p). Now the group G acts transitively on $V \times U$ by

$$(a, w)(x, l) = (ax + w, \mathrm{Ad}_a^{\#} l) \tag{42.15}$$

and it is easy to check that this action preserves ω_U. Thus G acts transitively on the quotient symplectic space $\mathscr{C}$; that is, $\mathscr{C}$ is a symplectic homogeneous space for G. On the principle that you can't get something for nothing, we should expect that $\mathscr{C}$ must be the original orbit $\mathcal{O}$. Let us prove this by computing the kernel of null foliation through the point $(x, l) = (x, (S, p))$. Write a tangent vector $\xi \in T(V \times U)_{(x,l)}$ as

$$\xi = \xi_V + \tilde{\eta},$$

where $\xi_V \in V$ is tangent to V and $\tilde{\eta}$ is the vector tangent to U corresponding to $\eta \in h$. Let h_p denote the isotropy algebra of p in h. Then

$$i(\xi)\omega_U = i(\tilde{\eta})\Omega_{|U} + \hat{\eta}_p \cdot dv - dp \cdot \xi_V,$$

where $\hat{\eta}_p$ denotes the tangent vector to $p \in V^*$ corresponding to η. From this we see that the equation $i(\xi)\omega_U = 0$ is equivalent to the conditions

$$\eta \in h_p \tag{42.16}$$

and

$$dp \cdot \xi_V = i(\tilde{\eta})\Omega_U \qquad \text{as a linear function on } h. \tag{42.17}$$

If we compare these two equations with (42.12) we see that these are precisely the conditions that the element (η, ξ) of g belong to the isotropy algebra of (S, p). Thus, *the null foliation at $(v, (S, p))$ coincides with the space* *of vectors* $\xi_V + \eta_{(S,p)}$, *where* $(\eta, \xi) \in g$ *ranges over the isotropy algebra of* (S, p). In particular, if the isotropy group $G_{(S,p)}$ of (S, p) is connected, we may identify $\mathcal{O}$ and $\mathcal{C}$ with $G/G_{(S,p)}$ showing that $\mathcal{O} = \mathcal{C}$ as homogeneous spaces. In any event, we have the G map $\Phi: V \times U \to \mathcal{O}$ given by

$$\Phi(v, (S, p)) = (S + p \odot v, p)$$

and we have

$$\Phi^*\Omega = \omega_U,$$

as can be checked from (42.14) by looking at vector fields coming from g.

Notice that we end up with the same symplectic manifold $\mathcal{C}$ no matter which H suborbit $U \subset \mathcal{O}$ we choose. Since the different suborbits will, in general, be of varying dimensions, this reflects itself in the fact that the dimensions of the null foliation will vary. Let $h_p(S) \subset h^*$ denote the set of all elements ζ_S, where $\zeta \in h_p$ and ζ_S denotes the tangent vector at S corresponding to ζ (considered as an element of h^*). We can write equation (42.17) as

$$\zeta_S = p \odot \xi. \tag{42.18}$$

Let $p \odot V \subset h^*$ denote the space of all elements of the form $p \odot v$. From (42.18) it is clear that *if* we can choose (S, p) so that

$$h_p(S) \cap p \odot V = \{0\}, \tag{42.19}$$

then the null foliation has minimal dimension and is spanned by vectors ζ in V satisfying

$$p \odot \zeta = 0. \tag{42.20}$$

(We should point out that our assertion that $\mathcal{C}$ depends only on $\mathcal{O}$ and is independent of the choice of U is not necessarily true in any case other than the flat case of the trivial bundle.)

Let us now specialize to the case that V is a real vector space with a scalar product and that $H = SO(V)$ is the connected component of the orthogonal group for this scalar product. (We soon specialize further to the case that $V = \mathbb{R}^{1,3}$ is Minkowski space and H is the proper Lorentz group.) We may then identify V and V^* using the scalar product as well as h and h^*. We may also identify h with $\wedge^2 V$, where $u \wedge w$ is identified with the linear operator $A_{u \wedge w}$ given by

$$A_{u \wedge w} v = (w, v)u - (u, v)w.$$

Under these identifications it is easy to check that

$$p \odot v = p \wedge v \tag{42.21}$$

and

$$\zeta_S = [\zeta, S]. \tag{42.22}$$

The orbits of H acting on V are the connected components of the (d-1)-dimensional "pseudospheres" $\| p \|^2 = \text{const.}$ together with the trivial orbit $\{0\}$. Let us examine condition (42.19) for the nontrivial orbits. We claim that there is an essential difference between the cases $\| p \|^2 = 0$ and $\| p \|^2 \neq 0$. It is easy to see that,

$$h_p = \wedge^2(p^\perp) \tag{42.23}$$

where $p^\perp$ denotes the subspace of V consisting of all vectors orthogonal to p and $\wedge^2(p^\perp)$ is the subspace of $\wedge^2(V) = h$. Now if $\| p \|^2 \neq 0$, then $p \in p^\perp$ and we have the direct sum decomposition

$$\wedge^2(V) = \wedge^2(p^\perp) \oplus p \odot V. \tag{42.24}$$

(This decomposition is known as the decomposition of angular momentum into "intrinsic" and "orbital" components.) In particular, for any orbit $\mathcal{O}$ of G sitting over the H orbit we are considering, we can, by an appropriate translation, "get rid of the orbital component," that is, move to an H suborbit whose points (S, p) satisfy (42.19). For such suborbits, we know according to (42.21) that (42.20) becomes

$$p \wedge v = 0, \tag{42.25}$$

which has the one-dimensional space of solutions $v = rp$. For this choice of U the null foliation is one-dimensional, and the projection onto M of the leaf of the null foliation through $(x, (S, p))$ consists of all points of the form

$$x + rp$$

a worldline of a free particle with momentum p. We should remark that condition (42.19) has a number of equivalent reformulations valid when $\| p \|^2 \neq 0$:

$$S \in h_p,$$
$$S_p = 0,$$
$$*S \wedge p = 0,$$

where in this last formulation, we have used the star operator mapping $\bigwedge^2(V) \sim \bigwedge^{d-2}(V)$ so that $*S \wedge p \in \bigwedge^{d-1}(V)$.

In particular, let us consider the G orbit passing through $(0, p)$ that is $2(d-1)$-dimensional and its H suborbit U passing through $(0, p)$ that is $(d-1)$-dimensional. For any Riemannian or Lorentzian manifold M, we can form a bundle of orthonormal frames P_H and the associated bundle $U(P_H)$, which has (total) dimension $2d-1$, where d is the dimension of M. It is clear from its definition that the bundle $U(P_H)$ can be identified with the "sphere bundle" $S^*_{\|p\|}(M)$ consisting of all covectors of length $\|p\|$. Thus we may identify $U(P_H)$ as a subbundle of T^*M. We may take the standard symplectic form on T^*M and restrict to $U(P_H)$ and it is easy to check that this coincides with the presymplectic form ω_U. (This verification depends on the fact that the torsion of the Levi-Civita connection vanishes.) The null foliation of this presymplectic form gives precisely the geodesic flow on $S^*_{\|p\|}(M) = U(P_H)$. In this way our group-theoretical construction recovers the standard equations for geodesics.

For $\|p\|^2 = 0$ the situation is quite different because now $p \in p^\perp$. Then (42.24) no longer holds so that we cannot arrange, by an appropriate translation, that $S \in h_p$, for an arbitrary orbit. Furthermore, even if we could arrange that $S \in h_p$, if $S \neq 0$, this will not imply (42.19) and the null foliation will be higher-dimensional corresponding to nonlocalizability of the particle. Let us illustrate this for the case of the Poincaré group. Let e_0, e_1, e_2, e_3 be an orthonormal basis of $\mathbb{R}^{1,3} = V$. There are two nontrivial H orbits with $\|p\|^2 \neq 0$, the "forward light cone," which is the orbit through $e_0 + e_3$ and the "backward light cone," which is the orbit through $-e_0 + e_3$. With no loss of generality, we can concentrate on the forward light cone and $p = e_0 + e_3$. Let us set $u = e_0 - e_3$ so that $(u, p) = 2$ and write

$$V = \mathbb{R}p \oplus \mathbb{R}u \oplus W, \quad \text{where} \quad W = u^\perp \cap p^\perp \quad \text{is spanned by } e_1, e_2.$$

We can write the most general element of h as

$$S = u \wedge w + se_1 \wedge e_2 + p \wedge v, \quad \text{where} \quad w \in W \quad \text{and} \quad v \in V.$$

By translation we can get rid of the last term so we may choose (S, p) on $\mathcal{O}$ such that

$$S = u \wedge w + se_1 \wedge e_2. \tag{42.26}$$

Note that

$$\|S \wedge p\|^2 = -4\|w\|^2,$$

so the first term in (42.26) will vanish or not according as $\|S \wedge p\|^2$ vanishes or not. If $w \neq 0$, the orbit $\mathcal{O}$ is eight-dimensional and does not correspond to known particles. Let us examine the case $w = 0$ with $\mathcal{O}$ six-dimensional and $s \neq 0$. Then $h_{e_0+e_3}$ is the two-dimensional Euclidean algebra $e(2)$ spanned by

$$(e_0 + e_3) \wedge e_1, \qquad (e_0 + e_3) \wedge e_2, \qquad e_1 \wedge e_2.$$

The H orbit of $(se_1 \wedge e_2, e_0 + e_3)$ is five-dimensional; the points in the H orbit sitting over $(e_0 + e_3)$ consist of

$$(S, p) = (se_1 \wedge e_2 + (e_0 + e_3) \wedge w, e_0 + e_3), \qquad w \in W.$$

From this we see that the null foliation is three-dimensional and is spanned by the images of the elements

$$(0, e_0 + e_3), \qquad ((e_0 + e_3) \wedge e_1, -e_1), \qquad ((e_0 + e_3) \wedge e_2, -e_2),$$

of g. Thus the leaf of the null foliation through $(x, (S, p))$ will be the three-dimensional affine space consisting of all points

$$x + r(e_0 + e_3) + a_1 e_1 + a_2 e_2.$$

We can think of this as a plane in space parallel to the e_1-e_2 plane passing through x and moving with the velocity of light in the direction e_3. We can "interpret" this plane as a wave front, but we cannot localize the particle on this wave front in a relativistically invariant manner.

Suppose we break the relativistic invariance by choosing some space-time splitting, which amounts to the choice of some fixed vector e_0. We can then cut down the bundle of frames to an $O(3)$ subbundle and also look at the submanifold $U' \subset U$ consisting of those (S, p) satisfying

$$Se_0 = 0.$$

Then U' is three-dimensional and

$$\mathscr{E} = U'(P_{O(3)})$$

is a submanifold of $U(P_H)$ such that the restriction of ω_U has a one-dimensional null foliation. We obtain in diagram (42.5) the projection onto M of the leaf of the null foliation passing through $(x, (S, p))$ in $\mathscr{E}$ as the "light ray" consisting of all points $x + rp$. Thus the process of space localization involves breaking the global symmetry. The subgroup $O(3)$ does not act transitively on the forward light cone. However, for the mass-zero particles, we can enlarge the Lorentz group by throwing in the scale transformations; that is, replace H by $H^+ = H \times \mathbb{R}^+$ and G by $G^+ =$ semidirect product of

H^+ with the translations. The group $O(3)^+ = O(3) \times \mathbb{R}^+$ acts transitively on the forward light cone and all the spaces in diagram (42.5) become homogeneous spaces. In fact, (42.5) becomes

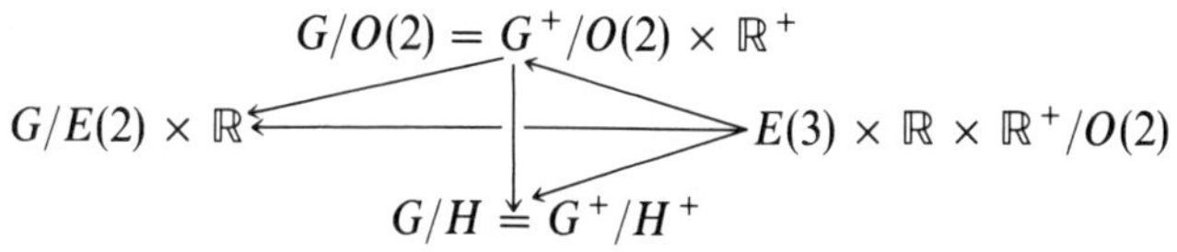

$$\begin{array}{ccc} & G/O(2) = G^+/O(2) \times \mathbb{R}^+ & \\ G/E(2) \times \mathbb{R} & & E(3) \times \mathbb{R} \times \mathbb{R}^+/O(2) \\ & G/H \doteq G^+/H^+ & \end{array}$$

At the end of Section 20 the equations for a spinning particle in general relativity (in the absence of magnetic moment) were obtained by considering, over the cotangent bundle T^*M, the bundle associated to a four-dimensional orbit of the Lorentz group characterized by $S \circ S = 0$ and $S \cdot S = s^2$. This gives rise to a twelve-dimensional symplectic manifold. The condition $*S \wedge p = 0$, $p^2 = m^2$ were imposed, which defined a nine-dimensional submanifold. The restriction of the symplectic form from the twelve-dimensional symplectic manifold to its nine-dimensional submanifold induced a presymplectic structure with one-dimensional null foliation on the latter. It is easy to check that the equations so obtained coincide with the equations written down above for massive spinning particles.

Finally we should point out that there is a slightly different way of writing the equations of motion derived in this section, due to Duval (1976) and Duval and Horvathy (1982) which is sometimes more convenient for the purposes of computation, and which works for the case that Q is a G orbit $\mathcal{O}$ in g^*. Let us fix a point α in U and consider that map

$$j_\alpha : P_H \to P_H \times U, \qquad j_\alpha(z) = (z, \alpha).$$

Then

$$j_\alpha^* \sigma_Q = d\langle \alpha, \Theta \rangle. \tag{42.27}$$

Let

$$r_\alpha = \rho \circ j_\alpha,$$

where ρ denotes the projection of $P_H \times U$ onto $U(P_H)$. Then r_α is a submersion and we have

$$r_\alpha^* \omega_U = d\langle \alpha, \Theta \rangle. \tag{42.28}$$

The expression on the right of (42.28) is quite simple and amenable to computations. Of course, it will have a larger null foliation, existing on the principal bundle rather than the associated bundle.

43. The moment map and kinetic theory

We close this chapter with a brief sketch of some recent exciting results of Marsden and Weinstein (1981, 1982) and their co-workers on the equations of plasma physics. The ideas involved would seem applicable in many other areas of kinetic theory. There are several observations, each of which is elegant and important in its own right:

(1) Let g be a Lie algebra having an invariant bilinear form that is nondegenerate. This allows an identification of the dual space g^* with g. Let H be any function on g^* and let $T: g^* \to g$ be our identification. Then H induces a vector field on g^*. Using T, the vector field can be written as

$$\tilde{H}(\alpha) = T^{-1}[T\alpha, L_H(\alpha)], \tag{43.1}$$

since we are using T to identify the adjoint and coadjoint representations. We shall want to apply this observation to the situation where M is a symplectic manifold, g is the Lie algebra of smooth functions on M under Poisson bracket, and the invariant bilinear form is

$$b(f_1, f_2) = \int_M f_1 f_2 \Omega, \qquad \text{where} \qquad \Omega = \omega^n, 2n = \dim M \tag{43.2}$$

is the Liouville measure. The space g^* is identified with the space of distribution densities whose elements we can write symbolically as

$$f\Omega, \tag{43.3}$$

where f is a generalized function and the corresponding map T would be given by

$$T(f\Omega) = f. \tag{43.4}$$

Of course, as it stands, (43.2) and (43.4) do not make sense; for instance, the integral in (43.2) need not converge. We might take g to consist of all smooth functions, in which case the f occurring in (43.3) would have compact support, but still (43.4) is not well defined as a map into g, since f is only a generalized function and not a function. However, we can consider the subspace $g^\# \subset g^*$ consisting of all $f\Omega$ with f smooth and of compact support. It is easy to check (by integration by parts) that $g^\#$ is an invariant subspace of g^* under the coadjoint action of g. We then replace g^* by $g^\#$ in (43.4) and in (43.1) (where we need only assume that H is defined on $g^\#$). (In specific instances, one might want to have slightly different definitions of g and $g^\#$, perhaps replacing compact support by some sort of decrease at infinity, etc.)

Marsden and Weinstein (1981) and Gibbons (1982) consider the following example. Take $M = T^*\mathbb{R}^3$ and consider f as an (unnormalized) probability density on phase space. Thus

$$\Omega = dp\,dx, \qquad dp = dp_1\,dp_2\,dp_3 \qquad dx = dx_1\,dx_2\,dx_3$$

and

$$\rho_f = \int f\,dp \tag{43.5}$$

as the associated matter density on $\mathbb{R}^3$. The total kinetic energy associated with such a system would be

$$\tfrac{1}{2}\int \frac{p^2}{m} f\,dp\,dx;$$

and if the particles were in an external potential field, the external potential energy would be given by

$$\int V(x) f(x, p)\,dx\,dp.$$

Suppose that we imagine a two-particle interaction with "Green's function" $G(x, x')$. For example, for electrostatic (Coulomb) interaction, we might take (up to constants depending on unit convention)

$$G(x, x') = \frac{q}{\|x - x'\|},$$

so that

$$\varphi_\rho = \int G(x, x')\rho(x')\,dx'$$

is the solution of the Poisson equation

$$\Delta\varphi_\rho = q\rho.$$

Writing G for the integral operator corresponding to the function $G(x, x')$ the "self-interaction" energy corresponding to f is then given by

$$\int fGf\,dx.$$

Hence, the total energy thought of as a function on $g^{\#}$ is given by

$$H(f\Omega) = \int\left(\frac{p^2}{2m} + V(x)\right) f(x, p)\,dx\,dp$$

$$+ \iint G(x, x') f(x, p) f(x', p')\,dx\,dp\,dx'\,dp' \tag{43.6}$$

This is an (inhomogeneous) quadratic function, and it is clear that

$$L_H(f\Omega)(x, p) = \frac{1}{2m} p^2 + V(x) + (G\rho_f)(x). \tag{43.7}$$

In equation (43.1) the [,] become Poisson brackets and (43.1) becomes

$$\frac{\partial f}{\partial t} = \{f, L_H(f\Omega)\}. \tag{43.8}$$

For the electromagnetic case (in the absence of V) this is known as the Poisson–Vlasov equation.

A second crucial observation made by Marsden and Weinstein is that the operation of integrating over the momentum variable, that is, the passage from f to ρ_f in (43.5), can be regarded as a momentum map. Indeed, if $h \to g$ is a homomorphism of Lie algebras, the corresponding momentum map is just the dual map $g^* \to h^*$. If X is a differentiable manifold the projection $\pi: T^*X \to$ induces a map $\pi^*: F(X) \to F(T^*X)$ from functions on X to functions on T^*X. We can regard $F(X)$ as a commutative Lie algebra and π^* as a homomorphism into the Lie algebra of all functions on T^*X under Poisson bracket. The dual spaces are the (generalized) densities and the dual map to the "pull-back" π^* is just the "push-forward" π_* which is just integration over the fiber (cf., e.g., Guillemin and Sternberg, 1976, Chap. VI).

Let $\mathscr{A}$ denote the vector space of all smooth linear differential forms on X. The operator d gives a map form $F(X) \to \mathscr{A}$ and hence a corresponding (translation) action of $F(X)$ on $\mathscr{A}$:

$$(f, A) \to A + df.$$

(We can think of $F(X)$ as the Lie algebra of the gauge group of a trivial circle or line bundle over X and of $\mathscr{A}$ as the space of all its connections; then this action is just the usual (infinitesimal version of) the action of the gauge group on the space of connections.) Assuming X to be oriented, we can identify $\mathscr{A}^*$ with the space of generalized $n - 1$ forms on X ($n = \dim X$) with the pairing given by integration over X:

$$\langle A, Y \rangle = \int_X A \wedge Y.$$

(As usual, appropriate compact support assumptions must be made.) The action of $F(X)$ on $\mathscr{A}$ induces a Hamiltonian action of $F(X)$ on $T^*\mathscr{A} = \mathscr{A} + \mathscr{A}^*$

$$(f;(A, Y)) \to (A + df, Y).$$

We can identify the dual space $F(X)^*$ with the space of generalized n-forms on X. The moment map $\Phi: T^*\mathscr{A} \to \bigwedge^n(X)_{\text{gen}}$ is then given by

$$\langle \Phi(A, Y), f \rangle = \langle df, Y \rangle = \int df \wedge Y,$$

or, by Stokes' theorem,

$$\Phi(A, Y) = -dY. \tag{43.9}$$

We can consider the Marsden–Weinstein reduction of $T^*\mathscr{A}$ at some point ρ: Thus $\Phi^{-1}(\rho)$ consists of all pairs (A, Y) with

$$dY = -\rho \tag{43.10}$$

Since the gauge group is commutative, the isotropy group of ρ is the whole gauge group. So dividing by the isotropy group means identifying A with $A + df$ for any function f. This, of course, amounts to replacing A by $dA = B$, where

$$dB = 0. \tag{43.11}$$

If X were simply connected, the Marsden–Weinstein reduced space T^*X_ρ could then be identified with the set of all pairs (B, E), where $E = -*Y$ satisfying (43.11) and

$$d*E = \rho. \tag{43.12}$$

This shows that we should think of E and B as the electric and magnetic fields. Indeed, Marsden and Weinstein consider the case where $X = \mathbb{R}^3$ and take the Hamiltonian

$$H(A, Y) = \tfrac{1}{2}\int_{\mathbb{R}^3} (\|dA\|^2 + \|Y\|^2)\, dx. \tag{43.13}$$

This quadratic Hamiltonian is clearly invariant under the action of the "gauge group" $F(X)$ and so induces a flow on the Marsden–Weinstein reduced space. It is easy to check that the equations for the corresponding reduced flow, together with (43.11) and (43.12) are precisely Maxwell's equations.

Finally, they consider the action of $F(X)$ on the product

$$F(T^*\mathbb{R}^3) \times T^*\mathscr{A}.$$

The corresponding moment map is given by

$$\Phi((f\Omega, (A, Y)) = \rho_f \, dx + dY.$$

The Hamiltonian

$$H(f, A, Y) = \tfrac{1}{2} \int \| p - A(x) \|^2 f\Omega + \tfrac{1}{2} \int (\| Y \|^2 + \| dA \|^2) \, dx \qquad (43.14)$$

is invariant under the action of the "gauge group." The corresponding equations of motion are verified to be the "Maxwell–Vlasov" equations of plasma physics, when expressed in terms of the Marsden–Weinstein reduced space at zero.

IV

Complete integrability

In this chapter we will give some illustrations of the use of group-theoretical methods, and particularly of the moment map, in establishing the complete integrability of various examples of mechanical systems. Our treatment will not be exhaustive or complete. There have been a number of important research and survey articles on this subject as well as treatments in standard texts, and we must refer the reader to the literature for the complete study. On the other hand, we felt that we must include some discussion of this elegant subject, and so have settled for an anecdotal account. We begin with a general discussion of the notion of complete integrability and of "action angle variables." (Again here we refer the reader to standard texts, such as Abraham and Marsden (1978) or Arnold (1978) for all the details, in particular, for explicit integration methods. Our treatment will be rather structural and abstract.) We will then discuss the various group-theoretical methods and their applications.

44. Fibrations by tori

We begin with some examples. The harmonic oscillator in one dimension has as its Hamiltonian the function

$$\tfrac{1}{2}(p^2/m + kq^2),$$

where m is the mass of the point particle and k is the restoring force (the spring constant). By making the symplectic change of variables

$$p = (mk)^{1/4}x, \qquad q = (mk)^{-1/4}y$$

this becomes

$$\tfrac{1}{2}\lambda(x^2 + y^2) \qquad \text{with} \qquad \lambda = (k/m)^{1/2}.$$

A system of n uncoupled harmonic oscillators is thus given by the Hamiltonian

$$\mathscr{H} = \sum \lambda_i(x_i^2 + y_i^2)$$

on $\mathbb{R}^{2n}$. Let $I_1, \ldots, I_n$ denote the functions

$$I_i = (x_i^2 + y_i^2)$$

and

$$\rho = (I_1, \ldots, I_n)$$

so that ρ maps $\mathbb{R}^{2n}$ onto the positive "2^n-tant" in $\mathbb{R}^n$. For any point in the inverse image of an interior point, the map ρ is a submersion, and the inverse image of any such interior point is an n-dimensional torus that is Lagrangian. The Hamiltonian $\mathscr{H}$ can be written as

$$\mathscr{H} = \rho^*h = h \circ \rho, \tag{44.1}$$

where

$$h = \sum \lambda_i I_i$$

is a function on $\mathbb{R}^n$. The flow determined by $\mathscr{H}$ is a straight-line flow on each torus.

In Section 30 we discussed the spherical pendulum. The phase space there was T^*S^2, and we defined a map $\rho: T^*S^2 \to \mathbb{R}^2$ such that (44.1) holds. There as well, after excluding a certain critical set (consisting of points that map onto the boundary of the image and one point in the interior), the map ρ is a submersion and the fibers are Lagrangian tori.

We want to study this situation in general:

$$\begin{aligned} &M \text{ is a symplectic manifold (of dimension } 2n); && \text{(i)} \\ &\rho: M \to B \text{ is a fibration (where } B \text{ is } n\text{-dimensional)}; && \text{(ii)} \\ &\text{the fibers of } \rho \text{ are compact connected Lagrangian submanifolds of } M. && \text{(iii)} \end{aligned} \tag{44.2}$$

For this we must first review some facts about *affine tori*. Let V be a (finite-dimensional real) vector space thought of as a commutative Lie group. Let X be a compact, connected smooth manifold with $\dim X = \dim V$ and suppose that we are given an action of V on X. Recall that for each $x \in X$ we have the map $\psi_x: V \to X$ given by the action

$$\psi_x(v) = \text{the image of } x \text{ under } v.$$

The action is called locally transitive if, for all x, $d(\psi_x)_0 : V \to TX_x$ is an isomorphism. We claim that under our hypotheses *the action is locally transitive if and only if it is transitive.*

Proof. Local transitivity implies that ψ_x maps a neighborhood of the origin in V onto a neighborhood of x. This means that $\psi_x(V)$, the orbit through x, is open. Since X is compact, a finite number of orbits cover X, and orbits by definition are disjoint. Hence the connectedness of X implies that there is only a single orbit, that is, that the action is transitive. Now assume that the action is transitive. Suppose that $d(\psi_x)_0$ is singular for some x. Since $\psi_x(v) = v \cdot \psi_x(0)$, this implies that $d(\psi_x)_v = dv \circ d(\psi_x)_0$ is singular at all v. Thus all points of V are critical points for the map ψ_x, and hence, by Sard's theorem (Sternberg, 1983) the image $\psi_x(V)$ must have measure 0, contradicting the transitivity, which says that $\psi_x(V) = X$. Q.E.D.

We say that X is an *affine torus* if the V action is transitive. Then the isotropy group of any two points of X must be conjugate, and since V is Abelian they must be the same. Let L denote this common isotropy group. The local transitivity implies that there is some neighborhood of the origin for which $\psi_x(v) = x$ implies $v = 0$; thus L is a discrete subgroup of V. Thus L is a lattice, and $V/L = X$ is compact, so L is a maximal lattice and X is a torus. Of course, the identification of X with V/L depends on the choice of a base point x. If $V \to B$ is a vector bundle and $X \to B$ is a fiber bundle with compact connected fibers, and we are given an action of V on X, then the previous discussion carries through, where now L is a subbundle of V, whose intersection with each fiber is a lattice. The identification of X with V/L becomes a little more problematical: One must choose a section of X in order to make this identification, and there may be some topological obstruction to the existence of such a section. All we can say is that if a section exists, then a choice of such a section gives rise to an identification of X with V/L.

(Without going into the details we remind the reader of the nature of the topological problem, referring to the standard texts in algebraic topology for details: One triangulates B so that the fibration is trivial over each simplex. One then attempts to define a section inductively over the k skeleton, assuming that a section has already been defined over the k-1 skeleton. Thus for each simplex, one has a map of its boundary into the fiber F (due to the local triviality) and thus an element of the homotopy group $\pi_{k-1}(F)$. This defines an element of $H^k(B, \underline{\pi_{k-1}(F)})$, where $\underline{\pi_{k-1}(F)}$ denotes the locally constant sheaf of homotopy groups of the fiber. The vanishing of this class is a necessary and sufficient condition for the extendability of the section from the k-1 skeleton to the k skeleton. In the case at hand, the fiber F is a torus, so $\pi_1(F) = \mathbb{Z}^n$ and $\pi_k(F) = 0$ for $k > 1$. There is only one obstruction to the existence of a global section and it is an element of $H^2(B, \pi_1(F))$ and is known as the Chern class of the fibration.)

Let us now go back to our more specific concern, where M is a symplectic manifold and $\rho: M \to B$ is a foliation by compact, connected Lagrangian submanifolds. Let x be a point of M and $b = \rho(x)$ the corresponding base point in B, and let F denote the fiber through x. We then have the exact sequence

$$0 \to TF_x \to TM_x \xrightarrow{d\rho_x} TB_b \to 0. \tag{44.3}$$

The dual map $d\rho_x^*$ sends TB_b^* into those covectors that vanish on TF_x; that is,

$$d\rho_x^*: TB_b^* \simeq NF_x^*. \tag{44.4}$$

But since F is Lagrangian, the symplectic form gives an identification of NF_x^* with TF_x. In this way an element v of TB_b^* gives rise to a tangent to the fiber at each $x \in F$, that is, to a vector field $\hat{v}$ along the fiber F. Suppose that we choose some function h on B such that $dh_b = v$. Then $d(\rho^*h)_x = d\rho_x^*(dh_b)$, and hence it follows from the above definitions that

$$\xi_{\rho^*h}(x) = \hat{v}(x) \qquad \text{if} \qquad v = dh_b. \tag{44.5}$$

In particular,

$$\xi_{\rho^*h} \text{ is tangent to the fibration;} \tag{44.6}$$

and hence

$$\xi_{\rho^*h_1}\rho^*h_2 = 0. \tag{44.7}$$

Thus

$$[\hat{v}_1, \hat{v}_2] = 0 \qquad \forall\, v_1 \qquad \text{and} \qquad v_2 \text{ in } TB_b^*. \tag{44.8}$$

We thus have an infinitesimal action of the commutative Lie algebra TB_b^* on F. Since F is compact and TB_b^* is connected, this exponentiates to an action of the commutative group TB_b^* on F. Since (44.4) is an isomorphism, this action is locally transitive and hence transitive. We thus get a transitive action of the vector bundle T^*B on M.

Let α be a linear differential form on B, so we can view α as a section of T^*B. Then α gives rise to a vertical vector field $\hat{\alpha}$ on M, and hence a flow

$$\kappa_{t,\alpha} = \exp t\hat{\alpha} \tag{44.9}$$

and hence a diffeomorphism

$$\kappa_\alpha = \kappa_{1,\alpha}. \tag{44.10}$$

The flow $\kappa_{t,\alpha}$ is vertical, so

$$\rho \circ \kappa_{t,\alpha} = \rho, \tag{44.11}$$

and the vector field $\hat{\alpha}$ is determined by the equation

$$i(\hat{\alpha})\omega = \rho^*\alpha, \tag{44.12}$$

where ω is the symplectic form of M. Therefore

$$D_{\hat{\alpha}}\omega = i(\hat{\alpha})\, d\omega + di(\hat{\alpha})\omega = d\rho^*\alpha. \tag{44.13}$$

Since

$$\frac{d}{dt}\kappa_{t,\alpha}^* = \kappa_{t,\alpha}^* D_{\hat{\alpha}}$$

and

$$\kappa_\alpha^* = \mathrm{id} + \int_0^1 \frac{d}{dt}\kappa_{t,\alpha}^*\, dt$$

it follows from (44.11)–(44.13) that

$$\kappa_\alpha^*\omega = \omega + d\rho^*\alpha. \tag{44.14}$$

We can summarize our results in the following:

Theorem 44.1. *Let* (M, ω) *be a symplectic manifold and let* $\rho: \mathrm{M} \to \mathrm{B}$ *be a fibration whose fibers are compact connected Lagrangian manifolds. Then there is a transitive action of* $\mathrm{T^*B}$ *on* M *and the fibers are tori. Each linear differential form* α *on* B *defines an automorphism* κ_α *of the fibration and (44.14) holds.*

Notice that if the fibers are not assumed to be compact, then the infinitesimal action of T^*B on M need not exponentiate to a global action. If it does, then most of Theorem 44.1 remains true. The fibers are no longer tori, but are of the form $T^k \times \mathbb{R}^{n-k}$, a torus cross a Euclidean space. For example, we could consider the fibration $T^*B \to B$, giving the standard (vector bundle) action of T^*B on itself. We shall make use of (44.14) for this case.

Now let us make the assumption that a section $s: B \to M$ of the fibration exists. Then $s^*\omega$ is a 2-form on B. Let us assume that

$$s^*\omega = 0, \tag{44.15}$$

that is, that the section s defines a Lagrangian submanifold.

(Actually, if a section s exists, the form $s^*\omega$ defines a deRham cohomology class $[s^*\omega] \in H^2(B, \mathbb{R})$. Suppose that we merely assume that this cohomology class vanishes, so that

$$[s^*\omega] = 0 \tag{44.16}$$

holds. Thus $s^*\omega = -d\alpha$ for some linear differential form α on B. Then replacing s by $\kappa_\alpha \circ s$ and using (44.14) we get a new section that satisfies (44.15). So if we start with a section that satisfies (44.16) we can always modify it so as to obtain a section satisfying (44.15). In particular, if ω itself is exact (i.e., $[\omega] = 0$ on M), then (44.16) holds for any section, so we can arrange for (44.15) to hold. This is frequently the case in examples in classical mechanics where ω is given as the (negative of) exterior derivative of the action form. Similarly, if $H^2(B, \mathbb{R}) = \{0\}$ we can always arrange that our section satisfy (44.15).)

In any event, let us assume (44.15). This then defines a map $\chi: T^*B \to M$, where

$$\chi(v) = -v \cdot s(x), \qquad v \in TB_b^*,$$

where the right-hand side denotes the action of v on the element $s(x)$ that lies in the fiber of M above b. It is clear that χ is a fiber map, that is, the diagram

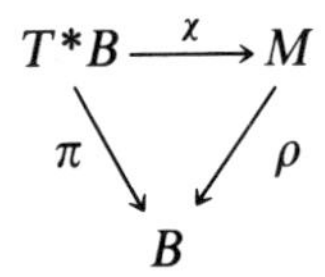

commutes, where π denotes the standard projection of $T^*B \to B$. It is also clear that χ carries the standard (group) action of T^*B on itself into the action of T^*B on M:

$$\chi(v + w) = v \cdot \chi(w).$$

The main observation of this section is

Theorem 44.2. *The map χ is (locally) symplectic; that is, $\chi^*\omega$ is the standard symplectic form on* T*B.

Proof. Let B_0 denote the zero section of T^*B, so that $\chi(B_0) = s(B)$. We claim that it is enough to prove that χ is symplectic at all points of B_0. Indeed, if v is any point of T^*B, we can write $v = \alpha(b)$, where α is a linear differential form on B, so $v = \hat{\kappa}_\alpha(0)$, where 0 is the zero element of T^*B_b, and where $\hat{\kappa}_\alpha$ denotes the diffeomorphism of T^*B given by α, just as κ_α denotes the corresponding diffeomorphism of M. Since

$$\chi \circ \hat{\kappa}_\alpha = \kappa_\alpha \circ \chi$$

and since (44.14) (and its analog for $\hat{\kappa}_\alpha$) both hold, if χ is a symplectic map at 0, then it is a symplectic map at v.

So we must prove that χ is symplectic at points of B_0. Let b be a point of B and $(b, 0)$ the corresponding point of B_0. The tangent space to B_0 in T^*B is a Lagrangian space, and it is mapped by $d\chi$ into the tangent space to $s(B)$ at $s(b)$, and this is assumed to be symplectic. The tangent space to the fiber of T^*B is a Lagrangian subspace, and thus is mapped by $d\chi$ into the tangent space to the fiber of M that is Lagrangian.

These two spaces span the tangent space of T^*B. So what we must show is the following: Let u be a tangent to the fiber at $(b, 0)$. We can think of u as an element of T^*B_b. Let w be a tangent of the zero section B_0. We can think of w as an element of TB_b. We must show that

$$\omega_0(u, w) = \omega\big(d\chi(u), d\chi(w)\big),$$

where ω_0 denotes the symplectic form on T^*B (and where we have written $d\chi$ instead of $d\chi_{(b,0)}$). But

$$\omega_0(u, w) = -\langle u, w\rangle,$$

the evaluation of $u \in T^*B_b$ on $w \in TB_b$. Also, $d\chi(u) = -\hat{u}\big(s(b)\big)$ and $d\chi(w) = ds_b(w)$. But, by definition of $\hat{u}$,

$$\begin{aligned}\omega\big(\hat{u}, ds_b(w)\big) &= \langle d\rho_b^* u, ds_b(w)\rangle \\ &= \langle u, d\rho_b \circ ds_b(w)\rangle \\ &= \langle u, w\rangle,\end{aligned}$$

since $\rho \circ s = id$. This proves Theorem 44.2.

Let us now return to the general case of Theorem 44.1, where we do not assume the existence of a section satisfying (44.15). We have the action of T^*B on M and hence the subbundle L of elements that act trivially. The intersection $L \cap TB_b^*$ of L with each fiber is a lattice in the vector space TB_b^*. We claim that

Theorem 44.3. *The subbundle* L *is a Lagrangian submanifold of* T*B.

Proof. Since the assertion of the theorem is local, we can assume that we have chosen a section satisfying (44.15), since the choice of such a section is always possible locally. But then

$$L = \chi^{-1}(s(B)), \tag{44.17}$$

and since $s(B)$ is Lagrangian and χ is symplectic, L is Lagrangian.

Suppose we pick a point b of B and a basis $\beta_1(b), \ldots, \beta_n(b)$ of $L \cap TB_b^*$. Then there are unique differential forms $\beta_1, \ldots, \beta_n$ defined in some neighborhood of b that take on the given values at b and that form a local basis of L. Each of the β_i defines a Lagrangian section of T^*B by Theorem 44.3; in other words, $d\beta_i = 0$. Thus, in a neighborhood of b, we can write

$$\beta_j = 2\pi\, dI_j,$$

where $I_1, \ldots, I_n$ are functions on B. The fact that the β_i form a basis of L over each point means, in particular, that they are independent. Thus we can use the functions $I_1, \ldots, I_n$ as coordinates near b. Let $\theta_1, \ldots, \theta_n$ be the corresponding dual variables on T^*B near $\pi^{-1}B$. In terms of the local coordinates $I_1, \ldots, I_n; \theta_1, \ldots, \theta_n$, the lattice subbundle is described by

$$\theta_j \in 2\pi\mathbb{Z}, \qquad j = 1, \ldots, n.$$

Thus

$$(I_1, \ldots, I_n; \theta_1, \ldots, \theta_n), \qquad \theta_j \bmod 2\pi, \tag{44.18}$$

form a system of local coordinates on M known as *action angle coordinates*, and the symplectic form on M is locally given as

$$\omega = d\sigma, \tag{44.19}$$

where

$$\sigma = \sum I_i d\theta_i \tag{44.20}$$

is a well-defined linear differential form on M. Note that

$$I_i(c) = \frac{1}{2\pi} \int_{\gamma_i(c)} \sigma, \tag{44.21}$$

where $\gamma_i(c)$ is the curve above c given by $0 \leqslant \theta_i \leqslant 2\pi$, $\theta_j = 0, j \neq i$. We have the theorem of Arnold, Jost, Markus, and Meyer (see Markus and Meyer, 1974).

Theorem 44.4. *Under the hypotheses of Theorem 44.1, local action angle coordinates exist. If σ is a linear differential form on* M *such that (44.19) holds, and if γ_i(c) are smoothly varying curves in the fiber above* c *whose* *homotopy classes* $[\gamma_i(c)]$, $i = 1, \ldots, n$ *form a basis of the fundamental group of the fiber above* c *for each* c, *then the functions* I_i *defined by (44.21) give action variables, whose dual variables give the angle variables.*

We must prove the second assertion. Fix some point b. We can parametrize the curve $\gamma_i(b)$ by $0 \leqslant t \leqslant 2\pi$. This lifts to a unique curve $\tilde{\gamma}_i$ in TB_b^* and the values $\tilde{\gamma}_i(2\pi)$ form a basis of the lattice $L \cap TB_b^*$. Choose these as the $\beta_1(b), \ldots, \beta_n(b)$ above and take the unique local extension $\beta_1, \ldots, \beta_n$ and introduce the angle action variables as above. The curves $\gamma_i(c)$ are then homotopic in each fiber to the curves described just after (44.21) in each fiber for c near b. But the fibers are Lagrangian; that is, $d\sigma$ vanishes when restricted to each fiber. So by Stokes' theorem we may assume that the curves are in fact the curves $0 \leqslant \theta_i \leqslant 2\pi$, $\theta_j = 0$, $j \neq i$, given above. Now if σ were given by (44.20) we would get back exactly our action coordinates described in the proof of the first part of the theorem. So suppose that

$$\sigma = \sum I_i d\theta_i + \mu.$$

Then $d\mu = 0$. Let $\bar{\mu}$ be the pull-back of μ to T^*B. Then $\bar{\mu}$ is a linear differential form on T^*B, which is periodic (relative to the lattice L), and, since $d\bar{\mu} = 0$ and (we can assume locally that) T^*B is simply connected, $\bar{\mu} = df$.

Then

$$f(I_1, \ldots, I_n; \theta_1 + 2\pi, \theta_2, \ldots, \theta_n) - f(I_1, \ldots, I_n; \theta_1, \ldots, \theta_n)$$

is a function whose differential $= 0$ (i.e., is a constant), say, k_1, and hence

$$\int_{\gamma_1} \sigma = I_1 + k_1,$$

and similarly the remaining I_j defined by (44.21) differ from the original action variables by constants. Since constants are irrelevant to the definition of the action variables, this completes the proof of the theorem.

A Hamiltonian H given by (44.1) is expressed in terms of the action variables as $H = h(I_1, \ldots, I_n)$, and hence the corresponding differential equations are given by

$$\frac{dI_i}{dt} = 0$$

$$\frac{d\theta_i}{dt} = -\frac{\partial h}{\partial I_i}. \qquad (44.22)$$

The flow is thus given by a straight-line flow on each torus.

The question of the existence of *global* action angle variables depends on a number of topological considerations. We have already discussed the problem of the existence of a section of the bundle $\rho: M \to B$. Assuming that a section satisfying (44.15) exists, we still have the following "monodromy" problem: Our choice of the action coordinates depended (up to a constant) on the choice of a basis in the lattice over a point b. As we continue along any curve in B, this will determine a continuous choice of lattice basis. If we go around a closed curve and come back to b, we get a possibly different lattice basis, one that differs from the original one by a nonsingular integral matrix. Duistermaat (1980) shows that this determines a homomorphism of $\pi_1(B, b)$ into $Gl(n, \mathbb{Z})$, which is the only remaining obstruction to the existence of global action angle coordinates. He also computes this homomorphism for the case of the spherical pendulum and shows that it is not trivial.

In the discussion so far, we have been assuming that we have a Lagrangian foliation defined on all of M, which is a fibration with compact, connected fibers. In the two examples mentioned, there were some singular points where these conditions do not hold. Usually, the Lagrangian foliation is given locally by a collection of functions $f_1, \ldots, f_n$, which Poisson-commute with one another, the foliation being given by the simultaneous level surface of the functions. This does indeed define a foliation at those points where the functions are functionally independent. This is, of course, an open subset M_r the regular points of M. If $\{H, f_i\} = 0$, $i = 1, \ldots, n$, then M_r is clearly invariant under $\exp t\xi_H$. Let us now assume that we have in fact a fibration with Lagrangian fibers, and let M_c denote the set of points p in M such that the connected component of the fiber through p is compact. We claim that M_c is an open subset of M. Indeed, suppose not. Then there would exist a point p in M_c and a sequence of points $p_i \notin M_c$ with $p_i \to p$. Let U be an open neighborhood of p that contains the connected component of the fiber through p, and such that $\bar{U}$ is compact and does not intersect any of the other components of the fiber through p. We claim that for p_i close enough to p, ∂U does not intersect the fiber through p_i. If this were not so, we could choose a point p_i' on the fiber through p_i and lying in ∂U, and by passing to a subsequence if necessary, arrange that these points converge to some point p'. But, by continuity, p' would have to lie in the fiber through p, contradicting the assumption that ∂U has no points in common with this fiber. Thus, for p_i close enough to p, the connected component of the fiber through p_i lies entirely in U and hence in $\bar{U}$ and hence is compact – contradicting the assumption that $p_i \notin M_c$.

In this way we arrange, at least locally, that the conditions of Theorem 44.1 is met. As far as the compactness is concerned, we should also point out the following observation: Suppose that H is a function that is constant on the leaves of the foliation, for example, the foliation is given by functions $f_1, \ldots, f_n$ and $\{H, f_i\} = 0$. Suppose, as is frequently the case, that the set $H \leqslant c$ is compact for any c. Then any closed leaf is automatically compact.

45. Collective complete integrability

Let Y be a Poisson manifold. That is, we assume that the space of all smooth functions on Y has the structure of a Poisson algebra. For example, Y could be a symplectic manifold, or, like g^*, an important case for us, it could be a union of symplectic manifolds. A submanifold C of Y is called coisotropic if the Poisson bracket of two functions vanishing on C again vanishes on C. As usual, this is a local condition and needs to be checked only for local generators of the ideal of functions vanishing on C. For example, suppose that Y is the union of symplectic manifolds $\mathcal{O}$, with the Poisson bracket evaluated at a point of $\mathcal{O}$ coming from the symplectic structure of $\mathcal{O}$ as in (28.6). Then each $\mathcal{O}$ is coisotropic in this sense.

Let X be a second Poisson manifold and let $\Psi: X \to Y$ be a Poisson map; that is, a smooth map such that the induced mapping $\Psi^*: C^\infty(Y) \to C^\infty(X)$ is a homomorphism of Poisson algebras. Then it follows from the definitions that if Ψ has transversal or clean intersection with C, then $\Psi^{-1}(C)$ is coisotropic if C is. For example, if $Y = g^*$ and $\Phi: M \to g^*$ is a moment map, then Φ is a Poisson map and so, as we know, $\Phi^{-1}(\mathcal{O})$ is a coisotropic submanifold of M.

If we are given a coisotropic foliation L of Y, that is, a foliation of Y whose leaves are all coisotropic submanifolds, then under the appropriate cleanness hypotheses, Ψ^*L, the foliation of X whole leaves are the inverse images of the leaves of L, will be a coisotropic foliation of X. (As usual, L may have some singularities (lower-dimensional leaves) and Ψ may not be everywhere transversal to the leaves of L, which may require replacing X or Y or both by open subsets.) If the foliation on Y is (locally) given by the simultaneous level surfaces of functions $f_1, \ldots, f_r$ with $\{f_i, f_j\}_Y = 0$, then $\Psi^*(L)$ is locally given by the level surfaces of $\Psi^*f_1, \ldots, \Psi^*f_r$. We have $\{\Psi^*f_i, \Psi^*f_j\}_X = 0$, since Ψ^* is a homomorphism.

We shall be particularly interested in the case where $X = M$ is a symplectic manifold, and where the coisotropic foliation Ψ^*L is Lagrangian.

Thus each leaf of Ψ^*L is a Lagrangian submanifold. In particular, any subspace of the tangent space to a leaf will be an isotropic subspace. Since $\ker d\Psi_m$ is always tangent to the leaf of Ψ^*L, we see that a necessary condition for Ψ^*L to be Lagrangian is that

$$\ker d\Psi_m \qquad \text{be isotropic at all } m. \tag{45.1}$$

Suppose we are given a Hamiltonian action of a Lie group G on a symplectic manifold M with moment map $\Phi: M \to g^*$. A function of the form Φ^*f, where f is a function on g^*, was called *collective* in Section 28. If H is any G-invariant function on M and Φ^*f is a collective function, then we know from Section 28 that

$$\{H, \Phi^*f\}_M = 0. \tag{45.2}$$

Similarly, if L is a coisotropic foliation on g^*, with Φ intersecting the leaves of L cleanly, we shall call Φ^*L a collective foliation on m. (More generally, we need only assume that $\Phi(M)$ is a submanifold with L defined on $\Phi(M)$ and with the appropriate cleanness assumptions.) Locally such a foliation will be given by the level surfaces of collective functions $\Phi^*f_1, \ldots, \Phi^*f_r$, which Poisson-commute with one another. Equation (45.2) then implies that H is constant along the null foliation of each leaf of the foliation Φ^*L. In particular, we will be interested in the case where Φ^*L is Lagrangian, so that H is constant along the leaves of Φ^*L and we have a completely integrable system. In this case we say that we have *collective* complete integrability. Now (45.1) provides us with a necessary condition on the moment map Φ for the existence of a collective Lagrangian foliation of M. Indeed, (26.3) asserts

$$\ker d\Phi_m = g_M(m)^\perp,$$

so (45.1) says that

$$g_M(m)^\perp \qquad \text{is isotropic} \tag{45.3}$$

or, what amounts to the same thing, that

$$g_M(m)^\perp \subset g_M(m). \tag{45.4}$$

In other words,

$$\textit{the G orbits are coisotropic.} \tag{44.4'}$$

Let $\alpha = \Phi(m)$ and let $\mathcal{O}$ denote the G orbit through α. Assuming that Φ intersects $\mathcal{O}$ cleanly, (26.11) says that

$$d\Phi_m^{-1}(T\mathcal{O}_\alpha) = g_M(m) + g_M(m)^\perp.$$

Thus (45.4) is equivalent to

$$d\Phi_m^{-1}(T\mathcal{O}_\alpha) = g_M(m), \tag{45.5}$$

which under our cleanness assumptions is the same as

$$G^0 \textit{ acts transitively on the connected components of } \Phi^{-1}(\mathcal{O}_\alpha) \tag{45.6}$$

or to

$$G_\alpha^0 \textit{ acts transitively on the connected components of } \Phi^{-1}(\alpha), \tag{45.7}$$

where G^0 denotes the connected component of G.

By Theorem 26.5 this is the same as saying that

The Marsden–Weinstein reduced spaces, $M_\mathcal{O}$ are points. (45.8)

In general the coisotropic manifold $\Phi^{-1}(\mathcal{O}_\alpha)$ quotiented by its null foliation is a symplectic manifold (if it is a manifold) which we may call the KKS reduced space, cf. Section 26. Then the preceding are equivalent to

The KKS reduced space of $\mathcal{O}$ is $\mathcal{O}$. (45.8′)

Finally, let us consider the algebra of G-invariant functions on M. By (45.4′) (generically) the common level surfaces of these functions are coisotropic. So the G-invariant functions Poisson commute. Thus (45.1)–(45.8′) are all equivalent to

The algebra of all G-invariant functions on M is commutative under Poisson bracket. (45.9)

Condition (45.9) is the "classical analog" of a well-known "quantum" property: If we are given a representation of a group, then the algebra of G-invariant operators (those which commute with the G action) is commutative if and only if the representation is multiplicity-free, that is, if and only if every irreducible of G occurs in the representation with multiplicity at most 1. In other words, the quantum action of G is multiplicity-free if and only if the algebra of G-invariant *quantum* observables is commutative. The classical analog would thus be that the algebra of G-invariant *classical* observables be commutative. This is exactly (45.9).

For this reason we will call a Hamiltonian G space *multiplicity-free* if any (and hence all) of the equivalent conditions hold. (This name was suggested to us by Joe Wolf.) We study these multiplicity-free spaces in more detail in Guillemin and Sternberg (1984a).

We have thus proved:

If the Hamiltonian G space, M, admits a collective completely integrable system then it must be multiplicity-free. (45.10)

We will now investigate various examples where this collection of equivalent conditions holds (or holds almost everywhere).

The simplest example is where M is itself an orbit $\mathcal{O}$ in g^*: The conditions then hold trivially, with Φ the inclusion of $\mathcal{O}$ in g^*. We have seen that a number of interesting mechanical systems, for example, constrained harmonic motion on a sphere or the geodesic flow on an ellipsoid, can be viewed as equations on G orbits. We shall see in Section 47 that these systems are collectively completely integrable. Similarly, one key step in the solution of a collective Hamiltonian system involves solving a Hamiltonian system on a coadjoint orbit (cf. the discussion in Section 28).

The invariant functions on g^* restrict to constants on each coadjoint orbit and hence give nothing interesting in this case. But still we may look for families of noninvariant functions on g^* that Poisson-commute with one another and whose restriction to a coadjoint orbit $\mathcal{O}$ give a completely integrable system. If $\mathcal{O}$ is of relatively low dimension, we will not need so many functions, and the method has some chance of success. We shall discuss methods of this type in Section 48.

Suppose that $N = G/K$ is a homogeneous space for G and that $M = T^*N$. Let us investigate our necessary conditions in this situation. By homogeneity, it suffices to check our conditions for points of $T^*N_{n_0}$, where $n_0 = K$ is the base point of N. Now the G action on N gives identification of

$$TN_{n_0} \qquad \text{with} \qquad g/k$$

and hence the dual identification of

$$TN^*_{n_0} \qquad \text{with} \qquad k^0, \text{ the annihilator of } k \text{ in } g^*.$$

Under this identification, the restriction of Φ to $TN^*_{n_0}$ is just the injection of k^0 as a subspace of g^*. Now K is still free to act, and condition (45.6) is clearly fulfilled if

$$K \qquad \text{acts transitively on} \qquad \mathcal{O} \cap k^0. \tag{45.11}$$

Actually, we only need the infinitesimal version of (45.11) for all orbits, which says that

$$g \cdot l \cap k^0 = k \cdot l \qquad \forall l \qquad \text{in} \qquad k^0, \tag{45.12}$$

where $\circ$ denotes the coadjoint action. In other words, if, for l in k^0 and ξ in g, $\xi \cdot l$ lies in k^0, then $\xi \cdot l = \eta \cdot l$ for some η in k.

As an example of a group pair G, K where this condition certainly holds, consider the case where G is the semidirect product of K with a vector space V on which K is represented. Thus $G/K = V$, and we can identify k^0 with the set of all elements of the form $(0, p)$ in $g^* = k^* + V^*$. Now we know the structure of all coadjoint orbits of a semidirect product. They are all fibered over K orbits in V^* and the ones that intersect $k^0 = V^*$ are just the cotangent bundles of these K orbits in V^*, the intersection being the zero section of these K orbits. Hence K acts transitively on these orbits, so condition (45.11) is fulfilled.

Here is another collection of important examples where we can verify (45.12). Suppose that g has a bilinear form invariant under the adjoint representation, which is nonsingular, so we may identify g^* with g and the coadjoint representation with the adjoint representation, and

$$k^0 \qquad \text{with} \qquad p = k^{\perp} \subset g.$$

Then p is a linear space complement to k, stable under the adjoint action of k in g. Condition (45.12) becomes

$$[p, l] \cap p = [k, l] \qquad \forall l \in p. \tag{45.13}$$

For example, if X is a symmetric space, then it is a standard theorem, see [Helgason (1978)] that there is such an invariant symmetric bilinear form and that

$$[p, p] \subset k, \tag{45.14}$$

so the left-hand side of (45.13) is $\{0\}$ and hence (45.13) is trivially verified. Thus

condition (45.12) is verified for the case that $M = T^*X$, where X is a symmetric space. (45.15)

For a general study of homogeneous spaces for which (45.11) holds, see Kraemer (1979). See also Planchart (1982).

We now want to consider the following example: Suppose that G_1 and G_2 are Lie groups, together with a (smooth) homomorphism $j: G_1 \to G_2$. This gives a homomorphism $\iota: g_1 \to g_2$ and hence a dual map

$$\iota^*: g_2^* \to g_1^*.$$

We get a Hamiltonian (Poisson) action of G_1 on g_2^* whose corresponding moment map is exactly ι^*. In particular, if $\mathcal{O}$ is a G_2 orbit in g_2^*, we get a Hamiltonian action of G_1 on $\mathcal{O}$, whose moment map is the restriction of ι^* to $\mathcal{O}$. So we want to investigate our conditions (45.3)–(45.8) for the case that $G = G_1$ and $M = \mathcal{O}$.

For example, suppose that $G_1 = U(n)$ and $G_2 = U(n+1)$. We may use the Killing form to identify the Lie algebras $u(n)$ and $u(n+1)$ with their dual spaces. Also, we may (after dividing by i) identify these spaces with the spaces of $(n+1) \times (n+1)$ and $n \times n$ Hermitian matrices. The projection $\iota^* : u(n+1)^* \to u(n)^*$ assigns to each $(n+1) \times (n+1)$ matrix its $n \times n$ lower right corner. We claim that for each $U(n+1)$ orbit $\mathcal{O}$, (45.7) holds almost everywhere. To verify this, it is enough to check that it holds on almost all $U(n)$ suborbits of $\mathcal{O}$, and for this it is enough to check it on cross sections to the $U(n)$ suborbits.

Now an orbit $\mathcal{O}_\lambda$ in $u(n+1)^* \sim u(n+1)$ consists of all self-adjoint matrices with given eigenvalues $\lambda = (\lambda_0, \ldots, \lambda_n)$, where

$$\lambda_0 \leqslant \lambda_1 \leqslant \cdots \leqslant \lambda_n.$$

If $A \in \mathcal{O}_\lambda$, the minimax principle says that the eigenvalues μ of $\Phi(A)$ must interlace those of A; that is, that

$$\lambda_0 \leqslant \mu_1 \leqslant \lambda_1 \leqslant \cdots \leqslant \mu_n \leqslant \lambda_m. \tag{45.16}$$

Conversely, every self-adjoint $n \times n$ matrix with eigenvalues satisfying (45.1) lies in the image of $\mathcal{O}_\lambda$ under Φ. We may choose our cross section to the $U(n)$ orbits in $\Phi(\mathcal{O}_\lambda)$ to consists of diagonal matrices. Let diag μ denote the diagonal matrices with (nondecreasing) μ along the diagonal.

Then $\Phi^{-1}(\operatorname{diag} \mu)$ consists of all matrices of the form

$$A = \begin{pmatrix} a_0 & a_1 & \cdots & a_n \\ \bar{a}_1 & \mu_1 & & 0 \\ \vdots & 0 & \ddots & \\ \bar{a}_n & & & \mu_n \end{pmatrix}, \tag{45.17}$$

with eigenvalues $\lambda_1, \ldots, \lambda_n$. The characteristic polynomial of A of the form (45.17) is

$$\prod_j (\lambda - \mu_j)(\lambda_0 - a_0) - \sum |a_i|^2 \prod_{j \neq i} (\lambda - \mu_j). \tag{45.18}$$

For any choice of λ and for generic μ in the image of $\Phi(O_\lambda)$ the values of a_0 and $|a_i|^2$, $i = 1, \ldots, n$, are determined by μ and the condition that A lie in $\Phi^{-1}(\mu)$.

For example, suppose that

$$\lambda_0 < \lambda_1 < \cdots < \lambda_n. \tag{45.19}$$

Then a generic μ will have

$$\lambda_0 < \mu_1 < \lambda_2 \cdots \mu_n < \lambda_n, \tag{45.20}$$

and, in particular, no two of the μ's equal. Then we can write (45.18) as

$$\prod_j (\lambda - u_j)\left[a_0 - \lambda - \sum \frac{|a_i|^2}{\lambda - u_i}\right]. \tag{45.21}$$

(It is clear from (45.17) that if we start with μ and the a_i's then (45.21) must vanish at all roots of the characteristic polynomial.) Since the product in front of (45.21) does not vanish, the condition that we obtain on the a's is

$$a_0 - \lambda_j = \sum_{i=1}^{n} \frac{|a_i|^2}{\lambda_j - \mu_i}, \qquad j = 0, \ldots, n. \tag{45.22}$$

We can solve these equations for a_0 and the $|a_i|^2$. This then determines $\Phi^{-1}(\operatorname{diag}\mu)$. Let us denote the isotropy group of diag μ by G_μ. Then G_μ in this case is just the n-torus consisting of diagonal matrices and it acts on the a_i's, $i = 1, \ldots, n$, by multiplying each a_i by an independent phase factor. In particular, it acts transitively on $\Phi^{-1}(\mu)$ verifying (45.7).

In case some of the λ's are equal, then (45.7) still holds at generic μ in $\Phi(O_\lambda)$. The group G_μ is larger but certain a_i must vanish and the net effect is as before. For example suppose $n = 4$ and

$$\lambda_1 = \lambda_2 = \lambda_3 < \lambda_4 < \lambda_5.$$

Then the μ that are generic in $\Phi(O_\lambda)$ will have

$$\mu_1 = \mu_2 = \lambda_3 < \mu_3 < \lambda_4 < \mu_5 < \lambda_5.$$

Differentiating (45.21) twice and setting $\lambda = \lambda_1$ (which is a triple root) gives

$$|a_1|^2 = |a_2|^2 = 0.$$

Although G_μ contains a $U(2)$ factor, it acts trivially on $\Phi^{-1}(\mu)$ as expected from the discussion preceding Theorem 26.5. The remaining a's are determined by equation (45.22) in fewer variables and the leftover torus, $G_\mu/U(2)$ acts transitively on $\Phi^{-1}(\mu)$. A similar result holds when we take $G_1 = O(n+1)$ and $G_2 = O(n)$. As the computations are almost identical, we illustrate them for the special case of $O(5)$ and $O(4)$, leaving the general discussion to the reader. The Lie algebra $o(n)$ and its dual space $o(n)^*$ can be identified with the space of all real antisymmetric matrices. The eigenvalues of a 5×5 antisymmetric matrix will be $0, \pm i\lambda_1, \pm i\lambda_2$, where $0 \leqslant \lambda_1 \leqslant \lambda_2$. A choice of λ_1, λ_2 determines an orbit $\mathcal{O}$ in $o(5)^*$. The eigenvalues of an antisymmetric 4×4 matrix will be of the form $\pm i\mu_1, \pm i\mu_2$ with $0 \leqslant \mu_1 \leqslant \mu_2$. The condition that an antisymmetric 4×4 matrix lie in $\Phi(\mathcal{O})$ is that

$$0 \leqslant \mu_1 \leqslant \lambda_1 \leqslant \mu_2 \leqslant \lambda_2.$$

We can choose matrices of the form

$$\alpha_\mu = \begin{pmatrix} 0 & -\mu_1 & 0 & 0 \\ \mu_1 & 0 & 0 & 0 \\ 0 & 0 & 0 & -\mu_2 \\ 0 & 0 & \mu_2 & 0 \end{pmatrix}$$

to give a cross section. For fixed μ_1, and μ_2, $\Phi^{-1}(\alpha_\mu)$ consists of all matrices of the form

$$\begin{pmatrix} 0 & x_1 & x_2 & x_3 & x_4 \\ -x_1 & 0 & -\mu_1 & 0 & 0 \\ -x_2 & \mu_1 & 0 & 0 & 0 \\ -x_3 & 0 & 0 & 0 & -\mu_2 \\ -x_4 & 0 & 0 & \mu_2 & 0 \end{pmatrix}$$

with the appropriate eigenvalues. Computing the characteristic polynomial yields (if $0 < \mu_1 < \lambda_1 < \mu_2 < \lambda_2$)

$$\lambda(\lambda^2 + \mu_1^2)(\lambda^2 + \mu_2^2)\left(1 - \frac{x_1^2 + x_2^2}{(\lambda^2 + \mu_1^2)} - \frac{x_3^2 + x_4^2}{(\lambda^2 + \mu_2^2)}\right).$$

The vanishing of this polynomial for $\lambda = \pm i\lambda_1$ and $\lambda = \pm i\lambda_2$ gives a pair of linear equations that determine $x_1^2 + x_2^2$ and $x_3^2 + x_4^2$. The isotropy group consist of all rotations in the x_1x_2 and x_3-x_4 planes and, hence, (45.7) holds. We have thus verified a theorem of Heckman (1980):

If $\mathrm{G} = \mathrm{U(n)}$ *and* M *is a coadjoint* $\mathrm{U(n+1)}$ *orbit, or if* $\mathrm{G} = \mathrm{O(n)}$ *and* M *is a coadjoint* $\mathrm{O(n+1)}$ *orbit, then* (45.7) *holds.* (45.23)

Heckman also shows that if G_1 and G_2 are a pair of compact semisimple Lie groups such that (45.7) holds for $G = G_2$ and M is a *generic* coadjoint orbit of G_1, then $(G_1, G_2) = (U(n+1), U(n))$ or $(O(n+1), O(n))$.

We can now put the preceding results together by an ingenious method of Thimm (1981) to give a sufficient condition for complete integrability. (See also Guillemin and Sternberg, 1983c.) Suppose we are given a chain of subgroups $G = G_1 \supset G_2 \supset \ldots$, with $G_n = \{e\}$, and such that the necessary conditions (45.3)–(45.9) hold almost everywhere for almost all orbits for the pair (G_i, G_{i+1}) (or at least for those orbits in g_i^* in the image of the moment maps from g_{i-1}^*). Then, starting with the trivial foliation of $g_n^* = \{0\}$ and working our way back, we get a foliation L, which is Lagrangian on each orbit. In particular, this applies to the chain $U(n) \supset U(n-1) \ldots$ and the chain $O(n) \supset O(n-1) \ldots$. Now suppose that we are given a Hamiltonian action of G on M that satisfies (45.3)–(45.9) almost everywhere. Then any G-invariant Hamiltonian on M is completely integrable. In particular, we obtain the results of Thimm (1980a, b, 1981)

Any U(n)- *or* O(n)-*invariant Hamiltonian system on a symmetric space is completely integrable. In particular, the geodesic flow on real or complex Grassmannian is completely integrable.*

For any Lie group G there is an alternative method of constructing a family of Poisson-commuting functions on g^*. It consists of taking the G-invariant functions on g^* and considering all possible shifts

$$f(\cdot + ta),$$

where a is in g^* and t is any real number. Mishchenko and Fomenko prove that the collection of these functions all Poisson-commute. We shall present this result in Section 47. Furthermore, if G is semisimple and a is generic, they prove (1978) that these provide a completely integrable system on *generic* orbits. Dao Chong Tkhi (1978) claims to prove that this system is also completely integrable on lower dimensional orbits. But there seem to be some problems with his proof, cf. the comments of Mishchenko (1982). If the results of Dao Chong Tkhi are correct, this would imply that the geodesic flow for a symmetric space of any semisimple Lie group is a collective completely integrable system.

46. Collective action variables

Suppose we are given a Hamiltonian action of G on M that is collectively completely integrable. How do we go about finding action angle variables for this system? In view of our discussion in Section 44, our problem is to find functions $H_1, \dots, H_n$ on g^* such that $\Phi^*H_1, \dots, \Phi^*H_n$ give our collective completely integrable system and so that the vector fields $\xi_{\Phi^*H_i}$ all generate flows that are periodic with the same period, say, with period one. Among the functions H_i will be invariant functions whose corresponding vector fields on M lie in the null foliation of $\Phi^{-1}(\mathcal{O})$ for each g^* orbit $(\mathcal{O})$ and the remaining functions define a Lagrangian foliation on (almost) every orbit $\mathcal{O}$. Our problem thus splits into two parts:

(1) Find enough invariant functions H on g^* whose associated flows on M are periodic with period one;
(2) Suppose that H is a function on g^* whose flow on each g^* orbit is periodic with period one. Find an invariant functions $\bar{H}$ such that the vector field $\xi_{\Phi^*(H+\bar{H})}$ generates a flow which is periodic with period one.

Then, if we had functions $H_1, \ldots, H_k$ on g^* that defined action coordinates on almost every orbit, we could modify them by (2) and supplement them by (1) so as to obtain collective action coordinates on M. For the rest of this section we will assume that the group G is compact.

Let T be a maximal torus (Cartan subgroup) of G and let t be its Lie algebra. Let $\hat{t} \subset g^*$ denote the space of T-fixed elements of g^*. The space $\hat{t}$ is nonsingularly paired with t under the pairing of g^* with g (and hence we may identify $\hat{t}$ with t^*). In $\hat{t}$ there is a convex cone t^*_+ (called the positive Weyl chamber) with the property that every orbit in g^* intersects t^*_+ in exactly one point. (We may choose an invariant scalar product on g that allows us to identify g^* with g. Then $\hat{t}$ becomes identified with t and the above assertions are standard facts about the adjoint representation of compact groups. (See, for example, Helgason, 1983, or Jacobson, 1979.) If $G = U(n)$, for example, we may identify g^* with the space of (skew)-Hermitian matrices, take T and hence $\hat{t}$ to be diagonal matrices, and t^*_+ to consist of diagonal matrices with decreasing eigenvalues.)

Let $\beta: g^* \to t$ be the map assigning to each orbit its point of intersection with t^*_+. This mapping is a smooth fiber mapping over Int t^*_+ and is continuous on all of g^*.

Now let $\xi \in t$. Because of the pairing between t and t^*, ξ defines a linear function l_ξ on t^* and, by restriction, a function, l^+_ξ, on t^*_+. Let $H_\xi = l^+_\xi \circ \beta$. This function is a continuous function on all of g^* and is smooth on the set of regular elements g_{reg} of g^*. (By definition g_{reg} is the preimage of Int t^*_+ with respect to β.) We will now compute the trajectories of the Hamiltonian flow associated with $H_\xi \circ \Phi$. We will have to confine ourselves, of course, to those points for which this flow is defined, that is, those points, $p \in X$ for which $\Phi(p) = \alpha$ is in g_{reg}. Since $H_\xi \circ \Phi$ is G-invariant we can, without loss of generality, assume that $\alpha \in$ Int t^*_+. We will show below that in this case, with $H = H_\xi$,

$$\mathscr{L}_H(\alpha) = \xi. \tag{46.1}$$

Let us defer the proof of (46.1) for the moment. Since H_ξ is G-invariant we can apply step (3) of the integration procedure of Section 28 with $\xi = L_H(\alpha) =$ constant. We conclude *that with $H = H_\xi$ then at all points $p \in X$ such that $\Phi(p) \in$ Int t^*_+ the trajectory of ξ_H through p is*

$$(\exp t\xi)p, \qquad -\infty < t < \infty. \tag{46.2}$$

Since T is a torus, the exponential mapping of t into T is a morphism of Abelian groups and its kernel Γ is a lattice subgroup of t. Thus if $\xi \in \Gamma$, the one-parameter subgroup $\exp t\xi$, $-\infty < t < \infty$, is periodic with period 1. From (46.2) we conclude

Theorem 46.1. *If* $\xi \in \Gamma$ *the Hamiltonian flow associated with* $H_\xi \circ \Phi$ *is periodic with period one.*

We must still prove (46.1). This is equivalent to proving that for all $v \in T_f g^* = g^*$

$$dH_f(v) = \langle \xi, v \rangle. \tag{46.3}$$

This is true by definition of $v \in t^*$. On the other hand if v is tangent to the orbit $\mathcal{O}$ through f, the left-hand side of (46.3) is 0 since H is an invariant function; so we will be done if we can show that for v tangent to $\mathcal{O}$ at f, $\langle \xi, v \rangle = 0$. This, in turn, will be true if we can show that the tangent space to $\mathcal{O}$ at f is annihilator of t in g^*. Since f is a regular element of g^* the isotropy algebra of f is t, so this is a special case of (26.4) with $M = \mathcal{O}$ and $\Phi = id$.

We now turn to step (2) in our program. We wish to prove

Theorem 46.2. *Let* H *be a smooth function defined on an open subset of* g^*. *Suppose that for every coadjoint orbit* $\mathcal{O}$, *the restriction of* H *to* $\mathcal{O}$ *generates a Hamiltonian flow that is periodic of period one. Then there exists an* $(Ad\ G)^*$*-invariant function* H^0 *on* g^* *with the following universal property. Let* $H' = H - H^0$. *Then, for every Hamiltonian* G*-space* X *with moment mapping* $\Phi: X \to g^*$, *the Hamiltonian flow associated with* $H' \circ \Phi$ *is periodic of period one.*

Proof. Suppose that H is a function with the property described in Theorem 46.2; that is, for every coadjoint $\mathcal{O}$ the restriction of H to $\mathcal{O}$ generates a Hamiltonian flow that is periodic of period one. Consider the cotangent bundle of G, T^*G, as a Hamiltonian G-space by letting G act on it on the left, and let Φ_L: $T^*G \to g^*$ be the associated moment map. Let ξ_H be the Hamiltonian flow associated with $H \circ \Phi$ and let f: $T^*G \to T^*G$ be the mapping obtained by evaluating the trajectories of this flow at time one; that is, $f = \exp \xi_H$. It is clear that f is symplectic. We will prove

Proposition 46.1. f *commutes both with the left action of* G *on* T^*G *and with the right action of* G *on* T^*G.

Proof. The left action of G on T^*G commutes with the right action; so *all* collective Hamiltonians $H \circ \Phi_L$ commute with the right action of G on T^*G.

We claim that to show that f commutes with the left action of G on T^*G it is enough to show that $\Phi_L \circ f = \Phi_L$. Indeed if this is the case, then for every $\xi \in g$, $f^*\phi_\xi = \phi_\xi$, where $\phi_\xi = \langle \Phi_L, \xi \rangle$. Let $\hat{\xi}$ be the Hamiltonian vector field on T^*G associated with the infinitesimal left action of G on T^*G. Since f is symplectic and preserves ϕ_ξ, it preserves $\hat{\xi}$. (Recall that $\hat{\xi}$ is the Hamiltonian vector field associated with ϕ_ξ.) Thus f commutes with the infinitesimal left action of G on T^*G; and, therefore, since G is connected, with the global left action of G.

We still must prove that $\Phi_L \circ f = \Phi_L$. Let $p \in T^*G$ and let $\alpha = \Phi(p)$. Let $\mathcal{O}$ be the coadjoint orbit through α, let $H_\mathcal{O}$ be the restriction of H to $\mathcal{O}$, and let $\xi_\mathcal{O}$ be the Hamiltonian vector field defined by $H_\mathcal{O}$. Since the moment map is equivariant

$$\Phi_L[(\exp t\xi_H)(p)] = (\exp t\xi_\mathcal{O})(\alpha).$$

Since $\exp t\xi_\mathcal{O}$, $-\infty < t < \infty$, is periodic of period one and $f = \exp \xi_H$, this proves that $\Phi_L \circ f = \Phi_L$. Q.E.D.

Thus the proof of Theorem 46.2 comes down to the problem of determining all bi-invariant symplectic mappings on T^*G. Let T be the Cartan subgroup of G. Let T_e^* be the cotangent space of the identity element of G, and let t^* be the subspace of T_e^* which is left fixed by the adjoint action of T on T_e^*. If we identify T_e^* with g^*, t^* gets identified with the dual of the Lie algebra of T. Now every point in T^*G can be conjugated to a point in T_e^* by the left action of G, and every point in T_e^* can be conjugated to a point in t^* by the adjoint action of G; so we have proved that *f is completely determined by its restriction to t^**. Next consider the set of points in T^*G that are fixed by the adjoint action of T. We know from Section 30 that the set of fixed points of T is always a symplectic submanifold of the ambient space. In our case this manifold consists of the set of left translates by T of t^* (regarded as a subspace of T_e^*) and can be identified with the cotangent bundle of the group T that is, $T^*(T)$. Since f commutes with the adjoint action of T on T^*G, it leaves this set fixed, and hence indices a symplectic mapping $\tilde{f}$ of $T^*(T)$ into itself. By our previous remark, f is completely determined by $\tilde{f}$. Notice also that the restriction of Φ_L to $T^*(T)$ is just the moment mapping Φ_T associated with the canonical action of T on $T^*(T)$; so $\Phi_T \circ \tilde{f} = \Phi_T$. Finally, notice that $\tilde{f}$ commutes with the adjoint action of the Weyl group on $T^*(T)$. We will show how to determine all symplectic mappings of $T^*(T)$ with these two properties. Identify $T^*(T)$ with $T \times t^*$ by means of the canonical T-invariant trivialization and identify T with t/Γ, where Γ is the integral lattice in t.

Then we have a covering map

$$T^*(T) \stackrel{\pi}{\leftarrow} t \times t^*,$$

which is symplectic providing we give $t \times t^*$ its standard linear symplectic structure. Therefore, $\tilde{f}$ lifts to a symplectic mapping $\hat{f}$ such that

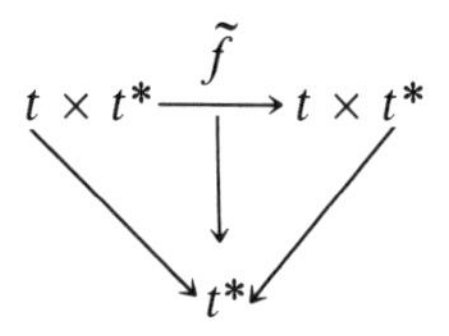

commutes. But for every such symplectic mapping one can find a function H^0 on t^* such that for all $(v, \alpha) \in t \times t^*$

$$\hat{f}(v, \alpha) = (v + \mathscr{L}_{H^0}(\alpha), \alpha), \tag{46.4}$$

$\mathscr{L}_{H^0}$ being the Legendre transform. In our case H^0 has to be Weyl group invariant, so it extends to an $(\mathrm{Ad}\, G)^*$-invariant function on g^* by means of β. Consider now the collective Hamiltonian $H^0 \circ \Phi_L$. This commutes with $H \circ \Phi_L$ by Proposition 46.1, and if ξ_{H_0} is the Hamiltonian vector field associated with H_0, $\exp \xi_{H_0} = f$ on the fixed point set of T by 46.4 and hence on *all* of T^*G because of the fact that both $\exp \xi_{H_0}$ and f are bi-invariant. Thus if we set $H' = H - H^0$ and let $\xi_{H'}$ be the Hamiltonian vector field on X associated with $H' \circ \Phi_L$, then $\exp \xi_{H'} = (\exp \xi_H)(\exp \xi_{H^0})^{-1} = \mathrm{id}$. This proves that the Hamiltonian flow associated with $H' \circ \Phi_L$ on T^*G is periodic of period one. However, if we compute the trajectory of this flow, with initial point lying in the fiber above e in T^*G, using the recipe from Section 28; we see that the curves $g(t)$ solving (28.4) are periodic of period 1. Therefore, by step (3) of Section 28, the flow associated with H' is periodic of period one for *all* Hamiltonian G-spaces.

For a further discussion of these points see Guillemin and Sternberg 1983d.

47. The Kostant–Symes lemma and some of its variants

If I is an invariant function on g^* then $\{I, f\}_{g^*} = 0$ for all functions on g^*. In particular, the invariants Poisson-commute with one another. In themselves, they do not give an interesting collection of Poisson-commuting functions, because they are constant when restricted to each coadjoint orbit. However, we can use these invariants to construct other functions that Poisson-commute and that are more interesting from the point of view of complete integrability. We have already seen one example of this kind of procedure in section 45: If g_2 is a subalgebra of g_1, then we have a projection $\iota^*: g_1^* \to g_2^*$ and the invariant functions on g_2^* pull back to give a collection of nontrivial Poisson-commuting functions on g_1^*. Indeed, we saw that taking a chain $u(n) \supset u(n-1) \supset \cdots$ or $o(n) \supset o(n-1) \supset \cdots$ provides a completely integrable system on each $u(n)$ or $o(n)$ coadjoint orbit. Another method consists of passing from a larger algebra g to a smaller one via a "splitting." This is the method of Kostant and Symes. Suppose that we have a vector space decomposition of the Lie algebra g into a direct sum of two subalgebras:

$$g = a + b, \qquad [a, a] \subset a, \qquad [b, b] \subset b \tag{47.1}$$

For example we could let $g = gl(n)$ be the algebra of all $n \times n$ matrices, let a consist of all strictly lower triangular matrices and b consist of all upper triangular matrices.

The decomposition $g = a + b$ gives the corresponding decomposition

$$g^* = b^0 + a^0, \tag{47.2}$$

which allows an identification

$$b^0 \sim a^*, \qquad a^0 \sim b^*. \tag{47.3}$$

The following proposition was proved by Kostant and Symes independently when $\lambda = 0$, and by Kostant with general λ.

Proposition 47.1. *Suppose that* $\lambda \in \mathrm{g}^*$ *satisfies*

$$\langle \lambda, [\mathrm{a}, \mathrm{a}] \rangle = 0 \tag{47.4}$$

and

$$\langle \lambda, [\mathrm{b}, \mathrm{b}] \rangle = 0. \tag{47.5}$$

For any function f, *on* g^*, *let* $\mathrm{f}_{\mathrm{a},\lambda}$ *be the function on* $\mathrm{b}^0 \sim \mathrm{a}^*$ *given by*

$$\mathrm{f}_{\mathrm{a},\lambda}(l) = \mathrm{f}(l + \lambda), \qquad l \in \mathrm{b}^0,$$

and similarly

$$\mathrm{f}_{\mathrm{b},\lambda}(l) = \mathrm{f}(l + \lambda), \qquad l \in \mathrm{a}^0.$$

Then, if f *and* h *are invariants,*

$$\{f_{\mathrm{a},\lambda}, \mathrm{h}_{\mathrm{a},\lambda}\}_{\mathrm{a}^*} = 0.$$

Proof. For any $\xi \in g$, we let ξ^a and ξ^b describe its a and b components in the decomposition (47.1). Then for any $l \in b^0$, we have by (28.5)

$$\begin{aligned}\{f_{a,\lambda}, h_{a,\lambda}\}_{a^*}(l) &= \langle l, [L_{f_{a,\lambda}}(l), L_{h_{a,\lambda}}(l)]\rangle \\ &= \langle l, [L_f(l+\lambda)^a, L_h(l+\lambda)^a]\rangle \\ &= \langle l+\lambda, [L_f(l+\lambda)^a, L_h(l+\lambda)^a]\rangle\end{aligned}$$

by (47.4). Now f is an invariant so

$$\langle l+\lambda, [L_f(l+\lambda), \eta]\rangle = 0$$

for any $\eta \in g$. Since $L_f(l+\lambda) = L_f(l+\lambda)^a + L_f(l+\lambda)^b$, we have

$$\begin{aligned}\langle l+\lambda, [L_f(l+\lambda)^a, L_h(l+\lambda)^a]\rangle &= -\langle l+\lambda, [L_f(l+\lambda)^b, L_h(l+\lambda)^a]\rangle \\ &= \langle l+\lambda, [L_f(l+\lambda)^b, L_h(l+\lambda)^b]\rangle,\end{aligned}$$

since h is also an invariant. But b is a subalgebra and $l \in b^0$ so $\langle l, [b,b]\rangle = 0$, and $\langle \lambda, [b,b]\rangle = 0$ by hypothesis. Hence, the last expression above vanishes, proving the proposition.

Let us now concentrate on the case where $\lambda = 0$. Let π_a and π_b denote the projections onto the summands in (47.1) and π_{b^0} and π_{a^0} the projection onto the summands in (47.2) Any η in a, can be regarded as an element of g and hence has a coadjoint action on g^* which we shall for present purposes denote by $\mathrm{ad}_g^\# \eta$; similarly, it has a coadjoint action on a^* which we shall denote by $\mathrm{ad}_a^\# \eta$. Under the identification of a^* with $b^0 \subset g^*$ given by (47.2), the relation between these two representations is clearly

$$\mathrm{ad}_a^\# \eta = \pi_{b^0}\mathrm{ad}_g^\# \eta. \tag{47.6}$$

If f is a function on g^* and L_f is the corresponding map of $g^* \to g$, we can also consider the restriction, $f_{|b^0}$ of f to b^0. If we identify b^0 with a^* we then get a map, $L_{f_{|b^0}}: b^0 \to a$. The relation between these two maps (a fact that we have already implicitly used in the proof of Proposition 47.1) is clearly

$$\pi_a(L^f)_{|b^0} = L_{(f|_{b^0})}. \tag{47.7}$$

Combining (47.6) and (47.7) we see that for any l in b^0, the vector field associated to $f_{|b^0}$ is given by the rule assigning to the point l the value

$$\pi_{b^0}\,\mathrm{ad}_g^\#\,(\pi_a L_f(l)). \tag{47.8}$$

A particularly important case of this formula is when f is an invariant function, so that its corresponding vector field on g^* vanishes; that is,

$$\mathrm{ad}_g^\#\,(L_f(l)) = 0$$

for all l. Then

$$\mathrm{ad}^{\#}_{g}\,(\pi_a L^f(l)) = -\mathrm{ad}^{\#}_{g}\,(\pi_b L^f(l))$$

and $\mathrm{ad}^{\#}_{g}\,\zeta$ leaves b^0 invariant for any ζ in b. So, substituting into (47.8) we can drop the π_{b^0}. Thus,

Proposition 47.2. If f *is an invariant function on* g^*, *then the vector field on* $\mathrm{a}^* \sim b^0$ *assigns to each point* l *the value*

$$-\mathrm{ad}^{\#}_{\mathrm{g}}\,(\pi_{\mathrm{b}}\mathrm{L}_{\mathrm{f}}(l)). \tag{47.9}$$

For example, let $g = sl(n)$, let $a =$ all lower triangular matrices of trace 0, and let $b = o(n)$. If we use the bilinear form $b(A, B) = \frac{1}{2}\operatorname{tr} AB$ to identify g with g^*, then

$$b^0 = \text{all symmetric traceless matrices,} \tag{47.10}$$

while

$$a^0 \text{ consists of all strictly lower triangular matrices.} \tag{47.11}$$

If we write any matrix A as a sum of its strictly upper triangular diagonal and strictly lower triangular parts,

$$A = A_+ + A_0 + A_-, \tag{47.12}$$

then

$$\pi_{b^0}(A) = A_+ + A_0 + A^{\mathrm{t}}_+. \tag{47.13}$$

when we think of A as an element of g^*. If we think of A as an element of g, then we have

$$\pi_b(A) = A_+ - A^{\mathrm{t}}_+. \tag{47.14}$$

Let f be the invariant function on $g^* \sim g$ defined by

$$f(A) = \tfrac{1}{2}\operatorname{tr} A^2. \tag{47.15}$$

Then, under our identification of g^* with g, we have

$$L_f(A) = A. \tag{47.16}$$

Thus, on the space b^0 of all symmetric matrices, the flow given by (47.9) is

$$\dot{\mathrm{A}} = [A, B], \qquad B = A_+ - A^{\mathrm{t}}_-. \tag{47.17}$$

We can also consider the equations (47.17) restricted to specific coadjoint orbits in $a^* \sim b^0$. For example, suppose we take the element

$$\begin{pmatrix} 0 & 1 & & & \\ 1 & 0 & & & \\ & & \ddots & \ddots & \\ & & \ddots & \ddots & \\ & & & & 1 \\ & & & 1 & 0 \end{pmatrix}.$$

The bracket of this matrix with any lower triangular matrix will have no nonzero entries on any diagonal two or more positions above the principal diagonal, and can have arbitrary entries on the principal diagonal (except that the trace must be 0) and in the diagonal immediately above the principal diagonal. Hence using the projection (47.13) we see that the coadjoint orbit can be identified with the space of traceless Jacobi matrices

$$A = \begin{pmatrix} a_1 & b_1 & 0 & \cdots & 0 \\ b_1 & a_2 & b_2 & \cdots & 0 \\ 0 & b_2 & a_3 & \cdots & 0 \\ \vdots & \vdots & \vdots & & b_{n-1} \\ 0 & 0 & 0 & b_{n-1} & a_n \end{pmatrix}, \qquad b_k > 0, k = 1, \ldots, n-1, \operatorname{tr} A = 0. \tag{47.18}$$

(This identification of the space of Jacobi matrices with a coadjoint orbit of the group of lower triangular matrices is due to Kostant. The identification of the corresponding equations of motion (47.16) with those of the Toda lattice, to be described below, together with the group-theoretical proof of their complete integrability, is also due to Kostant (1978), and independently to Symes (1980a).)

The functions

$$f_k(A) = \operatorname{tr} A^k$$

are all $Gl(n)$-invariant. Hence by the Kostant–Symes lemma, Proposition 47.1, they give a family of Poisson-commuting functions when restricted to b^0, with $f = \frac{1}{2}f_2$. It is easy to check that on the orbit of all matrices of the form (47.18) these functions are independent almost everywhere, and hence the flow generated by f on this orbit is completely integrable.

In terms of the a's and b's in (47.18) the equation (47.17) becomes

$$\begin{aligned} \dot{b}_k &= b_k(a_{k+1} - a_k) \qquad & k &= 1, 2, \ldots, n-1, \\ \dot{a}_k &= 2(b_k^2 - b_{k-1}^2) \qquad & k &= 1, \ldots, n, \\ &\text{with} & b_0 &= b_n = 0. \end{aligned} \tag{47.19}$$

If we make the change of variables

$$b_k = \tfrac{1}{2}e^{(x_k - x_{k+1})/2}, \qquad a_k = -\tfrac{1}{2}y_k, \tag{47.20}$$

the first two equations in (47.19) go over into the equations

$$\dot{x}_k = H_{y_k}, \qquad \dot{Y}_k = -H_{x_k}, \tag{47.21}$$

with

$$H = \tfrac{1}{2}\sum y_k^2 + \sum e^{(x_k - x_{k-1})} \tag{47.22}$$

with the last equations in (47.19) going over into the (formal) boundary conditions $x_0 = -\infty$ and $x_{n+1} = +\infty$. The Hamiltonian in (47.22) is known as the (finite version of) the Hamiltonian of the Toda lattice. The change of variables (47.20) and the realization that the equations (47.19) can be written in matrix form (47.17) is due to Flashka (1974). A detailed discussion of the asymptotic behavior of the trajectories can be found in Moser (1975). An explicit solution of these equations using a "collective motion" method due to Symes (1980) will be presented in the next section. A generalization of these equations, together with their complete explicit solutions in terms of the data of representation theory can be found in Kostant (1979).

We now turn to some variants of the Kostant–Symes lemma. One, due to Miscenko and Fomenko (1978a), deals with translated invariants and no splitting, with conditions (47.4, 5) unnecessary.

Proposition 47.3 (Miscenko and Fomenko). *Let* f *and* h *be invariant functions on* g^* *and define the functions* $\mathrm{f}_{s\lambda}$, $\mathrm{h}_{t\lambda}$ *by*

$$\mathrm{f}_{s\lambda}(l) = \mathrm{f}(l + s\lambda),$$

$$h_{t\lambda}(l) = \mathrm{h}(l + t\lambda),$$

where s *and* t *are real numbers and* λ *is any element* g^*. *Then* $\mathrm{f}_{s\lambda}$ *and* $\mathrm{h}_{t\lambda}$ *Poisson-commute.*

Proof. It suffices to prove the theorem when $s \neq t$ since it then follows for $s = t$ by continuity. Set

$$l = a(l + s\lambda) + b(l + t\lambda), \qquad a = \frac{t}{t-s}, \qquad b = \frac{s}{s-t}.$$

Then

$$\begin{aligned}\{f_{s\lambda}, h_{t\lambda}\}(l) &= \langle l, [L_f(l + s\lambda), L_h(l + t\lambda)]\rangle \\ &= a\langle l + s\lambda, [L_f(l + s\lambda), L_h(l + t\lambda)]\rangle \\ &\quad + b\langle l + t\lambda, [L_f(l + s\lambda), L_h(l + t\lambda)]\rangle.\end{aligned}$$

The first term vanishes because f is an invariant and the second term vanishes because h is an invariant, proving the proposition.

In view of Proposition 47.3 it would seem that (47.4) should not be necessary in Proposition 47.1, or, in any event, the hypotheses of Proposition 47.1 are too stringent. The proper setting is not clear to us at present.

A useful variant of this result is due to Ratiu: Let h be Lie algebra, and consider the adjoint representation of h on itself. and let

$$g = h + h$$

the corresponding semidirect product. The Lie bracket on this algebra is given by

$$[(\xi_1, \eta_1), (\xi_2, \eta_2)] = ([\xi_1, \xi_2], [\xi_1, \eta_2] + [\xi_2, \eta_1]). \tag{47.23}$$

Let λ be a fixed element of h^*, let f be a function defined on h^* and t a real number. Define the function f_t on g^* by

$$f_t(\xi, \eta) = f(t\xi + \eta + t^2\lambda) \tag{47.24}$$

so that $L_{f_t}: g^* \to g$ is given by

$$L_{f_t}(\xi, \eta) = (t\, df(x), df(x)) \qquad x = t\xi + \eta + t^2\lambda. \tag{47.25}$$

We then have

Proposition 47.4 (Ratiu, 1981b). *Let* $\mathrm{g} = \mathrm{h} + \mathrm{h}$ *as above, with* $\lambda \in \mathrm{h}^*$. *Let* t *and* s *be real numbers and let* f_1 *and* f_2 *be invariant differentiable functions on* h^* *(under the coadjoint action of* H*). Define the functions* f_{1t} *and* f_{2s} *as in (47.24). Then* f_{1t} and f_{2s} *Poisson-commute.*

Proof. As in the proof of Proposition 47.3, we may assume that $s \neq t$. Then by the formula (28.5) for the Poisson bracket and by (47.22) we have

$$\begin{aligned}\{f_{1t}, f_{2s}\}_{g^*}(\xi, \eta) &= \langle (\xi, \eta), df_{1t}(\xi, \eta), df_{2t}(\xi, \eta)\rangle \\ &= \langle st\xi + (s + t)\eta, [df_1(x), df_2(y)]\rangle,\end{aligned}$$

where $x = t\xi + \eta + t^2\lambda$ and $y = s\xi + \eta + s^2\lambda$. Write

$$st\xi + (s + t)\eta = \frac{s^2}{s - t}x - \frac{t^2}{s - t}y;$$

and the fact that f_1 and f_2 are h^*-invariant implies the vanishing of the Poisson bracket.

Ratiu combines the preceding proposition with the following one to explain the complete integrability of the C. Neumann system and the geodesic flow on the ellipsoid and the n-dimensional Lagrange top:

Proposition 47.5. *Let* g *be a Lie algebra,* k *a subalgebra and* p *a complementary subspace to* k *that is* k*-invariant, so*

$$\mathrm{g} = \mathrm{k} + \mathrm{p}, \qquad [\mathrm{k},\mathrm{k}] \subset \mathrm{k}, \qquad [\mathrm{k},\mathrm{p}] \subset \mathrm{p},$$

with π_{k} *and* π_{p} *the corresponding projections. Let* f_1 *and* f_2 *be two smooth functions on* g^* *that Poisson commute. Suppose that for every* $\mathrm{l} \in \mathrm{p}^0$

$$\langle \mathrm{l}, \pi_{\mathrm{k}}[\pi_{\mathrm{p}}\, \mathrm{df}_1(\mathrm{l}), \pi_{\mathrm{p}}\, \mathrm{df}_2(\mathrm{l})]\rangle = 0 \tag{47.26}$$

Then $\mathrm{f}_{1\mathrm{p}^0}$ *and* $\mathrm{f}_{2\mathrm{p}^0}$ *commute in the Poisson bracket of* $\mathrm{k}^* \sim \mathrm{p}^0$.

Proof. The Poisson bracket on p^0 is given by

$$\begin{aligned}\{f_1, f_2\}_{p^0}(l) &= \langle l, [\pi_k\, df_1(l), \pi_k\, df_2(l)]\rangle \\ &= \langle l, [df_1(l), df_2(l)]\rangle - \langle l, [\pi_p(l), \pi_k(l)]\rangle \\ &\quad - \langle l, [\pi_k(l), \pi_p(l)]\rangle + \langle l, [\pi_p(l), \pi_p(l)]\rangle.\end{aligned}$$

The first term on the right-hand side of this equation vanishes because it equals $\{f_1, f_2\}_{g^*}(l)$, which vanishes by assumption. The second and third term vanish because $[k, p] \subset p$ and $l \in p^0$. Also, the last term equals the right-hand side of (47.26) since $l \in P^0$. Q.E.D.

Two cases where (47.26) clearly hold are

$$df_i(l) \in k \qquad \text{for} \qquad i = 1, 2 \qquad \text{so} \qquad \pi_p\, df_i(l) = 0 \tag{47.26$'$}$$

and

$$p \text{ is a subalgebra, so } \pi_k[\xi, \eta] = 0 \text{ for and } \xi, \eta \in p. \tag{47.26$''$}$$

Ratiu applies the last two propositions to prove the complete integrability of various kinds of Euler–Poisson equations as follows: (We refer to the notation of Section 29, especially equation (29.19) with the following changes in notation: We shall denote our semidirect product Lie algebra as

$g = L + V$ instead of $g = h + V$, as appears there.) Now suppose that as a vector space, we can identify $L + V$ as the underlying vector space of a Lie algebra h, such that L is a subalgebra and the action of L on V is given by the adjoint representation. For example, in the case of the geodesic flow on the ellipsoid discussed in Section 30, we would take

$$L = o(n),\ V = \mathrm{symm}\,(n), \qquad \text{(all } n \times n \text{ symmetric matrices)}$$

and so

$$h = gl(n).$$

Notice that in this case, and we shall so assume in general, there is a nonsingular invariant bilinear form ($\mathrm{tr}\, AB$ for $gl(n)$) under which $V = L^{\perp}$ and also

$$[V, V] \subset L.$$

Then we may identify $(L + V)^*$ with $L + V$ and all the operations $\hat{\alpha}p$, $p \odot x$, and $\hat{\alpha} \cdot \alpha$ occurring in (29.19) become identified with Lie bracket. Thus equations (29.19) become

$$\begin{aligned} \frac{d\alpha}{dt} &= [\hat{\alpha}, \alpha] + [p, x], \\ \frac{dp}{dt} &= [\hat{\alpha}, p]. \end{aligned} \tag{47.27}$$

Here $\hat{\alpha} = T\alpha$, where T is some linear transformation of L into itself, and x is a constant element of V. Ratiu's ingenious method is as follows: Suppose that we can find some constant element y of h such that

$$[\hat{\alpha}, y] - [x, \alpha] = 0, \qquad \forall\, \alpha \in L. \tag{47.28}$$

Then equations (47.25) are equivalent (by comparing coefficients of powers of the parameter s) to

$$\frac{d(p + s\alpha + s^2 y)}{dt} = [\hat{\alpha} - sx, p + s\alpha + s^2 y]. \tag{47.29}$$

Equation (47.29) clearly implies that if f is any ad-invariant function on h, then the functions f_s on $L + V$ given by

$$f_s(\alpha, p) = f(p + s\alpha + s^2 y)$$

must be constant along trajectories of our system (47.24). In particular, if f is an analytic function, each of the coefficients to the Taylor expansion of f_s

is a constant along the trajectories of (47.25), that is, an invariant of the Euler–Poisson Hamiltonian. By Proposition 47.3, if f_1 and f_2 are two invariants then f_{1s} and $f_{2s'}$ Poisson-commute on $h + h$ for any values of s and s'. Now in Proposition 47.4, we can take

$$g = h + h,$$
$$k = L + V,$$
$$p = V + L.$$

If we can verify (47.24) for coefficients of various powers of s in f_{1s} and f_{2s} we will get commuting integrals of motion.

Example. Take $L = o(n)$ and $V = \mathrm{symm}(n)$ so $h = gl(n)$, as above. Let F_k be the function on $gl(n)$ given by $F_k(B) = \mathrm{tr}\, B^{k+1}$. We let f_k be the coefficient of s^{2k} in $F_k(p + s\alpha + s^2 y)$ so up to a constant

$$f_k(\alpha, p) = \mathrm{tr}\left(\sum_0^k y^i p y^{k-i} + \sum y^i \alpha y^j \alpha y^m\right), \qquad i + j + m = k - 1 \quad (47.30)$$

and, up to constants,

$$df_k(\alpha, p) = \left(\sum_0^{k-1} y^i \alpha y^{k-i-1}, y^k\right). \quad (47.31)$$

Notice that if $y \in \mathrm{symm}(n)$ then the right-hand side belongs to k; that is, (47.26)′ holds, so the f_k Poisson-commute. For the case of the C. Neumann problem cf. in Section 30, we had

$$\hat{\alpha} = \alpha \qquad \text{and} \qquad x = A$$

[cf. (30.8)]. Thus we can take $y = -A$ in (47.28).

For the geodesic flow on the ellipsoid we had

$$\hat{\alpha} = A^{-1}\alpha A^{-1} \qquad \text{and} \qquad x = A^{-1},$$

so again we can take $y = -A$ in (47.28). The coadjoint orbits for the two problems are the same. Thus the functions f_k given by (47.30) with $y = -A$ form a family of Poisson-commuting functions for both problems. It still requires a bit of work to show that these functions $f_1, \ldots, f_{n-1}$ are functionally independent almost everywhere on this orbit. For this, we refer the reader to Ratiu (1981b). We also refer the reader to Ratiu (1981a) or to Ratiu and van Moerbeke (1982) for an application of the same method to the n-dimensional free or Lagrange top.

48. Systems of Calogero type

Suppose that we are given a Hamiltonian action of G on some symplectic manifold M, and H is a G-invariant Hamiltonian. We know that H then induces a Hamiltonian system, with Hamiltonian $H_{\mathcal{O}}$ on each of the Marsden–Weinstein reduced spaces $M_{\mathcal{O}}$ (whenever they are defined for g^* orbits $\mathcal{O}$). In the most standard examples of classical mechanics, this is typified by "passing to the center-of-mass frame" or "elimination of nodes" – a reduction of the number of variables that arise in the differential equations; which might be regarded as a "simplification" of the problem of solving the original equations of motion on M.

However, it is conceivable that quite the reverse might be the case – that the original equations of motion are quite transparent – but that the equations of motion of the quotient system appear more complicated. Indeed, we shall show in this section that the Calogero system of n particles on the line moving under the inverse-square potential, and the corresponding Sutherland system of n particles on the circle moving under the $\sin^{-2}$ potential are examples of mechanical systems that arise as quotients of much simpler looking mechanical systems. Our discussion follows very closely that of Kahzdan, Kostant, and Sternberg (1978).

Suppose that we are given a linear representation of a Lie group G on some vector space V. We then get the corresponding Hamiltonian action of G on $T^*V = V + V^*$ and its moment map $\Phi: V + V^* \to g^*$ is given by

$$\Phi(v, p) = p \odot v,$$

where, as usual

$$\langle p \odot v, \xi \rangle = \langle p, \xi v \rangle.$$

In particular, we shall consider the case where $v = g$ carries the adjoint representation and where g has an invariant nondegenerate bilinear form allowing the identification of g with g^*. Then as we know, $\odot$ gets identified with $[\,,]$ so

$$\Phi(\beta, \zeta) = [\beta, \zeta]. \tag{48.1}$$

For any G action on X, the invariant functions on X pull back to give a Poisson-commuting family of G-invariant functions on T^*X and hence a Poisson-commuting family of functions on any Marsden–Weinstein reduced space of T^*X. In the case at hand, the collection of ad-invariant functions on g thus give rise to a family of Poisson-commuting functions on

$(g + g)_{\mathcal{O}}$ for any orbit $\mathcal{O}$ in $g^* \sim g$. If the Marsden–Weinstein reduced space $(g + g)_{\mathcal{O}}$ has low enough dimension, this might give rise to a completely integrable system.

Let us apply these considerations to the case $G = U(n)$. (It will be slightly more convenient to work with the reductive group $U(n)$ than with the simple group $SU(n)$. Of course, all the results are really taking place in $SU(n)$. It is notationally more convenient to use $U(n)$.) Then g consists of skew-adjoint matrices, or by multiplying by $-i$, we may identify g as a vector space with the space of $n \times n$ self-adjoint matrices and identify g with g^* via the bilinear form tr AB. The orbits of minimum positive dimension are orbits of translates of the rank one self-adjoint matrices by multiples of the identity matrix. Thus, for example, if

$$\alpha = \lambda I + v^* \otimes v,$$

then the orbit through α consists of all $\lambda I + w^* \otimes w$ with $\|w\| = \|v\|$. This set has (real) dimension $2(n - 1)$. By (48.1), $\Phi^{-1}(\alpha)$ consists of all pairs of self-adjoint matrices A and B such that

$$\frac{1}{i}[A, B] = \lambda I + v^* \otimes v. \tag{48.2}$$

Notice that if A and B and CAC^{-1}, CBC^{-1} both lie in $\Phi^{-1}(\alpha)$ then $Cv^* \otimes Cv = v^* \otimes v$ so that $Cv = e^{i\theta}v$ for some θ. Since multiplying C by $e^{-i\theta}$ does not affect conjugation by C we can assert that if A and B and A', B' in $\Phi^{-1}(\alpha)$ are conjugate by some element of G then they are conjugate by some C satisfying $Cv = v$. To fix the ideas, let us take $\lambda = -1$ and v the column vector with all entries $= 1$. Then α is the matrix with 0's on the diagonal and 1's in all other positions. Suppose that $(1/i)\,[A, B] = \alpha$. Let us choose some unitary matrix C such that $CAC^{-1} = D$ is diagonal. Let $E = CBC^{-1}$ and $w = Cv$. Then $(1/i)\ [D, E] = -I + w^* \otimes \mathrm{w}$ has 0's on the main diagonal, since D is a diagonal matrix. This implies that $\bar{w}_j w_j = 1$ $(j = 1, \ldots, n)$, that is, that $w_j = \exp(i\theta_j)$ for suitable θ_j. Multiplying C on the left by the diagonal matrix whose entries along the diagonal are $\exp(-i\theta_j)$ gives a new matrix C' with $C'v = v$ and $C'AC'^{-1}$ still equal to D. Thus we can always diagonalize A by a matrix that preserves v. Let $x_1, \ldots, x_n$ be the eigenvalues of A (i.e., the diagonal entries of D). Since $[D, E] = \alpha$ has 1's in all the off-diagonal positions, we conclude that $x_i \neq x_j$ for $i \neq j$ and that the off-diagonal entries of E are $i/(x_i - x_j)$. The diagonal entries of E are, of course, arbitrary. Let us denote them by $y_1, \ldots, y_n$. Thus we can use $(x_1, \ldots, x_n; y_1, \ldots, y_n)$ as coordinates on the

Marsden–Weinstein reduced space, which we shall denote here by $\mathcal{O}'$, so

$$(g + g)_{\mathcal{O}} = \mathcal{O}' = \Phi^{-1}(\alpha)/G_{\alpha},$$

with the understanding that the x's and the y's are determined only up to permutation. We claim that the x's and y's form a symplectic coordinate system for $\mathcal{O}'$, that is, we claim that symplectic form of $\mathcal{O}'$ is $\sum dy_i \wedge dx_i$. To prove this it suffices to prove that

$$\{x_i, x_j\} = 0, \qquad \{x_i, y_j\} = \delta_{ij}, \qquad \{y_i, y_j\} = 0.$$

To verify these equations at some particular point $\bar{p}$ of $\mathcal{O}'$ it is enough to find functions f_i and g_j defined near some point p in the inverse image of p in $g + g$ that agree with the pull-back of the functions x_i and y_j up to second order at p and whose Poisson brackets have the desired properties. Suppose that we have arranged the eigenvalues $x_1, \ldots, x_n$ in increasing order. If D is a diagonal matrix with diagonal entries $x_1, \ldots, x_n$ and if F is a matrix with 0's along the diagonal then the ith eigenvalue of $D + \epsilon F$ is $x_i + O(\epsilon^2)$. Thus, if a_{ii} is the function on $g + g$ assigning to the pair A, B the ith diagonal entry of A and b_{ii} the function assigning the ith diagonal entry of B, we may take $p = D, E$ and $f_i = a_{ii}, g_j = b_{jj}$. These functions clearly give the desired Poisson brackets.

We can now consider the ring of invariant polynomials in B as giving a completely integrable system on $\mathcal{O}'$. Thus, for example $\operatorname{tr} B = y_1 + \cdots + y_n$ corresponds to the "total momentum" in the x-y coordinate system. Similarly,

$$\tfrac{1}{2}\operatorname{tr} B^2 = \tfrac{1}{2}\sum y_j^2 + \tfrac{1}{2}\sum_{i \neq j} (x_i - x_j)^{-2} \tag{48.3}$$

is the Hamiltonian of the Calogero system of points, moving on a line with r^{-2} potential. Notice that the various integrals,

$$I_k = \frac{1}{k!}\operatorname{tr} B^k \tag{48.4}$$

are polynomials in the y_j and $(x_j - x_k)^{-1}$. If we consider the Hamiltonian $H = I_2$ to be defined on the full original space $M = g + g^*$, it is easy to see explicitly that the corresponding flow sends the pair of matrices (A, B) into $(A + tB, B)$ at time t. (Observe that $[A + tB, B] = [A, B]$ so that the set $\Phi^{-1}(\alpha)$ is preserved by the flow.) In terms of the $x - y$ coordinates that we have introduced on $\mathcal{O}'$, the flow looks somewhat complicated, but we can introduce more suitable coordinates as follows: Instead of using the I_k, which are coefficients of the characteristic polynomial as integrals, we could equally well use the eigenvalues λ_k. Suppose that we write them in increasing order, so $\lambda_1 \leqslant \lambda_2 \leqslant \cdots \leqslant \lambda_n$. We thus diagonalize the matrix B instead of the matrix A! If we choose the unitary matrix U such that UEU^{-1} is diagonal and $U\alpha U^{-1} = \alpha$, then since A and B enter antisymmetrically into the definition of $\mathcal{O}'$, we conclude that UDU^{-1} has off-diagonal entries $-i(\lambda_j - \lambda_k)^{-1}$ and diagonal entries

$$\mu_i = \sum_j |u_{ij}|^2 x_j. \tag{48.5}$$

The μ's and λ's form a symplectic coordinate system in terms of which the Hamiltonian takes the form $H = \frac{1}{2}\sum \lambda_i^2$ so that the flow is simply given as $(\vec{\mu}, \vec{\lambda}) \to (\vec{\mu} + t\vec{\lambda}, \vec{\lambda})$. Since the potential (in the $x - y$ system) is repulsive, it is easy to see that as $t \to \infty$ the particles move apart, so that $(x_j - x_k)^{-1} \to 0$ and E approaches a diagonal matrix and thus $y_i^{(+\infty)} = \lambda_i$, since the velocities y_i must satisfy $y_1 < y_2 <$, etc. as the particles are moving apart, cf. Moser (1975). As $t \to \infty$, therefore, the doubly stochastic matrix $(|u_{ij}^2|)$ approaches the identity and hence $\mu_j - x_j \to 0$. Thus we can think of the transformation from (x, y) to (μ, λ) coordinates as a sort of "wave operator," where the (μ, λ) coordinates give the free state at $+\infty$. As $t \to -\infty$, the condition that the particles move apart requires that $y_1 > y_2 > \cdots$ show that E approaches a diagonal matrix whose eigenvalues are still the λ_i, but in reverse order: $y_i(-\infty) = \lambda_{n-i+1}$. There is thus an "exchange of velocities" as a result of the interaction. Furthermore, as $t \to -\infty$ the matrix U and hence the doubly stochastic matrix $(|u_{ij}|^2)$ approaches the matrix with 1's on the antidiagonal and 0's elsewhere. Thus $\mu_j - x_{n-j+1} \to 0$. This, together with the form of the flow in (μ, λ) coordinates clearly implies that

$$x_k(t) - x_{n-k+1}(-t) - 2\dot{x}(+\infty)t \to 0 \tag{48.6}$$

(i.e., the absence of "phase shifts"). We have thus given a complete description of the classical scattering theory for the Calogero system [cf. Moser (1975)].

As a second illustration of how to simplify a system by "dereducing" it, consider the conjugation action of a Lie group G on itself and the corresponding induced Hamiltonian action of G on T^*G. It follows from (28.8) and (28.9) that the moment map for this action is given by

$$\Psi(c, \beta) = c\beta - \beta. \tag{48.7}$$

In particular, if we take $G = U(n)$ and α the element we used to describe the Calogero system above, we can argue just as above to derive a normal form for the elements of $\Psi^{-1}(\alpha)$: We may assume that c is diagonal with eigenvalues $\exp(i\theta_j)$ and hence that all the off-diagonal entries of β are $(\exp[i(\theta_j - \theta_k)] - 1)^{-1}$. Then if $y_1, \ldots, y_n$ are the diagonal entries of β

$$\operatorname{tr}\beta^2 = y_j^2 + \sum_{i \neq j} (\sin\tfrac{1}{2}(\theta_j - \theta_k))^{-2}, \tag{48.8}$$

and we obtain the complete integrability of the Sutherland system (see Sutherland, 1972). To verify that the θ's and y's form a symplectic coordinate system on $\mathcal{O}'$ we need a formula for the symplectic structure on T^*G in terms of the (left) trivialization that we have chosen. Let us derive this formula in general. Let Θ denote the canonical 1-form on T^*G. If $\tilde{\eta}$ is a left-invariant vector field on G, then, under the identification $T^*G = G \times g^*$, and for any $\beta \in g^*$ (thought of as a tangent vector to g^*) we have

$$\Theta_{(c,\alpha)}(\tilde{\eta}, \beta) = \langle \alpha, \eta \rangle,$$

where $\langle\, , \rangle$ denotes the evaluation map between g^* and g. We can write this equation as

$$\Theta_{(c,\alpha)} = \langle \alpha, c^{-1}\, dc \rangle,$$

where α is thought of as the function from $G \times g^*$ to g^* given by projection onto the second component, and where $c^{-1}\, dc$ denotes the 1-form on G (considered as a 1-form on $G \times g^*$) that assigns to each tangent vector at c the associated vector in g, thus $(c^{-1}\, dc)(\tilde{\eta}) = \eta$ so that $c^{-1}\, dc$ is a 1-form with values in g. The symplectic structure is given by $d\Theta$, and we see from the preceding equation that

$$d\Theta_{(c,\alpha)} = \langle d\alpha \wedge c^{-1}\, dc \rangle - \langle \alpha, d(c^{-1}\, dc) \rangle,$$

or, since $d(c^{-1}\, dc) = -c^{-1}\, dc \wedge c^{-1}\, dc$, we see that if (η_1, β_1) and (η_2, β_2) are elements of $g + g^*$ (with the β thought of as tangent vectors to g^*) then

$$d\Theta_{(c,\alpha)}((\eta_1, \beta_1), (\eta_2, \beta_2)) = \langle \alpha, [\eta_1, \eta_2] \rangle + \langle \beta_2, \eta_1 \rangle - \langle \beta_1, \eta_2 \rangle. \tag{48.9}$$

(It is interesting at this point to check that this formula is consistent with our preceding formula for $f_\eta = \langle \alpha, \eta \rangle$. That is, let us check that $\{f_\eta, f_{\eta'}\} = f_{[\eta,\eta']}$. For this purpose we must find the vector field associated to the function f_η by the symplectic structure. Now

$$df_\eta(\eta', \beta') = \langle \beta', \eta \rangle$$

for any tangent vector (η', β'). Thus, if we denote the vector field associated to f_η by $(\bar{\eta}, \bar{\beta})$ we must have

$$d\Theta_{(c,\alpha)}((\bar{\eta}, \bar{\beta}), (\eta', \beta')) = \langle \beta', \eta \rangle$$

or

$$\langle \alpha, [\bar{\eta}, \eta'] \rangle + \langle \beta, \bar{\eta} \rangle - \langle \bar{\beta}, \eta' \rangle = \langle \beta', \eta \rangle.$$

This implies that $\eta = \bar{\eta}$, as expected, and that $\bar{\beta}$ is given as the linear function

$$\langle \bar{\beta}, \eta' \rangle = \langle \alpha, [\eta, \eta'] \rangle.$$

Now the function f_η depends only on the g^* component since $f_\eta(c, \alpha) = \langle \alpha, \eta \rangle$. Thus, in applying the vector field $(\eta, \bar{\beta})$ to $f_{\eta'}$, we need only apply the $\bar{\beta}$ component. Thus

$$\begin{aligned} \{f_\eta, f_{\eta'}\} &= (\eta, \bar{\beta}) f_{\eta'} \\ &= \bar{\beta} \langle \cdot, \eta' \rangle \\ &= \langle \bar{\beta}, \eta' \rangle \\ &= \langle \alpha, [\eta, \eta'] \rangle \\ &= f_{[\eta, \eta']}(c, \alpha), \end{aligned}$$

as was to be proved.) Getting back to the Sutherland system, we observe that $[\eta_1, \eta_2] = 0$ for diagonal matrices, and thus the argument that we have the correct symplectic structure is the same as in the Calogero case.

As a third illustration of simplification by "dereduction," we explain the method of Symes for solving a Toda-like system. Suppose that we have a (global) decomposition $G = LK$, where L and K are Lie subgroups of the Lie group G. This means that every element a of G can be written uniquely as $a = cb$, where $c \in L$ and $b \in K$. The corresponding decomposition of Lie algebras is $g = l + k$, with the consequent identification

$$l^* = k^0. \tag{48.10}$$

Let f be an invariant function on g^*. The restriction, $f_{|k^0}$ defines a function on l^*, which is not in general L-invariant. The problem is to integrate the corresponding equations of motion on l^*. We first observe that any function f on g^* determines a left-invariant function F on $T^*G \sim G \times g^*$ collective for the moment map of right multiplication:

$$F(a, \alpha) = f(-\alpha),$$

(cf. (28.9) and the general discussion in Section 28). As indicated there, since f is invariant, the solution curves for the Hamiltonian system on T^*G given by F will all be orbits (under right multiplication) of one-parameter groups: The orbit through the point (a, α) will be

$$(a \exp(-t\xi), \alpha), \qquad \text{where} \qquad \xi = L_f(-\alpha), \tag{48.11}$$

since $\exp t\xi\alpha = \alpha$. The function F on T^*G is bi-invariant, that is, invariant under the actions induced by both right and left multiplication, in particular, under the action induced by right multiplication by K. The moment map $\Phi^K: T^*G \to k^*$ for this action is given by

$$\Phi^K(a, \alpha) = -\pi_k{}^*\alpha,$$

where $\pi_k{}^*: g^* \to k^*$ is the projection dual to the injection of k in g as a subalgebra. Therefore F induces a function on each of the Marsden–Weinstein reduced spaces for this K action. In particular,

$$\Phi^{K-1}(0) = \{(a, \alpha), \alpha \in k^0\} \tag{48.12}$$

and, as we know from Section 29, the corresponding Marsden–Weinstein reduced space at 0 is

$$\Phi K^{-1}(0)/K \sim T^*(G/K). \tag{48.13}$$

In the presence of a group decomposition $G = LK$ we may identify G/K with L and hence $T^*(G/K)$ with T^*L. This means that in (48.13) we represent the equivalence class of

$$(a, \alpha) \text{ by } (c, b\alpha), \qquad \text{where} \qquad a = cb. \tag{48.14}$$

Notice that $b \in K$, so that $b\alpha \in k^0$ if $\alpha \in k^0$. The function F_L induced by F on T^*L is thus given by

$$F_L(c, \beta) = f(-\beta).$$

It is thus collective for the right action of L on T^*L and corresponds to the function f_{k^0}. Thus the solution curves of f_{k^0} are just the images of solution curves of F_L on T^*L, and these are just the projections onto T^*L (regarded as the Marsden–Weinstein reduced space at 0) of the solution curves of F on T^*G. Putting together (48.11) and (48.14) we see that the solution through $\alpha \in k^0$ of f_{k^0} is obtained as follows (by taking $(e, -\alpha)$ as the initial point in (48.11)):

Let $\xi = L_f(\alpha)$ and write

$$\exp(-t\xi) = c(t)b(t). \tag{48.15}$$

Then

$$b(t)\alpha \tag{48.16}$$

is the desired solution curve.

Let us show how this method gives the "scattering theory" of the aperiodic Toda lattice. Our treatment follows the ideas of Symes (1982) with some minor corrections, and was worked out jointly with Kostant. So we will take $G = Sl(n, \mathbb{R})$ and L to be the group of upper triangular matrices (a minor notational change from Section 47), K to be $O(n)$ and $f(\alpha) = \frac{1}{2}\operatorname{tr}\alpha^2$ so $\xi = \alpha$. Thus, identifying the adjoint and coadjoint representation in this case and writing the adjoint representation as matrix conjugation, the solution curve through ξ is given by

$$b(t)\xi b(t)^{-1},$$

where $b(t)$ is given by (48.15). We claim that for ξ belonging to the Toda lattice orbit as described in Section 47,

$$\lim_{t\to\infty} b(t)\xi b(t)^{-1} = \Delta_+, \tag{48.17}$$

where Δ_+ is the diagonal matrix whose entries are the (distinct) eigenvalues of ξ arranged in decreasing order. In fact, the following two general facts are true:

(a) Let ξ be any symmetric matrix with distinct eigenvalues, and let $v_1, \ldots, v_n$ be the eigenvalues of ξ and let $\delta_1, \ldots, \delta_n$ be the standard basis of $\mathbb{R}^n$. Then if

$$\begin{gathered}(v_1, \delta_1) \neq 0,\\ (v_1 \wedge v_2, \delta_1 \wedge \delta_2) \neq 0,\\ \vdots\\ (v_1 \wedge \cdots \wedge v_{n-1}, \delta_1 \wedge \cdots \wedge \delta_{n-1}) \neq 0,\end{gathered} \tag{48.18}$$

then (48.17) holds when $b(t)$ is given by (48.15).

Indeed, to prove (48.17) it is enough to show that, in the limit, $b(t)^{-1}$ carries the basis $\delta = (\delta_1, \ldots, \delta_n)$ to the basis $v = (v_1, \ldots, v_n)$. Since b is orthogonal, this is the same as showing that the flag, $\boldsymbol{\delta}$ generated by δ (the collection of nested subspaces spanned by δ_1, by δ_1, δ_2, by $\delta_1, \delta_2, \delta_3$, etc.) tends to the flag $\mathbf{v}$ generated by $(v_1, \ldots, v_n)$. Now $c\boldsymbol{\delta} = \boldsymbol{\delta}$ for any lower triangular matrix, so

$$b(t)^{-1}\boldsymbol{\delta} = b(t)^{-1}c(t)^{-1}\boldsymbol{\delta} = e^{t\xi}\boldsymbol{\delta}.$$

So we must prove that

$$\lim_{t\to\infty} e^{t\xi}\boldsymbol{\delta} = \mathbf{v}. \tag{48.19}$$

Now if we expand δ_1 in terms of the orthonormal basis v we have, by the first inequality in (48.18),

$$\delta_1 = a_1 v_1 + \cdots, \qquad a_1 \neq 0.$$

Thus

$$e^{t\xi}\delta_1 = e^{t\lambda_1}a_1 v_1 + e^{t\lambda_2}a_2 v_2 + \cdots,$$

where $\lambda_1 > \lambda_2 \cdots > \lambda_n$ are the eigenvalues of ξ. The line through $e^{t\xi}\delta_1$ tends to the line through v_1, since the first exponential term dominates. In $\bigwedge^2(\mathbb{R}^n)$ the eigenvalues of ξ are $\lambda_i + \lambda_j$ $(i \neq j)$. If we expand $\delta_1 \wedge \delta_2$ in terms of the orthonormal basis $v_i \wedge v_j$ we have

$$\delta_1 \wedge \delta_2 = a_{12}v_1 \wedge v_2 + \cdots, \qquad a_{12} \neq 0,$$

so

$$e^{t\xi}(\delta_1 \wedge \delta_2) = \exp[(\lambda_1 + \lambda_2)t]a_{12}v_1 \wedge v_2 + \cdots .$$

Since $\lambda_1 + \lambda_2$ is the maximal eigenvalue of ξ on $\bigwedge^2$, we see that the line through $e^{t\xi}\delta_1 \wedge e^{t\xi}\delta_2$ in $\bigwedge^2$ tends to the line through $v_1 \wedge v_2$, or what amounts to the same thing, the plane spanned by $e^{t\xi}\delta_1$ and $e^{t\xi}\delta_2$ tends to the plane spanned by v_1 and v_2. Proceeding in this way proves (48.18).

(As a matter of fact, if ξ is any symmetric matrix, not necessarily satisfying (48.18), we will always have

$$b(t)\xi b(t)^{-1} \to \Delta, \tag{48.20}$$

where Δ is some diagonal matrix whose entries are the eigenvalues of ξ, but not necessarily in decreasing order. The reason for this is as follows: Let M denote the manifold of all flags in $\mathbb{C}^n$. This is a Kaehler manifold and $U(n)$ acts as automorphisms. Thus $i\xi$ is a Hamiltonian vector field for the symplectic form that is the imaginary part of the Kaehler form. That is

$$\mathrm{Im}\langle i\xi_{M^\times}(p), v\rangle_p = df(v) \qquad \forall\, v \in TM_p.$$

But this is the same as

$$\mathrm{Re}\langle \xi_M(p), v\rangle_p = df(v)$$

or

$$\xi_M = \mathrm{grad}\, f,$$

where the gradient is with respect to the Riemann metric Re $\langle,\rangle$. As ξ_M is a gradient field, the flow it generates will carry any point to a critical point (of f or what is the same, of ξ). The critical points are easily seen to be the flags of eigenvectors and this proves (48.19). Generically, points will flow to the maximum critical point. The role of (48.18) is to ensure that $\boldsymbol{\delta}$ is generic in this sense.)

(b) Let ξ be any semisimple matrix with nonzero elements on the diagonal below the main diagonal and with all lower diagonals identically zero. Then ξ has distinct eigenvalues and (48.18) holds for *any* arrangement of the eigenvectors. To prove the first of these assertions, observe that the only matrices that commute with the matrix

$$\begin{pmatrix} 0 & \cdots & \cdots & \cdots & 0 \\ 1 & \ddots & & & \\ \vdots & & & & \\ \vdots & & 1 & & \\ \vdots & & & \ddots & \\ 0 & & & 1 & 0 \end{pmatrix} = P$$

are linear combinations of the identity matrix and

$$\begin{pmatrix} 0 & \cdots & \cdots & 0 \\ 1 & \ddots & & \vdots \\ 1 & & & \vdots \\ \vdots & \ddots & & \vdots \\ 0 & 1 & 0 & 0 \end{pmatrix}, \begin{pmatrix} 0 & \cdots & \cdots & \cdots & \cdots & 0 \\ 0 & 0 & & & & \vdots \\ 1 & 0 & \ddots & & & \vdots \\ 0 & 1 & & \ddots & & \vdots \\ \vdots & & \ddots & & \ddots & \vdots \\ \vdots & & & \ddots & & \ddots \; \vdots \\ 0 & \cdots & \cdots & 1 & 0 & 0 \end{pmatrix}, \ldots, \begin{pmatrix} 0 & & & \\ \vdots & & & \\ 0 & & & \\ 1 & 0 & \cdots & 0 \end{pmatrix}.$$

So the matrix P is regular in the sense that the dimension of the space of matrices commuting with P is n. Now if

$$\xi = \begin{pmatrix} 0 & & & \\ b_1 & \ddots & & \\ & \ddots & & \\ & & b_{n-1} & 0 \end{pmatrix}, \qquad b_i \neq 0.$$

Then conjugating ξ by diag $(1, b_1^{-1}, b_1^{-1} b_2^{-1}, \ldots)$ brings ξ to the form $P + X$, where X is upper triangular. Further conjugation by $\operatorname{diag}(\lambda, \lambda^2, \ldots, \lambda^n)$ yields $\lambda P + X(\lambda)$, where $X(\lambda)$ is bounded. This is clearly regular for large enough λ, and so ξ is regular. For semisimple ξ regularity implies distinct eigenvalues.

Let us now prove that (48.18) holds. Let R denote the change of basis matrix so that

$$R\xi = \Delta R.$$

Let $\pi_k\colon \mathbb{R}^n \to \mathbb{R}^n$ denote projection onto the first k components. Conditions (48.18) are the same as asserting that

$$\pi_k R \pi_k \text{ has rank } k$$

or that

$$\text{no nonzero vector of the form } x = \begin{pmatrix} x_1 \\ \vdots \\ x_k \\ 0 \\ \vdots \\ 0 \end{pmatrix}$$

lies in the kernel of $\pi_k R$. Suppose this were not true and choose k smallest with a nontrivial such kernel, so we may assume that $x_k \neq 0$. Now

$$\pi_k RAx = \pi_k \Delta Rx = \pi_k \Delta \pi_k Rx = 0$$

since $\pi_k \Delta = \pi_k \Delta \pi_k$. Thus Ax lies in ker $\pi_k R$. But

$$Ax = \begin{pmatrix} \vdots \\ a_k x_k \\ 0 \\ \vdots \end{pmatrix} \neq 0 \qquad \text{since} \qquad a_k \neq 0$$

and similarly $x, Ax, \ldots, A^{n-k}x$ are all linearly independent and lie in ker $\pi_k R$. But since R is nonsingular, $\pi_k R$ has rank k, so its kernel cannot contain $n - k + 1$ independent vectors. This proves (48.19).

In particular, for Jacobi matrices, the matrices on the Toda lattice orbits, (48.17) holds.

If we now look at (47.11) (with A instead of ξ) and at (47.20) we see that (up to a sign) the eigenvalues which by (48.17) are the limits of the diagonal elements, are the limiting momenta of the system as $t \to +\infty$. As $t \to -\infty$ the momenta tend to the eigenvalues, but in reverse order. The scattering is thus described by an exchange of momenta.

49. Solitons and coadjoint structures

In the next few sections we wish to sketch how some of the ideas of this chapter have been applied to the study of nonlinear partial differential equations. We will only touch on the formal aspects of the theory, and even here only insofar as it relates to coadjoint structures, following in the main the expository article by Iacob and Sternberg (1980). There is now a vast and growing literature on the subject, with many deep results. It would take a whole book to begin to do justice to the subject. So we must refer the interested reader to the literature for study of the subject in depth.

The concept of a solitary wave, now called a soliton, was first introduced into the science of hydrodynamics in August 1834 when it was observed by Scott-Russell. He described his observations in his "Report on waves." It has been customary in expositions of this subject to quote from his description, and we shall follow the custom:

> I was observing the motion of a boat which was rapidly drawn along a narrow channel by a pair of horses, when the boat suddenly stopped – not so the mass of water in the channel which it had put in motion; it accumulated round the prow of the vessel in a state of violent agitation, then suddenly leaving it behind, rolled forward with great velocity, assuming the form of a large solitary elevation, a rounded, smooth and well-defined heap of water, which continued its course along the channel apparently without change of form or diminution of speed. I followed it on horseback, and overtook it still rolling on at a rate of some eight or nine miles an hour, preserving its original figure some thirty feet long and a foot to a foot and a half in height. Its height gradually diminished, and after a chase of one or two miles I lost it in the windings of the channel. Such, in the month of August 1834, was my first chance interview with that singular and beautiful phenomenon ....

In 1895 Korteweg and de Vries provided a simple analytic foundation for the study of solitary waves by developing an equation for shallow-water waves that includes both nonlinear and dispersive effects, but ignores dissipation. A more abstract derivation of this equation, showing its central importance was presented by Manin (1979) and we follow his exposition.

The linear one-dimensional wave equation with speed of propagation c is

$$u_{tt} - c^2 u_{xx} = 0.$$

Its general solution is a sum of two waves of arbitrary form $u = f(x + ct) + g(x - ct)$, one of which moves to the left and the other to the right with constant speed c. We consider the equation $u_t + cu_x = 0$, which distinguishes waves moving to the right. Among its solutions are the harmonic waves $u = \exp i(\omega t - kx)$, where the frequency ω and the wave number k are related by $\omega = ck$, or for waves of both types by $\omega^2 = c^2k^2$, where c is characteristic of the medium.

If the wave equation remains linear but includes derivatives of higher order, then the relation between the frequency and the wave number of a harmonic wave may have the more general form

$$\omega^2 = f(k^2),$$

where f can be a general, not necessarily linear, function. Such a relation between frequency ω and wave number k is called a *dispersion relation*. In the approximation of long waves (i.e., small k), we may restrict attention to the first two terms of the Taylor series for f and write $\omega^2 \approx c^2k^2 + \epsilon k^4$ or

$$\omega \approx ck + \frac{1}{2}\frac{\epsilon k^3}{c}.$$

Waves with this dispersion relation are described by the equation

$$u_t + cu_x - \frac{\epsilon}{2c} u_{xxx} = 0.$$

On the other hand, the simplest nonlinearity enters if it is assumed that the speed depends on the amplitude u. For waves of small amplitude it may be assumed that the dependence is linear, and the equation can be written in the form $u_t + (c + \alpha u)u_x = 0$. The dependence of the speed on the amplitude for suitable sign of α may cause the crest of the wave to move faster than the trough; that is, curling of the front occurs with subsequent formation of breakers and decay of the wave.

Simultaneous consideration of dispersion and nonlinearity leads to the equation $u_t + cu_x - \alpha uu_x - (\epsilon/2c)u_{xxx} = 0$. If we go over to a system of coordinates moving to the right with speed c, the term cu_x drops out, and we arrive at the Korteweg–de Vries equation

$$u_t = 6uu_x - u_{xxx}, \tag{49.1}$$

up to a normalization constant that can be changed by scaling u, x, and t. Manin's derivation has the advantage that it nowhere appeals to hydrodynamics and indicates the universal applicability of the Korteweg–de Vries equation to one-dimensional media where the essential features are only weak dispersion and weak nonlinearity. We can find a traveling-wave solution of (49.1) as follows: Write

$$u(x, t) = U(x - vt),$$

where the function U is called the waveform and v is a constant speed (we recall v is really the speed by which the traveling wave exceeds the wave speed in the simplest approximation $u_t + cu_x = 0$).

For U we obtain the equation $-vU' = 6UV' - U'''$. Integrating we obtain $-vU = 3U^2 - U'' + a$, where a is a constant. Multiplying by U' and integrating again, we find $-v(U^2/2) = U^3 - \frac{1}{2}U'^2 + aU + b$, where b is a new constant, or $U'^2 = 2U^3 + vU^2 + aU + b$. Up to a normalization constant, the general solution of this equation is the Weierstrass $\mathscr{P}$ function $U(x - vt) = c_1\mathscr{P}(x - vt) + c_2$, the periods of which are the periods of the elliptic curve $\Gamma: Y^2 = 2X^3 + vX^2 + aX + b$.

The soliton is obtained for the curve with a double point at the origin: $Y^2 = 2X^3 + vX^2$, $a = b = 0$. The explicit formula for it has the form $U(x - vt) = -(v/2)\,\mathrm{ch}^{-2}[(\sqrt{v}/2)(x - vt)]$. This is the solitary wave (in the present normalization it is rather a "solitary well") with trough at the point $x = vt$. The depth of the well is proportional to its speed, which may be arbitrary.

Since solitons decrease at infinity and large solitons move faster than small ones, we may attempt to consider the solution of the Cauchy problem for which $u(x, 0)$ is the sum of two widely separated solitons of which the left is larger than the right and therefore begins to move almost independently of the right soliton and strives to overtake it. Of course, when the solitons approach each other, we expect complicated behavior. But as a result of computer studies it was discovered that after a long period of time the original solitons emerged with their shape and velocities unchanged. Indeed it was discovered that for (49.1) there are certain families of solutions, the N-soliton solutions. These solutions were first found numerically, then verified analytically. They have a number of truly remarkable properties. For large negative time, a solution looks like the superposition of a number of isolated waves, spatially separated. In time these come together, interact, and then, for large positive time, again emerge as spatially separated waves, moving with different velocities. For each fixed time, the set of functions of x can be parametrized by $2N$ real parameters (related to the locations and amplitudes of the waves) and the time evolution, in terms of these parameters, is given by the solution of a Hamiltonian system of ordinary differential equations. The manifold of solutions at each time can be described as extremals of some "conserved quantity" of X. This last property, that a manifold of extremals of some F gives a class of solutions evolving according to some Hamiltonian systems, can be understood in terms of the calculus of variations. Suppose that $u^t = u^t(x)$ is an actual solution of the evolution equation, with $u^0 = u$ and F is some function of u and its x derivatives. Then

$$\frac{dF}{dt}(u^t, u^t_x, \ldots)_{|t=0} = \frac{\partial F}{\partial u} u_t + \frac{\partial F}{\partial u_x}\frac{du_t}{dx} + \ldots$$

$$= \frac{\partial F}{\partial u} X + \frac{\partial F}{\partial u_x}\frac{dX}{dx} + \ldots.$$

The operator

$$\hat{X} = X\frac{\partial}{\partial u} + \left(\frac{d}{dx}X\right)\frac{\partial}{\partial u_x} + \left(\frac{d^2}{dx^2}X\right)\frac{\partial}{\partial u_{xx}} + \cdots \tag{49.2}$$

can be applied to any F, whether or not we know about the solubility of the equation.

Here

$$\frac{d}{dx} = \frac{\partial}{\partial x} + u_x \frac{\partial}{\partial u} + u_{xx} \frac{\partial}{\partial u_x} + \cdots \tag{49.3}$$

is the operation of total derivative with respect to x. (See Section 51 for a more detailed and invariant description.) We say that F is invariant under X if

$$\hat{X}F = 0. \tag{49.4}$$

We can think of F as defining a "functional" on the space of functions u by assigning the "value"

$$\int F(u, u_x, \ldots)\, dx$$

to u. Of course, to get an actual functional, we need to specify the domain of integration, and if infinite, worry about convergence. But the variational equations (i.e., the Euler–Lagrange equations),

$$\frac{\delta F}{\delta u} = 0, \tag{49.5}$$

where

$$\frac{\delta F}{\delta u} = \sum_0^\infty (-1)^i \left(\frac{d}{dx}\right)^i \frac{\partial L}{\partial u^i} \quad \text{(finite sum since } L \text{ depends only on finitely many derivatives of } u) \tag{49.6}$$

associated to F are well defined locally.

The operators of the form $\hat{X}$ form a Lie algebra under commutator. We shall present a more invariant study of them in Section 51. We shall also prove the following:

Theorem 49.1. *Suppose that the matrix of partial derivatives* $\partial^2 \mathrm{F}/\partial \mathrm{u}^{(n)}\, \partial u^{(n)}$ *is nonsingular. Then the space* M *of solutions of (49.5) is a symplectic manifold of dimension* 2dn *(where* d *is the number of* u *variables). If* $\hat{\mathrm{X}}\mathrm{F} = 0$ *then* X *induces a flow* ξ_{X} *on* M *that is Hamiltonian and a solution curve* u(x, t) *of this vector field solves the partial differential equation*

$$\mathrm{u}_t = \mathrm{X}(\mathrm{u}). \tag{49.7}$$

If X *and* Y *both satisfy* (49.4) *then*

$$[\xi_{\hat{\mathrm{X}}}, \xi_{\hat{\mathrm{Y}}}] = \xi_{[\hat{\mathrm{X}}, \hat{\mathrm{Y}}]}. \tag{49.8}$$

Thus we are led to search for invariant F's. If also, we can find other $\hat{Y}$'s that commute with $\hat{X}$ and satisfy $\hat{Y}F = 0$, then this would help in the solution of the mechanical system given by $\xi_{\hat{X}}$. A method that has led to the discovery of such invariants and commuting vector fields was the introduction of a Poisson bracket structure by Gardner (1971) into the space of functions F, for the one-dependent-variable case of the Korteweg–de Vries equation. Recall that a Poisson structure is a rule that assigns to each function F a vector field $\hat{X}_F$, so that one has the Poisson bracket

$$\{F, G\} = \hat{X}_F G$$

defining a Lie algebra operation on functions, and such that

$$[\hat{X}_F, \hat{X}_G] = \hat{X}_{\{F,G\}}.$$

Gardner's Poisson bracket was given by assigning to F the evolutionary equation $u_t = X_F$ where

$$X_F = \frac{d}{dx}\frac{\delta F}{\delta u}. \tag{49.9}$$

If we take

$$F(u) = u^3 + \tfrac{1}{2}u^2$$

then

$$\frac{\delta F}{\delta u} = 3u^2 - u_{xx}$$

and

$$\frac{d}{dx}\frac{\delta F}{\delta u} = 6uu_x - u_{xxx},$$

so the corresponding evolution equation is precisely the Korteweg–de Vries equation.

It turns out that the various N-soliton solutions of the Korteweg–de Vries equation are in fact given by the method of (49.7) for various different choice of F (cf. Lax, 1975, 1976).

We must check that the assignment (49.9) does indeed define a Poisson bracket. This can be done directly. We shall give a more instructive interpretation, due to Adler and Lebedev and Manin in the next section. The Gardner–Poisson bracket (49.9) was greatly generalized by Gel'fand and Dikii in an important series of papers. Their calculation and the formalism that they had to introduce were extremely involved. It was the remarkable discovery of Adler (1979) and Lebedev and Manin (1978) that the Gel'fand–Dikii symplectic structures are obtained from the dual space of a certain infinite-dimension Lie algebra, the algebra of formal pseudodifferential operators of negative degree. We will explain this circle of ideas in the next section.

50. The algebra of formal pseudodifferential operators

We begin by recalling the formula for the composition of two differential operators in $\mathbb{R}^n$. Let $x = (x_1, \ldots, x_n)$ and $\xi = (\xi_1, \ldots, \xi_n)$. For each $\alpha = (\alpha_1, \ldots, \alpha_n)$, with the α_i integers, we let ξ^α denote the monomial

$$\xi^\alpha = \xi_1^{\alpha_1} \cdots \xi_n^{\alpha_n}.$$

We let

$$D_j = \frac{1}{i}\frac{\partial}{\partial x_j}, \qquad i = \sqrt{-1},$$

and

$$D^\alpha = D_1^{\alpha_1} \cdots D_n^{\alpha_n}.$$

We let

$$\xi \cdot x = \xi_1 x_1 + \cdots + \xi_n x_n$$

so that

$$D^\alpha e^{i\xi \cdot x} = \xi^\alpha e^{i\xi \cdot x}.$$

A differential operator of degree (at most) m is an operator of the form

$$P(x, D) = \sum_{|\alpha| \leqslant m} a_\alpha(x) D^\alpha, \qquad \alpha_j \geqslant 0,$$

where the a_α are C^∞ functions of x and $|\alpha| = \alpha_1 + \cdots + \alpha_n$. Thus

$$P(x, D)e^{i\xi \cdot x} = e^{i\xi \cdot x} P(x, \xi),$$

where

$$P(x, \xi) = \sum a_\alpha(x) \xi^\alpha.$$

If u is any C^∞ function that vanishes sufficiently rapidly at infinity, we can write

$$u(x) = (2\pi)^{-n} \int e^{i\xi \cdot x} \hat{u}(\xi)\, d\xi,$$

where the Fourier transform $\hat{u}$ of u is given by

$$\hat{u}(\xi) = \int e^{-i\xi \cdot y} u(y)\, dy.$$

If the a_α do not grow too fast at infinity (for example, if they are bounded), then we can pass the P under the integral sign and write

$$Pu(x) = (2\pi)^{-n} \int e^{i\eta \cdot (x-y)} P(x, \eta) u(y)\, dy\, d\eta.$$

This formula allows us to compute the effect of conjugating P by multiplication by $e^{i\xi \cdot x}$. Let M_ξ denote this operation of multiplication, so that

$$(M_\xi u)(x) = e^{i\xi \cdot x} u(x).$$

Then

$$\begin{aligned}
[(M_\xi^{-1} P M_\xi) u] u(x) &= (2\pi)^{-n} e^{-i\xi \cdot x} \int e^{i\eta \cdot (x-y)} P(x, \eta) e^{i\xi \cdot y} u(y)\, dy\, d\eta \\
&= (2\pi)^{-n} \int e^{i(\eta - \xi) \cdot (x-y)} P(x, \eta) u(y)\, dy\, d\eta \\
&= (2\pi)^{-n} \int e^{i\eta \cdot (x-y)} P(x, \eta + \xi) u(y)\, dy\, d\eta \\
&= P(x, \xi + D) u.
\end{aligned}$$

This last expression, or rather the next to the last integral, is to be interpreted as follows: Take the Taylor expansion of $P(x, \cdot)$ about the point ξ:

$$P(x, \xi + \eta) = \sum \frac{1}{\alpha!} \partial_\xi^\alpha P(x, \xi) \eta^\alpha,$$

which is a polynomial in η for each fixed x and ξ. Substitute D for η. Here $\alpha! = \alpha_1!, \ldots, \alpha_n!$ and

$$\partial_\xi^\alpha = \left(\frac{\partial}{\partial \xi_1}\right)^{\alpha_1} \cdots \left(\frac{\partial}{\partial \xi_n}\right)^{\alpha_n}.$$

Let P and Q be two differential operators. Then

$$\begin{aligned}
(P \circ Q) e^{i\xi \cdot x} &= P[e^{i\xi \cdot x} Q(x, \xi)] \\
&= e^{i\xi \cdot x} (M_\xi^{-1} P M_\xi) Q(x, \xi) \\
&= e^{i\xi \cdot x} P(x, \xi + D) Q(x, \xi) \\
&= e^{i\xi \cdot x} \sum \frac{1}{\alpha!} \partial_\xi P(x, \xi) D^\alpha Q(x, \xi).
\end{aligned}$$

In this last expression, the D^α acts as a differential operator on Q, thought of as a function of x for fixed ξ. This gives the formula for the composition of two differential operators. We have proved it under some restrictions on the growth of the coefficients of P and Q at infinity. But clearly the formula for the composition of two differential operators is purely local; and hence the above formula is valid without restriction. We can think of it as defining a new kind of multiplication on functions $P = P(x, \xi)$ that are polynomials in ξ with coefficients that are C^∞ functions of x:

$$(P \circ Q)(x, \xi) = \sum \frac{1}{\alpha!} \partial_\xi^\alpha P(x, \xi) D^\alpha Q(x, \xi). \tag{50.1}$$

In particular,

$$(\xi^\beta \circ Q)(x, \xi) = (\xi + D)^\beta Q(x, \xi). \tag{50.2}$$

Now suppose that P and Q are formal series of the form

$$P = \sum_{-\infty}^{m} P_j(x, \xi),$$

$$Q = \sum_{-\infty}^{m'} Q_j(x, \xi),$$

where the P_j and Q_j are C^∞ functions of x, and ξ is defined for $\xi \neq 0$, which are homogeneous of degree j in ξ. We can now define the composition of P and Q by (50.1), where we apply (50.1) to P_j and Q_k and collect all terms of a fixed degree of homogeneity. Since any partial derivative with respect to ξ lowers the degree of homogeneity by 1, there will be only a finite number of each fixed degree of homogeneity, and hence $P \circ Q$ will be well defined as a formal sum. The operation $\circ$ is associative. (In the case that P and Q are "symbols" of pseudodifferential operators, $P \circ Q$ is the symbol of their composition. The notion of pseudodifferential operators and the multiplication of their symbols was introduced by Kohn and Nirenberg, 1965. The ring of "formal pseudodifferential operators" described above was introduced by Guillemin, Quillen, and Sternberg, 1970.) We say that P is of degree m if the highest j such that $P_j \neq 0$ is $j = m$. If P is of degree m and Q is of degree m' then $P \circ Q$ is of degree $m + m'$. Also, the highest-order term in $P \circ Q$ is just the ordinary product of P_m and $Q_{m'}$. Thus, the commutator

$$[P, Q] = P \circ Q - Q \circ P$$

has degree $m + m' - 1$. In particular, the elements of negative degree form a Lie algebra that we shall denote by g. For future reference we record the following useful fact: For any P and Q we can write

$$[P,Q] = \sum \partial_{\xi_i} A_i + \sum D_j B_j \tag{50.3}$$

(i.e., as a sum of ξ derivatives and of x derivatives). Indeed, for any E and F, we have

$$\partial_{\xi_j} E D_j F = \partial_{\xi_j}(E D_j F) - E \partial_{\xi_j} D_j F$$

and

$$D_j E \partial_{\xi_j} F = D_j(E \partial_{\xi_j} F) - E D_j \partial_{\xi_j} F.$$

Since D_j and ∂_{ξ_j} commute, this implies that

$$\partial_{\xi_j} E D_j F - \partial_{\xi_j} F D_j E = \partial_{\xi_j}(E D_j F) - D_j(E \partial_{\xi_j} F).$$

This establishes (50.3) for the terms in $[P,Q]$ that involve only first derivatives in (50.1). Repeated application of this argument gives (50.3) for derivatives of any order.

Let us now restrict attention to the case of one independent variable, $n = 1$. We shall also restrict ξ to be positive. We can write an element X of g as

$$X = \sum_1^\infty c_{-j} \xi^{-j} = \sum_0^\infty (\xi + D)^{-k-1} b_k$$

since we can use the binomial expansion of $(\xi + D)^{-k-1}$ to recursively solve for the b's in terms of the c's and vice versa. Thus

$$c_{-1} = b_0, \qquad c_{-2} = b_1 - Db_0, \qquad \text{etc.}$$

We shall use the representation in terms of the b's to put a topology on g: Each b_k is a C^∞ function, and we topologize the space of C^∞ functions on $\mathbb{R}$ in the standard way. Thus an element of g is a sequence of elements in a topological vector space, so we give g the weak-product topology; thus an element A of g^* (i.e., a continuous linear function on g) is a finite sequence $A = (a_0, \ldots, a_m)$, where each a_j is a distribution of compact support and

$$\langle A, X \rangle = \langle a_0, b_0 \rangle + \cdots + \langle a_m, b_m \rangle.$$

Let $g^\# \subset g^*$ denote the subspace of smooth distributions, that is, those A for which all the a_j are smooth functions of compact support. Then we can write

$$\langle A, X \rangle = \int (a_0 b_0 + \cdots + a_m b_m)\, dx.$$

We now make the key observation due to Adler. Let us write A as a differential operator

$$A = a_0 + a_1\xi + \cdots + a_m\xi^m.$$

Then

$$\begin{aligned} A \circ X &= (\textstyle\sum a_k\xi^k)\circ(\sum(\xi + D)^{-j-1}b_j) \\ &= \sum_{k,j} a_k(\xi + D)^k(\xi + D)^{-j-1}b_j \\ &= \sum_{k,j} a_k(\xi + D)^{k-j-1}b_j. \end{aligned}$$

Let us look for the coefficient of ξ^{-1} in this last expression. If $k > j$, then $(\xi + D)^{k-j-1}$ is a polynomial in ξ and D and so contains no ξ^{-1} terms. If $k < j$ then the expansion of $(\xi + D)^{k-j-1}$ starts with ξ^{k-j-1} and goes down in powers of ξ, and $k - j - 1 \leqslant -2$. Thus contributions only come from $k = j$, and in these we have $(\xi + D)^{-1} = \xi^{-1}$ plus lower powers of ξ. Thus

$$(A \circ X)_{-1} = a_0b_0 + \cdots + a_mb_m,$$

where $(\ )_{-1}$ denotes the coefficient of ξ^{-1}. We thus obtain Adler's formula

$$\langle A, X\rangle = \int (A \circ X)_{-1}\,dx. \tag{50.4}$$

Next we have Adler's lemma.

Lemma 50.1. *For any* P *and* Q, $([\mathrm{P},\mathrm{Q}])_{-1}$ *is always a total derivative; that is,* $([\mathrm{P},\mathrm{Q}])_{-1} = \mathrm{Db}$ *for some function* b. *Thus, if* P *or* Q *has compact support then* $\int([\mathrm{P},\mathrm{Q}])_{-1}\,\mathrm{dx} = 0$.

Proof. By (50.3) we know that $[P,Q] = \partial_\xi A + DB$. But ξ^{-1} is not the ξ derivative of any power of ξ, hence the first term can not contribute to the coefficient of ξ^{-1} and hence $([P,Q])_{-1} = (DB)_{-1} = DB_{-1}$, proving the lemma with $b = B_{-1}$.

Corollary. *For any* P, *let* P_+ *be the nonnegative part of* P, *so that* $\mathrm{P}_+ = \sum_{\mathrm{j}\geqslant 0}\mathrm{P}_\mathrm{j}$ *if* $\mathrm{P} = \sum \mathrm{P}_\mathrm{j}$. *We can now use the lemma to derive Adler's formula for the coadjoint action: For any* X *and* Y *in* g,

$$\langle \mathrm{A}, [\mathrm{X},\mathrm{Y}]\rangle = \langle [\mathrm{A},\mathrm{X}]_+, \mathrm{Y}\rangle. \tag{50.5}$$

Proof. Write $A \circ (X \circ Y - Y \circ X) = (A \circ X - X \circ A) \circ Y + X \circ (A \circ Y) - (A \circ Y) \circ X$. Since A has compact support, so does $A \circ Y$ and so, when taking the integral of the ξ^{-1} component, the second term vanishes by the lemma. This proves (50.5). Thus the coadjoint action of $X \in g$ acting on $A \in g^*$ is given by

$$X \circ A = [X, A]_+. \tag{50.6}$$

Remembering that X is of negative degree, we see that if $A = a_0 + \cdots + a_m \xi^m$, the terms of degree m and m^{-1} in $[X, A]_+$ vanish. Thus the affine space given by

$$a_m = a, \qquad a_{m-1} = b,$$

where a and b are fixed functions, is invariant under the coadjoint action.

Now let f be a function on $g^\#$. We wish to construct the Legendre transformation associated to f. We shall restrict attention only to those f that are *local*, that is, to f which are of the form

$$f(A) = \int F\big(x, a_0(x), a_0'(x), \ldots, a_m(x), \ldots a_m^{(k)}(x)\big)\, dx,$$

where F is some smooth function of the variables $x, u_0, u_0', \ldots, u_0^{(k)}, u_1, \ldots, u_1^{(k)}, \ldots, u_m, \ldots, u_m^{(k)}$, compactly supported in x. Thus, F is a function defined on the space of k-jets of the A's. Here each fixed F depends on only finitely many a_i and their derivatives up to some finite order. But the number m of the a_i and the number k of the derivative may depend on the f. Now the expression for $df_A(B)$ will be an integral of a sum involving various partial derivatives of F, evaluated at the a's and their derivatives, multiplied by the values of the b's and their derivatives. Since the b's are of compact support, we can integrate by parts so as to eliminate all the expressions involving derivatives of the b's. The resulting expression will be

$$df_A(B) = \sum \int \frac{\delta F}{\delta a_i} b_i \, dx.$$

The expression $\delta F/\delta a_i$ is called the variational derivative of f with respect to a_i. It is to be evaluated, of course, at the points given by the values of the a_j and their derivatives. The important point about the above expression is that since the $\delta F/\delta a_i$ under the integral are smooth functions of x, the formula extends immediately to the case where the b's are allowed to be distributions. In other words, by the definition of $\langle B, X \rangle$ with $X = L_f(A)$

we see that

$$L_f(A) = (\xi + D)^{-1}\frac{\delta F}{\delta a_0} + (\xi + D)^{-2}\frac{\delta F}{\delta a_1} + \cdots, \tag{50.7}$$

$$X_f(A) = L_f(A) \circ A = [A, L_f(A)]_+, \tag{50.8}$$

and

$$\{f, h\} = X_f h = \langle A, [L_f(A), L_h(A)]\rangle$$
$$= \int (A \circ [L_f(A), L_h(A)])_{-1}\, dx. \tag{50.9}$$

The important point about formulas (50.7) and (50.8) is that they are local – the value of $L_f(A)$ at a point x (i.e., the values of each of the coefficients of the expansion of $L_f(A)$ at the point x) depends only on the values of the a's and their derivatives to sufficiently high order at x, and various partial derivatives of F (which enter into the expression for the variational derivatives of F) evaluated at these values. In particular, the right-hand side of (50.7) makes sense without any compactness assumptions on either F or A, and we can take it as a definition of the left-hand side. (Actually, in the absence of compactness assumptions on F, the "function" f is not defined since it involves an integration. So we should write $L_F(A)$ on the left of (50.7). On the other hand, if F and F' are two functions such that

$$\frac{\delta(F - F')}{\delta a_i} = 0 \qquad \forall\, i, \tag{50.10}$$

then clearly the right-hand side of (50.7) gives the same answer when evaluated on F or F'. We could then take (50.10) as defining an equivalence relation and let f denote the equivalence class of F. It can be shown (see Gel'fand and Dikii, 1975, 1976, 1977) that (50.10) holds if and only if $F - F'$ is a "total derivative" and so, if F and F' have compact support, they define the same honest function f.) Similarly, (50.8) is a local formula, and we can regard X_f as a vector field

$$X_f = (X_f)_0\frac{\partial}{\partial a_0} + (X_f)_1\frac{\partial}{\partial a_1} + \cdots + (X_f)_{m-2}\frac{\partial}{\partial a_{m-2}} \tag{50.11}$$

where the coefficients $(X_f)_i$ depend on the a's and their derivatives up to some high order (the explicit expression involving the values of various derivatives of F at these points) and so again do not depend on whether or not A or F is compact. Now (50.9) does involve an integration, and so requires some slight reformulation. If $h(A) = -\int H(\cdots)\, dx$ and X is a vector field like (50.11), then Xh is also given by a density easily computable in terms of X and H. Rather than clutter up the formula with indices, we illustrate how this computation goes when $m = 2$ and we write a for a_0, X for X_f, and assume that H only depends on a and its first two derivatives (i.e., on a, a', and a''). Then

$$\frac{\delta H}{\delta a} = \frac{\partial H}{\partial a} - \frac{d}{dx}\frac{\partial H}{\partial a'} + \left(\frac{d}{dx}\right)^2 \frac{\partial H}{\partial a''},$$

where

$$\frac{d}{dx} = \frac{\partial}{\partial x} + a'\frac{\partial}{\partial a} + a''\frac{\partial}{\partial a'} + a^{(3)}\frac{\partial}{\partial a''} + \cdots$$

is the total derivative in the x direction. Writing $X = X_0(\partial/\partial a)$, where X_0 can depend on a and its various derivatives, we have

$$\begin{aligned}(Xh)(A) &= \int X_0 \frac{\delta H}{\delta a}\, dx \\ &= \int\left[X_0\frac{\partial H}{\partial a} + \left(\left(\frac{d}{dx}\right)X_0\right)\frac{\partial H}{\partial a'} + \left(\left(\frac{d}{dx}\right)^2 X_0\right)\frac{\partial H}{\partial a''}\right] dx \\ &= \int \hat{X} H\, dx,\end{aligned}$$

where $\hat{X}$ is the vector field given by

$$\hat{X} = X_0\frac{\partial}{\partial a} + \left(\frac{d}{dx}X_0\right)\frac{\partial}{\partial a'} + \left(\left(\frac{d}{dx}\right)^2 X_0\right)\frac{\partial}{\partial a''} + \cdots. \tag{50.12}$$

In general, it is clear that any vector field of the form (50.11) gives rise to a vector field $\hat{X}_f$ obtained from it by formula (50.12) (applied to all the a_i) and the density corresponding to $X_f h$ is $\hat{X}_f H$. The content of (50.9) is that

$$\int \hat{X}_f H\, dx = \int (A \circ [L_f(A), L_h(A)])_{-1}\, dx.$$

Now the integrands on both sides of this equation are local expressions in A, F, and H, and the equation is to hold identically in F and H when they have compact support. This can only happen if the reason that the two integrals are equal is because of some (possibly complicated) integration by parts (i.e., that the two integrands differ by a total derivative). So we can write

$$\hat{X}_f H \doteq (A \circ [L_f(A), L_h(A)])_{-1}, \tag{50.13}$$

where $\doteq$ means equality up to a total derivative.

It is easy to check that the vector fields of the form (50.11) form a Lie algebra under Lie bracket and that they all commute with d/dx; in particular, they carry total derivatives into total derivatives. So, if h denotes the equivalence class of H, we can use (50.13) to define $\{f, h\}$; this Poisson bracket is well defined on the equivalence classes and from general considerations we know that

$$[\hat{X}_f, \hat{X}_g] = \hat{X}_{\{f,g\}}. \tag{50.14}$$

Let us illustrate the computation of this Poisson structure for the simplest case, where $m = 2$ and we fix $a_2 = 1$ and $a_1 = 0$. We write a for a_0 so $A = a + \xi^2$ is the time-independent Schrödinger operator with potential a. Then

$$L_f(A) = (\xi + D)^{-1}\frac{\delta F}{\delta a} = \frac{\delta F}{\delta a}\xi^{-1} - D\frac{\delta F}{\delta a}\xi^{-2} + \cdots$$

$$L_h(A) = (\xi + D)^{-1}\frac{\delta H}{\delta a} = \frac{\delta H}{\delta a}\xi^{-3} + \cdots,$$

$$[L_f(A), L_h(A)] = -\left(\frac{\delta F}{\delta a}D\frac{\delta H}{\delta a} - \frac{\delta H}{\delta a}D\frac{\delta F}{\delta a}\right)\xi^{-3} + \cdots,$$

so

$$\{f, h\} = (A \circ [L_f(A), L_h(A)])_{-1} = -\left(\frac{\delta F}{\delta a}D\frac{\delta H}{\delta a} - \frac{\delta H}{\delta a}D\frac{\delta F}{\delta a}\right)$$

is the Gardner–Poisson bracket. Also

$$\begin{aligned} X_f(A) &= [A, L_f(A)]_+ = [\xi^2 + a, L_f(A)]_+ \\ &= \left[\xi^2, \frac{\delta F}{\delta a}\xi^{-1}\right]_+ \\ &= 2D\frac{\delta F}{\delta a}, \end{aligned}$$

all other terms being of negative degree. If $F = i(a^3 + \frac{1}{2}a_x^2)$ then

$$X_f = (6aa_x + a_{xxx})\frac{\partial}{\partial a}$$

is the evolution field of the Korteweg–de Vries equation, where we have written a_x instead of a' to conform to standard usage. Let us give a heuristic sketch, following Adler (1979) of how the Kostant–Symes lemma,

Proposition 47.1, implies the Gel'fand–Dikii construction of integrals for evolution equations. (Our discussion is only heuristic because we don't want to go into the technical problems of giving a precise definition of the complex powers of a pseudodifferential operator or the proof that the appropriate component of the complex power is a local function.) "Define"

$$f_\sigma(A) = \int (A^\sigma)_{-1}\, dx.$$

Then

$$df_{\sigma A} = \int (\sigma A^{\sigma-1} \circ dA)_{-1}\, dx.$$

(Here we have to use Adler's lemma (50.1). For example, $dA^2 = A \circ dA + (dA) \circ A$.) But

$$\int (dA \circ A)_{-1}\, dx = \int (A \circ dA)_{-1}\, dx$$

so

$$\int (dA^2)_{-1}\, dx = 2 \int (A \circ dA)_{-1}\, dx.$$

We can decompose the algebra of all pseudodifferential operators into a vector space sum of $a + b$ where a is the algebra of differential operators and b is the algebra of pseudodifferential operators of negative degree that we have been considering. Thus, on the full algebra of all pseudodifferential operators we have

$$L_f(A) = \sigma A^{\sigma-1},$$

and the corresponding vector field (given by (50.8) without the subscript $+$) is $[A, \sigma A^{\sigma-1}] = 0$ since all powers of A commute. Thus f_σ is an invariant. The restriction of these invariants to $b^* = a^\perp$ give the Gel'fand–Dikii family of commuting integrals.

For example, if $d = 1$, we can find $(\xi^2 + u)^{1/2}$ by writing

$$(\xi^2 + u)^{1/2} = \xi + \tfrac{1}{2}u\xi^{-1} + b_{-2}\xi^2 + b_{-3}\xi^{-3} + \cdots$$

and then solving recursively for the b's in the equation

$$(\xi^2 + u)^{1/2} \circ (\xi^2 + u)^{1/2} = \xi^2 + u.$$

Thus, for example,

$$b_{-2} = \tfrac{1}{4}u'$$
$$b_{-3} = \tfrac{1}{8}(-u^2 + u'')$$
$$b_{-4} = \tfrac{1}{8}(3uu' - \tfrac{1}{2}u''')$$

etc. Having calculated the b's as far as necessary, we can then calculate the -1 coefficient of $(\xi^2 + u)^{N+1/2}$. These then give the desired integral (where we identify two functions differing by a total derivative).

51. The higher-order calculus of variations in one variable

In this section we present an invariant formulation of the results that we needed in Sections 49 and 50. We refer the reader to Goldschmidt and Sternberg (1973) for a study of the first-order calculus of variations in an arbitrary number of independent variables and to the important studies by Takens (1979), Tulczyjew (1977), Olver (1980), and others. Our treatment follows Sternberg (1978b) verbatim.

Let X be a differentiable manifold, for the time being of arbitrary dimension. (We shall, unfortunately, at the crucial juncture specialize to the case $\dim X = 1$.)† We let $Y \to X$ denote a fibered manifold over X. We shall let $J \to X$ denote the bundle of infinite jets of sections of Y, regarded as an "inverse limit" of the manifolds J_k of k-jets of sections of Y. All functions, differential forms, etc. are pull-backs of functions and forms defined on the finite jet bundles J_k. A vector field on J will be a derivation of the ring of functions and hence will have infinitely many "components" in a local coordinate system.

Any section s of Y over X gives rise to a section js of J over X. But not every section of J is the infinite jet of a section of Y. There is simple differential condition on sections u of J that determines whether or not $u = js$ for some section s of Y. Indeed (see Goldschmidt and Sternberg, 1973), there exists a "connection" ω, that is, a section of $\mathrm{Hom}(TJ, TJ)$, defined on J such that

$$\omega(\zeta) = \zeta \qquad \text{if} \qquad \zeta \text{ is a vertical tangent vector} \tag{51.1}$$

and

$$u^*\omega = 0 \qquad \text{iff} \qquad u = js. \tag{51.2}$$

We obtain an ideal I in the ring of differential forms on J consisting of

† For a discussion of the general case, see Garcia and Muñoz (1982).

all $\omega \bar{\wedge} \Omega$ where Ω is an arbitrary form. This deal I satisfies $dI \subset I$ and so defines a "foliation" on J whose "leaves" are of dim X. Of course there are no actual leaves but each vector field on X defines a unique vector field on J that is annihilated by ω. If $x^1, \ldots, x^n$ are coordinates on X so that $\partial/\partial x^1, \ldots, \partial/\partial x^n$ is a local basis for the vector fields on X, we shall denote the corresponding vector fields on J by $d/dx^1, \ldots, d/dx^n$. In particular,

$$\left[\frac{d}{dx^i}, \frac{d}{dx^j}\right] = 0.$$

We let V denote the algebra of vertical vector fields ξ that preserve I. Thus $\xi \in V$ iff $D_\xi \Omega \in I$ for all $\Omega \in I$. This implies that

$$\left[\xi, \frac{d}{dx^j}\right] = a_1 \frac{d}{dx^1} + \cdots + a_n \frac{d}{dx^n}.$$

If f is a function on X and $\rho^* f$ denotes the corresponding function on J then $(d/dx^i)\rho^* f = \rho^*(\partial f/\partial x^i)$, while $\xi \rho^* f = 0$ for any vertical vector field ξ. Taking $f = x^k$ in the preceding equation shows that $a_k = 0$. Thus

Proposition 51.1. *For any vertical vector field ξ, we have $\xi \in \mathrm{V}$ if and only if*

$$\left[\xi, \frac{\mathrm{d}}{\mathrm{dx}^{\mathrm{i}}}\right] = 0 \qquad \forall \mathrm{i}.$$

We now reduce to the case $n = \dim X = 1$. We let x be a coordinate on X and u, v, etc., fiber coordinates. (We shall usually write the formulas as if we had only one fiber coordinate, but this is irrelevant for much of what follows.) Coordinates on J will be x, u, v, u', v, u'', etc. The ideal I is generated by the forms

$$du - u'dx, \qquad du' - u''dx, \qquad \text{etc.}$$

Proposition 51.2. *The ideal* I *contains no nontrivial exact one forms; that is, if* $\mathrm{df} \in \mathrm{I}$ *then* $\mathrm{df} = 0$.

Proof. Write $df = a_0(du - u'dx) + \cdots + a_n(du^{(n)} - u^{(n+1)}dx)$, where a_n is the highest nonvanishing coefficient. Then $\partial f/\partial u^{(k)} = 0$ for $k > n$ so $\partial a_i/\partial u^{(n+1)} = \partial^2 f/\partial u^{(i)} \partial u^{(n+1)} = 0$ for all i. But then $\partial^2 f/\partial x\, \partial u^{(n+1)} = -a_n = 0$, contradicting the assumption that $a_n \neq 0$.

We call a tangent vector ζ *very vertical* if $d\pi(\zeta) = 0$, where π denotes the projection of J onto Y. Thus $\partial/\partial u'$, $\partial/\partial u''$, etc., very vertical vector fields.

Theorem 51.1. *For any one form ω there exists a unique one form θ_ω such that*

$$\theta_\omega \equiv \omega \bmod I \tag{i}$$

and

$$\mathrm{i}(\mathrm{v})\theta_\omega = 0 \tag{ii}$$

for all very vertical vector fields.

Proof (*uniqueness*). Suppose that $\theta = a_0(du - u'dx) + \cdots + a_n(du^{(n)} - u^{(n+1)}dx)$. Then

$$i(\partial/\partial u^{(n+1)})d\theta \equiv -a_n dx \bmod I.$$

Thus for the right-hand side to lie in I we must have $a_n = 0$. Proceeding recursively we see that $i(\zeta)d\theta \equiv 0 \bmod I$ and $\theta \equiv 0 \bmod I$ implies that $\theta = 0$.

(*Existence*). Let us write

$$\theta = Ldx + p_0(du - u'dx) + \cdots + p_n(du^{(n)} - u^{(n+1)}dx),$$

where

$$\omega \equiv Ldx \bmod I$$

and the p_i are to be determined. Then

$$i(\partial/\partial u^{(i)})\, d\theta \equiv (\partial L/\partial u^{(i)})\, dx - dp_i - p_{i-1}dx.$$

For $i > 0$ we want the right-hand side to belong to I, which is the same as saying that the right-hand side is annihilated by the vector field d/dx. Applying this vector field we obtain the recursion relations

$$\frac{dp_i}{dx} + p_{i-1} = \partial L/\partial u^{(i)},$$

which have the explicit solution

$$p_i = \sum_{j=0}^{\infty} (-1)^j \left(\frac{d}{dx}\right)^j \frac{\partial L}{\partial u^{(i+j+1)}}.$$

Since L depends on only finitely many of the variables $u^{(k)}$, the sum on the right is finite and vanishes for sufficiently large i. This proves the existence of θ_ω. If $\omega = Ldx \bmod I$, we define

$$\frac{\delta L}{\delta u} = i\left(\frac{d}{dx}\right) i\left(\frac{\partial}{\partial u}\right) d\theta_\omega.$$

From the above formula for θ_ω we obtain

$$i(\partial/\partial u)\, d\theta_\omega = (\partial L/\partial u)\, dx - dp_0$$

so

$$\frac{\delta L}{\delta u} = \sum_{j=0}^{\infty} (-1)^j \left(\frac{d}{dx}\right)^j \frac{\partial L}{\partial u^{(i+j)}}.$$

It follows as an immediate corollary of the theorem that

$$\theta_{df} = df, \qquad \text{by uniqueness and } d(df) = 0,$$

and, conversely, $d\theta = 0$ iff $\theta = df$ (locally), since we may apply the Poincaré lemma on the finite-dimensional manifold $J_k(Y)$. Thus

$$\theta_{\omega_1} = \theta_{\omega_2} \qquad \text{iff} \qquad \omega_1 \equiv \omega_2 \bmod I$$

and

$$d\theta_{\omega_1} = d\theta_{\omega_2} \qquad \text{iff} \qquad \omega_1 \equiv \omega_2 + df \bmod I \text{ for some } f.$$

Notice that if s and r are sections of Y over X that agree outside some region of compact support K, then

$$\int_D (jr)^*\omega_1 - \int_D (js)^*\omega_1 = \int_D (jr)^*\omega_2 - \int_D (js)^*\omega_2$$

if $\omega_1 \equiv \omega_2 + df \bmod I$, and if $K \subset \operatorname{int} D$. For each compact set K, we call s_t a K variation of s if s_t is a one-parameter family of sections of Y over X such that $s_0 = s$ and $s_t = s$ outside K. A section s is called an extremal of ω if for any domain D and any $K \subset \operatorname{int} D$ we have

$$\frac{d}{dt}\int (js_t)^*\omega = 0 \qquad \text{at} \qquad t = 0$$

for all K variations. From the preceding remarks we see that the set of extremals of ω depends only on $d\theta_\omega$. Suppose that s_t is a K variation of s. Then $(ds_t/dt)|_{t=0}$ is a vector field along s (i.e., a smooth map which assigns a tangent vector in $TY_{s(x)}$ to each $x \in X$) and this tangent vector is vertical, that is, has zero projection onto X. We will denote this vector field by ξ. Similarly, the one-parameter family of sections js_t of J gives rise to a vector field along js that we denote by ξ. Thus $d\pi(\xi_{js(x)}) = \xi(s(x))$, where π denotes the projection of J onto Y. Notice that $\xi_{s(x)} = 0$ if $x \notin K$. For any D containing K in its interior we have

$$\begin{aligned}\frac{d}{dt}\int_D (js_t)^*\omega = \frac{d}{dt}\int_D (js_t)^*\theta_\omega &= \int_D \{(js)^*i(\xi)d\theta_\omega + d(js)^*i(\xi)\theta_\omega\} \\ &= \int_D (js)^*i(\xi)\, d\theta_\omega,\end{aligned}$$

where all derivatives are evaluated at $t = 0$. In view of the defining properties of θ this last integral depends only on $\hat{\xi}$ and depends linearly on $\hat{\xi}$. Thus, for example, where $n = 1$ we can define the variational differential δL of any function L on J by

$$\delta L(\hat{\xi}) = i\left(\frac{d}{dx}\right) i(\xi)\, d\theta_{Ldx}.$$

If V denotes the vertical tangent vectors to Y considered as a vector bundle over J, then δL is well defined as a section of V^* over J, and, if $\omega \equiv L\,dx$ mod I we can write the last integral as

$$\int_D (js)^* \delta L(\hat{\xi})\, dx.$$

Since $\hat{\xi}$ can be an arbitrary vertical vector field of compact support, we conclude that $\delta L(js) = 0$ along extremals. Let F be an $(n-1)$-form (function) on J such that

$$dF \equiv i(\xi) d\theta_\omega \text{ mod } I$$

for some vertical vector field ξ on J. If s is an extremal of ω then

$$d(js)^* F = (js)(^*i(\xi) d\theta_\omega) = 0.$$

Thus $F \circ s$ is a constant. Thus each such function F defines a function f on the set of extremals. The collection of functions F satisfying the above equation is clearly an algebra and the restriction map $F \to f$ is a homomorphism into the algebra of functions on the set of extremals. In general, this homomorphism need not be surjective onto the algebra of "smooth" functions on the set of extremals. For example, if $d\theta_\omega = 0$, then every section is an extremal, but the only functions F satisfying the above equation are the constants. However, under suitable nondegeneracy assumptions, we can conclude that the set of extremals is a finite-dimensional symplectic manifold whose symplectic structure is induced from $d\theta_\omega$, in a way that we shall make precise below, and that the above homomorphism is surjective.

For example, suppose that the fiber has dimension 1 with local coordinate u, so that $x, u, u', u'', \ldots$ are local coordinates on J. Suppose that $L = L(x, u, u', u'', \ldots, u^{(n)})$. Then in the formula for $\delta L/\delta u$, the term involving the highest derivative of u will come from

$$(-1)^n \left(\frac{d}{dx}\right) \frac{n\partial L}{\partial u^{(n)}}$$

Now

$$\frac{d}{dx}\frac{\partial L}{\partial u^{(n)}} = \left(\frac{\partial}{\partial x} + u'\frac{\partial}{\partial u} + \cdots + u^{(n+1)}\frac{\partial}{\partial u^{(n)}} + \cdots\right)\frac{\partial L}{\partial u^{(n)}}$$

$$= u^{(n+1)}\frac{\partial^2 L}{(\partial u^{(n)})^2} + (\text{terms involving } u^{(k)} \text{ with } k < n+1).$$

Continuing to apply d/dx, we see that

$$\frac{\delta L}{\delta u} = (-1)^n u^{(2n)}\frac{\partial^2 L}{(\partial n^{(n)})^2} + (\text{terms involving } u^{(k)} \text{ with } k < 2n).$$

If we make the nondegeneracy assumption

$$\frac{\partial^2 L}{(\partial u^{(n)})^2} \neq 0,$$

then the equation $\delta L/\delta u = 0$ is a nonsingular $2n$th-order ordinary equation and we may use the "initial values" $u, u', \ldots, u^{(2n-1)}$ at some particular x to parametrize the space of extremals. Let us write

$$\theta_{Ldx} = Hdx + p_0 du + p_1 du' + \cdots + p_n du^{(n)},$$

where the p_i are given as above and where $H = L - p_0 u' - p_1 u'' - \cdots - p_n u^{(n+1)}$. We wish to show that the matrix whose i, jth entry is

$$i(\partial/\partial u^{(i)})i(\partial/\partial u^{(j)})\, d\theta_{Ldx}, \qquad i, j, = 0,1,\ldots,2n-1,$$

is nonsingular. In evaluating this matrix we can ignore $dH \wedge dx$, and we should also observe that

$$i(\partial/\partial u^{(2n-1)})\, d\theta_{Ldx} = \frac{\partial p_0}{\partial u^{(2n-1)}}\, du = (-1)^{n-1}\frac{\partial^2 L}{(\partial u^{(n)})^2}\, du,$$

$$i(\partial/\partial u^{(2n-2)})\, d\theta_{Ldx} = \frac{\partial p_0}{\partial u^{(2n-2)}}\, du + \frac{\partial p_1}{\partial u^{(2n-2)}}\, du;$$

$$\text{where} \qquad \frac{\partial p_1}{\partial u^{(2n-2)}} = (-1)^{n-2}\cdot\frac{\partial^2 L}{(\partial u^{(n)})^2}$$

etc. Thus the (antisymmetric) matrix has the form

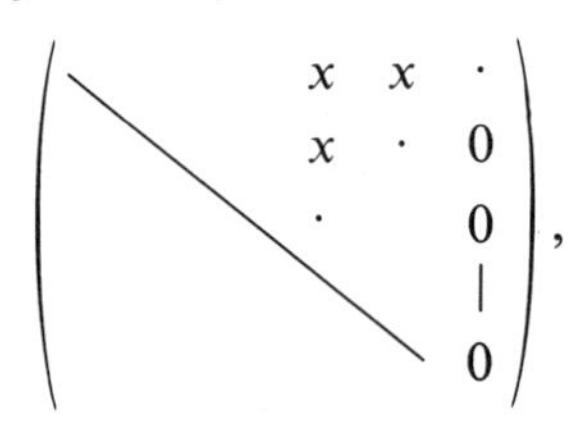

,

where the entry at the positions marked $\cdot$ is $\pm\partial^2 L/(\partial u^{(n)})^2$ and so the matrix is nonsingular.

It is clear that the same arguments work if the fiber dimension is greater than 1 and if we replace the nondegeneracy assumption by the assumption Hess $L \neq 0$, where Hess L is defined by

$$\text{Hess } L = \left(\frac{\partial^2 L}{\partial u_i^{(n)} \partial u_j^{(n)}}\right),$$

where n is the smallest integer such that L is defined on $J_n Y$. We call n the order of L. We have proved the following:

Theorem 51.2. *Assume that Hess* L $\neq 0$, *where* L *is of order* n *and where the fiber dimension is* l. *Then the set of all points of the form* js, *where* s *is an extremal of* Ldx *is a submanifold of* J *of dimension* $2nl + 1$, *call it* S_{Ldx}. *The form* $d\theta_{Ldx}$ *when restricted to this manifold has rank* 2nl and its *null direction is spanned by the vector field* d/dx. *We let* E_{Ldx} *denote the quotient of this* $(2nl + 1) =$ *dimensional manifold by the (one-dimensional) foliation spanned by* d/dx. *(Locally* E_L *looks like a manifold of dimension* 2nl *and also globally, if we make some completeness assumptions about the ordinary differential equals* $\delta L/\delta u = 0$.*) If* f *is a function that is defined on* E_L *we can think of* f *as a function defined on* S *such that* $df/dx = 0$. *Then* $df = i(\zeta)\, d\theta_{Ldx}$ *on* S, *where* ζ *is a vector field tangent* S *and determined up to a multiple of* d/dx.

$$\begin{array}{l} S \rightarrow J \\ \downarrow \\ E \end{array}$$

We claim that we can choose the vector field ζ in the theorem so that $[\zeta, d/dx] = 0$. In fact on S we have, for any choice of ζ,

$$0 = \frac{df}{dx} = D_{d/dx} i(\zeta)\, d\theta_{Ldx}) = i\left(\left[\frac{d}{dx}, \zeta\right]\right) d\theta_{Ldx} + i(\zeta) d\left(i\left(\frac{d}{dn}\right) d\theta_{Ldx}\right)$$

so

$$\left[\frac{d}{dx}, \zeta\right] = a\frac{d}{dx}$$

since d/dx is the generator of the null line of $d\theta_{Ldx}$ on S. We now simply choose a function b on S such that $db/dx = a$ and replace ζ by $\zeta - b(d/dx)$.

Now the set of vertical vector fields on J that preserve I can be identified with the set of sections of $V(Y)$ over J. In particular, we can regard ζ as a section of $V(Y)$ over S. Given any such section we can extend it to a section of $V(Y)$ over J. We have thus proved the following:

Proposition 51.3. *Suppose that $d\theta_\omega$ induces a symplectic structure on the space of extremals, for example suppose that Hess* L $\neq 0$. *Then for any Hamiltonian vector field, ζ on the space of extremals, we can find a vector field ξ in the algebra $\mathscr{V}$ such that ξ is tangent to the space of extremals and ξ coincides with ζ when restricted to this space.*

The last two assertions are the key results that we needed in Sections 49 and 50 for the applications to Poisson brackets and nonlinear evolution equations. These results were obtained in a more algebraic setting, with the above applications in mind, by Gel'fand and Dikii (1975, 1976, 1977).

Any vertical vector field ξ that commutes with d/dx determines a section, $\hat{\xi}$ of $V(Y)$, and, conversely, any section of $V(Y)$ gives rise to a unique vector field commuting with d/dx. We can write the equation

$$i(\xi)\, d\theta_{Ldx} = df \qquad \text{on} \qquad S$$

as

$$\delta L(\hat{\xi})\, dx = df \qquad \text{on} \qquad S,$$

since the value of $i(\xi)d\theta$ mod I depends only on $\hat{\xi}$.

Suppose that we start with some vector field $\xi \in \mathscr{V}$ satisfying

$$i(\xi)\, d\theta_{Ldx} \equiv dF \bmod I.$$

Now $\theta_{Ldx} = Ldx + \alpha$, where $\alpha \in I$ and hence $D_\xi \theta_{Ldx} = D_\xi(Ldx) + D_\xi \alpha'$, where, for any form β we have $D_\xi \beta = i(\xi)\, d\beta + d\big(i(\xi)\beta\big)$. Since $\xi \in \mathscr{V}$, we know that $D_\xi \alpha \in I$ and hence that

$$D_\xi(Ldx) \equiv dF \bmod I,$$

which implies that ξ is tangent to the manifold S and induces a Hamiltonian vector field on S.

On the other hand, suppose that we start with a function f on S and assume that S is a manifold of dimension $2nl + 1$ and that $d\theta_{Ldx}$ has rank $2nl$. Let us also assume that there is some k so that the projection of J onto J_k induces an immersion on S, and such that $d\theta_{Ldx}$ is defined on J_k and has constant rank there. (For example, in the case of Lagrangian of order n with Hess $L \neq 0$, we can take $k = 2n$ and the projection of J onto J_{2n} induces a diffeomorphism of S with J_{2n}, while $d\theta_{Ldx}$ has exactly one singular direction at every point.) We obtain a foliation of J_k consisting of the singular direction of $d\theta_{Ldx}$, and we can extend f to be a function F defined on all of J_k that is constant along the leaves of this foliation. As before, we can find $\hat{\xi}$ which induces a vector field ξ commuting with d/dx and such that $i(\xi)\, d\theta_{Ldx} = dF$. Thus the set of all vector fields satisfying $i(\xi)\, d\theta_{Ldx} \equiv dF$ mod I maps surjectively onto the algebra of Hamiltonian vector fields on S.

V

Contractions of symplectic homogeneous spaces

In Section 17 we gave two examples of "deformations" or "contractions" of one Lie algebra into another – the contraction of the Euclidean algebra $e(2)$ into the "free Schrödinger algebra" of Gaussian optics, and the Poincaré algebra into the Galilean algebra. In this chapter we want to study the corresponding deformations of the associated homogeneous symplectic spaces. This is of central physical importance because many basic advances in physics have involved the realization that the "true" group of physics is one that contracts into the group of an earlier theory. In fact (see Guillemin and Sternberg, 1977) the famous mass-energy equation $E = mc^2$ of Einstein occurs as an example of a deformation formula: In the notation of Section 17, we have introduced a 10-dimensional space g consisting of all vectors of the form $\xi = (A, v, x, t)$ and with a bracket $[,]_\epsilon$ depending on $\epsilon = c^{-2}$. For $\epsilon > 0$, the Lie algebra with this bracket is isomorphic to the Poincaré algebra, whereas for $\epsilon = 0$, it is the Galilean algebra. We can write a typical element of the dual space, g^*, as $(\tau, l, p, E) = (\Gamma, P)$, where $\Gamma = (\tau, l)$ and where $P = (p, E)$ is the energy-momentum vector. For each ϵ we have a map

$$d_\epsilon : g^* \to \wedge^2(g^*), \qquad (d_\epsilon \beta)(\xi_1, \xi_2) = \langle \beta, [\xi_1, \xi_2]_\epsilon \rangle$$

depending continuously on ϵ, and, more generally, maps

$$d_\epsilon : \wedge^k g^* \to \wedge^{k+1} g^*$$

given by the formulas of Section 24 with the $[,]$ depending on the parameter ϵ. In particular, from the formula for $[,]_\epsilon$ given in Section 17,

$$d_\epsilon(0, 0, 0, 1) = \epsilon\alpha,$$

where

$$\alpha(\xi_1, \xi_2) = v_1 \cdot x_2 - v_2 \cdot x_1.$$

It follows that $d_\epsilon \alpha = 0$ for $\epsilon > 0$, and hence, by continuity, also for $\epsilon = 0$. We shall see in Section 53 that the second cohomology of the Galilean algebra H^2 is one-dimensional and is generated by $[\alpha]$, the cohomology class of α. The most general second cohomology class can thus be written as $m[\alpha]$. The physical significance of the parameter m is that it determines the change in linear momentum effected by a velocity transformation, and hence can be identified with the mass. For $\epsilon > 0$ (i.e., for the Poincaré group) we shall see that $H^2 = 0$, but the mass enters as one of the parameters describing a coadjoint orbit. The mass of a point (τ, l, p, E) is given by $m^2 = E^2 - c^2p^2$. The equation

$$d_\epsilon(0,0,0,1) = \epsilon\alpha$$

or

$$d_\epsilon(0,0,0,E) = Ec^{-2}\alpha$$

shows that for a system of mass m we must choose $E = mc^2$. In other words, for a system of vanishingly small momentum, the inertial mass (when we replace the Poincaré group by the "limiting" Galilean group) is related to the energy by Einstein's famous formula.

Most of the mathematical results presented here on contractions of homogeneous symplectic spaces are due to Don Coppersmith and appeared in his (unpublished) Harvard thesis (1975). They are reproduced here with his kind permission. Some of the examples and physical applications come from Sternberg (1975, 1976) and Sternberg and Ungar (1978). For the convenience of the reader, we begin this chapter with two sections presenting standard facts about Lie algebras that allow the computation of $H^1(g)$ and $H^2(g)$ in many important cases.

52. The Whitehead lemmas

The purpose of this section is to prove that for a semisimple Lie algebra g (we give a definition of semisimple below) $H^1(g) = H^2(g) = \{0\}$. This is a special case of a more general set of facts, known as the Whitehead lemmas, which we will formulate in a moment. These are all basic to the theory of Lie algebras and our treatment follows the classical presentation of Jacobson (1979).

We begin with a generalization of the notion of Lie algebra cohomology that we presented in Section 24. Let g be a Lie algebra and let V be a g module, that is, a vector space on which we are given a representation of g. Let $C^k(g, V)$ denote the space of antisymmetric k linear maps of g in V. Define the operator

$$\delta : C^k(g, V) \to C^{k+1}(g, V)$$

by

$$\delta f(\xi_0, \ldots, \xi_k) = \sum_{i=0}^{k} (-1)^i \xi_i f(\xi_0, \ldots, \hat{\xi}_i, \ldots, \xi_k)$$
$$+ \sum_{i<j=1}^{k+1} (-1)^{i+j} f([\xi_i, \xi_j], \xi_0, \ldots, \hat{\xi}_i, \ldots, \hat{\xi}_j, \ldots, \xi_k). \quad (52.1)$$

Thus, for $k = 0$ f is an element of V and the last term does not occur so

$$\delta f(\xi) = \xi f. \quad (52.2)$$

For $k = 1$, f is a linear map from g to V and

$$\delta f(\xi_0, \xi_1) = \xi_0 f(\xi_1) - \xi_1 f(\xi_0) - f([\xi_0, \xi_1]). \quad (52.3)$$

For $k = 2$ we have

$$\delta f(\xi_0, \xi_1, \xi_2) = \xi_0 f(\xi_1, \xi_2) - \xi_1 f(\xi_0, \xi_2) + \xi_2 f(\xi_0, \xi_1)$$
$$- f([\xi_0, \xi_1], \xi_2) + f([\xi_0, \xi_2], \xi_1) - f([\xi_1, \xi_2], \xi_0). \quad (52.4)$$

Note that if V is a trivial g module, the first terms in (52.1) disappear. Then (52.2) and (52.3) reduce to (24.1) and (24.2). It is not difficult to check that $\delta^2 = 0$. (We shall need this fact for the cases $k = 0$ and 1 and the reader is advised to verify these cases.) We can then define $Z^k(g, V) \subset \wedge^k(g, V)$ as $\ker \delta$ and $B^k(g, V) \subset Z^k(g, V)$ as $\operatorname{im} \delta$ with

$$H^k(g, V) = Z^k(g, V)/B^k(g, V).$$

In what follows, all Lie algebras and modules will be understood to be finite-dimensional. To understand the significance of the first cohomology group let us prove the following:

Proposition 52.1. *A Lie algebra* g *has the property that* $\mathrm{H}^1(\mathrm{g}, \mathrm{V}) = \{0\}$ *for all* V *if and only if every representation of* g *is completely reducible.*

Proof. Suppose that $H^1(g, V) = \{0\}$ for all V. Let us be given a representation of g on some vector space W and let U be a g-invariant subspace. We want to prove the existence of a g-invariant complement U'. Let $P: W \to U$ be a projection. Any other projection Q from W onto U can be written as

$$Q = P + D,$$

where D is a linear mapping satisfying

$$D(W) \subset U \qquad \text{and} \quad D_{|U} = 0.$$

Conversely, given any such D, $Q = P + D$ is a projection onto U. To find an invariant complement to U is the same as to find a projection Q that commutes with all the $\rho(\xi)$, $\xi \in g$, where ρ is the representation. Let V denote the space of all D satisfying the above equations, and let σ be the representation on V given by

$$\sigma(\xi)D = [\rho(\xi), D].$$

(It is easy to check that this is well defined, i.e., that the right-hand side belongs to V.) Let $f: g \to V$ be defined by

$$f(\xi) = [\rho(\xi), P].$$

The right-hand side does belong to V. Indeed, since P maps W into U and U is g-invariant, the right-hand side maps W into U. Applied to an element x of U, $Px = x$ so $[\rho(\xi), P]x = \rho(\xi)Px - P\rho(\xi)x = \rho(\xi)x - \rho(\xi)x = 0$. On the other hand, if we think of P as an element of $\mathrm{Hom}(W, W)$ and think of $\mathrm{Hom}(W, W)$ as a g module under the commutator action, the definition of f is just $f = \delta P$, relative to the δ operator on the module $\mathrm{Hom}(W, W)$. For this reason, or just by direct verification, $\delta f = 0$. Since $H^1(g, V) = \{0\}$ by hypothesis, we can find some $E \in V$ such that $f = \delta E$, which means that

$$[\rho(\xi), E] = [\rho(\xi), P] \qquad \forall\, \xi \in g.$$

But this implies that

$$[\rho(\xi), Q] = 0, \qquad \text{where} \qquad Q = P - E$$

so Q is the desired projection along an invariant complement.

Conversely, let us assume that every representation of g is completely reducible. In particular, this implies that the adjoint representation of g is completely reducible. An invariant subspace of g for the adjoint representation is just an ideal, and so g must be a direct sum of simple ideals: $g = \bigoplus I_j$. For each of these ideals we must have either $[I_j, I_j] = I_j$ or $[I_j, I_j] = \{0\}$ by the irreducibility of the adjoint representation on I_j. Now any commutative Lie algebra has representations that are not completely reducible: It suffices to observe that the one-dimensional Lie algebra does have such representations since there exist linear transformation that are not completely reducible, and we can, by simply choosing a subspace of codimension 1, find a homomorphism from a commutative Lie algebra onto a one-dimensional one, and hence a noncompletely reducible representation of the commutative Lie algebra. Any representation of one of the ideals I_j of g extends to a representation of g by sending all the other ideals to 0. Hence g contains no commutative ideals. This implies that $[g, g] = g$, which, from Section 24, implies that $H^1(g) = \{0\}$ (i.e., the vanishing of H^1 for the trivial module).

Now let V be any g module. Define the representation $\xi \to D_\xi$ on $C^k(g, V)$ by

$$(D_\xi f)(\xi_1, \ldots, \xi_k) = \xi f(\xi_1, \ldots, \xi_k) - \sum f(\xi_1, \ldots, [\xi, \xi_i], \ldots, \xi_k).$$

It is easy to check that this is indeed a representation (we shall only use this fact for $k = 0, 1, 2, 3$) and that

$$D_\xi \delta = \delta D_\xi.$$

Notice that for $k = 1$ we have the formula

$$D_\xi f = \delta f(\xi, \cdot) + \delta(f(\xi)).$$

(There is, of course, a similar formula for all k, which is just the natural generalization of the fundamental formula of the differential calculus, (22.1) translated into Lie algebra terms.) In particular, $Z^1(g, V)$ is a representation space for g and $B^1(g, V)$ and invariant subspace. If all representations of g are completely reducible, then $B^1(g, V)$ must have an invariant complement in $Z^1(g, V)$. We must show that this complement is $\{0\}$. If f belongs to this complement, then by invariance, so must $D_\xi f$. Since $\delta f = 0$, the preceding equation says that $D_\xi f = \delta(f(\xi))$, and since $D_\xi f$ is to be in a space complementary to the space of all boundaries, this implies that $D_\xi f = 0 = \delta(f(\xi))$. But $\delta(f(\xi))(\eta) = \eta f(\xi)$ by (52.2) (with the constant value $f(\xi)$ substituted for f in (52.2)). Hence, the first two terms on the right of (52.2) vanish and

$$0 = \delta f(\xi_0, \xi_1) = -f([\xi_0, \xi_1]).$$

Since $[g, g] = g$, this last equation implies that $f = 0$, completing the proof of the proposition. By the way, a similar argument proves

Proposition 52.2. *Any Lie algebra satisfying the hypothesis of Proposition 52.1 has* $\mathrm{H}^2(\mathrm{g}) = \{0\}$.

Indeed, as we are now looking at cohomology with values in a trivial g module the first three terms in (52.4) vanish and we have an action of g on $C^3(g)$ given by

$$\begin{aligned}(D_\xi f)(\xi_1, \xi_2) &= -f([\xi, \xi_1], \xi_2) - f(\xi_1, [\xi, \xi_2]) \\ &= \delta f(\xi, \xi_1, \xi_2) - f(\xi, [\xi_1, \xi_2]) \\ &= \delta f(\xi, \xi_1, \xi_2) - \delta(f(\xi, \cdot))(\xi_1, \xi_2).\end{aligned}$$

If f lies in an invariant complement to $B^2(g)$ in $Z^2(g)$ we must have $D_\xi f = \delta(f(\xi,\cdot)) = 0$, or $f(\xi,[\xi_1,\xi_2]) = 0$ for all ξ, ξ_1, and ξ_2. Since $[g,g] = g$, again this implies that $f = 0$.

In fact, a more delicate argument [cf. Jacobson (1979) for example] will show that $H^2(g,V) = \{0\}$ for any module V. As we will not need this result, we shall not prove it here. A Lie algebra is called *semisimple* if it contains no commutative ideals. We now want to prove the converse to the result established earlier, that is, we want to prove that any semisimple algebra satisfies the equivalent hypotheses of Proposition 52.1. The crucial fact to be used for this proof is

Cartan's criterion for semisimplicity. *If* g *is a semisimple Lie algebra over the real or complex numbers (or any field of characteristic zero) and* (V,ρ) *is a one-to-one (finite-dimensional) representation of* g *then the trace form* b_ρ *where*

$$b_\rho(\xi,\eta) = \operatorname{tr}\rho(\xi)\rho(\eta)$$

is nondegenerate. If the trace form of the adjoint representation of a Lie algebra g *(called the Killing form) is nondegenerate, then* g *is semisimple.*

The last assertion in Cartan's criterion for semisimplicity is easy to prove: Suppose that I is an Abelian ideal in g. Then if we choose a basis for g whose last elements span I, every element of g has, in the adjoint representation, the matrix form $\begin{pmatrix} - & 0 \\ - & - \end{pmatrix}$, while every element of I has the matrix form $\begin{pmatrix} 0 & 0 \\ - & 0 \end{pmatrix}$. The trace of the product of any two such matrices is clearly 0. Thus every element of I is orthogonal, under the Killing form, to all of g; that is, the Killing form is degenerate. The first part of Cartan's criterion, which is the one we need to use, requires a careful study of solvable Lie algebras, which we will postpone momentarily. We will first use Cartan's criterion to prove that the hypotheses of Proposition 52.1 are satisfied. By the way, it follows immediately from Cartan's criterion that the property of semisimplicity for a Lie algebra is independent of the base field: If we pass from the real to the complex numbers (or from any field of characteristic 0 to its algebraic closure) the property of nondegeneracy of a bilinear form, and hence semisimplicity is not changed.

If g has no commutative ideals, then g cannot contain any elements whose bracket with all other elements is 0, since these would constitute a commutative ideal. Hence the adjoint representation must be one-to-one, and hence by Cartan's criterion, the Killing form must be nondegenerate. If I is any ideal of g, its orthogonal complement with respect to the Killing form must be a complementary ideal, and the restriction of the Killing form to it is nondegenerate; hence this complementary ideal is itself semisimple as a Lie algebra. In particular, if ρ is any representation of g, $\ker \rho$ is an ideal and its orthocomplement with respect to the Killing form is a semisimple Lie algebra such that the restriction of ρ to it is one-to-one. Hence, in proving that the hypotheses of Proposition 52.1 hold, we may as well assume that ρ is one-to-one. Let $b = b_\rho$ denote the associated trace form and let us write ξ instead of $\rho(\xi)$ (that is, we regard g as a subalgebra of $\mathrm{Hom}(V, V)$ via ρ). Thus

$$b(\xi, \eta) = \operatorname{tr} \xi\eta$$

and

$$b([\xi, \eta], \zeta) + b(\eta, [\xi, \zeta]) = 0 \qquad \forall\, \xi, \eta, \zeta.$$

By Cartan's criterion b is nondegenerate so we can use b to identify $g \otimes g$ with $g \otimes g^* = \mathrm{Hom}(g, g)$. The identity element in $\mathrm{Hom}(g, g)$ then corresponds to an element of $g \otimes g$ that is invariant under the tensor product of the adjoint representation with itself. Since g is regarded as a subspace of $\mathrm{Hom}(V, V)$ we have a multiplication map (the restriction of the operator multiplication of $\mathrm{Hom}(V, V)$) sending $g \otimes g$ into $\mathrm{Hom}(V, V)$ and the image of this invariant element must be an element C of $\mathrm{Hom}(V, V)$ that commutes with the action of g. Explicitly, this element can be constructed as follows: Choose a linear basis $\xi_1, \ldots, \xi_n$ of g, and let $\eta_1, \ldots, \eta_n$ be the dual basis with respect to b; that is

$$b(\xi_i, \eta_j) = \delta_{ij}.$$

Then

$$C = \sum \xi_i \eta_i$$

so that

$$\operatorname{tr} C = \sum b(\xi_i, \eta_i) = n = \dim g.$$

The definition of C (the Casimir element of the representation) is independent of the choice of basis. We now wish to show that $H^1(g, V) = \{0\}$. Decompose V into

$$V = V_0 + V_1$$

such that C is nilpotent on V_0 and nonsingular on V_1. Since C commutes with g, both of these spaces are g-invariant and we can write every function $f : g \to V$ as $f = f_0 + f_1$ with $\delta f_0 = \delta f_1 = 0$ if $\delta f = 0$. We may prove that $H^1(g, V) = \{0\}$ by induction on dim V and hence we may assume that either $V = V_0$ or $V = V_1$. If $V = V_0$ then $\operatorname{tr} C = 0 = \dim g$ so $g = \{0\}$ and there is nothing to prove. So we may assume that C is non-singular. Let

$$v = \sum \xi_i f(\eta_i),$$

where $\xi_1, \ldots, \xi_n$ and $\eta_1, \ldots, \eta_n$ are the dual bases used above. Then

$$\begin{aligned}\zeta v &= \sum \xi_i \zeta f(\eta_i) + \sum [\zeta, \xi_i] f(\eta_i) \\ &= \sum \xi_i \eta_i f(\zeta) + \sum \xi_i f([\zeta, \eta_i]) + \sum [\zeta, \xi_i] f(\eta_i)\end{aligned}$$

since $\delta f = 0$ (see (52.3)). Now since the ξ's and η's form a dual basis for the invariant form b, the matrix of ad relative to the ξ basis is the negative transpose of its matrix relative to the η basis, and, since f is linear, the last two terms in the above expression cancel, while the first term is just $Cf(\zeta)$. Thus

$$f(\zeta) = \zeta u, \qquad u = C^{-1} v,$$

or

$$f = \delta u$$

as was to be proved.

We must still prove Cartan's criterion. As was already mentioned, we may assume that the field is algebraically closed, for example, that we are over the complex numbers.

(The next few pages are based on Hausner and Sternberg (1957).) We begin with the definition of nilpotent and solvable Lie algebras:

Definition 52.1. *A* nilpotent *Lie Algebra* L *is one such that there exists an* N *with* $[\ldots[[\xi_1, \xi_2]\xi_3]\ldots\xi_N] = 0$ *for any* ξ_i *in* L.

Notation. $C(L) = [L, L]$ = algebra generated by elements of the form $[\xi_1, \xi_2]$.

Definition 52.2. *A* solvable *Lie algebra* L *is one such that* $L \supset C(L) \supset C(C(L)) \supset \cdots$ *terminates with* 0 *after a finite number of steps.*

Note: Clearly a nilpotent algebra is solvable.

Lemma 52.1a (on matrices). *There exist constants* a^r_{jk} *such that for any matrices* T, S:

$$T^r S = S T^r + \sum a^r_{jk} T^j [S, T] T^k,$$

where $j + k + 1 = r$.

Proof. The proof is a trivial induction on r, using $TS = ST - [ST]$.

Theorem 52.1. *Let* L *be a solvable Lie algebra, and let* R *by any representation of* L. *Then* R *may be (simultaneously) put into triangular form.*

Proof. We may assume R is irreducible. For if we have proven the theorem for irreducible R, we may prove for arbitrary R by induction on the dimension of the representation. Thus, if

$$R(l) \sim \left(\begin{array}{c|c} R_1(l) & \sim \\ \hline 0 & R_2(l) \end{array}\right),$$

by placing R_1 and R_2 in triangular form, we have the result. The proof consists of an induction on the dimension of L, the result being clear for dimension 1. We have $[L, L] \subset L$ (strict inclusion). Take K of dimension one less than L with $[L, L] \subset K \subset L$. Then $[K, K] \subset [L, L] \subset K$ and hence K is a solvable algebra. Hence R on K may be put into triangular form and, in particular, there is a common eigenvector $v \in V$ with

$$R(k)v = \alpha(k)v \qquad \alpha \text{ a scalar.}$$

Let $x \notin K$. We contend that the space generated by $v, R(x)v, \ldots, R^j(x)v, \ldots$ is invariant under R, and hence is all of V. For it is certainly invariant under $R(x)$. As for $R(k)$, we shall show that

$$R(k)R^n(x)v = \text{linear combination of } R^j(x)v, j \leqslant n.$$

We do this by induction on n. It is true for $n = 0$. In general,

$$R(k)R^n(x) = R^n(x)R(k) + \sum a^n_{jk} R^j(x)[R(k), R(x)]R^k(x)(j + k = n - 1).$$

Hence

$$\begin{aligned} R(k)R^n(x)v &= R^n(x)R(k)v + \sum a^n_{jk} R^j(x)R[k, x]R^k(x)v \\ &= \alpha(k)R^n(x)v + \sum c_j R^j(x)v \qquad j < n, \end{aligned}$$

since $[k, x] \in [L, L] \subset K$.

We further observe from the above, that

$$\operatorname{tr} R(k)R^n(x) = \text{a multiple of } \alpha(k).$$

If $k \in [L, L]$, $\operatorname{tr} R(k) = 0$ (as the commutator of matrices), and hence

$$\alpha(k) = 0 \qquad \text{for} \qquad k \in [L, L].$$

Thus,

$$R[L, L]v = 0.$$

If W = space of w such that $R[L, L]w = 0$, we have $v \in W$. We show that W is invariant under R, and hence $W = V$. For,

$$R([l, m])R(x)w = R(x)R[l, m]w + R[[l, m], x]w = 0.$$

Hence $R[L, L] = 0$ (on all of V). Hence the matrices $R(L)$ commute. But an irreducible space for commuting matrices is one-dimensional, and the result is proven.

Note 1. In particular, this result applies to nilpotent Lie algebras.

Note 2. In particular, there is a common eigenvector for any representation of a solvable algebra.

Lemma 52.1b (on matrices). *Let* S, T *be matrices, with* $\mathrm{S}_1 = [\mathrm{S}, \mathrm{T}], \ldots, \mathrm{S}_{n+1} = [S_n, \mathrm{T}], \ldots$ *Let* $\mathrm{V}_\lambda(\mathrm{T})$ *be the space of vectors* v *such that* $(\mathrm{T} - \lambda \mathrm{I})^\nu \mathrm{v} = 0$ for suitable ν. ($\mathrm{V}_\lambda(\mathrm{T})$ *is the largest space on which* $\mathrm{T} - \lambda\mathrm{I}$ *is nilpotent.*) *Then if* $\mathrm{S}_k = 0$, V_λ *is invariant under* S.

Proof. By replacing T by $T - \lambda I$, it suffices to consider only the case $\lambda = 0$. We use induction on k. For $k = 1$, $S_1 = 0$, we have

$$T^r S = S T^r.$$

Hence if $v \in V_0(T) = V_0$, $Sv \in V_0$. Now assume $S_{k+1} = 0$. By the induction hypothesis V_0 is invariant under $S_1 = [S, T]$. By Lemma 52.1a,

$$T^r S = S T^r + \sum c^r_{jk} T^j S_1 T^k,$$

where $j + k = r - 1$. For r sufficiently large, all terms on the right vanish on V_0, since eventually T^j or T^k vanishes on V_0.

Theorem 52.2. *Let* R *be any representation of a nilpotent algebra* N *on a vector space* V. *Then* V *decomposes into a direct sum of subspace* V_i, $i = 1, \ldots, k$, *with corresponding linear complex-valued functions* α_i *on* N *such that*

(1) V_i *is invariant under* R;
(2) $\mathrm{R}(n) - \alpha_i(n)\mathrm{I}$ *is nilpotent on* V_i, *for all* $n \in \mathrm{N}$.

Proof. By induction on dim V. For each $n \in N$, we consider the spaces V^n_λ as in Lemma 52.1b for the linear transformation $R(n)$. We choose n_0, λ_0 so that dim $V^{n_0}_{\lambda_0} = \min_n$ dim V^n_λ. Then $V = V^{n_0}_{\lambda_0} + V^{n_0}_{\lambda_1} + \cdots$. Since any of the transformations $R(n)$ satisfy the hypotheses of Lemma 52.1b, $V^{n_0}_{\lambda_i}$ is invariant under $R(n)$ for all n. If $V^{n_0}_{\lambda_0} = V$, then each $R(n) = \lambda_0(n)I +$ nilpotent, and we have the result. If $V^{n_0}_{\lambda_0} \neq V$, we apply the induction to $V^{n_0}_{\lambda_0}$ and to $V^{n_0}_{\lambda_1} + \cdots$ and the theorem follows. Moreover $\alpha_i(n)$ is clearly linear since it is the unique eigenvalue on V_i of $R(n)$ and by Theorem 52.1, note 2, we have a common eigenvector in V_i.

Remark 1. By combining the V_i's corresponding to identical α_i's, we may arrange to have the α_i's distinct. From now on, we assume that this has been done.

Remark 2. The spaces V_i and functions $\alpha_i(n)$ are uniquely determined by R. For if $v = \sum v_i (v_i \in V_i)$ is such that $R(n) - \beta(n)I$ is nilpotent on v for all n, then all but one of the $v_i = 0$. Hence, if $V = \sum W_i$, $R(n) - \beta_i(n)$ nilpotent on W_i, we have each $W_i \subseteq$ some V_j, and the decompositions are identical.

Definition 52.3. *The* $\alpha_i(n)$ *are called the* weights *of the representation* R.

Remark. By Theorem 52.2, there exist k distinct weights α_i. The spaces V_i on which $R(n) - \alpha_i(n)$ is nilpotent are invariant under R, and V is the direct sum of V_i.

We apply the preceding to the case where N is a nilpotent subalgebra of a Lie algebra L. The representation of N will be the adjoint representation.

Definition 52.4. *The adjoint representation of a subalgebra* N *of* L *is defined by* $(ad\, n)(x) = [n, x]$ *and is a representation of* N *in* L. *In this case, the weights are called roots.*

Note. In this case the roots α of N relative to the representation ad breaks up L into sub (vector) spaces L_α. We note that $\alpha = 0$ is a root and in fact $L_0 \supset N$. For $(\mathrm{ad}\, n)^r m = \pm[[[m, n], n] \ldots] = 0$ for n, $m \in N$, since N is nilpotent.

Theorem 52.3. $[L_\alpha, L_\beta] \subset L_{\alpha+\beta}$.

Note. In particular, if $\alpha + \beta$ is not a root $[L_\alpha, L_\beta] = 0$.

Proof. Let $x_\alpha \in L_\alpha, y_\beta \in L_\beta$. Then

$$\begin{aligned}(\operatorname{ad} n - \alpha(n)I - \beta(n)I)[x_\alpha, x_\beta] \\ &= [n,[x_\alpha, x_\beta]] - [\alpha(n)x_\alpha, x_\beta] - [x_\alpha, \beta(n)x_\beta] \\ &= [x_\alpha[n, x_\beta]] + [[n, x_\alpha]x_\beta] - [\alpha(n)x_\alpha, x_\beta] - [x_\alpha, \beta(n)x_\beta] \\ &= [x_\alpha, (\operatorname{ad} n - \beta(n)I)x_\beta] + [(\operatorname{ad} n - \alpha(n)I)x_\alpha, x_\beta].\end{aligned}$$

Iterating this formula sufficiently, we finally have $\operatorname{ad} n - \alpha(n)I - \beta(n)I$ is nilpotent on $[x_\alpha, x_\beta]$, and hence $[x_\alpha, x_\beta] \in L_{\alpha+\beta}$.

Corollary: $\mathrm{L}_0 \supset \mathrm{N}$ *is a subalgebra of* L.

Definition 52.5. N *is a* Cartan subalgebra *of* L *if* $\mathrm{L}_0 = \mathrm{N}$.
Our immediate aim is to prove the existence of Cartan subalgebras of L. This is done with the heip of the following.

Lemma 52.2 (Engel's theorem). *Let* L *be a finite-dimensional Lie algebra of matrices, each of which is nilpotent. Then* L *is a nilpotent Lie algebra.*

Proof. We prove more: In fact any N products (associative multiplication) of elements in L vanishes. Let M be a maximal sub- (Lie) algebra of L that is (associative) nilpotent. We show $M = L$. If $M \neq L$, choose $y \notin M$. If $M^r = 0$, then $[[y, M], M] \ldots M] = 0 \subset M$ $(2r - 1$ factors.)

Let k be the least integer such that $[yM] \ldots M] \subset M$ (k factors). Then there exist elements $m_i \in M$ such that $X = [y, m_1] \ldots m_{k-1}] \notin M$, but $[X, M] \subset M$. Hence there exists an element $X \notin M$ such that $[X, M] \subset M$. Then the Lie algebra generated by X and M will satisfy the conclusion of the theorem. For, any N products of elements in M vanish, and $X^r = 0$ by hypothesis.

Using $[X, m_1] \in M$, we have

$$Xm_1 = m_1 X + m^*.$$

Thus, any product of elements of M and X's vanishes if N or more elements of M appear in it. But, if restricted to $N - 1$ elements of M in a product, and no X^r appearing, the product can clearly have at most $(N - 1) + N(r - 1)$ factors. Thus any product of Nr factors vanish, which gives the result.

Theorem 52.4. *For any Lie algebra* L *there exists a Cartan subalgebra* N.

Proof. We shall construct an algebra N on which $\operatorname{ad}_N(n)$ is nilpotent for all $n \in N$. Since, by the above lemma, $\operatorname{ad}_N N$ will be nilpotent, this will show that N is nilpotent: Any $h \in L$ generates a (one-dimensional) nilpotent algebra and hence yields in the decomposition, the algebra $L_0(h)$. Choose h minimizing the dimension of $L_0(h)$ and set $N = L_0(h)$. Now let $h^* \in N$, and we assert that $\mathrm{ad}_N h^*$ is nilpotent. For, we consider $\mathrm{ad}(ah + h^*)$ on $L_\alpha(h)$; $a \neq 0$. Since h is nonsingular on L_α, $\mathrm{ad}(ah + h^*)$ is nonsingular with a finite number of exceptions. Hence the nilpotent space of $\mathrm{ad}(ah + h^*) \subset L_0$. But by the minimality choice of h, $\mathrm{ad}(ah + h^*)$ is nilpotent on L_0 for infinitely many a's. But, if $aA + \mathrm{B}$ is nilpotent for infinitely many a's, B is nilpotent; for this can be reduced to a polynomial condition on a (with matrix coefficients). Thus, $\mathrm{ad}_N h^*$ is nilpotent, and N is an nilpotent algebra. But the space on which all of N acts nilpotently $(L_0(N))$ is contained in the space on which a particular element $h \in N$ acts nilpotently. Thus $L_0(N) \subset L_0(h) = N \subset L_0(N)$. Thus, N is a Cartan subalgebra of L.

Lemma 52.3. *Let* N *be a Cartan subalgebra of* L *and* (ρ, V) *a representation of* L. *Suppose that* α *is a root of* L *such that* $-\alpha$ *is also a root. Let* e_α *and* $\mathrm{e}_{-\alpha}$ *be nonzero elements of* L_α *and* $\mathrm{L}_{-\alpha}$ *and set* $\mathrm{h}_\alpha = [\mathrm{e}_\alpha, \mathrm{e}_{-\alpha}]$. *Then* $\beta(\mathrm{h}_\alpha)$ *is a rational multiple of* $\alpha(\mathrm{h}_\alpha)$ *for any weight* β *on* N *in* V.

Proof. Consider the functions of the form $\beta + j\alpha$ on $N, j = 0, \pm 1, \pm 2, \ldots$, which are weights, and let W be the direct sum of the corresponding weight spaces. Clearly, if w is a weight vector with weight γ then $e_\alpha w$ is a weight vector with weight $\gamma + \alpha$ and similarly $e_{-\alpha} w$ is a weight vector with weight $\gamma - \alpha$. Hence the space W is invariant under the three-dimensional algebra spanned by $e_\alpha, e_{-\alpha}$, and h_α. Since $\rho(h_\alpha)$ is a commutator,

$$0 = \mathrm{tr}\, \rho(h_\alpha) = \sum n_{\beta + j\alpha}\big(\beta(h_\alpha) + j\alpha(h_\alpha)\big),$$

where $n_{\beta + j\alpha}$ is the dimension of the space of weight vectors with weight $\beta + j\alpha$. Collecting coefficients, and the coefficients of $\beta(h_\alpha)$ are all positive integers, gives the desired rational relation. Using the lemma we can now prove:

Cartan's criterion for solvable algebras. *Let* L *be a Lie algebra and* (ρ, V) *a representation of* L *such that*

(*1*) *ker* ρ *is solvable;*

and

(*2*) *tr* $\rho(\xi)^2 = 0 \;\forall \xi \in \mathrm{C(L)}$.

Then L *is solvable.*

Proof. It suffices to prove that $C(L)$ is strictly contained in L. For the restriction of ρ to $C(L)$ satisfies all the conditions and hence, by induction

$C(L)$ and therefore L is solvable. So we must derive a contradiction from the assumed properties and the assertion that $C(L) = L$. Let N be a Cartan subalgebra of L and L_α the corresponding root spaces. Then by Lemma 52.3, we can conclude that

$$N \cap C(L) = \sum [L_\alpha, L_{-\alpha}],$$

where the sum is taken over all α (including possibly $\alpha = 0$) such that $-\alpha$ is also a root. The assertion that $C(L) = L$ says that every element of N can be written as a sum of terms of the form $[e_\alpha, e_{-\alpha}] = h_\alpha$. For any weight β, the element $\rho(h_\alpha)$ has the unique eigenvalue $\beta(h_\alpha)$ on the space of weight vectors with weight β, and hence $\operatorname{tr} \rho(h_\alpha)^2$ on this subspace is just $n_\beta(\beta(h_\alpha))^2$. Thus

$$\operatorname{tr} \rho(h_\alpha)^2 = \sum n_\beta(\beta(h_\alpha))^2 = \sum n_\beta r_\beta^2(\alpha(h_\alpha))^2,$$

where the r_β are the rational numbers provided by the lemma. Since the n_β are positive integers, this implies that

$$\alpha(h_\alpha) = 0.$$

By the lemma again, this implies that $\beta(h_\alpha) = 0$ and since the h_α span all of N by hypothesis, we see that $\beta = 0$. So 0 is the only weight, and hence every element $\rho(h_\alpha)$ is nilpotent. Since $e_\alpha v$ has weight α if v has weight 0, we conclude that $e_\alpha v = 0$; that is, all the L_α with $\alpha \neq 0$ must be contained in ker ρ. Thus $\rho(L) = \rho(N)$ and hence $\rho(L)$ is nilpotent and hence solvable, being the homomorphic image of a nilpotent algebra. But if ker ρ and im ρ are both solvable, this implies that L is solvable (and in particular $C(L) \neq L$).

Let g be any Lie algebra. Suppose that I and J are two solvable ideals in g. Then $(I + J)/J$ is isomorphic to $I/(I \cap J)$ and hence is solvable. If

$$0 \to L_1 \to L \to L_2 \to 0$$

is an exact sequence of Lie algebras with L_1 and L_2 solvable, then clearly L is also solvable. Taking $L = I + J$, $L_1 = I$ and $L_2 = J$ shows that $I + J$ is solvable. Therefore, any Lie algebra g has a maximal solvable ideal, called the radical of g. If L is an ideal in g then so is $C(L)$. So if L is a solvable ideal in g, then the last nonzero $C^k(L)$ will be a commutative ideal. Thus an algebra g is semisimple (has no commutative ideals) if and only if it has no solvable ideals (i.e., its radical is $\{0\}$).

We are finally in a position to prove Cartan's criterion for semisimplicity. Suppose that ρ is a one-to-one representation of g with trace form $b = b_\rho$. Then $g^\perp$, the null space of this trace form is an ideal in g, and every element of $g^\perp$ satisfies $b(\xi, \xi) = \operatorname{tr} \rho(\xi)^2 = 0$. Hence, by Cartan's criterion for solvability, $g^\perp$ is solvable. If g is semisimple, this means that $g^\perp = \{0\}$, that is, that b is nondegenerate.

We have already verified that the algebra $sp(2n)$ is simple (i.e., has no ideals). Hence it is semisimple and its first and second cohomology groups vanish. It is equally easy to check that the algebras $sl(n)$ and $o(n)$ (for $n > 2$) are simple, or to verify directly that the Killing form, which is up to a scalar multiple, given by $f(\xi, \eta) = \operatorname{tr} \xi\eta$, is nondegenerate. Hence for these algebras as well, $H^1(g) = H^2(g) = \{0\}$. These are all the "classical" simple finite-dimensional Lie algebras over the complex numbers, and have various compact real forms, for example $o(p, q)$, etc. In addition, there are the "exceptional" simple algebras (see for example Jacobson, 1979, for a discussion of the Cartan classification). For all of these, the first two cohomology groups vanish.

53. The Hochschild–Serre spectral sequence

In this section we wish to show how to compute $H^1(g)$ and $H^2(g)$ for a more general class of Lie algebras. In particular, we will see that the first two cohomology groups of the Poincaré group vanish, whereas they are one-dimensional for the Galilean group. The method is one relating the cohomology of g to that of a subalgebra, and, in reality, is a special case of a general construction known as the Hochschild–Serre spectral sequence (see Hochschild and Serre, 1953). We shall not present this general theory, but rather carry out the computations in detail for the first and second cohomology groups – the ones of interest to us in symplectic geometry. The computations are based on Sternberg (1975b) and, follow our treatment in Guillemin and Sternberg (1976) almost verbatim.

We will assume that there are two subspaces k and p of g such that

$$g = k + p, \qquad k \cap p = \{0\}, [k, k] \subset k, \quad \text{and} \quad [k, p] \subset p.$$

Thus we are assuming that k is a subalgebra of g and that p is a supplementary subspace to k that is stable under the action of k. We do not make any further assumptions at the moment about p. Thus $[p, p]$ will have both a k and a p component, which we denote by r and s, respectively: For η and η' in p we have $[\eta, \eta'] = r(\eta, \eta') + s(\eta, \eta')$, where $r(\eta, \eta') \in k$ and $s(\eta, \eta') \in p$. Jacobi's identity implies some identities on r and s. It is easy to check that these are

$$\mathfrak{S} r(s(\eta, \eta'), \eta'') = 0,$$

$$\mathfrak{S}\{s(s(\eta, \eta'), \eta'') + [r(\eta, \eta'), \eta'']\} = 0,$$

where $\mathfrak{S}$ denotes cyclic sum. Also

$$[\xi, r(\eta, \eta')] = r(\xi \cdot \eta, \eta') + r(\eta, \xi \cdot \eta'),$$

where $\xi \in k$ and $\eta, \eta' \in p$ and we have written $\xi \cdot \eta$ for $[\xi, \eta]$, thinking of k acting on p. We also have the equation

$$\xi \cdot s(\eta, \eta') = s(\xi \cdot \eta, \eta') + s(\eta, \xi \cdot \eta').$$

In addition, we have the identity asserting that k acts as a Lie algebra of linear transformations on p and Jacobi's identity for k. Conversely, starting from any action of a Lie algebra k on a vector space p together with r and s satisfying the above identities it is clear that $g = k + p$ becomes a Lie algebra. Let us give some illustration of this situation:

(A) $r = s = 0$. In this case p is a supplementary Abelian ideal, and k acts as linear transformations on p. In other words, g is the semidirect product of k and p, where k is a Lie algebra with a given linear representation of k on p. Any such linear representation of k gives rise to a Lie algebra g, which is called the associated affine algebra.

(B) $r = 0$. Here all that is assumed is that p is a supplementary ideal to k. An important illustration of this situation is the case of the Galilean group. Recall that the Galilean group can be regarded as the group of all 5×5 matrices of the form

$$\begin{bmatrix} A & v & x \\ 0 & 1 & t \\ 0 & 0 & 1 \end{bmatrix},$$

where $A \in O(3)$, $v \in \mathbb{R}^3$, $x \in \mathbb{R}^3$ and $t \in \mathbb{R}$. Such a matrix carries the space-time point (x_0, t_0) into the space-time point $(Ax_0 + x + t_0 v, t + t_0)$. The corresponding Lie algebra consists of all matrices of the form

$$\begin{bmatrix} a & v & x \\ 0 & 0 & t \\ 0 & 0 & 0 \end{bmatrix},$$

where $a \in o(3)$ and v, x, t as before. Here we can take $k \sim o(3)$ to consist of the subalgebra with $x = v = t = 0$ and p to be the seven-dimensional subalgebra with $a = 0$. Denoting an element of p by (v, x, t) we see that $[(v, x, t), (v', x', t')] = s\big((v, x, t), (v', x', t')\big) = (0, t'v - tv', 0)$ and $\xi \cdot (v, x, t) = (\xi \cdot v, \xi \cdot x, 0)$, where $\xi \cdot v$ denotes the usual action of $\xi \in o(3)$ on $v \in \mathbb{R}^3$ and similarly for $\xi \cdot x$.

(C) The case where g is semisimple and k, p corresponds to a Cartan decomposition. Here $s = 0$.

(D) The case where k is an ideal. Here the action of k on p is trivial. For example, in the case of the Heisenberg algebra we can take k to be the center. For this case p is a symplectic vector space, $k = \mathbb{R}$ acts trivially on p, and r is the symplectic 2-form, while $s = 0$.

Let $f \in \bigwedge^2 g^*$ be a 2-form. Identifying $\bigwedge^2 g^*$ with

$$\bigwedge^2 k^* \oplus k^* \otimes p^* \oplus \bigwedge^2 p^*$$

allows us to write $f = a + b + c$ so that

$$f(\xi + \eta, \xi' + \eta') = a(\xi, \xi') + b(\xi, \eta') - b(\xi', \eta) + c(\eta, \eta').$$

Now $df \in \bigwedge^3 g^*$ is given by $df(\chi, \chi', \chi'') = \mathfrak{S} f([\chi, \chi'], \chi'')$, where $\mathfrak{S}$ denotes cyclic sum. Writing $\chi = \xi + \eta$, etc., the equation $df = 0$ becomes

$$\begin{aligned}\mathfrak{S}\{a([\xi, \xi'] + r(\eta, \eta'), \xi'') &+ b([\xi, \xi'] + r(\eta, \eta'), \eta'') \\ &- b(\xi'', \xi \cdot \eta' - \xi' \cdot \eta + s(\eta, \eta')) \\ &+ c(\xi \cdot \eta - \xi' \cdot \eta + s(\eta, \eta'), \eta'')\} = 0.\end{aligned}$$

We now derive various identities for a, b, and c by considering special cases of this identity:

(i) $\xi = \xi' = \xi'' = 0$. In this case the identity becomes

$$\mathfrak{S}\{b(r(\eta, \eta'), \eta'') + c(s(\eta, \eta'), \eta'')\} = 0. \tag{53.1}$$

For the case of the affine algebra, this identity is vacuous. If p is a subalgebra so that $r = 0$, only the identity involving c remains. For example, a direct computation in the case of the Galilean group shows that (53.1) reduces to the condition $c((0, x, 0), (0, x', 0)) = 0$. For the case of the Cartan decomposition, only the identity involving b remains. Similarly for the case of the Heisenberg algebra.

(ii) $\xi = \xi' = 0$, $\eta'' = 0$. In this case the identity becomes

$$a(r(\eta, \eta'), \xi'') - b(\xi'', s(\eta, \eta')) + c(\xi'' \cdot \eta, \eta') + c(\eta, \xi'' \cdot \eta') = 0. \tag{53.2}$$

For the case of the affine algebra both r and s vanish and this identity becomes

$$c(\xi \cdot \eta, \eta') + c(\eta, \xi \cdot \eta') = 0, \tag{53.2a}$$

which asserts that the antisymmetric form c is invariant under the action of k. For example, in the case of the Poincaré algebra where $k = o(3, 1)$ and $p = \mathbb{R}^4$ there is no invariant antisymmetric form so we conclude that $c = 0$.

In the case that we only assume that p is a subalgebra so that r vanishes the identity becomes

$$c(\xi \cdot \eta, \eta') + c(\eta, \xi \cdot \eta') = b(\xi, s(\eta, \eta')). \tag{53.2b}$$

For example, in the case of the Galilean algebra, if we apply this identity to $\eta = (v, x, 0)$ and $\eta' = (v', x', 0)$ the right-hand side vanishes and we conclude that c, when restricted to $(\mathbb{R}^3 + \mathbb{R}^3) \wedge (\mathbb{R}^3 + \mathbb{R}^3)$ is invariant under the action of $o(3)$, which acts diagonally on $\mathbb{R}^3 + \mathbb{R}^3$. There is obviously only one such invariant (up to scalar multiples) and it is given by

$$c\big((v, x, 0), (v', x', 0)\big) = m(\langle v, x'\rangle - \langle v', x\rangle),$$

where $\langle\, ,\rangle$ denotes the Euclidean scalar product. If we take $\eta = (0, x, 0)$ and $\eta' = (0, 0, t)$ the right side of (53.2b) still vanishes. On the left the term $\xi \cdot \eta'$ vanishes and $\xi \cdot x$ is arbitrary. We conclude that

$$c\big((0, x, 0), (0, 0, t)\big) = 0.$$

Thus

$$c\big((v, x, t), (v', x', t')\big) = m(\langle v, x'\rangle - \langle v', x\rangle) + \langle l, t'v - tv'\rangle$$

for some $l \in \mathbb{R}^3$, where (53.2b) implies that

$$\langle l, \xi \cdot v\rangle = b\big(\xi, (0, v, 0)\big).$$

In the case of a Cartan decomposition, or, more generally when $s = 0$ the identity (53.2) becomes

$$c(\xi \cdot \eta, \eta') + c(\eta, \xi \cdot \eta') = a\big(\xi, r(\eta, \eta')\big). \tag{53.2c}$$

For the case where k is an ideal, (53.2) becomes

$$a\big(\xi, r(\eta, \eta')\big) + b\big(\xi, s(\eta, \eta')\big) = 0. \tag{53.2d}$$

(iii) $\xi = 0$, $\eta' = \eta'' = 0$. In this case neither a nor c contributes and we obtain the identity

$$b([\xi', \xi''], \eta) + b(\xi'', \xi' \cdot \eta) - b(\xi', \xi'' \cdot \eta) = 0. \tag{53.3}$$

This identity says that the map from k to p^* sending $\xi \rightsquigarrow b(\xi, \cdot)$ is a cocycle. If k is semisimple, then Whitehead's lemma asserts that b must be a coboundary, that is, that there exists a $\theta \in p^*$ such that

$$b(\xi, \eta) = \theta(\xi \cdot \eta). \tag{53.3a}$$

Suppose that instead of assuming that k is semisimple we assume that k contains an element in its center that acts as the identity transformation on p. Taking ξ to be this element and ξ'' to be an arbitrary ξ in (53.3), we see that (53.3a) holds with $\theta(\eta) = b(\xi', \eta)$. Thus

if either k is semisimple or k contains an element in its center acting as the identity transformation on p then (53.3a) holds.

For example, in the case of the Galilean algebra, we see that the bilinear form b is given by

$$b\big(\xi, (v, x, t)\big) = \langle l', \xi \cdot v\rangle + \langle l, \xi \cdot x\rangle,$$

where l' and l are elements of $\mathbb{R}^3$.

(iv) $\eta = \eta' = \eta'' = 0$. In this case we simply obtain the identity that asserts that a is a cocycle in $\wedge^2 k^*$. Again, if k is semisimple we can conclude that a must be a coboundary.

In the case of the Galilean algebra we have thus established that the most general cocycle can be written as

$$f((\xi, v, x, t), (\xi', v', x', t')) = \tau([\xi, \xi']) + \langle l', \xi v' - \xi' v\rangle + \langle l, \xi x' - \xi' x + t' v - t v'\rangle + m(\langle v, x'\rangle - \langle v', x\rangle),$$

where $\tau \in o(3)^*$. Now the sum of the first three terms can be written as $\theta([(\xi, v, x, t), (\xi', v', x', t')])$, where $\theta = (\tau, l', l, 0) \in g^*$, that is, as a coboundary. On the other hand, it is clear that the last term is definitely not a coboundary. We have thus recovered a result first proved by Bargmann (1954).

If G is the Galilean group then $H^2(g)$ is one-dimensional and, up to coboundaries, a cocycle can be written as $f = m\alpha$, where

$$\alpha((\xi, v, x, t), (\xi', v', x', t')) = (\langle v, x'\rangle - \langle v', x\rangle).$$

In order to understand the significance of the cocycle α, we do the following computation: Let T_u denote the "velocity transformation" with velocity u, so that T_u corresponds to the matrix

$$\begin{pmatrix} 1 & 0 & 0 \\ u & 1 & 0 \\ 0 & 0 & 1 \end{pmatrix}$$

and

$$\mathrm{Ad}\, T_u(v, A, x, T) = (v - Au, A, x + tu, t).$$

Then, if

$$\xi_1 = (v_1, A_1, x_1, t_1) \qquad \text{and} \qquad \xi_2 = (v_2, A_2, x_2, t_2),$$

$$\begin{aligned} \alpha(\mathrm{Ad}\, T_u^{-1}\xi_1, \mathrm{Ad}\, T_u^{-1}\xi_2) &= (v_1 + A_1 u)\cdot(x_2 - t_2 u) \\ &\quad - (v_2 + A_2 u)\cdot(x_1 - t_1 u) \\ &= v_1 \cdot x_2 - v_2 \cdot x_1 \\ &\quad - u\cdot(A_1 x_2 + t_2 v_1 - A_2 x_1 - t_1 v_2) \\ &= \alpha(\xi_1, \xi_2) - \langle \beta_u, [\xi_1, \xi_2]_0\rangle, \end{aligned}$$

where

$$\beta_u = (0, 0, u, 0)$$

In other words, if we set $\mathrm{Ad}^{\#}\, T_u\alpha(\xi_1, \xi_2) = \alpha(\mathrm{Ad}\, T_u^{-1}\xi_1, \mathrm{Ad}\, T_u^{-1}\xi_2)$, we can write the above equation as

$$\mathrm{Ad}^{\#}\, T_u\alpha = \alpha + d\beta_u.$$

Now let M be any symplectic homogeneous space for the Galilean group. By the construction of Section 25, we know that there is a map Ψ from $M \to Z^2(g)$ (given by pulling back the symplectic form of M to G). Let σ denote this map. We know that

$$\Psi(ax) = \mathrm{Ad}_a^{\#}\, \Psi(x), \qquad a \in G \text{ and } x \in M,$$

where $\mathrm{Ad}^{\#}$ denotes the representation of G on $Z^2(g)$. On the other hand, we can write

$$\Psi = d_0(l, \tau, p, E) + m\alpha,$$

where l, τ, p, and m are well-defined functions on M. It follows from the preceding computation that m is a constant (since it is unchanged under T_u and under rotations – it is obviously unchanged under translations) and that

$$\Psi \circ T_u = d_0(l, \tau, p + mu, E - p \cdot u) + m\alpha.$$

Thus the parameter m tells us how the momentum p changes when we apply the velocity transformation with velocity u. We see that the ratio of the change in momentum to the velocity is m. For this reason m can be identified with the "mass" of the system.

Let us consider our "deformed Poincaré algebra" with parameter $\epsilon = c^{-2}$. For $\epsilon \neq 0$, we know that α is a coboundary of d_ϵ. In fact,

$$d_\epsilon(0, 0, 0, 1) = \epsilon\alpha$$

as can be seen by direct computation. Therefore

$$d_\epsilon(0, 0, 0, E) = m\alpha$$

if we take $E = \epsilon^{-1} m$, that is,

$$E = mc^2.$$

Notice that $(0, 0, 0, E)$ is dual to time translation, and hence corresponds to an energy.

For $\epsilon > 0$, we know that g_ϵ is isomorphic to the Poincaré algebra and hence the most general element (l, τ, p, E) in g_ϵ^* can be written as $(l, \tau, p, E) = (\Gamma, P)$, where $\Gamma = (l, \tau) \in O(1, 3)^*$ and $P = (E, p) \in \mathbb{R}^{1,3^*}$. We know that $\|P\|^2 = E^2 - c^2p^2$ is an invariant of the orbit through (Γ, P). Thus, as indicated in the introduction to this chapter we can summarize the results of the preceding computation (and the interpretation of the famous Einstein mass-energy formula) as follows: Within the context of the Galilean group the mass enters as a parameter describing the element of $H^2(g_0)$ associated with a given symplectic homogeneous space of the Galilean group. Its physical significance is that it determines the change in momentum effected by a velocity transformation. Within the framework of the Poincaré group the mass enters as one of the parameters describing an orbit in g^*. The equation $d_\epsilon(0,0,0,1) = \epsilon\alpha$ shows that if we restrict ourselves to small velocity transformations, while the speed of light is regarded as a large number, then for an element of vanishingly small momentum, the inertial mass (when we replace the Poincaré group by the "limiting" Galilean group) is related to the energy by Einstein's formula $E = mc^2$.

Note that our procedure to compute cohomology shows that if V is any vector space carrying a nondegenerate scalar product, and if $\dim V > 2$, then the corresponding Euclidean algebra, $E(V) = O(V) \circledS V$ has vanishing first and second cohomology. In particular, all its symplectic homogeneous spaces are (up to covering) coadjoint orbits, and we know how to find all of these since our group is a semidirect product.

In the next section, we will give a detailed description of the symplectic homogeneous spaces for the Galilean and Poincaré groups, together with a "physical interpretation." From the mathematical point of view, in addition to the phenomenon of "creation of cohomology" as we contract from the Poincaré group to the Galilean, there is another interesting phenomenon: A single Poincaré orbit can break up, in the limit, into a continuum of Galilean orbits. (Physically this corresponds to the fact that the wavelength of light is a Galilean, but not a Poincaré, invariant.) We shall present the general theory of these phenomena, due to Coppersmith, in the section after next. As the computations in the next section might get tedious, some readers might prefer to go directly to the general theory.

The computations of the next section fall into the following framework (a special case of the one discussed in Chapter III): We are given a "group of nature," G (say the Galilean or Poincaré group). We are also given a "world geometry" that is a homogeneous space $N = G/L$ (for example, Galilean or Lorentzian space-time, so, for the Poincaré group L is the Lorentz group and N is M Minkowski space). We wish to construct, for each homogeneous symplectic G manifold S, an "evolution space" E that is simultaneously fibered over S and N by G-equivariant fibrations:

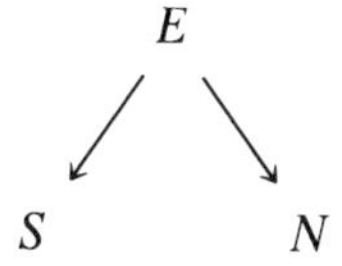

The fibers of the S fibration then project onto submanifolds in N describing the evolution of the system. In certain favorable cases, the fibers will be one-dimensional, and we can think of the image curves in N as "trajectories" but in general this is not possible.

54. Galilean and Poincaré elementary particles

Let us now describe the symplectic homogeneous spaces for the Galilean group. We can subdivide them into two cases according to whether the mass m is zero or not:

$$m \neq 0.$$

By applying a velocity transformation we can choose our point x_0 on M so that the value of Ψ at that point is given by

$$\sigma = \Psi(x_0) = d_0\beta + m\alpha,$$

where the p component of β is zero. We can then apply a space translation so that the l component of β is zero. (These two choices are known as "moving to a frame in which the particle is at rest" and "choosing the center of mass as the origin of the coordinate system.") Thus $\beta = (0, \tau, 0, E_0)$. We can still apply any element of $SO(3)$, so that $\|\tau\| = s$ is a further invariant. We thus have two different kinds of cases according to whether $s = 0$ or $s > 0$.

$m \neq 0, s = 0$. In this case we can choose our representative point with $\sigma = d_0(0, 0, 0, E_0) + m\alpha$. The subalgebra

$$h_\sigma = \{\xi | i(\xi)\sigma = 0\}$$

consists of all ξ of the form $\xi = (0, A, 0, t)$, that is, infinitesimal rotations and time translations. The corresponding group, $H = SO(3) \times \mathbb{R}$, acts on the four-dimensional space $\mathbb{R} \times \mathbb{R}^3$ as rotation about the origin in $\mathbb{R}^3$ and translation in the time direction. Thus G/H is six-dimensional and consists of all lines in $\mathbb{R} \times \mathbb{R}^3$ that are not parallel to the subspace $\mathbb{R}^3$, that is, those that projection surjectively onto the time axis. Any such line consists of the set of all points of the form $(t, x + tv)$, and we could use (x, v) as coordinates. To describe the symplectic structure it is convenient to proceed as follows: On $T^*(\mathbb{R}^3) \times \mathbb{R}$ we have the linear differential form

$$\kappa_m = p \cdot dq - (1/2m)p^2\, dt.$$

This form is clearly invariant under the Euclidean group and time translations, whereas if velocity transformations T_u act by $T_u(q, p, t) = (q + tu, p + mu, t)$ then

$$T_u^* \kappa_m = \kappa_m + \tfrac{1}{2}mu^2\, dt + mu \cdot dq$$

and thus the form $d\kappa_m$ is preserved by all Galilean transformations. The lines $s \to (q + (p/m)s, p, t + s)$ are the solutions of the field of singular directions for the form $d\kappa_m$, and the set of such lines, which can be identified with the set of lines in $\mathbb{R} \times \mathbb{R}^3$ described above (with $x = q$ and $v = p/m$) carries an induced symplectic structure that is Galilean invariant. It is easy to check that these are exactly the symplectic manifolds with $m \neq 0$ and $s = 0$. Upon restriction to the Euclidean group $E(3) \subset G$, these six-dimensional manifolds decompose into one three-dimensional orbit (corresponding to $v = 0$) and a family of five-dimensional orbits. The symplectic form, when restricted to the three-dimensional orbit, is identically zero, and hence this orbit corresponds to the zero-dimensional symplectic homogeneous space for the Euclidean group. The symplectic form, when restricted to the five-dimensional orbits, has a one-dimensional null direction at each point, and so each five-dimensional orbit is fibered over a four-dimensional symplectic homogeneous space of the Euclidean group, and each spin-0 orbit of the Euclidean group occurs once in this decomposition.

$m \neq 0$, $s \neq 0$. In this case h_σ is two-dimensional; if we choose $\tau = se_1 \wedge e_2$ and $\sigma = d_0(0, \tau, 0, 0) + m\alpha$, then h_σ consists of all $(0, A, 0, t)$, with A infinitesimal rotations about the e_3 axis and t time translations. The space G/H in this case is eight-dimensional and can be identified, as a manifold, with $\mathbb{R}^3 \times \mathbb{R}^3 \times S^2$, where S^2 is the two-dimensional sphere. To describe the symplectic structure (and also to deal with the mass-0 case), it is convenient for us to have an explicit expression for the action of G on g^* in terms of the basis of g^* that we have been using. A direct computation shows that if

$$a = \begin{pmatrix} 1 & 0 & r \\ w & B & y \\ 0 & 0 & 1 \end{pmatrix},$$

then

$$a^{\#}(l, \tau, p, E) = \big(B(l - rp), B\tau + Bl \wedge w + y \wedge Bp, E + Bp\big).$$

Now consider the nine-dimensional manifold consisting of all (x, v, t, u), where

$$[x, v, u \in \mathbb{R}^3, \|u\| = s, t \in \mathbb{R}].$$

We let G act on this nine-dimensional space by

$$a(x, v, t, u) = (Bx + y + tw, Bv + w, t + r, Bu),$$

and define a map $\hat{\Psi}$ of this nine-dimensional space into g^* by

$$\hat{\Psi}(x, v, t, u) = \big(m(x - vt), *u + mx \wedge v, mv, \tfrac{1}{2}mv^2\big).$$

Then

$$\hat{\Psi}\big(a(x, v, t, u)\big) = (l', \tau', p', E'),$$

where

$$\begin{aligned}
l' &= m(Bx + y - tBv - rBv - rw),\\
\tau' &= B * u + m\big(B(x \wedge v) + Bx \wedge w + y \wedge Bv + tw \wedge Bv + y \wedge w\big),\\
p' &= Bmv + mw,\\
E' &= \tfrac{1}{2}mv^2 + Bmv \cdot w + \tfrac{1}{2}mw^2,
\end{aligned}$$

while

$$a^{\#}\hat{\Psi}(x, v, t, u) = (l'', \tau'', p'', E''),$$

where

$$\begin{aligned}
l'' &= Bm(x - tv) - rBmv && = l' - m(y - rw),\\
\tau'' &= B *u + m\big(B(x \wedge v) + B(x - vt) \wedge w + y \wedge Bv\big) && = \tau' - m(y \wedge w),\\
p'' &= Bmv && = p' - mw,\\
E'' &= \tfrac{1}{2}mv^2 + w \cdot mBv && = E' - \tfrac{1}{2}mw^2.
\end{aligned}$$

Thus

$$\hat{\Psi}a = a^{\#}\hat{\Psi} + m\Theta(a),$$

where $\Theta : G \to g^*$ is defined by

$$\Theta(a) = (y - rw, y \wedge w, w, \tfrac{1}{2}w^2).$$

If $(v_1, A_1, x_1, t_1) = \xi_1$ is an element of $g = TG_e$, then $d\Theta_e(\xi_1) = (x_1, 0, v_1, 0)$

and so

$$d\Theta_e(\xi_1)(\xi_2) = x_1 \cdot v_2 - x_2 \cdot v_1 = \alpha(\xi_1, \xi_2).$$

Thus the image of $\hat{\Psi}$ is a homogeneous symplectic manifold corresponding to mass m. The set of points of the form $(x + rv, v, t + r, u)$ is the pre-image of $\Psi(x, v, t, u)$. We can thus think of the points on the symplectic manifold as describing trajectories of particles of mass m moving with velocity v with momentum $p = mv$ spinning in the direction of u, with spin velocity $s = \|u\|$. Under restriction to the three-dimensional Euclidean group, the orbit of a point (x, v, t, u) is six-dimensional if $v \neq 0$ and u is not collinear with v. These orbits are fibered over the four-dimensional symplectic manifolds of $E(3)$ with momentum $m\|v\|$ and "spin" $u \cdot v/m\|v\|$. The fibers are two-dimensional because the point x moves along a straight line and u can "precess" about v, maintaining a constant value of $u \cdot v$. If $v \neq 0$ and u and v are collinear, then the orbit through (x, v, t, u) under $E(3)$ is a five-dimensional symplectic manifold. If $v = 0$, the orbit is five-dimensional and the corresponding symplectic manifolds are two-dimensional.

We now examine the massless case,

$$m = 0.$$

We must look at orbits of G on g^*. From the explicit formula for the action, we see that $\|p\|$ and $\|l \wedge p\|$ are invariants. We thus subdivide into cases according to whether these invariants are zero or not.

$m = 0$, $\|p\| \neq 0$, $\|p \wedge l\| = 0$. By appropriate choice of r we can arrange that $l = 0$ and by appropriate choice of y that $\tau = k * p$. By a rotation we can arrange that $p = \|p\|e_3$. Thus we can arrange to choose, on each orbit in g^* of this type, a point of the form $(0, se_1 \wedge e_2, \|p\|e_3, E)$. We thus see that the only invariants of these kinds of orbits are $\|p\|$ and $s = k/\|p\|$. Such orbits are six-dimensional, consisting of all points of the form $(rp, (s/\|p\|)^*p + p \wedge q, p, E)$, where r and E are arbitrary real numbers, where $q \in \mathbb{R}^3$ and $p \in \mathbb{R}^3$ has length $\|p\|$. Upon restriction to the Euclidean group $E(3)$ we see that the orbit through each point is four-dimensional and corresponds to the Euclidean symplectic space with momentum p and spin s. We can relate these six-dimensional orbits of the Galilean group to the eight-dimensional orbits by a form of limiting process as follows: Suppose we go back to the nine-dimensional space of all (x, v, t, u) and map $\Psi = \Psi_m$ described above. Suppose that we let $m \to 0$ and $v \to \infty$ in such a way that $mv \to p$ is finite, while keeping x finite. Then $\Psi_m(x, v, t, u)$ tends to a point of one of the six-dimensional orbits that we have been describing. We can thus think of these orbits as describing trajectories of "particles of zero mass and infinite velocity." For spin ± 1 these would, for example, correspond to photons in Galilean relativity where the "speed of light" is assumed to be infinite.

$m = 0, p \neq 0, p \wedge l \neq 0$. Here, by suitable choice of y and w both $\perp p$ we can arrange that $\tau = 0$, by choice of r that $l \perp p$, and by suitable choices of y and w in the direction of p that $E = 0$ without destroying the normalization $\tau = 0$. By suitable rotation we can then arrange that our point on the orbit has the form $(ke_3, 0, k'e_1, 0)$. From this we see that $\|p\|$ and $\|p \wedge l\|$ are the only invariants. The isotropy group of a typical point is one-dimensional, consisting of translations in the direction of p, and so the orbits are eight-dimensional. We can get them as limits of orbits through points $\Psi_m(x, v, t, u)$, where now we let $m \to 0$, $v \to \infty$ so that $mv \to p$, but also let $x \to \infty$ in a direction orthogonal to v, with $mx \to l$. Thus these orbits would correspond to "particles at infinity with infinite velocity and mass zero," which may be why these orbits do not correspond to known particles.

$m = 0$, $p = 0$. Here l is an invariant, as $\|\tau \wedge l\|$ if $\|l\| \neq 0$ and $\|\tau\|$ if $l = 0$. Also E is an invariant. These orbits are four-dimensional if $l \neq 0$ and two-dimensional if $l = 0$. A four-dimensional orbit projects onto the whole family of two-dimensional orbits of $E(3)$ while the two-dimensional orbit gives rise to the corresponding two-dimensional $E(3)$ orbit. Again, neither of these orbits correspond to physical particles. (In terms of the limit description given above they would correspond to "particles of zero mass and finite velocity at infinity.") Finally, we have the zero dimensional orbits of the form $(0, 0, 0, E)$.

We now describe the homogeneous symplectic manifolds for the Poincaré group, which, as we have seen, amounts to describing the orbits of the Poincaré group acting on the dual of its Lie algebra. Let us write the most general element of the dual of the Lie algebra as $(\Gamma, P) = (l, \tau, p, E)$, where $\Gamma = (l, \tau) \in o(1, 3)^*$ and $P = (p, E) \in \mathbb{R}^{1,3*}$. From our general considerations we know that $\|P\|^2$ and $\|\Gamma \wedge P\|^2$ are invariants. Here

$$\|P\|^2 = E^2 - c^2 p^2$$

and

$$\|P \wedge \Gamma\|^2 = E^2(\tau + l \wedge p)^2 - c^2(\tau \wedge p)^2.$$

(Here an expression such as x^2 means the length of the vector in three-dimensional Euclidean space, or the length of a 2-vector, etc.) If $\|P\|^2 \geqslant 0$, sign E is also an invariant. For reasons that will become clear below, we call the second-order invariant $\|P\|^2$ the mass² and denote it by M^2c^4. We now describe the orbits in terms of these invariants.

$$M^2 > 0, E > 0, \Gamma \wedge P \neq 0.$$

By an element of $SO(1,3)$ we can arrange that $P = (0, E)$ and then by a space translation that $l = 0$, and by a further rotation that $\tau = se_1 \wedge e_2$. Thus $M^2c^4 = E^2$ and $\|\Gamma \wedge P\|^2 = c^4M^2s^2$ determine the orbit completely. These orbits are eight-dimensional. We can describe these orbits as "trajectories" in the following way: Consider the nine-dimensional manifold consisting of all (X, P, U) where $X, P, U \in \mathbb{R}^{1,3}$, $\|P\|^2 = M^2c^4$, $\|U\|^2 = -s^2$, $P \cdot U = 0$. We let the Poincaré group act on X as *affine* transformations, while we consider P as an element of the dual space $\mathbb{R}^{1,3*}$ upon which the Poincaré group acts by *linear* transformations, and U as an element of $\mathbb{R}^{1,3}$ with the Poincaré group acting as linear transformations again; that is, the translation components act trivially on P and U. We let $* : \wedge^2(\mathbb{R}^{1,3}) \to \wedge^2(\mathbb{R}^{1,3})$ denote the star operator associated to the nondegenerate scalar product on $\mathbb{R}^{1,3}$. We can then map this nine-dimensional space into the dual of the Lie algebra of the Poincaré group by

$$(X, P, U) \rightsquigarrow (X \wedge P + *(U \wedge P), P).$$

This map commutes with the action of the group, and the image is the orbit with invariants M^2c^4 and $M^2c^4s^2$. Now P and U are determined by the image point, but we can modify X by replacing it by $X + \rho P$ for any real number, ρ. An element $\xi = (w, B, y, r) \in g$ thought of as the matrix

$$\begin{bmatrix} 0 & c^{-2}w^t & r \\ w & B & y \\ 0 & 0 & 0 \end{bmatrix},$$

assigns the tangent vector (ξ_X, ξ_P, ξ_U) where,

$$\text{if} \quad X = \begin{pmatrix} t \\ x \end{pmatrix}, \quad \text{then} \quad \xi_X = \begin{pmatrix} c^{-2}w \cdot x + r \\ tw + Bx + y \end{pmatrix};$$

$$\text{if} \quad P = (p, E), \quad \text{then} \quad \xi_P = (c^{-2}Ew, w \cdot p);$$

$$\text{if} \quad U = \begin{pmatrix} \mu \\ u \end{pmatrix} \quad \text{then} \quad \xi_U = \begin{pmatrix} c^{-2}w \cdot u \\ \mu w \end{pmatrix}.$$

We can relate the nine-dimensional space of (X, P, U) of the Poincaré group to the nine-dimensional space of (x, v, t, u) of the Galilean group by the maps

$$X \to \begin{pmatrix} t \\ x \end{pmatrix}.$$

$v = (c^2/E)p$, where $P = (p, E)$ and $E^2 = p^2c^2 + M^2c^4$, $U = \begin{pmatrix} \mu \\ u \end{pmatrix}$, where $\mu^2 = c^{-2}(u^2 - s^2)$. The tangent vector corresponding to ξ at v will then be given as follows: Let us write $\xi_P = (\xi_p, \xi_E) = (c^{-2}Ew, w \cdot p)$. Then

$$\begin{aligned} \xi_v &= (c^2/E)\,\xi_p - (c^2/E^2)(\xi_E)p \\ &= w - \big(w \cdot p/(p^2 + M^2c^2)\big)P. \end{aligned}$$

We thus see that in the limit we get the situation corresponding to the eight-dimensional symplectic manifolds for the Galilean group provided that we take $m = M$ if $E > 0$ and $m = -M$ if $E < 0$ and $c^2\mu^2 \to 0$. It is interesting to describe what is happening from the point of view of submanifolds in the one fixed ten-dimensional vector space g^*. For each finite c, we are considering the orbit through a point of the form $(0, *u, 0, MC^2)$. As $c \to \infty$, this orbit moves off to infinity, and the action of the group in a neighborhood of the point $(0, *u, 0, Mc^2)$ tends to the action of the Galilean group on a symplectic manifold having nontrivial cohomology class M. In this way we see that the orbits "at infinity" give rise to the symplectic manifold of the limiting group that correspond to cocycles that are not coboundaries. We shall discuss the general theory of this phenomenon in the next section.

$$M^2 > 0, \ \|p \wedge l\|^2 = 0.$$

From the preceding discussion we see that these correspond to massive particles with spin 0, the orbits being six-dimensional. All of the preceding discussion applies with the simplification that $u = 0$.

$$M^2 < 0.$$

These would correspond to *tachyons*, that is, particles traveling faster than the speed of light, which are not observed.

$$M^2 = 0, \ P \neq 0.$$

Here there are two possibilities depending on whether $\|\Gamma \wedge P\|^2 = 0$ or not.

$P^2 = 0, P \neq 0, E > 0, \|\Gamma \wedge P\|^2 = 0$. The group $SO(1, 3)$ acts transitively on the "forward light cone," that is, on the set of all vectors satisfying $\|P\|^2 = 0, E > 0$. Let us denote the vector $P = (0, 1)$ by e_0, and the vectors

$(e_i, 0)$ $(i = 1, 2, 3)$ by e_i, so that e_0, e_1, e_2, e_3 is a basis of $\mathbb{R}^{1,3}$. We may thus choose our point (Γ, P) on our orbit so that $P = e_0 + c^{-1}e_3$. By applying a translation we can add any linear combination of $(e_0 + c^{-1}e_3) \wedge e_i (i = 1, 2, 3)$ to Γ. Thus we may choose Γ so that

$$\Gamma = a(e_0 - c^{-1}e_3) \wedge e_1 + b(e_0 - c^{-1}e_3) \wedge e_2 + se_1 \wedge e_2.$$

Then

$$\Gamma \wedge P = c^{-1}2e_0 \wedge (ae_1 \wedge e_3 + be_2 \wedge e_3) + s(e_0 + c^{-1}e_3) \wedge e_1 \wedge e_2,$$

and the condition $\|\Gamma \wedge P\|^2 = 0$ implies that $a = b = 0$. Thus we may assume that

$$\Gamma = se_1 \wedge e_2.$$

We thus see that each of the orbits of the kind we are considering is completely determined by the one additional parameter s. Let us now calculate the isotropy algebra h of the element

$$(\Gamma, P) = se_1 \wedge e_2 + e_0 + c^{-1}e_3 = (0, se_1 \wedge e_2, c^{-1}e_3, 1) \in g^*.$$

A direct computation shows that h is four-dimensional and consists of the elements ξ_1, ξ_2, ξ_3, and ξ_4, where ξ_1 is infinitesimal rotation about the e_3 axis; i.e., that is,

$$\xi_1 = \begin{pmatrix} 0 & 0 & 0 \\ 0 & e_1 \wedge e_2 & 0 \\ 0 & 0 & 0 \end{pmatrix},$$

where ξ_2 is infinitesimal translation

$$\xi_2 = \begin{pmatrix} 0 & 0 & 1 \\ 0 & 0 & c^1 e_3 \\ 0 & 0 & 0 \end{pmatrix}$$

and where ξ_3 and ξ_4 are the elements

$$\xi_3 = \begin{pmatrix} 0 & c^{-2}e_2^t & 0 \\ e_2 & c^{-1}e_3 \wedge e_2 & se_1 \\ 0 & 0 & 0 \end{pmatrix}$$

and

$$\xi_4 = \begin{pmatrix} 0 & c^{-2}e_1 & 0 \\ e_1 & c^{-1}e_3 \wedge e_1 & -se_2 \\ 0 & 0 & 0 \end{pmatrix}.$$

Notice that $h \cap o(1,3)$ has dimension 1; and a straightforward computation shows that $h \cap \mathrm{Ad}_a o(1,3)$ is at most one-dimensional. This has the following consequence: Let H be the subgroup of the Poincaré group G, which fixes the point on our orbit, so that the orbit can be identified as $M = G/H$. Suppose, as in the preceding examples, we wish to regard M as the "space of trajectories" of some system N and we would like to be able to associate to each point of N some space-time position (i.e., some point in $\mathbb{R}^{1,3}$). In other words, we are looking for a manifold N together with a map $\pi: N \to M$ and a map $\rho: N \to \mathbb{R}^{1,3}$. We would like N to be a homogeneous space for G; that is, $N = G/K$ for some closed subgroup K, and we would like the maps π and ρ to commute with the action of G. We thus want to have the diagram

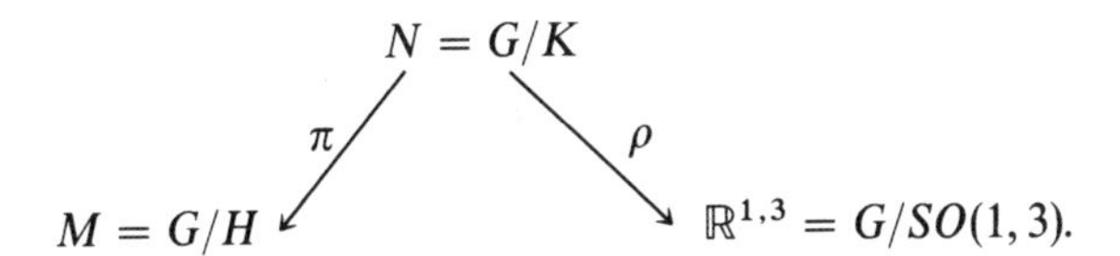

(More generally, we might want to replace $SO(1,3)$ by a conjugate subgroup.) But then $\pi^{-1}(H) = H \cdot K/K = H/K$ has dimension at least 3, as does $\rho(\pi^{-1}(H)) \subset \mathbb{R}^{1,3}$. We have discussed this phenomenon in general in Chapter III.

Explicitly, we can construct such a nine-dimensional manifold N as follows: We let N consist of all $(X, P, \bar{P})$ X, P, $\bar{P} \in \mathbb{R}^{1,3}$ with $P^2 = \bar{P}^2 = 0$, $P \cdot \bar{P} = -1$, where G acts as affine transformations of the X component and as linear transformations on the P and $\bar{P}$ components. We set $\rho(X, P, \bar{P}) = X$ and

$$\pi(X, P, \bar{P}) = \big(s^*(P \wedge \bar{P}) + X \wedge P, P\big).$$

This map is clearly equivariant. If we take $X = 0$, $P = (e_0 + c^{-1}e_3)$ and $\bar{P} = \frac{1}{2}(e_0 - c^{-1}e_3)$ then

$$\pi(X, P, \bar{P}) = (se_1 \wedge e_2, e_0 + c^{-1}e_3)$$

so the image is precisely the orbit we are considering. Then for any (Γ, P) the image $\rho\pi^{-1}(\Gamma, P)$ consist of points of the form $X + Y$, where $Y \perp P$. (It can be thought of as a 2-plane in space moving with the velocity of light, and thus as a plane wave.) Of course, while the Poincaré group acts transitively on the set of forward null vectors, the Galilean group does not, and so this last orbit corresponds to all the six-dimensional mass-0 orbits of the Galilean group when we let $c \to \infty$. This corresponds to the fact that the wavelength is an invariant of light in Galilean relativity but not in special relativity, due to the Doppler effect.

55. Coppersmith's theory

We first must define what we mean by a contraction of a (one-parameter family of) Lie algebra(s). For this purpose we introduce the concept of a filtered Lie algebra. (In contrast to the notation in Guillemin and Sternberg, 1964, it will be more convenient here for us to use an ascending filtration.)

Definition. *A* filtered Lie algebra *is a Lie algebra* g *together with a vector space filtration* $g_i \supset g_{i-1}, g_N = g, g_{-1} = 0$, *such that* $[g_i, g_j] \subset g_{i+j}, i, j \in \mathbb{Z}$.

Let

$$\mathrm{gr}\, g = \bigoplus \mathrm{gr}_i g, \qquad \mathrm{gr}_i g = g_i / g_{i-1}$$

and let

$$p_i : g_i \to \mathrm{gr}_i g$$

denote the natural projection. Then gr g is a graded Lie algebra under the bracket $[\,,]_{\mathrm{gr}}$ given by

$$[p_i(x), p_j(y)]_{\mathrm{gr}} = p_{i+j}([x, y]), \qquad x \in g_i,\ y \in g_j.$$

This bracket is independent of the choice of x and y, depending only on $p_i(x)$ and $p_j(y)$ and so is well defined.

For example, let $g = o(n) = g_1$ and $g_0 = o(n-1)$. (More generally let $g = k + p$ be any semisimple Lie algebra with a Cartan decomposition so that $[p, p] \subset k$, and take $g_0 = k$ and $g_1 = g$.) Then $\mathrm{gr}\, g = E(n-1)$, the semidirect product of $o(n-1)$ with the vector space $\mathbb{R}^n = o(n)/o(n-1)$. (Or more generally in the $k + p$ case, gr g is the semidirect product of k with the vector space p.)

As a second example, we can take $g = E(2)$, as in Section 17. Then g consists of all vectors $\xi = (v, z, q)$ with

$$[\xi_1, \xi_2] = (0, q_2 v_1 - q_1 v_2, x_1 v_2 - x_2 v_1).$$

Take $g = g_1$ and $g_0 = \{(0, 0, q)\}$. The corresponding graded algebra is the "algebra of free transmission in Gaussian optics" or the "free one-dimensional Schrödinger particle" as described in Section 17.

As a third example we can take g to be the Poincaré algebra so that g consists of all

$$\xi = (A, v, x, t)$$

with

$$[\xi_1, \xi_2] = ([A_1, A_2] + v_1 \wedge v_2, A_1 v_2 - A_2 v_1,$$
$$A_1 x_2 + t_2 v_1 - A_2 x_1 - t_1 v_2, v_1 \cdot x_2 - v_2 \cdot x_1).$$

We take $g_1 = g$ and g_0 to be the subalgebra spanned by all ξ with $v = x = 0$. We claim that the corresponding graded algebra is the Galilean algebra. To see this, let us choose the complement g^1 to g_0 to be given as

$$g^1 = \{(0, v, x, 0)\}.$$

We can identify g^1 with $\mathrm{gr}_1 g$ via p_1. Notice that under that Poincaré bracket we have

$$[g^1, g^1] \subset g_0.$$

Thus we see in the graded bracket that two elements of gr_1 bracket to 0 and $\mathrm{gr}\, g$ is the Galilean algebra.

Let us return to the general theory. We can define a continuous deformation or *contraction* of g into $\mathrm{gr}\, g$ as follows. For each k, choose a vector space complement g^k to g_{k-1} in g_k so

$$g_k = g^k \oplus g_{k-1} \qquad \text{(vector space direct sum)}.$$

Thus

$$g = \bigoplus g^k \qquad \text{(vector space direct sum)}.$$

For each $\epsilon \neq 0$ define the vector space automorphism $f_\epsilon : g \to g$ by

$$f_\epsilon \xi_i = \epsilon^{-i} \xi_i \qquad \text{for} \qquad \xi_i \in g^i.$$

Since f_ϵ is a vector space isomorphism, we can use it to define a new Lie bracket on g by "transport of structure":

$$[\xi, \eta]_\epsilon = f_\epsilon [f_\epsilon^{-1} \xi, f_\epsilon^{-1} \eta].$$

If $\xi_i \in g^i$ and $\eta_j \in g^j$, then since g is a filtered Lie algebra, we can write

$$[\xi_i, \eta_j] = \sum_{k \leqslant i+j} \zeta_k, \qquad \zeta_k \in g^k.$$

Then it is clear that

$$[\xi_i, \eta_j]_\epsilon = \sum_{k < i+j} \epsilon^{i+j-k} \xi_k.$$

This shows that the bracket $[\,,]_\epsilon$ depends continuously on ϵ up to and including $\epsilon = 0$, where $[\,,]_0 = \lim\, [\,,]_\epsilon$ as $\epsilon \to 0$. It also follows from the formula that $[\,,]_0$ can be identified with $[\,,]_{\mathrm{gr}}$ if we identify g^i with $\mathrm{gr}_i g$ under the projection p_i. In this way, we have "deformed" the original filtered algebra g into its associated graded algebra. The reader should check that this deformation for the Poincaré and $e(2)$ algebras coincides with the ones introduced earlier.

The bracket $[\,,]_\epsilon$ defines operators $d_\epsilon : \wedge^k(g^*) \to \wedge^{k+1}(g^*)$, (c.f. Section 24). The operators d_ϵ depend continuously on ϵ. But, as we have seen from the example of the Poincaré and Galilean algebras, even if g has vanishing (first and second) cohomology, it is possible that the algebra gr g has nonvanishing cohomology. Recall from Section 24 that for any Lie algebra, the first cohomology group $H^1(g)$ can be identified with $(g/[g, g])^*$. For this reason, it is useful to choose the complements g^k in the construction of our contraction a little more carefully, so as to be able to measure the "rate of acquisition of first cohomology." We define a bigrading on g as follows:

Set

$$g_{k,n} = g_k \cap \sum_{i+j=k+n} [g_i, g_j]$$

for $n = 0, 1, \ldots$. As $g_{k,(n-1)} + g_{(k-1),(n+1)} \subset g_{k,n}$ we may choose a vector space complement $g^{k,n}$, so that $g_{k,n} = g^{k,n} \oplus (g_{k,(n-1)} + g_{(k-1),(n+1)})$. If $[g, g] \neq g$, we can also define $g^{k,\infty}$ as a vector space complement to $(g_{k-1} + (g_k \cap [g, g]))$ in g_k. We have $g = \bigoplus_{k,n} g^{k,n}$, where k ranges through 0, 1, ..., N and n ranges through 0, 1, ..., $2N$, ∞. Set $g^k = \sum_n g^{k,n}$ and $g^{(n)} = \sum_k g^{k,n}$. The g^k are complements to the g_{k-1} in g_k and we will use them to define our contraction.

For example, a bigrading of the Poincaré algebra is given by

$$g^{0,0} = \{(0, 0, 0, 0)\},$$

$$g^{0,2} = \{(0, 0, 0, t)\},$$

$$g^{1,0} = \{(0, \sigma, x, 0)\},$$

and all other $g^{k,n} = \{0\}$.

We can write the bracket $[\,,]_\epsilon$ as

$$[\xi_i, \eta_j]_\epsilon = \sum_{\substack{k \leqslant i+j \\ n \leqslant i+j-k}} \epsilon^{i+j-k} \xi_{k,n}.$$

Thus $[g, g]_0 = g^{(0)}$. We can write the coefficient of $\xi_{k,n}$ in the above expression as $\epsilon^{n+(i+j-k-n)}$. Let ∂_ϵ denote the map $[\,,]_\epsilon$ of $\wedge^2 g \to g$. If U is a bounded neighborhood of 0 in $\wedge^2 g$, then the projection of $\partial_\epsilon U$ onto $g^{(n)}$ has size ϵ^n. Indeed, for $\xi^i \in g^{(i)}$ and $\eta^j \in g^{(j)}$ let us consider the projection of $[\xi^i, \eta^j]_\epsilon$ onto g^{kn}. If $i + j > k + n$, then the coefficient of the g^k component of $[\xi^i, \eta^j]_\epsilon$ will be ϵ^l, where $l = i + j - k > n$, and so is negligible in comparison with ϵ^n. If $i + j < n$ then (by the definition of the spaces $g^{p,q}$) the projection on $g^{k,n}$ of $[\xi^i, \eta^j]$, and hence of $[\xi^i, \eta^j]_\epsilon$, will be 0. Hence we need only consider the case where $i + j = k + n$. For these values the coefficient of the g^k component will be exactly ϵ^n.

A number of facts emerge from this analysis: First of all we see that

$$[g, g]_0 = g^{(0)}. \tag{55.1}$$

Second, if we consider the corresponding dual space decomposition

$$g^* = \bigoplus g^{(n)*}$$

then

$$d_\epsilon \text{ vanishes identically on } g^{(\infty)}. \tag{55.2}$$

Furthermore, for

$$\theta \in g^{(n)}, \qquad 1 \leqslant n < \infty$$

the limits

$$\lim_{\epsilon \to 0} \epsilon^{-n} d_\epsilon \theta$$

exist and the limiting values are cocycles for d_0. The cohomology classes corresponding to these various limits are linearly independent of one another for different n and the limit of $\epsilon^{-n} d_\epsilon$ is injective on each $g^{(n)*}$ for $1 \leqslant n < \infty$. Thus, by this limiting procedure each such $g^{(n)*}$ is mapped injectively into $H_0^2(g)$ and the classes so obtained are linearly independent.

For example, let us consider the contraction of the three-dimensional algebra $o(3)$ into the Euclidean algebra $e(2)$. From the considerations of Section 53 we know that $H^2(e(2))$ is one-dimensional. It is easy to describe the symplectic manifolds corresponding to the nonvanishing cohomology classes – they are just the Euclidean planes – the symplectic form being the oriented area form of the plane. The metric, and hence the area form is only determined up to scalar factor and this (together with the choice of orientation) accounts for a one-parameter family. In addition, we have the coadjoint orbits of $E(2)$. Since $E(2)$ is a semidirect product, we can compute its coadjoint orbits by the method of section. In terms of the standard basis X, Y, Z of $e(2)$, the coadjoint orbits are just the cylinders $X^2 + Y^2 = r^2$ for each $r > 0$ and the points on the $X = Y = 0$ axis. The algebra $o(3)$ is semisimple and hence $H^1(o(3)) = H^2(o(3)) = 0$. For $g = o(3)$, let us take $g_0 = o(2)$ generated by Z and $g_1 = g$. Then

$$g_{00} = g_{01} = 0 \qquad \text{and} \qquad g_{02} = g_0,$$

while

$$g_{10} = \{X, Y\}, \qquad g_{11} = g_1 = g.$$

Thus we take

$$g^{(2)} = g_{02} = \{Z\} = g^0 = g^{02};$$
$$g^{(0)} = g_{10} = \{X, Y\} = g^1 = g^{10}.$$

The orbits in $g^* = o(3)^*$ are just the spheres

$$S_r: X^2 + Y^2 + Z^2 = r^2.$$

Let us now attempt to follow these symplectic manifolds throughout the contraction process. The most naive approach would be to consider the maps $f_\epsilon^*: g^* \to g$, and, for r fixed, the manifolds $f_\epsilon^{*-1} S_r$ ($\epsilon > 0$). By transport of structure, these will be coadjoint orbits for the group whose algebra is given by $[\,,]_\epsilon$ and hence in the limit, we will get an invariant set (a union of coadjoint orbits) for the contracted algebra (in our present example $e(2)$). In our example, $f_\epsilon^{*-1}(S_r)$ is described by the

$$\epsilon^{-2}(X^2 + Y^2) + Z^2 = r^2.$$

For small ϵ this is an ellipsoid with principal axis going from $-r$ to r along the $X = Y = 0$ axis, with maximal radius $\epsilon^2 r^2$ in the X and Y directions. In the limit this will tend to the interval from $-r$ to r on the $X = Y = 0$ axis, that is, to a union of the zero-dimensional orbits of $E(2)$. It is clear that to get the cylinders we must consider the family of spheres $S_{\epsilon^{-1}r}$. Then $f_\epsilon^{*-1}(S_{\epsilon^{-1}r})$ is the ellipsoid

$$X^2 + Y^2 + \epsilon^2 Z^2 = r^2, \tag{55.3}$$

which tends to the cylinder of radius r^2 about the $X = Y = 0$ axis. However, this point set limit in (each bounded region of) g^* is not an accurate description of the limiting homogeneous symplectic manifolds for $E(2)$ associated with elongating ellipsoid (55.3). The point is that we can consider regions "near the top" of the elongating ellipsoid by letting Z grow at a rate of $\epsilon^{-2}c$ for any $c \leqslant r$. Such a region on the ellipsoid will tend, in the limit, to a cylinder of radius $r^2 - c^2$. Thus the homogeneous symplectic manifolds for $E(2)$ associated with the family of ellipsoids (55.3) will consist of all cylinders radius $\leqslant r$ (together with the one-point symplectic manifolds). This can be better visualized by examining what happens in Z^2. In our present example, it is easy to see that $Z_\epsilon^2(g)$ is all of $\wedge^2 g^*$ for all values of ϵ including $\epsilon = 0$. Let α, β, γ be the basis of g^* dual to the basis X, Y, Z. Then, since in our example the "volume element" $\alpha \wedge \beta \wedge \gamma$ is invariant, we see that we can identify $\wedge^2 g^*$ with g as representation spaces for all values of ϵ including $\epsilon = 0$. Here $\beta \wedge \gamma$ corresponds to X, etc. We know from Section 25 that the G orbits in $Z^2(g)$ parametrize the homogeneous symplectic G spaces. For $\epsilon = 1$ the orbits are spheres. As $\epsilon \to 0$ these spheres flatten out into oblate ellipsoids with the two large axes in the X and Y (i.e., $\beta \wedge \gamma$ and $\alpha \wedge \gamma$) directions. In the limit $\epsilon = 0$ the orbits are planes parallel to (but not equal to) the plane through $\alpha \wedge \gamma$ and $\beta \wedge \gamma$ (the $X = Y$ plane) and circles (and the origin) in this plane. The circular orbits correspond to the cylinders, while the planar orbits correspond the planes that are symplectic homogeneous spaces for $E(2)$, thus to symplectic homogeneous spaces with nonvanishing cohomology classes. The map $d_\epsilon : g^* \to \wedge^2 g^*$ is given by

$$d_\epsilon(x\alpha + y\beta + z\gamma) = x\beta \wedge \gamma - y\alpha \wedge \gamma + \epsilon^2 z\alpha \wedge \beta.$$

Thus the elongated ellipsoid (55.3) is mapped by d_ϵ into the oblate ellipsoid

$$A^2 + B^2 + \epsilon^{-2}C^2 = r^2, \tag{55.4}$$

where A, B, C are the dual basis to $\beta \wedge \gamma$, $\alpha \wedge \gamma$, $\alpha \wedge \beta$. In the limit, this degenerates into the union of circles or radius $\leqslant r^2$ in the $X - Y$ plane. We can now see how to obtain the planar symplectic homogeneous spaces for $E(2)$ (which correspond to nonvanishing cohomology) as limits of $O(3)$ orbits: Take the spheres of radius $\epsilon^{-2}r$. Then $f_\epsilon^{*-1}(S_{\epsilon^{-2}r})$ is the ellipsoid

$$X^2 + Y^2 + \epsilon^2 z^2 = \epsilon^{-2} r^2. \tag{55.5}$$

These ellipsoids are not only getting elongated, they are also getting larger, and, seemingly, disappearing off to infinity in g^*. However, the tops (and bottoms) of these ellipsoids are flattening out, and the action of the Lie algebra near the tops approaches the action of $e(2)$ on a Euclidean plane, as can be checked by direct verification. The operator d_ϵ carries the ellipsoid (55.5) into the ellipsoid

$$A^2 + B^2 + \epsilon^{-2}C^2 = \epsilon^{-2}r^2, \tag{55.6}$$

which approaches the planes $C = r$ and $C = -r$.

From this example it is clear that it is relatively simple to follow the deformation of families of orbits in $\wedge^2 g^*$. In order to have a description of the deformation of the symplectic manifolds, it is convenient to have a procedure for describing the symplectic homogeneous spaces associated to arbitrary orbits in $Z^2(g)$. This can be done as follows: Given any Lie algebra g, we can choose a basis $c_1, \ldots, c_k$ for $H^2(g)$ and cycles $z_1, \ldots, z_k$ representing the c's. Then a Lie algebra structure is given to the space $H^2(g) + g$ by defining

$$[(v, x), (w, y)] = \left(\sum z_j(x, y)c_j, [x, y]\right)$$

The algebra so obtained is a central extension of g by $H^2(g)$. If $\theta_i \in (H^2(g) + g)^*$ is given by $\langle \theta_i, (v, x) \rangle = a_i$, where $v = \sum a_i c_i$, then it is clear that $d\theta_i = z_i$. In this way every cocycle in $Z^2(g)$ becomes a coboundary in the extended algebra, and the G orbit through θ is the symplectic manifold associated with $z = d\theta$.

Let us now return to our general problem of contraction of homogeneous symplectic spaces. So g is a filtered algebra as above. Let M be the topological closure in $H^2(g) + g^*$ of the union of a class of symplectic homogeneous spaces for G. This class might be
(A) a single G orbit;
(B) all nonzero scalar multiples of a single G orbit; or
(C) a "Dixmier class": all G orbits in $H^2(g) + g^*$ having the (conjugacy class of) isotropy subalgebras.
From M we get a G-invariant subset L of $Z^2(g)$. The isomorphism f_ϵ, for $\epsilon > 0$, induces an isomorphism of $Z^2_\epsilon(g)$ with $Z^2(g)$ and so we get a subset L_ϵ of $Z^2_\epsilon(g)$ corresponding to L. The limiting set L_0 will be a subset of $Z^2_0(g)$ that is invariant under the simply connected group corresponding to gr g; let us call this group G'. We thus get a parametrization of the limiting G' symplectic manifolds. In order to trace the symplectic manifolds themselves, we must first choose the central extensions corresponding to $[\,,]_\epsilon$ in a consistent manner. We do so as follows: For $z \in \wedge^2 g^*$ define Rk as $\min i + j \,|\, z(g_i \wedge g_j) \neq 0$. When we choose the basis of $H^2(g)$ and representatives z_k of the basis elements in $Z^2(g)$, we do so to maximize $\sum Rk(z_j)$. Then set

$$z_{j\epsilon} = \epsilon^{-Rk(z_j) f_\epsilon^{*-1}}(z_j)$$

and

$$y_{j\epsilon} = [z_{j\epsilon}]_\epsilon,$$

where $[\;]_\epsilon$ denotes cohomology class relative to d_ϵ. One checks that the classes $y_{j,\epsilon}$ remain finite and linearly independent in the limit, that is, that the $z_{j\epsilon}$ tend to limiting cocycles z_{j0}, whose cohomology classes y_{j0} are linearly independent. The y_{j0} will not, in general, form a basis of $H^2(\text{gr } g)$ – the spaces $g^{(n)}$ will contribute new limiting classes as we have shown above, and there might be additional linearly independent classes as well.

Following Coppersmith, we shall introduce the following notation: We shall let g_ϵ denote a copy of g for each positive ϵ and $v_\epsilon : g \to g_\epsilon$ the vector space identity map. We shall also let $v_0 : g \to \text{gr } g$ denote the vector space isomorphism between these two spaces (coming from the identification of $\text{gr}_i g$ with g^i). We thus think of f_ϵ as a Lie algebra isomorphism between g

and g_ϵ for positive ϵ and write, for example, $H^2(g_\epsilon)$ instead of $H^2_\epsilon(g)$. For each positive ϵ define the spaces g_ϵ^+ by

$$g_\epsilon^+ = (H^2(g_\epsilon) + g_\epsilon)^* + A,$$

where

$$A = \bigoplus_{1 \leqslant n < \infty} g^{(n)^*} \otimes P^n$$

and P^n is the (one-dimensional) space of polynomials in one variable with homogeneous degree n, and write g^+ for g_1^+.

Define the map $v_\epsilon^+ : g^+ \to g_\epsilon^+$ $(\epsilon > 0)$ by

$$\begin{aligned} v_\epsilon^+ &= v_\epsilon^{*-1} \qquad \text{on } g^*, \\ &= \text{id} \qquad \text{on } A, \end{aligned}$$

and

$$v_\epsilon^+(y_j) = y_{j\epsilon},$$

where the y's are the classes constructed above.

By construction, one sees that v_ϵ^+ remains nonsingular as $\epsilon \to 0$; let g_0^+ be the limit of the spaces g_ϵ^+, and let v_0^+ be the limiting map.

Define the map $E_\epsilon : g_\epsilon^+ \to H^2(g_\epsilon^*) + g_\epsilon^*$. E_ϵ acts as the identity on $H^2(g_\epsilon^*) + g_\epsilon^* \subset g_\epsilon^+$. For $\alpha \otimes p$ in A, α in $g^{(n)^*}$, p in P^n, we set $E_\epsilon(\alpha \otimes p) = p(\epsilon^{-1})\alpha = \epsilon^{-n} p(1)\alpha$ in g^*. $\varphi : H^2(g^*) + g^* \to \bigwedge^2 g^*$ is the map defined by

$$\varphi(\theta) = d^*\theta, \qquad \theta \text{ in } g^*;$$

$$\varphi(y_j) = z_j, \qquad y_j \text{ a basis element of } H^2(g^*).$$

Analogously, we define maps $\varphi_\epsilon : H^2(g_\epsilon^*) + g_\epsilon^* \to \bigwedge^2 g_\epsilon^*$ and the map

$$\psi_0 : H^2(\text{gr } g) + \text{gr } g^* \to \bigwedge \text{gr g}^*$$

by:

$$\varphi_\epsilon(\theta) = d_\epsilon^*\theta, \qquad \theta \text{ in } g_\epsilon^*;$$

$$\varphi_\epsilon(y_{j\epsilon}) = z_{j\epsilon};$$

$$\varphi_0(\theta) = d_0^*\theta;$$

$$\varphi_0(y_j') = z_j'.$$

The maps v_ϵ^* and v_0^* are obtained from the vector isomorphisms v_ϵ and v_0; these give us a basis for comparison of $\bigwedge^2 g_\epsilon^*$, $\bigwedge^2(\text{gr } g^*)$, and $\bigwedge^2 g^*$.

Note that although the map E_ϵ becomes unbounded in the limit $\epsilon \to 0$

(since it contains ϵ^{-n}), the composition with φ_ϵ remains bounded, since by definition of $g^{(n)}$, the operator d_ϵ^* is of order $O(\epsilon^n)$ on $g^{(n)*}$. Thus, denoting the composite map

$$g^+ \xrightarrow{v_\epsilon^+} g_\epsilon^+ \xrightarrow{E_\epsilon} H^2(g_\epsilon^*) + g_\epsilon^* \xrightarrow{\varphi_\epsilon} \wedge^2 g_\epsilon^* \xrightarrow{v_\epsilon^*} \wedge^2 g^*$$

by $F_\epsilon : g^+ \to \wedge^2 g^*$, we find F_ϵ remains bounded as $\epsilon \to 0$. Thus we may define $F_0 : g^+ \to \wedge^2 g^*$ by $F_0 = \lim F_\epsilon$. We choose the map $\Phi : g_0^+ \to H^2(\operatorname{gr} g^*) + \operatorname{gr} g^*$ to satisfy $F_0 = v_0^* \varphi_0 \Phi v_0^+$.

The map Φ can be gotten more directly and less mysteriously. Let $C = \Phi v_0$. C sends $g^* \subset g^+$ to $\operatorname{gr} g'^*$ via v_0^{*-1}. For $\alpha \otimes p$ in A, α in $g^{(n)*}$, p in P^n, $C(\alpha \otimes p)$ is the class in $H^2(\operatorname{gr} g)$ of the two-cocycle

$$\lim p(\epsilon^{-1}) d_\epsilon^* \alpha = \lim \epsilon^{-n} p(1) d_\epsilon^* \alpha = \lim p(1) D_\epsilon \alpha = p(1) D_0 \alpha;$$

in particular, the basis element $\alpha_i \otimes p_i$ is sent to z_i. For y_j a basis element of $H^2(g^*)$, $C(y_j) = y'_j$, the class in $H^2(\operatorname{gr} g^*)$ of the two-cocycle $\lim \epsilon^{-Rk(Z_j)} f_\epsilon^{*-1} z_j$.

The point of this whole exercise is that elements of A give rise, via Φ to the "new" elements of $H^2(\operatorname{gr} g)$, not obtainable as limits of elements of $H^2(g_\epsilon^*)$.

Set $M_\epsilon = f_\epsilon^{*-1} M \subset H^2(g_\epsilon^*) + g_\epsilon^*$; M_ϵ is again the closure of a union of symplectic manifolds for g_ϵ.

Now let $N_\epsilon = (E_\epsilon v_\epsilon^+)^{-1} M_\epsilon \subset g^+$; this represents the given class of symplectic manifolds, together will all possible "renormalizations" in A. Set $N_0 = \lim N_\epsilon$, and set $M_0 = \Phi v_0^+ N_0 \subset H^2(g'^*) + g'^*$.

We thus have the (limit commutative) diagram (see Figure 55.1).

The maps v are vector space isomorphisms, which remain nonsingular as $\epsilon \to 0$ but do not preserve Lie structure. The maps E, φ, and f are related to the Lie structure but may become singular or infinite in the limit $\epsilon \to 0$.

We claim that M_0 is a union of symplectic homogeneous spaces for $\operatorname{gr} g$. Since the action of G' is the limit of the actions of G_ϵ, M_0 is closed under the action of G', which is all that is required.

We claim also that the symplectic manifolds in M_0 are precisely those that can be obtained as the renormalized limits of the symplectic manifolds of M_ϵ. This is just a tautology – a point α in M_0 is the image of a point in N_0, which is the limit of points in N_ϵ, that is, points in M_ϵ together with points in A. Thus we have for each ϵ (or at least for a sequence of ϵ's coverging to 0) a point of X_ϵ (a symplectic manifold in M_ϵ) and a point A. Then the orbit $\operatorname{Ad}^\#(G')\alpha$ is part of the limit of the symplectic manifolds X_ϵ, "renormalized" by the given sequence of points in A. (Note

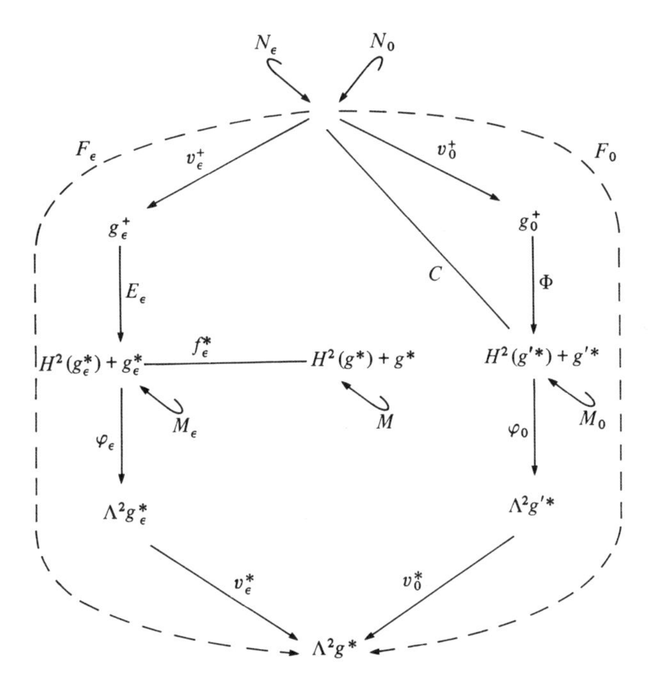

Figure 55.1.

well that this renormalized limit may decompose into a family of smaller-dimensional orbits, but that all these orbits are fully contained in the limit.)

The map $C = \Phi v_0^+$ is an injection. One can check that C is injective on each component A, $H^2(g^*)$, and g^* of g^+, and that the images are disjoint (CA and $CH^2(g^*)$ give disjoint subspaces of $H^2(\text{gr g})^*$, and $Cg^* = g'^*$.) It is not a surjection in general: $\text{Ker}\, d_0^{2*} \subset \wedge^2(\text{gr g})^*$ may be strictly larger than $\text{Ker}\, d_\epsilon^{2*} \subset \wedge^2 g_\epsilon^*$, and $\text{Im}\ C$ is by construction $\varphi_0^{-1}(\lim \text{Ker}\, d_\epsilon^{2*}) \subset \varphi_0^{-1} \text{Ker}\, d_0^{2*} = H^2(\text{gr g})^* + (\text{gr g})^*$. This is unfortunate because the only symplectic manifolds we can get by our procedure are those contained in Im C.

We give here some examples of our procedure, mostly dealing with the Poincaré algebra g and the Galilean algebra $g' = \text{gr}\ g$.

Example. Let M be a single symplectic manifold (Case A) corresponding to a photon with nonzero spin s; this is a six-dimensional subvariety of $g^* = (H^2(g) + g)^*$. Each M_ϵ is also six-dimensional. $N_\epsilon = (E_\epsilon v_\epsilon^+)^{-1} M_\epsilon$ is seven-dimensional, since $E_\epsilon v_\epsilon^+$ is a submersion from $g^+(\dim = 11)$ to $g^*(\dim = 10)$. It turns out that the limit N_0 is also seven-dimensional, as is $M_0 = CN_0$, by injectivity of C. We find that $N_0 \subset g^*$ (the A component vanishes), and thus $CN_0 \subset g'^* \subset (H^2(g') + g')^*$. On general principles the dimension of a limit orbit is less than or equal to that of the original orbit; in this case, a Galilean photon has six dimensions, as does a relativistic (Poincaré) photon. Thus M_0 is a one-parameter (momentum) family of six-dimensional orbits. This gives us an idea of how we can gain dimension in the limiting process. Physically, what we have done is to associate a single spinning relativistic photon to a whole family of spinning classical photons, parametrized by the absolute value of their momenta. (M_0 also contains some lower-dimensional orbits; for example, four-dimensional orbits obtained earlier in the "naive approach.")

Example. Let M be the class of all scalar multiples of the orbit associated with a spinning photon (Case B); this is also the entire Dixmier class of the spinning photon (Case C), as one can verify. M is a one-parameter (spin) family of six-dimensional orbits, thus has seven dimensions.

The preceding example predicts everything correctly; we just add one dimension to everything. Thus dim M = dim $M_\epsilon = 7$, dim N_ϵ = dim N_0 = dim $M_0 = 8$, and the variety M_0 breaks into a two-parameter (spin, momentum) family of classical spinning photons, plus lower-dimensional things as usual.

Example. Suppose we want to obtain (as part of our limit M_0) a massive spinning Galilean particle. If we start with Case A, M being the orbit of a single massive spinning relativistic particle (an eight-dimensional orbit), we find that the particle represented by M_ϵ, corresponding to deformation parameter ϵ, has mass only $m\epsilon$ (where m/c is the mass of the original particle); thus computing N_ϵ, N_0, and M_0, every point in the limiting variety M_0 has mass 0, so that we cannot get the required orbit of a (nonzero) massive spinning classical particle. If we start with Case B, M being the union of all scalar multiples of this orbit of a single massive spinning relativistic particle, we still cannot get the required orbit: Choosing points on M_ϵ with mass m, the spin s_ϵ would go off to infinity as ϵ^{-1}, so that there is no way we could get both mass and spin to be finite and nonzero in the limit. On the other hand, starting with Case C, M is (the closure of) a Dixmier class of all orbits "equivalent" to the orbit of a massive spinning particle. One checks that all massive spinning orbits are equivalent (as long as mass

$\neq 0$, spin $\neq 0$), so that this Dixmier class is dense in g^*. Thus in Case C, M is all of $g^* = H^2(g^*) + g^*$, and M_0 is all of Im C (which in this case is all of $H^2(g'^*) + g'^*$), including the desired massive spinning Galilean particle. We had to choose our orbit X_ϵ in M_ϵ to be "equivalent to," but *not* a scalar multiple of, the image of the original orbit, in order to achieve a limiting case where spin and mass were both finite and nonzero.

References

Books and articles are referred to in the text by author and year. In case two or more articles by the same author, published in the same year occur in the references, these are distinguished by letters a, b, and so forth, which appear in this bibliography at the end of the individual references.

Abraham, R., and Marsden, J., *Foundations of Mechanics*. Benjamin/Cummings, Menlo Park, Calif., 1978.

Adler, M., On a trace functional for formal pseudodifferential operators and the symplectic structure for the Korteweg–de Vries type equations. *Invent. Math. 50* (1979), 219–48.

Adler, M., and van Moerbeke, P., Completely integrable systems, Kac–Moody Lie algebras, and curves. *Adv. Math. 38(3)* (1980), 267–317. (a)

– Linearization of Hamiltonian systems, Jacobi varieties, and representation theory. *Adv. Math. 38(3)* (1980), 318–79. (b)

Arms, J., Marsden, J., and Moncrief, V., Symmetry and bifurcations of momentum mappings. *Commun. Math. Phys. 78* (1980), 455–78.

Arnold, V., Sur le geometrie differentielle des groupes de Lie de dimension infinite et ses applications a l'hydrodynamique des fluids parfaits. *Ann. Inst. Fourier (Grenoble) 16(1)* (1966), 319–61.

– *Mathematical Methods of Classical Mechanics*. Springer Graduate Texts in Mathematics, No. 60. Springer-Verlag, New York, 1978.

Atiyah, M. F., Convexity and commuting Hamiltonians. *Bull. London Math. Soc. 14* (1982), 1–15.

Atiyah, M. F. and Bott, R. The Lefschetz fixed-point formula for elliptic complexes: I, *Ann. of Math. 86* (1967), 374–407.

Atiyah, M. F., and Bott, R. Yang–Mills and bundles over algebraic curves. *Proc. Ind. Acad. Sci. Math. Sci.* 90 (1981) No. 1, 11–20.

Atiyah, M. F., and Singer, I. M., The index of elliptic operators, III. *Ann. Math. 87* (1968) 546–604.

Auslander, L., and Kostant, B., Polarizations and unitary representations of solvable Lie groups. *Invent. Math. 14* (1971), 255–354.

Balachandran, A. P., Salomonson, P., Skagerstam, B.-S., and Winnberg, J.-O., Classical description of a particle interacting with a non-Abelian gauge field. *Phys. Rev. Lett. D 15* (1977).

Bargmann, V. On unitary ray representations of continuous groups. *Ann. of Math. 59* (1954), 1–46.

Bargmann, V., Michel, L., and Telegdi, V. L., Precession of the polarization of particles moving in a homogeneous electromagnetic field. *Phys. Rev. Lett. 2* (1959), 435.

Berline, N., and M. Vergne, Zeros d'un champ de vecteurs et classes characteristiques equivariantes. *Duke. Math. Jour.* 50 (1983) 539–49.

– Fourier transforms of orbits of the coadjoint representation (to appear).

Bohr, A., and Mottleson, B. R., *Nuclear Structure*, Vol. II. Benjamin, New York, 1976.

Bott, R., Nondegenerate critical manifolds. *Ann. Math. 60(2)* (1959), 248–61.

Bredon, G. E., *Introduction to Compact Transformation Groups*. Academic Press, New York, 1972.

Buck, R., Biedenharn, L. C., and Cusson, R. Y., *Nucl. Phys. A 37* (1979), 205–41.

Calogero, F., Solution of the one-dimensional n-body problems with quadratic and/or inversely quadratic pair potentials. *J. Math. Phys. 12* (1971), 419–36.

Calogero, F., and Marchioro, C., Exact solution of a one-dimensional three-body scattering problem with two and/or three body inverse square potential. *J. Math. Phys. 15* (1974), 1425–30.

Chapline, M., *Nucl. Phys. B 184* (1981), 391.

Chernoff, P., Mathematical obstructions to quantization. *Hadronic J. 4* (1981), 879–98

Chu, R. Y., Symplectic homogenous spaces. *Trans. Amer. Math. Soc. 197* (1975), 145–59.

Coppersmith D., Ph. D. Thesis, Harvard University, 1977.

Corwin, L. Nèeman, Y., and Sternberg, S., Graded lie algebras in mathematics and physics (Bose-Fermi symmetry). *Rev. Mod. Phy. 47* (1975) 573–603.

Duflo, M., and Vergne, M., Une propriété de la représentation coadjointe d'une algèbre de Lie. *C. R. Acad. Sci. Paris A-B 268* (1969), 583–5.

Duistermaat, J. J. On global action-angle coordinates. *Commun. Pure Appl. Math. 33* (1980), 687–706.

Duistermaat, J. J., and Heckman, G. J. On the variation in the cohomology of the Symplectic Form of the reduced phase space. *Invent. Math. 69* (1982), 259–68.

Duval, Ch., The general relativistic Dirac–Pauli particle: an underlying classical model. *Ann. Inst. Henri Poincaré A 25* (1976), 345–62.

Duval, Ch., and Horvathy, P., Particles with internal structure: The geometry of classical motions and conservation laws. *Ann. Phy.* 142 (1982) 10–33.

Ebin, D. G., The motion of slightly compressible fluids viewed as a motion with strong constraining force. *Ann. Math. 105* (1976), 141–200.

Ebin, D. G., and Marsden, J., Groups of diffeomorphisms and the motion of an incompressible fluid. *Ann. Math. 92* (1970), 102–63; *Bull. Amer. Math. Soc. 75* (1969), 962–7.

Ehresmann, C., "Les connexions infinitesimales dans un espace fibre differentiable," *Colloque de topologie*, Bruxelles, 1950.

Einstein, A., Infeld, L., and Hoffman, B., *The Gravitational Equations and the Problems of Motion*, Annals of Mathematics, Vol. 39, pp. 65–100. Princeton University Press, Princeton, N.J., 1938.

Faddeev, L., and Zakharov, V. E., Korteweg–de Vries equations as completely integrable Hamiltonian system. *Funk. Anal. Priloz. 5* (1971), 18–27. (In Russian.)

Flashka, H., *Phys. Rev. B 9* (1974), 1924.

Garcia, P. L., The Poincaré–Cartan invariant in the calculus of variations. In *Symposia Mathematics*, Vol. 14, pp. 219–46. Instituto Nazionale di Alta Matematica, Roma, 1974.

– Gauge algebras, curvature and symplectic structure. *J. Differential Geom. 12* (1977), 351–9.

Garcia, P. L., Muñoz, J., On the geometrical structure of higher order variational calculus. Lecture at the IUTAM–ISIMM Symposium on Modern developments in Analytical Mechanics, Acad. de la Scienze di Torino (1982).

Garcia, P. L., and Perez-Rendon, A., Symplectic approach to the theory of quantized fields. I. *Commun. Math. Phys. 13* (1969) 22–44; II. *Arch. Rational Mech. Anal. 43* (1969), 101–24.

Gardner, C. S., Korteweg–de Vries equation and generalizations. IV: The Korteweg–de Vries equation as a Hamiltonian system. *J. Math. Phys. 12* (1971), 1548–51.

Gardner, C. S., Greene, J. M., Kruskal, M. D., and Miura, R. M., Method for solving the Korteweg–de Vries equation. *Phys. Rev. Lett. 19* (1967), 1095–7.

– Korteweg–de Vries equation and generalizations VI: Methods of exact solution. *Commun. Pure Appl. Math 27* (1974), 97–133.

Gel'fand, I. M., and Cetlin, M. L., Finite dimensional representations of the group of unimodular matrices. *Dokl. Akad. Nauk SSSR 71* (1950), 825–8.

Gel'fand, I. M., and Dikii, L. A., Asymptotic behavior of the resolvent of Sturm–Liouville equations and the algebra of the Korteweg–de Vries equations. *Usp. Mat. Nauk 30 (5)* (1975), 67–100.

– Fractional powers of operators and Hamiltonian systems. *Funk. Anal. Priloz. 10 (4)* (1976).

– The resolvent and Hamiltonian systems. *Funk. Anal. Priloz. 11* (1977), 11–27.

Gel'fand, I. M., Manin, Yu. I., and Shubin, M. A., Poisson brackets and the kernel of a variational derivative in formal variational calculus. *Funk. Anal. Priloz. 10 (4)* (1976).

Gerrard, A., and Burch, J. M., *Introduction to Matrix Methods in Optics.* Wiley, New York, 1975.

Gibbons, J., Collisionless Boltzmann equations, and integrable moment equations. *Physica D 3* (1981), 503–11.

Goldschmidt, H., and Sternberg, S., The Hamilton–Jacobi formalism in the calculus of variations. *Ann. Inst. Fourier (Grenoble) 23* (1973), 203–67.

Gotay, J., and Nester, J. M. Generalized constant algorithms and special presymplectic manifolds. *Springer Lecture Notes in Mathematics*, No. 775, pp. 78–104. Springer-Verlag, New York, 1980.

Gotay, M. J. On coisotropic imbeddings of presymplectic manifolds, *Proc. A.M.S. 84* (1982) 111–113.

Greub, W., Halperin, S., and van Stone, R., *Connections, Curvature and Cohomology.* Academic Press, New York, 1972.

Groenwald, H. J., On the principles of elementary quantum mechanics. *Physica 12* (1946), 405–60.

Guillemin, V., and Pollack, A. *Differential Topology.* Prentice-Hall, England Cliffs, N.J., 1974.

Guillemin, V., Quillen, D., and Sternberg, S., The integrability of characteristics. *Commun. Pure Appl. Phys. 23* (1970), 39–77.

Guillemin, V., and Sternberg, S., An algebraic model of transitive differential geometry. *Bull. Amer. Math. Soc.* (1964), 16–47.

– *Geometric Asymptotics.* American Mathematical Society, Providence, R. I., 1977.

– On the equations of motion of a classical particle in a Yang–Mills field and the principle of general covariance. *Hadronic J. 1* (1978), 1–32.

– Homogeneous quantization and multiplicities of group-representations. *J. Functional Anal. 47* (1980), 344–80.

– The metaplectic representation, Weyl operators, and spectral theory. *J. Functional Anal. 42* (1981), 129–225.

– Convexity properties of the moment mapping. *Invent. Math., 67* (1982), 491–513. (a)

– Geometric quantization and multiplicities of group representations. *Invent. Math.*, 67 (1982) 515–38. (b)

– In the universal phase spaces for homogeneous principal bundles. *Lett. Math. Phys. 6* (1982), 231–2. (c)

– Moments and reductions. In *Proceedings of the Clausthal Summer School on Mathematical Physics*, 1980 (to appear). (a)

– Symplectic analogies. In *Conference on Differential Geometric Methods in Theoretical Physics*, World Scientific, Singapore (1983), 87–102.

– On collective complete integrability according to the method of Theorem. *Ergodic Theory* 1983.
– Multiplicity free spaces. *Journal of Diff. Geom.* (1983).
– The Gel'fand–Cetlin system and quantization of the complex flag manifolds. *J. Functional Anal. 52* (1983). (e)
Harnad, J., Shnider, S., and Tafel, J., Group actions on principal bundles and dimensional reduction. *Lett. Math. Phys. 4* (1980), 107–13.
Harnad, J., Shnider, S., and Vinet, J., *J. Math. Phys. 21* (*12*) (1980), 271.
Hausner, M., and Sternberg, S., Notes on Lie algebras. *RIAS* (1957).
Heckman, G., Projections of orbits and asymptotic behavior of multiplicities for compact Lie groups. Ph. D. thesis, University of Leiden (1980).
Helgason, S., *Differential Geometry and Symmetric Spaces.* Academic Press, New York, 1978.
Hermann, R., *Toda lattices, Cosymplectic Manifolds, Bäcklund Transformations, Kinks*; Part A, Vol. 15; Part B, Vol. 18, Interdisciplinary Mathematics, Math. Sci. Press.
Hochschild, G., and Serre, J. P., Cohomology of Lie Algebras. *Ann. Math. 57* (1953), 591–603.
Holmes, P., and Marsden, J., Horseshoes and Arnold diffusion for Hamiltonian systems on Lie groups. *Indiana Univ. Math. Jour. 32* (1983), 273–309.
Humphreys, J., *Introduction to Lie Algebras and Representation Theory.* Springer, New York, 1972.
Iacob, A., *Topological Methods in Mechanics.* Editura Acad. Repub. Social. Romania, Bucharest, 1973. (In Romanian.)
Iacob, A., and Sternberg, S., *Coadjoint Structures, Solitons, and Integrability.* Springer Lecture Notes in Physics, No. 120 Springer-Verlag, New York (1980).
Jacobson, N., *Lie Algebras.* Dover, New York, 1979.
Jost, R., Poisson brackets (an unpedagogical lecture). *Rev. Mod. Phys. 36* (1964), 572–9.
Kawasaki, T., The Riemann–Roch theorem for complex V-manifolds. *Osaka J. Math. 16* (1979), 151–9.
Kazhdan, D., Kostant, B., and Sternberg, S., Hamiltonian group actions and dynamical systems of Calogero type. *Commun. Pure Appl. Math. 31* (1978), 481–508.
Kempf, G., and Ness, L., *The Length of Vectors in Representation Space.* Springer Lecture Notes in Mathematics, No. 732. Springer-Verlag, New York, 1979.
Kirillov, A. A., Unitary Representations of Nilpotent Lie Groups. *Russ. Math. Surveys 17* (*4*) (1962), 57–101.
– *Elements of the Theory of Representations.* pp. 226–235, 290–292. Springer-Verlag, New York, 1976.
Kobayashi, and Shoshichi, On connections of Cartan. *Canad. J. Math. 8* (1956), 145–56.
– Theory of connections. *Ann. Mat. Pure Appl. 43* (*4*) (1957), 119–94.
Kohn, J. J., and Nirenberg, L., Pseudodifferential operators. *Commun. Pure Appl. Math. 18* (1965), 269–305.
Korteweg, D. J., and de Vries G., On the change of form of long waves advancing in a rectangular canal, and on a new type of long stationary waves. *Philos. Mag. 39* (1895), 422–43.
Kostant, B., Orbits, symplectic structures, and representation theory, In *Proceedings, US–Japan Seminar in Differential Geometry, Kyoto,* (*1965*). Nippon Hyoronsha, Tokyo, 1966.
– Quantization and unitary representations. In *Modern Analysis and Applications.* Springer Lecture Notes in Mathematics, No. 170, pp. 87–207. New York, Springer-Verlag, 1970.
– On Whittaker vectors and representation theory. *Invent. Math. 48* (*2*) (1978), 101–84.
– The solution to a generalized Toda lattice and representation theory. *Adv. Math. 34* (*3*) (1979), 195–338.

– and Sternberg, S., Symplectic Projective Orbits. In *New Directions in Applied Mathematics* (P. Hilton and G. S. Young, eds.). Springer-Verlag, New York, 1982.

Kowalewsky, S., Sur le problème de la rotation d'un corps solide autour d'un point fixe. *Acta Math 12* (1889), 177–232.

Kraemer, M., Spharische Untergruppen in Kompacten zusammenhangenden Lie Gruppen. *Compositio Math. 38* (1979), 129–53.

Kunzle, H. P., Canonical dynamics of spinning particles in gravitational and electromagnetic fields. *J. Math. Phys. 13* (1972), 729.

Kuperschmidt, B., and Vinogradov, A. M., *Usp. Mat. Nauk 32 (4)* (1977). 175–236.

– Periodic solutions of the KdV equations. *Commun. Pure Appl. Math. 28* (1975), 141–88.

– Almost periodic solutions of the Korteweg–de Vries equation, *SIAM Rev. 18* (1976), 351–375.

Lax, P., Integrals of nonlinear equations and solitary waves. *Commun. Pure Appl. Math. 21* (1968), 467–90.

Lebedev, D. R., and Manin, Yu. I., Gel'fand–Dikii Hamiltonian operator and coadjoint representation of Volterra group (to appear).

Loomis, L., and Sternberg, S., *Advanced Calculus.* Addison–Wesley, Reading Mass., 1968.

Mackey, G. W., *The Mathematical Foundations of quantum Mechanics.* Benjamin, New York, 1963.

– *Induced Representation.* Benjamin, New York, 1968.

Manakov, S. V., Note on the integration of Euler's equations of the dynamics of an n-dimensional rigid body. *Funct. Anal. Appl. 10 (4)* (1976), 328–9.

Manin, Yu., Algebraic aspects of nonlinear differential equations. *J. Sov. Math 11 (1)* (1979), 1–122.

Manton, N. S., A new six dimensional approach to the Weinberg–Salam model. *Nucl. Phys. B* (1979), 141–53.

Markus, L., and Meyer, K. R., Generic Hamiltonian systems are neither integrable nor ergodic. *Mem. Amer. Math. Soc. 144* (1974).

Marle, C.-M.

Actions de groupes et separations de variables dans ls systèmes Hamiltoniens, *J. Math. pures et appl. 59* (1980) 133–144.

Sous-variétés de rang constant et sous-variétés symplectiquement régulières d'une variété symplectique, C.R Acad. Sci. Paris *295* (1982) I, 119–122.

Sous-variétés de rang constant d'une variété symplectique, *Asterisque* 107–108 (1983) 69–86.

Poisson manifolds in mechanics, in *Bifurcation theory, Mechanics and Physics* (Bruter, C. P., Aragnol, A. and Lichnerowics, A. ed.) Reidel, Dordrecht 1983.

Marsden, J., *Lectures on Geometric Methods in Mathematical Physics.* CBMS-NSF Regional Conference Series, No. 37. Society for Industrial and Applied Mathematics (SIAM), New York, 1981.

– and Weinstein, A., Reduction of symplectic manifolds with symmetry. *Rep. Math. Phys. 5* (1974), 121–30.

– The Hamiltonian structure of the Maxwell–Vlasov equations. *Physica D 4* (1982), 394–406.

Mayer, M., *Fiber Bundle Techniques in Gauge Theory.* Springer Lecture Notes in Physics, No. 67. Springer-Verlag, New York, 1977.

Milnor, J. W., *Morse Theory*, Annals of Mathematical Studies, Vol. 51. Princeton University Press, Princeton, N. J., 1963.

Miscenko, A. C., Integrability of the geodesic flow on symmetric spaces. *Mat. Zamet. 31* (1982), 257–62. (In Russian.)

Mishehenko, A. S., and Fomenko, A. T., Euler equations on finite dimensional Lie groups, *Math. USSR Izv. 12* (3) (1978), 371–89. (a)

– Generalized Liouville method of integration of Hamiltonian systems. *Funct. Anal. Appl.* *12* (1978), 133–21. (b)

Miura, R. M., The Korteweg–de Vries equation: A survey of results. *SIAM Rev.* (1975), 412–458.

Miura, R. M., Gardner, C. S., and Kruskal, M. D., The Korteweg–de Vries equation and generalizations, II: Existence of conservation laws and constants of motion. *J. Math. Phys.* *9* (1968), 1204–9.

Montgomery, D., Samuelson, H., and Zippin, L., Singular points of compact transformation groups. *Ann Math.* (*2*) *G3* (1956), 1–9.

Moser, J., Three integrable Hamiltonian systems connected with isospectral deformations. *Adv. Math.* *16* (1975), 197–220.

– Various aspects of integrable Hamiltonian systems. Proceedings, CIME, Bressanone, Italy, June 1978. In *Progress in Mathematics* No. 8. Birkhäuser, 1980; In *Dynamical System*, (J. Guckenheimer, J. Moser, S. Newhouse, eds.), 233–89. (a)

– Geometry of quadrics and spectral theory. In *Proceedings, Chern Symposium, Berkeley 1979*, pp. 147–88. Springer-Verlag, New York, 1980. (b)

Mostow, G. D., Equivariant embeddings in euclidean space. *Ann. Math.* *65* (1957), 432–46. (a)

– On a conjecture of Montogmery. *Ann. Math.* *65* (1957), 513–16. (b)

– Compact transformation groups of maximal rank. *Bull. Soc. Math. Belg.* *11* (1959), 3–8.

Mumford, D., *Algebraic Geometry I. Complex Projective Varieties.* Springer-Verlag, New York, 1976.

– *Geometric Invariant Theory.* Ergebnisse der Mathematics, Vol. 34, 2nd ed. Berlin-Heidelberg-New York, Springer-Verlag, 1982.

Olshanetsky, M., and Perelomov, A., Completely integrable Hamiltonian systems connected with semisimple Lie algebras. *Invent. Math.* *37* (1976), 93–108.

Olver, P., On the Hamiltonian structure of evolution equations. *Math. Proc. Camb. Philos. Soc.* *88* (1980), 1, 71–88.

Palais, R. S., A global formulation of the Lie theory of transformation groups. *Mem. Amer. Math. Soc.* (1957). (a)

– Imbedding of compact differentiable transformation groups in orthogonal representations. *J. Math. Mech.* *6* (1957), 673–8. (b)

– The classification of G-spaces. *Mem. Amer. Math. Soc.* *36* (1960). (a)

– Slices and equivariant imbeddings. In Seminar on Transformation Groups. Annals of Math. Studies No. 46, Chapter VIII. Princeton Univ. Press, Princeton, N. J., 1960. (b)

– On the existence of slices for actions of non-compact Lie groups. *Ann. Math.* *73* (1961), 295–323.

Planchart, E. A., Analogies in symplectic geometry of some results of Cartan in Representation Theory. Ph. D. Thesis, University of California, Berkeley, 1982.

Ratiu, T., *Euler–Poisson Equations on Lie Algebras.* Ph. D. Thesis, University of California, Berkeley, 1980. (a)

– Euler–Poisson equations on Lie Algebras and the N-dimensional heavy rigid body. *Proc. Nat. Acad. Sci. USA* *78* (1981), 1327–8. (a)

– Euler–Poisson equations on Lie algebras and the N-dimensional heavy rigid body, *Amer. J. Math.* *104* (1982), 409–48.

– The motion of the free n-dimensional rigid body. *Indiana Univ. Math. J.* *29* (4) (1980), 609–29. (b)

– *Involution Theorems.* Springer Lecture Notes in Mathematics, No. 775 pp. 219–57. Springer-Verlag, New York, 1980. (c)

– The C. Neumann problem as a completely integrable system on an adjoint orbit. *Trans. Amer. Math. Soc.* *264* (*2*) (1981), 321–9. (b)

Ratiu, T. and van Moerbeke, P., The Lagrange rigid body motion. *Ann. Inst. Fourier* (Grenoble) (1982).

Reeb, G., *Sur certaines propriétés topologiques des variétés feulletées. Act. Sci. Ind. 1183* (1952), Hermann, Paris.
Reyman, A. G., Integrable Hamiltonian systems connected with graded Lie algebras, *Zap. Nauk Sem.* Leningrad O. D. T. L., Math. Inst. Steklov (L.O.M.I.), Vol. 95, 3–54.
Reyman, A. G., and Semenov-Tjan-Sanskii, M. A., Reduction of Hamiltonian systems, affine Lie algebras, and Lax equations. *Invent. Math. 54 (1)* (1979), 81–100.
Rosensteel, G., and Rowe, D. J., *Ann. Phys. (N. Y.) 96* (1976), 1–42.
– *Ann. Phys. (N. Y.) 104* (1977), 134–44. (a)
– *Group Theoretical Methods in Physics: Proceedings of the Fifth International Colloquium.* pp. 115–32, Academic Press, New York, 1977. (b)
Satake, I., On a generalization of the notion of manifold. *Proc. Nat. Acad. Sci. USA 42* (1956), 359–63.
Shnider, S., and Sternberg, S. Dimensional reduction from the infinitesimal point of view. *Lett. Nuovo Cimento 34* (1982), 459–63.
– Dimensional reduction and symplectic reduction. *Nuovo Cimento B 73* (1983), 130–8.
Simms, D. J., and Woodhouse, N. M. J., *Lectures on Geometric Quantization.* Springer Lecture Notes in Physics, No. 53, Springer-Verlag, New York, 1976.
Sniatycki, J., *Geometric Quantization and Quantum Mechanics.* Springer-Verlag, New York, 1980.
Souriau, J. M., *Structure des systemes dynamiques.* Dunod, Paris, 1970.
– Modele de particule a spin dans le champ electromagnetique et gravitationnel. *Ann. Inst. Henry Poincaré A 20* (1974), 315.
Sternberg, S., Ciclo de Conferencias sobre el Tema Particulas Elementales Clasicas, Univ. de Carabobo, Valencia, Venezuela, 1975. (a)
– Symplectic homogeneous spaces. *Trans. Amer. Math. Soc. 212* (1975), 113–30. (b)
– On minimal coupling and the symplectic mechanics of a classical particle in the presence of a Yang–Mills Field, *Proc. Nat. Acad. Sci. 74* (1977), 5253–4.
– On the influence of field theories on our physical conception of geometry. In *Proceedings of the Bonn Conference on Differencial Geometric Methods in Physics (1977)*, Springer Lecture Notes in Mathematics, No. 676. Springer-Verlag, Bedin, New York, 1978. (a)
– Some preliminary remarks on the formal variational calculus of Gel'fand–Dikii. Springer Lecture Notes in Mathematics, No. 676, pp. 399–407. Springer-Verlag, New York, 1978. (b)
– Review of *Foundations of Mechanics* (2nd ed., rev. and enlgd.), by Ralph Abraham and Jerrold E. Marsden. *Bull. Amer. Math. Soc. 2* (2) March 1980.
– *Lectures on Differential Geometry*, 2nd ed. Chelsea, New York, 1983.
Sternberg. S., and Ungar, T., Classical and prequantized mechanics without Lagrangians or Hamiltonians. *Hadronic J. 1* (1978), 33–76.
Sternberg. S., and Wolf, J., Hermitian Lie algebras and metaplectic representations. *Trans. Amer. Math. Soc. 238* (1978), 1–43.
Sutherland, B., Exact results for a quantum many-body problem in one dimension. II, *Phys. Rev. A 5* (1972), 1372–6.
Symes, W., Systems of Toda type, inverse spectral problems, and representation theory. *Invent. Math. 159* (1980), 13–51. (a)
– Hamiltonian group actions and integrable systems. *Physica D 1* (1980), 339–74. (b)
– The QR Algorithm and scattering for the finite nonperiodic Toda lattice. *Physica D 4* (1982), 278–80.
Takens, F. A global version of the inverse problem of the calculus of variations. *J. Differential Geom. 14* (1979), 543–62.
Thi, Dao Chong, Integrability of the Euler equations on homogeneous symplectic manifolds. *Math. USSR Sbornik, 34* (1978), 707–13.
Thimm, A., Integrabilität beim geodetisch Fluss. *Bonner Math. Schrift. 10B* (1978). (a)

– Dissertation, Universität Bonn, 1980. (b)
– Integrable geodesic flows. *Ergodic theory and Dynamical System 1* (1981), 495–517.
Toda, M., Wave propagation in Harmonic lattices. *J. Phys. Soc. Japan 23* (1967), 501–6.
– Studies of a non-linear lattice. *Phys. Rep. Phys. Lett. C 18* (1975), 1–125.
Tulczyjew, W., The Lagrange complex. *Bull. Soc. Math. France 105* (1977), 419–31.
Tulczyjew, W., and Kijowski, J., *A Symplectic Framework for Field Theory.* Lecture Notes in Physics No. 107. Springer-Verlag, New York, 1979.
Ui, I., *Prog. Theor Phys. (Kyoto) 44* (1970), 153–71.
Ungar, T., *Elementary Systems for Lie Algebra Bundle Actions.* Diff. Geom. Math. in Math. Phy. Proc. Clausthal, 1980; Springer Lecture Notes in Mathematics, No. 905. Springer-Verlag, New York, 1982.
van Hove, L. Sur certaines representations unitaries d'un groupe infini de transformations. *Mem. Acad Roy Belg. 26* (1951), 61–102.
Vinogradov, A. M., *Dokl. Akad. Nauk SSR 238* (5) (1978), 1028–31.
Wang, H., *Nagoya Math. J. 13* (1958), 1–14.
Weaver, L., and Biedenharn, L. C., *Nucl. Phys. A 185* (1972).
Weinstein, A., *Lectures on Symplectic Manifolds.* CBMS Lecture Notes, Soc. No. 29 (1977). (a)
– Symplectic V-manifolds, periodic orbits of Hamiltonian systems and the volume of certain Riemann manifolds. *Commun. Pure Appl. Math. 30* (1977), 265–71. (b)
– A universal phase space for particles in Yang–Mills fields. *Lett. Math. Phys. 2* (1978), 417–20.
– Symplectic geometry, *Bull. Amer. Math. Soc. 5* (1981), 1–13.
– Fat Bundles and symplectic manifold. *Adv. in Math. 37* (1980), 239–50.
– Neighborhood classification of isotropic embedding. *J. Differential Geom. 16* (1981), 125–8.
– The local structure of Poisson manifolds. (Preprint, Berkeley, 1982.)
Whittaker, E. T., *A Treatise on Analytical Dynamics of Particles and Rigid Bodies*, 4th ed. Cambridge University Press, New York, 1959.
Wong, S. K., Field and particle equations for the classical Yang–Mills field and particles with isotopic spin. *Nuovo Cimento A65* (1970), 689.
Woodhouse, N., *Geometric Quantization.* Oxford University Press, Oxford, 1980.

Index